APPLICATIONS OF
MICROBIAL ENGINEERING

APPLICATIONS OF MICROBIAL ENGINEERING

Editors

Vijai Kumar Gupta, PhD
Department of Biochemistry
School of Natural Sciences
National University of Ireland Galway
Galway, Ireland

Monika Schmoll, PhD
Health and Environment Department, Bioresources
Austrian Institute of Technology (AIT)
Konrad-Lorenz Strasse 24
3430 Tulln, Austria

Marcio Antonio Mazutti, PhD
Department of Chemical Engineering
Federal University of Santa Maria
Brazil

Minna Mäki, PhD
Program Leader, NAT
Orion Diagnostica Oy.
Espoo, Finland

Maria G. Tuohy, PhD
Department of Biochemistry
School of Natural Sciences
National University of Ireland Galway
Galway, Ireland

CRC Press
Taylor & Francis Group
Boca Raton London New York

CRC Press is an imprint of the
Taylor & Francis Group, an **informa** business

A SCIENCE PUBLISHERS BOOK

CRC Press
Taylor & Francis Group
6000 Broken Sound Parkway NW, Suite 300
Boca Raton, FL 33487-2742

First issued in paperback 2019

Cover photo credit: *Dr. Monika Schmoll*

ISBN-13: 978-1-4665-8577-5 (hbk)
ISBN-13: 978-0-367-37983-4 (pbk)

This book contains information obtained from authentic and highly regarded sources. Reasonable efforts have been made to publish reliable data and information, but the author and publisher cannot assume responsibility for the validity of all materials or the consequences of their use. The authors and publishers have attempted to trace the copyright holders of all material reproduced in this publication and apologize to copyright holders if permission to publish in this form has not been obtained. If any copyright material has not been acknowledged please write and let us know so we may rectify in any future reprint.

Library of Congress Cataloging-in-Publication Data

Applications of microbial engineering / editors, Vijai Kumar Gupta, Monika Schmoll, Minna Mäki.
       p. ; cm.
  Includes bibliographical references and index.
  ISBN 978-1-4665-8577-5 (hardcover : alk. paper)
  I. Gupta, Vijai Kumar, editor of compilation. II. Schmoll, Monika, editor of compilation. III. Maki, Minna, editor of compilation.
  [DNLM: 1. Genetic Engineering--methods. 2. Microbiological Phenomena. QU 450]

  QH442
  572.8'6--dc23
                                                          2013011775

Visit the Taylor & Francis Web site at
http://www.taylorandfrancis.com

and the CRC Press Web site at
http://www.crcpress.com

Foreword

Humans have used microbes for centuries in the preparation of food and drink. Over these centuries, strain improvements have come randomly, through spontaneous, chemical or radiological mutation. In many cases, we do not know the nature of the changes to an organism that increase (or decrease) its productivity. In a sense, the cell factory is largely an enigma—the input is controlled and the output is predictable—what happens between input and output is not well understood. Taking an engineering approach to controlling the metabolic output of microbes is a recent advance relative to the amount of time that humans have made use of domesticated microbes.

Today we use microbes for the production of enzymes, fuels, commodity chemicals, food, alcohol, crops and pharmaceuticals. Therefore, the field of microbial engineering covers a huge breadth of research whose applications touch our lives on a daily basis. Moreover, the wide variety of approaches used for the engineering of microbes is also massive. Functional genomics, metabolic engineering, systems biology and synthetic biology are all terms used to describe various ingredients needed to successfully engineer microbes. Thus, despite the importance of microbes, it is a daunting challenge for researchers, students and educators to grasp the breadth of endpoints for engineered microbes.

The chapters of this book survey topics across the landscape of microbial engineering. The applications of microbial engineering that are covered by chapters range from food and drink to plant-microbe interactions to biofuel and organic acid production. The phylogenetic space covered by the variety of microbes is equally broad. It is therefore, the broad group of researchers, students and educators whose interest converge on the topic of microbial engineering that will benefit from the information contained in this book.

Scott E. Baker, Ph.D.
Interim Lead for Biology
Environmental Molecular Science Laboratory (EMSL)
Pacific Northwest National Laboratory
902 Battelle Boulevard
P.O. Box 999, MSIN P8-60
Richland, WA 99352, USA

Preface

Microbial engineering technologies have been identified as an essential and important subject area of bio-engineering and applied biological sciences. Currently, the world economy needs innovative research for sustainable and environmentally safe industrial processes. Microbial engineering is significant, not only due to the fact that microbes are able to perform chemical conversions in a more environment friendly manner, but also because of the advantages in the production of compounds, over classical chemistry. Scientists and engineers, alike, are motivated to develop sustainable green technology to underpin the predicted industrial revolution that will increasingly depend on microorganisms. However it should be emphasized, that expertise in the use of microbes for industrial applications has to be further developed, either through genetic manipulations or conventional mutagenesis.

Microbial engineering encompasses diverse areas including biotechnology, chemical engineering, and alternative fuel development. The discipline seeks to exploit fungi, bacteria and algae as workhorses for industry. A microbial engineer works on the biological, chemical and engineering aspects of biotechnology, manipulating microbes and developing new uses for bacteria and fungi.

With both introductory and specialized chapters on recent research work, this book will contribute to a better understanding of industrially important microbial processes and can highlight the potential of diverse organisms for production of enzymes or chemicals, which in many cases have become indispensable in daily life. The contributions by specialists on the respective topics provide a profound scientific basis for further research.

The editors of this book are grateful for research funding from Enterprise Ireland and the Industrial Development Authority, through the Technology Centre for Biorefining and Bioenergy (TCBB), as part of the Competence Centre programme under the National Development Plan 2007–2013 and to the Austrian Science Fund (FWF, project V152-B20). The support of Mr. B. Bonsall, Technology Leader (TCBB) and Prof. V. O'Flaherty, Chair of Microbiology, School of Natural Sciences & Deputy Director of the Ryan Institute for Environmental, Marine and Energy Research at NUI Galway, Ireland, is gratefully acknowledged during the compilation of this project. Editors of the book are very much thankful to Dr. Monika Schmoll for designing the cover page of the book. The editors also express gratitude to their colleagues in the Molecular Glycobiotechnology Group, Discipline of Biochemistry, School of Natural Sciences, NUI Galway, Ireland; MITS University, Rajasthan, India; Vienna University of Technology, the Austrian Institute of Technology (AIT); University of Santa Maria— UFSM Av. Roraima, Brazil and Orion Diagnostica Oy., Finland for their co-operation.

Editors

Contents

Aspergillus: A Cell Factory with Unlimited Prospects

Markus R.M. Fiedler, Benjamin M. Nitsche, Franziska Wanka and *Vera Meyer**

ABSTRACT

The genus *Aspergillus* covers a diverse group of filamentous fungi including industrially important species like *A. niger, A. oryzae, A. awamori, A. sojae* and *A. terreus*. Species of this genus have been exploited in large scale industrial production processes for almost 100 years. As microbial cell factories, filamentous fungi are outstanding with respect to their tolerance of extreme cultivation conditions, their ability to grow on plant biomass, their high secretion capacities and versatile secondary metabolism. The array of *Aspergillus* products includes bulk chemicals, enzymes for food and feed processing, homologous and recombinant proteins as well as bioactive compounds. This chapter aims at providing a comprehensive overview of the advances made during the last decade to further establish and improve Aspergilli as industrial production platforms. It starts with a description of the molecular genetics toolbox that has been developed for rational strain improvement, followed by various genetic strategies that have been applied to improve production of heterologous proteins including optimization of transcription, translation, secretory fluxes, product

Berlin University of Technology, Institute of Biotechnology, Department of Applied and Molecular Microbiology, Gustav-Meyer-Allee 25, 13355 Berlin, Germany.
* Corresponding author: vera.meyer@tu-berlin.de

degradation and morphology. The second part of this chapter provides an overview of omics tools established for Aspergilli and highlights recent omics studies on *Aspergillus* as producer of organic acids, plant polysaccharide degrading enzymes and secondary metabolites. Finally, the future prospects of *Aspergillus* as a cell factory are discussed.

Introduction

The kingdom of fungi covers a large and diverse group of lower eukaryotes which includes about 100,000 known species and presumably a million yet to be described and characterized (Hawksworth 1991). Fungi range from unicellular (yeasts) to multicellular organisms (filamentous fungi) and are diverse in morphology, physiology and ecology. Among the group of filamentous fungi, the genus *Aspergillus* is of considerable importance for industrial biotechnology. Their ability to grow on rather simple and inexpensive substrates as well as their natural capacity to secrete high amounts of hydrolytic proteins into the environment combined with its ability to synthesize and secrete various organic acids have attracted considerable interest to exploit them as production organisms in biotechnology and food industry. Important industrial production hosts include *A. niger, A. awamori* (a subspecies of the *Aspergillus* section Nigri (Perrone et al. 2011)), *A. oryzae, A. sojae* and *A. terreus*. In general, most members of the genus *Aspergillus* are saprophytic and are of vital importance for nutrient cycling and the function of ecosystems. However, few *Aspergilli* are pathogenic causing detrimental effects on plants and humans such as *A. flavus, A. parasiticus* and *A. fumigatus*.

Aspergilli exploited at an industrial scale, have a long history of safe use and many of their products have acquired the GRAS status meaning that they are generally regarded as safe by the American Food and Drug Administration (Table 1). The groundwork for *Aspergillus* as microbial cell factory was laid at the dawn of the twentieth century that was accompanied by advances in microbiology, biochemistry and fermentation technology. The pioneering works of Jokichi Takamine (production of amylase from Japanese koji mold, *Aspergillus oryzae*, 1894), James Currie (development of fungal fermentation for citric acid production 1917) and Alexander Fleming (discovery of penicillin production by *Penicillium notatum*, 1928) stimulated scientists to further explore fungal metabolic capacities and, moreover, triggered engineers to develop large-scale production processes for filamentous fungi. The findings of James Curie, for example, led to the establishment of the first industrial scale production process with a filamentous fungus by Pfizer already in 1919. Improvements of fungal capacities to produce metabolites of interest were, however, mainly restricted to classical mutagenesis techniques followed by tedious selection strategies.

Table 1. Selected examples of industrially important compounds produced by *Aspergilli*.

Product	Host	Company
Organic acids		
Citric acid	*A. niger*	Adcuram, ADM, Anhui BBCA Biochemical, Cargill, Jungbunzlauer, Gadot Biochemical Industries, Iwata Chemical Co. Ltd., Tate & Lyle
Itaconic acid	*A. terreus*	Itaconix, Shandong Kaison Biochemical, Qingdao Langyatai Group
Kojic acid	*A. oryzae*	Chengdu Jinkai Biology Engineering Industry, MHC INDUSTRIAL CO., LTD., Sansyo Pharmaceutical Co. Ltd., Wuxi syder Bio-products Co. Ltd.
Enzymes		
α-Amylase	*A. oryzae*	Amano Enzyme Co. Ltd., Biocon, DSM, Novozymes, Dupont IB, Novo Nordisk, Hunan Hong-Ying-Xiang Bio-Chemistry, Shin Nihon Chemical Co. Ltd.
Arabinase	*A. niger*	DSM, Shin Nihon Chemical Co. Ltd.
Asparaginase	*A. niger,* *A. oryzae*	DSM, Novozymes
Catalase	*A. niger*	DSM, Dupont IB, Novozymes, Shin Nihon Chemical Co. Ltd.
Cellulase	*A. niger*	Biocon, DSM, Dyadic, Genencor INT, Haihang Industry, Shin Nihon Chemical Co. Ltd., TNO
Chymosin	*A. niger*	Christian Hansen
β-Galactosidase	*A. niger,* *A. oryzae*	Amano Enzyme Co. Ltd., DSM, Dupont IB, Genencor INT, Novozymes, Shin Nihon Chemical Co. Ltd.
Glucoamylase	*A. niger*	Amano Enzyme Co. Ltd., Cangzhou Kangzhuang Chemical, DSM, Dyadic, Novozymes, Dupont IB, Shandong Longda Bio-Products
Glucose oxidase	*A. niger,* *A. oryzae*	Amano Enzyme Co. Ltd., DSM, Dupont IB, Dyadic, Novozymes
Hemicellulase	*A. niger*	Amano Enzyme Co. Ltd., BASF, Biocon, DSM, Dupont IB, Genencor INT, Novozymes, Shin Nihon Chemical Co. Ltd.
Lactoferrin	*A. niger*	DSM, Agennix
Lipase	*A. niger,* *A. oryzae*	DSM, Dupont IB, Novozymes, Novo Nordisk
Pectinase	*A. niger*	Biocon, DSM, Dupont IB, Novozymes, Shandong Longda Bio-Products
Phytase	*A. niger,* *A. oryzae*	BASF, DSM, Novozymes, TNO
Proteases (Acid, Neutral, Alkaline)	*A. niger,* *A. oryzae,* *A. saitoi*	Amano Enzyme Co. Ltd., DSM, Novozymes, Mitsubishi Foods Co. Ltd., Shin Nihon Chemical Co. Ltd.
Tannase	*A. oryzae,* *A. ficuum*	ASA Spezialenzyme GmbH, Biocon, Kikkoman Corp.
Secondary metabolites		
Fumagillin	*A. fumigatus*	Merck
Lovastatin	*A. terreus*	Biocon, Merck

After (Ward 2011, Meyer 2008)

New classical genetic techniques such as (para) sexual processes and protoplast fusion became available around the 1950's and further advanced the productivity of industrial processes. The birth of molecular biology in 1941 with the demonstration of the 'one gene-one enzyme' relationship in the filamentous fungus *Neurospora crassa* (Beadle and Tatum 1941) and the development of recombinant DNA technologies for filamentous fungi, shown for the first time in 1979 for *N. crassa* (Case et al. 1979), has finally revolutionised *Aspergillus* biotechnology. Since then, it became possible to obtain insights into the molecular basis of product formation and to improve traditional fungal fermentations by rational genetic engineering approaches, not only allowing production of homologous proteins but also production of proteins from non-fungal origin. For example, Novozymes has been the first company in the world which commercialised a recombinant lipase using *A. niger* as production host in 1984. Nowadays, the growing demand for industrial enzymes and organic aicds is met by *Aspergilli* (Table 1), which can be cultivated in large-scale stirred tank reactors reaching volumes up to 300,000 litres (Elander 2003). The advantage of *Aspergilli* over other microbial cell factories of bacterial or yeast origin is that they can tolerate extreme cultivation conditions covering a broad spectrum of pH (2–10), temperature (10–50°C), salinity (0–34%) and water activity (0.6–1) (Meyer et al. 2011b). As they are also able to efficiently degrade plant-derived polysaccharides such as starch, cellulose, hemicellulose, pectin and inulin, the importance of *Aspergilli* and its hydrolytic enzymes might even rise in the near future. For example, the efficiency of the saccharification process of second-generation feedstocks used for bioethanol production might become improved by *A. niger* derived (hemi)cellulases (Rumbold et al. 2009, 2010, Pel et al. 2007, de Souza et al. 2011).

The challenge for current and future strain development programs aiming at full exploitation of *Aspergilli* as multi-purpose expression platform is the full understanding of molecular cell biology of these hosts, the identification of pathway limitations and the substantiate prediction of beneficial metabolic engineering strategies. The aim of this chapter is to explore the possibilities and limitations of *Aspergillus* as a cell factory for the production of platform chemicals, proteins and pharmaceuticals. We review the progress made gestrichen in recent years to implement new molecular genetic engineering tools for rational strain improvements and discuss current technologies for the determination and evaluation of transcriptomic, proteomic and metabolomics data from different industrial *Aspergilli*. We highlight representative systems biology approaches which have uncovered some key players and regulatory mechanisms involved in protein secretion and the formation of primary and secondary metabolites. We also summarise the current knowledge of compartmentalized product biosynthesis as well as transport and traffic phenomena in *Aspergillus* as this is key to fully understand the link between product formation, secretion and morphology in these versatile expression hosts.

The Molecular Genetic Toolbox for *Aspergillus*

The basis for every rational experimental approach that applies genetic modification in any organism is a well-equipped molecular toolbox including suitable vectors, selection markers and transformation protocols. Although several plasmids have been found in bacteria, yeast and filamentous fungi such as *N. crassa*, no naturally occurring plasmids are present in *Aspergillus* (Griffiths 1995). Nevertheless, it has been shown that artificial plasmids with replication sites targeting the *Aspergillus* replication machinery are able to autonomously replicate in *Aspergilli* and distribute during mitosis. The introduction of the autonomous maintenance in *Aspergillus* (AMA1) sequence from a genomic library of *A. nidulans* resulted in plasmids which displayed autonomous replication properties similar to plasmids with homologous sequences used in *S. cerevisiae* (Verdoes et al. 1994b, Khalaj et al. 2007, Gems et al. 1991, Carvalho et al. 2010). However, the mycelium is heterogenic even under selection pressure, tolerating hyphae devoid of plasmid without showing any phenotype or growth defects (Aleksenko and Clutterbuck 1997). In addition, AMA1-based plasmids can get lost during long-term cultivation, especially under non-selective pressure. Due to these reasons, protein overexpression approaches mainly target the genome of *Aspergilli*. However, AMA1-based vectors carrying auxotrophic (*pyrG*) or dominat (*hygB*) markers are very helpful for complementation approaches of gene deletion mutants (Carvalho et al. 2010).

Several selection markers are available for genetic modification of *Aspergillus*. Well established are nutritional and auxotrophic markers like *argB*, *pyrG*, *pyrE*, *trpC*, *amdS* and *niaD* as well as antibiotic resistance markers based on hygromycin B (*hygB*) and phleomycin (*phle*) resistance (Meyer et al. 2010b). The advantage of using *pyrG* or *pyrE* as selection marker is that *pyrG⁻* or *pyrE⁻* strains can easily be obtained by direct selection on 5'-fluoroorotic acid medium (FOA) without any mutagenic treatment. Another advantage of these markers is that they can be used repeatedly, i.e. after integration of a *pyrG*- or *pyrE*-containing plasmid into the genome, transformants can be cured from *pyrG* or *pyrE* by cultivating them on FOA plates. The *amdS* gene can be used, much like *pyrG* and *pyrE*, as a bidirectional marker by counterselecting for the loss of *amdS* with media containing the antimetabolite fluoroacetamide (FAA). Detailed protocols for obtaining *pyrG⁻ pyrE⁻* or *amdS⁻* recipient strains of *A. niger* have recently been published (Meyer et al. 2010b).

Other versatile selection markers have recently been added to this collection. For example, three mutant alleles encoding subunits of a succinate dehydrogenase (*sdhB, sdhC, sdhD*) have been isolated from a carboxin resistant strain of *A. oryzae*. One of them (*sdhB*) has been shown to provide resistance to carboxin in *A. parasiticus* (Shima et al. 2009). Another auxotrophic marker for *A. niger* is based on the *sC* gene encoding an ATP-sulfurylase homologous to the *Saccharomyces cerevisisae MET3* gene. Complementation with a fully functional copy of the *sC* gene from a wild type *A. niger* strain conferes cysteine prototrophy (Varadarajalu and Punekar 2005). Finally, an efficient selection system targeting arginine catabolism of *Aspergillus* has most recently been established. An *A. niger* strain deficient in arginase (*agaA*) and hence unable to grow on arginine as the sole nitrogen source can efficiently be complemented with an arginase expression plasmid (Dave et al. 2012).

Counterselection of bi-directional markers is an effective approach to re-use selection markers in *Aspergilli*. As discussed above, counterselection with antimetabolites such as FOA and FAA allow the re-use of markers such as *pyrG* or *amdS*. The rationale is that the antimetabolite used is intracellularly converted into a toxic substance when an intact copy of the marker gene *pyrG* or *amdS* is present in the genome and actively expressed. Such a selection pressure allows the isolation of natural mutants which accumulate loss-of-function mutations within *pyrG* or *amdS* and/or the isolation of induced mutants which have lost the selection marker due to homologous recombination between direct repeats flanking *pyrG* or *amdS*, respectively (Meyer et al. 2010b, Carvalho et al. 2010). An alternative system for marker recycling has been developed for filamentous fungi by adaptation of the *cre/loxP* system from the bacteriophage P1. Here, the marker gene is flanked by 34 bp long DNA sequences (*loxP*), which become specifically recognized by the recombinase Cre. This enzyme efficiently catalyses the excision of any DNA sequence located between both *loxP* sites, which was successfully demonstrated in *A. nidulans, A. fumigatus* and *A. oryzae* (Krappmann et al. 2005, Forment et al. 2006, Mizutani et al. 2012).

For gene targeting approaches in *Aspergillus*, two challenges have to be met. Firstly, any recombinant DNA has to pass the cell wall and cell membrane of the recipient strain. Secondly, the DNA introduced has to become targeted to the desired locus efficiently. Several transformation techniques have been established to transform *Aspergilli*, among which polyethylene glycol (PEG)-mediated transformation of protoplasts is the most frequently used method. Other transformation methods such as *Agrobacterium tumefaciens*—mediated transformation (Michielse et al. 2008, de Groot et al. 1998), electroporation or biolistic transformation have been established as well, although their transformation efficiencies are not as high compared to PEG-mediated transformation of protoplasts (Meyer et al. 2003, Meyer 2008). Detailed protocols for PEG-mediated transformation of *A. niger* providing step-by-step instructions and helpful advices on how to avoid potential pitfalls have recently been published (Arentshorst et al. 2012, Meyer et al. 2010b). Basically, young mycelium is incubated with cell wall degrading enzymes or enzyme mixtures including Lysing Enzyme® from *Trichoderma harzianum*, chitinase from *Streptomyces griseus* or β-glucuronidase from *Helix pomatia* to degrade the cell wall, thus releasing protoplasts (de Bekker et al. 2009). Protoplasts are suspended in buffers with high salt or sugar concentrations (0.7–1.2 M sucrose, sorbitol or NaCl), in order to protect them from burst due to osmotic imbalance. DNA uptake is mediated by calcium chloride (10–50 mM) in combination with high concentrations of PEG. Although PEG-mediated transformation of cells is a well-established method for bacteria, yeast and filamentous fungi since the 1980s, the actual mechanism on how PEG enables DNA to enter the cell membrane has remained cryptic for almost three decades. Recently, it was uncovered that high concentrations of PEG mediate the attachment of dissolved DNA to cells and protoplasts of *S. cerevisiae* and facilitate the uptake of DNA via endocytosis (Kawai et al. 2010, Zheng et al. 2005). However, compared to transformation efficiencies obtained with *Escherichia coli* or *S. cerevisiae*, the transformation efficiencies of *Aspergilli*, and in general of filamentous fungi, are considerably lower, reaching only about 10–100 transformants per µg DNA (Fincham 1989).

A second challenge, when working with *Aspergilli* is their low frequency of homologous recombination, which hampered efficient functional gene analyses for a long time. Only after observing that disruption of the non-homologous end-joining (NHEJ) pathway in *N. crassa* resulted in homologous recombination frequencies up to 100% (Ninomiya et al. 2004), respective mutants were established in various filamentous fungi, thereby allowing genome-wide functional genomics studies to become feasible (see Kück and Hoff 2010, Meyer 2008). In brief, the NHEJ pathway is a eukaryotic mechanism which bridges broken DNA ends by the joint activities of the Ku heterodimer (Ku70/Ku80-protein complex) and the DNA ligase IV-Xrcc4 complex (Dudásová et al. 2004, Krogh and Symington 2004). In eukaryotes, the NHEJ pathway competes with another repair mechanism, the homologous recombination (HR) pathway, which mediates interaction between homologous DNA sequences, whereas the NHEJ pathway ligates double-strand breaks without the requirement of any homology (Shrivastav et al. 2008). By deleting either *ku70*, *ku80* or *lig4* genes, the HR frequency is dramatically increased in *Aspergilli* (Table 2). Another advantage of this high efficiency of gene targeting is that essential genes can easily be identified by the so called heterokaryon rescue technique as shown for *A. nidulans* and *A. niger* (Nayak et al. 2006, Carvalho et al. 2010). However, several studies have shown that inactivation of the NHEJ pathway makes fungal strains vulnerable to DNA damaging conditions thus increasing their sensitivity towards UV, X-ray or chemical mutagens (Meyer et al. 2007a,

Table 2. Homologous recombination frequencies in NHEJ defective mutants of different *Aspergilli.*

Species	Length of homologous sequence (bp)	Homologous recombination frequency (%)	Reference
A. fumigatus	100	(75)[a]	(Krappmann et al. 2006)
	500	(84)[a]	
	1000	96	
	1500	96	
	2000	95	
A. nidulans	500	89	(Nayak et al. 2006)
	1000	92	
	2000	90	
A. niger	100	18	(Meyer et al. 2007a)
	200	33	
	500	88	
	1000	95	
	1500	98	
A. sojae	500	14.3	(Takahashi et al. 2006)
	1000	71	
	1400	75	
	2000	87	
Neurospora crassa	100	10	(Ninomiya et al. 2004)
	500	91	
	1000	100	

[a]Frequencies given in brackets might not be significant due to low numbers of obtained transformants

Malik et al. 2006, Kito et al. 2008, Snoek et al. 2009). To eliminate the risk that NHEJ deficiency influences or obscures phenotypic analyses, *A. nidulans* and *A. niger* strains were established being transiently silenced in NHEJ, and respective strains have proven to perform as efficient as constitutive silenced NHEJ strains with respect to gene targeting (Carvalho et al. 2010, Nielsen et al. 2008).

In summary, the molecular genetic toolbox for *Aspergillus* has considerably been extended in recent years and provides the research community with versatile tools to genetically modify this genus in a rational and user-specified way.

Genetic Strategies to Improve *Aspergillus* as Protein Producer

Aspergilli are extraordinary in their ability to secrete high amounts of proteins into the environment. Concentrations up to 20 g/l culture medium are no peculiarities for a wide range of host specific proteins (Finkelstein 1987). Secreted fungal proteins like amylases, lipases or proteases are produced for a wide range of applications including laundry, biorefinery pre-processing or food industry (Table 1) (Fleissner and Dersch 2010). Their outstanding secretion capacity in the case of homologous fungal proteins, their ability to post-translationally modify proteins and their long history of safe use fostered attempts to produce proteins of non-fungal origin in *Aspergilli*. Unfortunately, respective expression levels are considerably lower compared to homologous proteins (Conesa et al. 2001). This phenomenon is still one of the major challenges in *Aspergillus* biotechnology, although some progress has been made in the last decade. Successful and combined approaches include (i) the use of strong, constitutive or inducible promoters, (ii) carrier protein approaches, i.e., a genetic fusion of the heterologous protein of interest with a homologous protein ('carrier protein') which is secreted by *Aspergillus*, (iii) increased expression of chaperones and foldases by tackling the unfolded protein response (UPR) and the endoplasmatic reticulum associated degradation (ERAD) pathway and (iv) down-regulation of proteases secreted by *Aspergillus*. We will briefly summarise these strategies here, and also recommend reading the reviews by Lubertozzi and Keasling 2009, Ward 2011, Meyer 2008, Fleissner and Dersch 2010 for more detailed information.

Improving recombinant protein expression by optimising transcription

To produce homologous and heterologous proteins in *Aspergilli*, different promoter systems growth-dependent or inducible, have been established and validated (Table 3) (Fleissner and Dersch 2010). In general, inducible promoters are superior to growth-dependent promoters because they offer more flexibility for controlled protein production. For example, protein expression can be induced after cells reach the exponential growth phase at an optimum level thus lowering any potential toxic side effects to the host.

Among the growth-dependent promoters, the glyceraldehyde-3-phosphate dehydrogenase (*gpdA*) promoter is most widely used in *Aspergilli*. Other successfully applied promoters include the *adhA* and *tpiA* promoter of *A. nidulans* (Upshall et al. 1987), the *pkiA* promoter of *A. niger* (Storms et al. 2005), the *gdhA* promoter of *A. awamori* (Moralejo et al. 1999) and the *hlyA* promoter from *A. oryzae* which has been shown to be highly active under solid state fermentations. Recently, the citrate synthase *citA* promoter of *A. niger* has been evaluated showing high expression under different carbon sources such as glucose, sucrose, starch, glycerol, acetate or molasse (Dave and Punekar 2011). Furthermore, Blumhoff and colleagues selected six novel growth-dependent promoters (*mbfA, coxA, srpB, tvdA, mdhA,*

Table 3. Established and new promoters applicable for recombinant protein overexpression in *Aspergillus*.

Species	Promoter	Gene function	Inducible	Remark	Reference
A. awamori	*exlA*	Xylanase	Yes	High expression level Inductor: D-xylose Repressor: glucose	(Gouka et al. 1996)
	gdhA	Glutamate-dehydrogenase	No	Nitrogen-and growth-dependent	(Cardoza et al. 1998)
	glaA	Glucoamylase	Yes	High expression level Inductor: starch, maltose, maltodextrin, glucose	(Ward et al. 1995)
A. fumigatus	*niiA*	Nitrite reductase	Yes	Nitrogen-dependent Repressor: ammonium	(Hu et al. 2007)
A. nidulans	*alcA*	Alcohol dehydrogenase I	Yes	Medium expression level Inductor: ethanol, ethylmethylketone Repressor: glucose	(Gwynne et al. 1989)
	gpdA	Glycerinaldehyd-3- phosphat dehydrogenase	No	High expression level Growth-dependent	(Punt et al. 1990)
A. niger	*catR*	Catalase	Yes	Inductor: H_2O_2	(Sharma et al. 2012)
	citA	Citrate synthase	No	High expression level	(Dave and Punekar 2011)
	glaA	Glucoamylase	Yes	High expression level Inductor: starch, maltose, maltodextrin, glucose Repressor: xylose, sucrose Leaky	(Fowler et al. 1990)
	HERα	Human estrogen receptor-expression system	Yes	Tunable expression Inductor: estrogen Tight but weak	(Pachlinger et al. 2005)
	inuE	Exoinulinase	Yes	Inductor: inulin and sucrose	(Yuan et al. 2008)

	Gene	Description		Expression characteristics	Reference
	mbfA	Predicted multiprotein bridging factor 1	No	High expression level	(Blumhoff et al. 2012)
	pkiA	Pyruvate kinase	No	High expression level	(Storms et al. 2005)
	sucA	ß-fructo-furanosidase	Yes	High expression level; Inductor: sucrose, inulin; Tight	(Roth and Dersch 2010)
	Tet-On	Tetracycline-expression system	Yes	Tunable expression; Inductor: Doxycycline; Tight	(Meyer et al. 2011a)
	ttdA	Predicted transport docking protein	No	Medium expression level	(Blumhoff et al. 2012)
A. oryzae	amyA	Alpha-amylase	Yes	High expression level; Inductor: starch, maltose	(Tada et al. 1991)
	hlyA	Hemolysin	No	High expression level; No glucose repression	(Bando et al. 2011)
	sodM	Manganese superoxid dismutase	Yes	Controllable with H_2O_2	(Ishida et al. 2004)
	tef1	Translation elongation factor	No	Growth-dependent; No glucose repression	(Kitamoto et al. 1998)
	thiA	Thiamine	Yes	Tunable expression; Inductor: Thiamine; Inactive under alkaline pH	(Shoji et al. 2005)

manB), which display different activation levels, thus being potentially applicable for fine-tuned protein expression when low or medium levels of expression are aimed for, from transcriptomic data of *A. niger* (Blumhoff et al. 2012).

Among the inducible promoter systems, the starch inducible *glaA* and the ethanol inducible *alcA* promoters are most widely used. The gene *glaA* encodes a γ-glucoamylase of *A. niger* which is induced by starch, maltose and repressed in the presence of xylose (Fowler et al. 1990). Activation of *glaA* is mediated via the transcriptional regulator AmyR, which controls the expression of amylolytic genes upon growth on starch or maltose as carbon sources (Vinck et al. 2011, Vongsangnak et al. 2009). The alcohol dehydrogenase (*alcA*) promoter of *A. nidulans*, becomes strongly activated by ethanol (Nikolaev et al. 2002, Felenbok 1991), which is mediated by the AlcR transcription factor and depends on acetaldehyde as co-inducer (Nikolaev et al. 2002). Expression of both the *glaA* and the *alcA* gene are repressed by the catabolite repressor CreA in the presence of an easily metabolizable carbon source like glucose (Prathumpai et al. 2004). CreA binds competitively to both promoters, thereby hindering accession of AlcR or AmyR to their promoter binding sites. This competitive inhibition was shown to be bypassed by introducing multiple copies of AlcR binding sites to the *alcA* promoter (Gwynne et al. 1989). In addition, the number of AlcR molecules expressed can increase the activation level of the *alcA* promoter (Panozzo et al. 1997).

Other inducible promoter systems include the xylose inducible *exlA* promoter (Gouka et al. 1996), the sucrose and inulin inducible *sucA* promoter of *A. niger*, repressible by glucose or maltose (Roth and Dersch 2010), the inulin inducible *inuE* promoter of *A. niger*, repressible by glucose (Yuan et al. 2008) and the hydrogen peroxide and calcium carbonate inducible *catR* promoter of *A. niger* (Sharma et al. 2012). Most of the inducible promoter systems described above are metabolism-dependent; hence, the choice of the growth medium is limited. In order to overcome this constraint, different efforts were undertaken to establish and validate metabolism-independent inducible promoter systems for the genus *Aspergillus*: the thiamine promoter system (P*thiA*) in *A. oryzae* (Shoji et al. 2005), the human estrogen receptor (hERα) system in *A. nidulans* and *A. niger* (Pachlinger et al. 2005), and a system based on the *Escherichia coli* tetracycline-resistance operon (Tet-On) in *A. fumigatus* (Vogt et al. 2005) and *A. niger* (Meyer et al. 2011a). All systems drive gene expression in a inducer-dependent manner, however the *hER*α promoter system is either leaky but strong or tight and weak, whereas the P*thiA* system is inactive under alkaline conditions. The most promising expression system is the Tet-On system, which is tight under non-induced conditions, and responds within minutes after inducer addition and mediates high gene expression levels comparable to levels of the *gpdA* promoter in *A. niger* (Meyer et al. 2011a). The Tet-On system is composed of the reverse transactivator rtTA2^s-M2 which is under control of the *gpdA* promoter, thus being constitutively expressed. It can only bind to the *tetO* operator sequence, which is located upstream of a minimal promoter (Pmin), after addition of doxycycline (DOX). Consequently it activates transcription of the gene of interest (GOI) in a DOX-dependent manner (Fig. 1). Most importantly, expression levels of the gene of interest can be fine-tuned in a user-specified manner: the amount of DOX added to medium and the copy number of the GOI expression cassette present in the genome of *A. niger* define the amount of protein produced. In addition, the spectrum of time-dependent induction can also vary between hours (one copy, low amount of DOX, low amount of protein product) to only few minutes (multiple copies, high DOX, high amount of protein product) (Meyer et al. 2011a). These characteristics make the Tet-On system superior to the other inducible systems mentioned above and suggest that this system can be used as a general and excellent tool for user-defined protein overexpression in *Aspergillus*.

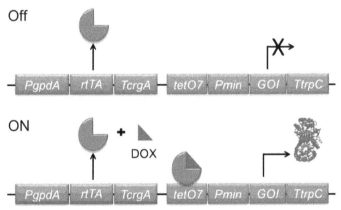

Fig. 1. Schematic representation of the Tet-On expression system for *A. niger* (Meyer et al. 2011a). The constitutively expressed reverse transactivator rtTA2^s-M2 (rtTA) is unable to bind to the operator sequence in the absence of DOX (the system is OFF). However, it undergoes a conformational change upon binding to DOX. Consequently, binding of rtTA2^s-M2 to the tetracycline operator, of which seven copies are present (*tetO7*), becomes possible and expression of the gene of interest (GOI) becomes induced (the system is ON). A prerequisite for induction of gene expression is the presence of a minimal promoter (Pmin) sequence which is located downstream of *tetO7*.

Improving recombinant protein expression by optimising translation

Production rates of recombinant proteins do not only depend on the amount of transcript produced but also on the amount of transcript translated and proteins channelled   through the secretory pathway before being released released into the cultivation broth. Codon bias has been known for a long time to be an important factor for efficient protein expression in pro- and eukaryotes. In highly expressed genes of *E. coli*, favoured codons correlate with high levels of the corresponding tRNAs (Ikemura 1981), which also holds true for *S. cerevisiae* (Ikemura 1982). The codon usage of *N. crassa* (Whittle et al. 2012, Edelmann and Staben 1994) and *A. nidulans* (Lloyd and Sharp 1991) have recently been revisited and compared with 10 food-related filamentous fungi, including *A. niger*, *A. oryzae* and *A. terreus*. This comparative study revealed that preferred nucleotides at the third position of a codon are pyrimidines (Chen et al. 2012). In case that a purine is present at the wobble position, guanine is the favoured nucleotide. That the presence/absence of abundant/rare codons can indeed affect protein translation has been proven for filamentous fungi two decades ago. It has been shown that the insertion of three rare codons into the highly expressed glutamate dehydrogenase of *N. crassa* decreases the protein level to 30% of the native gene while the transcription level was unaffected (Kinnaird et al. 1991). Several studies have confirmed this observation and higher product levels were indeed obtained by codon-optimization in filamentous fungi (Nelson et al. 2004, Cardoza et al. 2003). For instance, the rate of a recombinant aequorin expressed in *N. crassa* has increased 45 fold compared to the native gene (Nelson et al. 2004) and expression of a codon-optimised synthetic gene encoding an industrial relevant alpha-glucan phosphorylase in *A. niger* has been proved successful (Koda et al. 2005).

Other studies have shown that the codon usage can have an impact on polyadenylation and mRNA stability. For example, an optimal codon usage reduced the premature adenylation of a heterologous endoglucanase in *A. oryzae*, thus preventing the accumulation of abbreviated mRNA (Sasaguri et al. 2008). Likewise, optimization of the codon usage improved the mRNA stability of a heterologously expressed mite allergen in *A. oryzae*. The observed half-life of the codon optimized RNA was tripled to 43 minutes compared to its native mRNA, which is comparable to the half-life of the mRNA of the highly expressed *glaA* gene from *A. oryzae* (Tanaka et al. 2012). Most recent codon usage analyses of eight different *Aspergilli* (including the industrial strains *A. niger*, *A. oryzae* and *A. terreus*) made use of publically available genomic and transcriptomic data and confirmed that synonymous codon usage bias is indeed associated with expression levels in the analyzed species, since in all cases genes coding for ribosomal proteins and other highly expressed genes (e.g., translation elongation factors, enzymes of the Krebs cycle, tubulin) clustered together in correspondence analyses. This data was used to define, for each species, a set of "optimal codons" for highly expressed genes (Iriarte et al. 2012).

In addition to the optimisation of single codons, adaptation of codon pairs according to the host genomic distribution has been shown to influence mRNA folding stability and the accuracy and speed of the translation elongation process in *S. cerevisiae* (Kahali et al. 2011). Most recently, van Peij and co-workers have adapted this approach for *A. niger* by comparing the codon and codon pair usage of highly secreted endogenous and heterologous genes with those of poorly secreted proteins. Based on this data, the codon pair usage of poorly secreted proteins was redesigned, thereby considerably increasing the production yields, while maintaining the biological activity of the proteins produced (van Peij et al. 2012). Finally, machine learning algorithms were used to identify a large set of homologous (600) and heterologous (2,000) fungal genes (all being expressed from a standardized expression cassette at a specific genomic locus) those relevant DNA and protein sequence features which correlate with high/ low protein production levels in *A. niger*. Using this approach, van den Berg and colleagues uncovered that the amino-acid composition of the protein sequence is most predictive and that, for both homologous and heterologous gene expression, the same features are important: tyrosine and asparagine positively correlate with high-level production, whereas methionine and lysine composition contribute to unsuccessful production (van den Berg et al. 2012). The trained classifier algorithm that can be used as predictor for low/high protein level expression has been made available online at http://bioinformatics.tudelft.nl/hipsec.

Beside codon bias/codon pair optimisation, other strategies have been developed to increase protein production. In general, multiple copies of an expression construct do usually increase protein production levels (Verdoes et al. 1993, 1994a, Cardoza et al. 2003, Meyer et al. 2011a); however, this positive correlation is limited. For example, transcription factor titration and/or positional effects such as insertion of the expression cassette into silenced genomic regions or inactivation of surrounding genes which might cause pleiotropic effects, can interfere with protein expression levels (Verdoes et al. 1994c, 1994a, Kelly and Hynes 1987).

Although manifold rational genetic engineering strategies have been developed, such as codon bias/codon pair optimization, optimised expression cassettes with suitable promoters and multiple integration events, protein overexpression in *Aspergillus* remains an art and no generic solutions are available. However, changing the perspective from a protein-oriented to a host-oriented view might open new possibilities for substantial improvements. This means that it is not only important to adapt the characteristics of the protein of interest to the host's requirements but it is also mandatory to truly understand the capacities of *Aspergillus* to express and channel any protein of interest through the secretory pathway and at which cellular burden this is accomplished. Basically, induced overexpression of any recombinant protein in *Aspergillus* has to compete with the natural secretion capacity of homologous proteins thus being at the cost of other cellular processes or even at the cost of other homologous secretory proteins important for normal growth. In agreement, RNAi-mediated silencing of abundantly secreted endogenous amylases has been shown to improve the expression yield of recombinant chymosin in *A. oryzae* and silencing of an endogenous fungal cellobiohydrolase in *Trichoderma reesei* increased the amount of secreted lipase several fold compared to non-silenced strains (Nemoto et al. 2009, Qin et al. 2012). Hence, it is crucial to identify and understand all homeostatic control mechanisms, which modulate the secretory flux of *Aspergillus* so that the fungus can flexibly adapt its cellular capacities to the artificially induced burden and to its own cellular needs for fast growth.

Improving recombinant protein expression by optimising the secretory flux

An inventory of the Aspergillus secretion pathway

In general, *Aspergillus* proteins and enzymes destined for secretion into the extracellular space have to follow the secretory route (Fig. 2). Their journey starts with protein translocation into the lumen of the endoplasmic reticulum (ER) via a translocon that forms a channel through the ER (Römisch 1999). At the ER, several ER resident chaperons and foldases including the binding protein (BipA), protein disulfide isomerase (PdiA) and calnexin (ClxA) assist secretory proteins in their folding (Määttänen et al. 2010). Many secretory proteins also become glycosylated, via attaching oligosaccharides to their asparagine residues (N-glycosylation) or serine and threonine residues (O-glycosylation). After correct folding and glycosylation, secretory proteins are packed into vesicles and transported to and through the Golgi complex via coatamer protein complex vesicles (COPII) and are then delivered to the Spitzenkörper, which is localised at the hyphal tip, via a microtubule-mediated transport. At the Spitzenkörper, secretory vesicles become translocated to actin cables and transported to the plasma membrane, where the vesicles release their cargo into the extracellular space. Empty vesicles become recycled by endocytosis which takes place at subapical regions of the hyphal tip and are transported back to the Golgi to become reloaded with new cargo (Taheri-Talesh et al. 2008). Hence, protein secretion in *Aspergillus* should be viewed as a concerted and balanced activity of exo- and endocytotic mechanisms (Fig. 2).

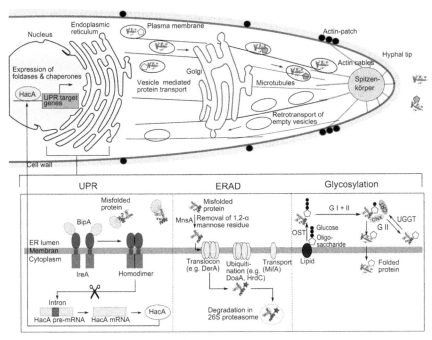

Fig. 2. The secretion pathway in *Aspergillus*.

High protein flux through the ER or expression of heterologous proteins in *Aspergillus* can result in the accumulation of misfolded proteins, which is recognized by a quality control system known as ER-associated degradation (ERAD). This control mechanism guides misfolded proteins to the cytosol where they become degraded by the proteasome (Fig. 2). In addition, another control mechanism becomes induced, named the unfolded protein response (UPR). The purpose of UPR is to refold misfolded proteins via induced expression of chaperones and foldases (Geysens et al. 2009). For this purpose, the UPR transcription factor HacA becomes activated via splicing of an unconventional 20-nt intron out of the *hacA* mRNA. This subsequently facilitates translation of the *hacA* mRNA and formation of HacA, which in turn induces transcription of a number of UPR target genes including *bipA* and *pdiA* (Mulder et al. 2004). Hence, both ERAD and UPR are key for the function of the secretory pathway—control mechanisms which are not only present in filamentous fungi but also in yeast and mammals (Kaufman 1999, Yoshida et al. 2001, Saloheimo et al. 2003, Saloheimo and Pakula 2012).

Improving secretory protein fluxes by manipulating quality control systems

There are several bottlenecks for the efficient production of recombinant proteins in *Aspergillus*. These are due to limitations within the secretory pathway (Nyyssönen and Keränen 1995, Jeenes et al. 1994, Gouka et al. 1997). Cullen and co-workers developed a protein carrier system for *A. niger* by genetically fusing the N-terminus of the abundantly secreted protein glucoamylase GlaA to the heterologous target protein chymosin (Cullen et al. 1987). They have shown that the fusion partner facilitated production of the recombinant protein by (i) increasing the translocation rates into the ER thus improving folding, maturation and stability (Bermúdez-Humarán et al. 2003, Le Loir et al. 2001) and (ii) masking the heterologous protein product thereby preventing proteolytic degradation (Gouka et al. 1997). To date, the GlaA protein is the most commonly used secretion carrier in *A. niger* and several proteins have been successfully produced using this approach (Martin et al. 2003, Ward et al. 1990, 1995, 2004). In addition to GlaA, other homologous and heterologous secretion carriers have been successfully used as carrier in different *Aspergillus* species, e.g., the α-amylase sequence of *A. oryzae* (Nakajima et al. 2006, Korman et al. 1990). The fusion carrier gets removed in the Golgi network prior to secretion when a Lys-Arg or Arg-Arg doublet is inserted between the carrier and the recombinant protein. These motifs are specifically recognized in the trans-Golgi network by the KexB protease, a furin-type endoprotease member of the kexin protease family, which removes the carrier protein (Jalving et al. 2000, Contreras et al. 1991, Punt et al. 2003). However, processing rates depend on the localization of the protease recognition site within the three dimensional structure of the fusion protein and can thus be inefficient in case that the site is inaccessible (Spencer et al. 1998).

When *Aspergillus* is forced to overexpress a heterologous protein, the ER protein folding capacity becomes overloaded, resulting in delayed or incorrect protein folding (Guillemette et al. 2007) and induction of the UPR (Kaufman et al. 2002, Punt et al. 1998, Wang et al. 2003). Hence, numerous studies have been conducted on *Aspergilli* in order to increase the level of chaperons and foldases to promote the synthesis of correctly folded heterologous proteins. For example, overexpression of the chaperone BipA in *A. awamori* resulted in a 2 to 2.5 fold increase of the sweet plant protein thaumatin (Lombraña et al. 2004). Similarly, insertion of multiple copies of the *pdiA* gene also led to an increased amount of thaumatin in *A. awamori* (Moralejo et al. 2001). In another approach, Valkonen and co-workers constitutively induced the UPR in *A. awamori* by removing the 20 bp intron of *hacA* leading to a 2.8 fold and 7.6 fold increase of a bovine chymosin and *Trametes vesicolor* laccase production, respectively (Valkonen et al. 2003).

However, whether recombinant protein production can be improved by up-regulation of the UPR is largely dependent on the target protein. For example, overexpression of BipA in *A. awamori* did not increase cutinase production (van Gemeren et al. 1998) and overexpression of PdiA did not improve expression levels of hen egg white lysozyme in *A. niger* (Ngiam et al. 2000).

To systematically explore the role of the UPR in protein secretion and the role of HacA as its central regulator, the transcriptome of *A. niger* expressing constitutively active form of HacA (HacA^{CA}), thus causing an overload of the ER with misfolded proteins, was compared with the transcriptome of a strain expressing the wild type copy of HacA (Carvalho et al. 2012). Several genes related to ER translocation, protein folding, protein glycosylation, vesicle-mediated transport between organelles, exo- and endocytosis and the ERAD pathway were up-regulated in the HacA^{CA} strain, including well known HacA targets such as *bipA, pdiA, tigA* and *prpA*, suggesting that HacA is a master regulator coordinating many processes within the secretory pathway. However, many genes involved in transcription and translation as well as in central catabolic pathways, including the AmyR regulon responsible for starch degradation, were down-regulated. This phenomenon, earlier reported for *A. niger* and the filamentous fungus *Trichoderma reesei* and known as the REpression under Secretion Stress (RESS) phenomenon (Pakula et al. 2003, Al-Sheikh et al. 2004), suggests that *Aspergillus* and other filamentous fungi respond to the protein overload in the secretion pathway by reducing the expression of genes encoding dispensable secretory proteins and by increasing the folding capacity of the cell. Overall, global mechanisms for energy generation and cell development seem to become arrested probably to prevent the entry and overload of newly synthesized proteins into the already "clogged" ER. Decreasing the metabolic activity reduces the influx of new proteins into the overloaded ER and extends the residence time of misfolded proteins within the ER to become successfully refolded.

The ERAD pathway is closely linked to the UPR and has been studied for a long time in *S. cerevisiae*. In case that correct folding of secretory proteins fails although expression of foldases and chaperones has been up-regulated by the UPR, misfolded proteins become targeted to the ERAD pathway (Fig. 2). In brief, protein degradation is initiated by the removal of an 1,2-α-mannose residue from the protein by the specific 1,2-α-mannosidase Mns1p in *S. cerevisiae* the respective homolog of which is MnsA in *A. niger* (Gonzalez et al. 1999, Tremblay and Herscovics 1999, Carvalho et al. 2011). Subsequently, the protein becomes retrotranslocated from the ER to the cytoplasm through the Sec61p translocon or the Der1p (DerA) translocon (Schäfer and Wolf 2009, Goder et al. 2008). Thereafter, the protein becomes ubiquitinylated by different complexes consisting of different proteins such as Hrd3p (HrdC) and Doa1p (DoaA) (Kostova et al. 2007). Mif1p (MifA) transports the labelled protein to the 26S proteasome (van Laar et al. 2001), where it becomes degraded in an ATP-dependent manner (Fischer et al. 1994). Carvalho and co-workers recently investigated the role of the ERAD pathway in *A. niger* and its potential link to the UPR. UPR-inducing conditions were achieved by induced expression of heterologous proteins from bacterial, metazoan and human origin or due to treatment of *A. niger* with dithiothreitol (DTT) or tunicamycin. These stress conditions were applied to different *A. niger* strains, a wild type strain or mutant strains carrying a single deletion in one of the ERAD components MnsA, DerA, DoaA, HrdC and MifA, respectively (Carvalho et al. 2011) Forced expression of heterologous proteins in *A. niger* indeed triggered up-regulation of BipA, PdiA, DerA and HrdC in the wild type background strain, thus proving a close link between the UPR and ERAD pathway in *A. niger*. Surprisingly, deletion of none of the ERAD proteins MnsA, DerA, DoaA, HrdC or MifA increased the susceptibility of *A. niger* towards DTT and tunicamycin which are both known to induce expression of the UPR genes *bipA* and *pdiA*. In agreement, deletion of ERAD components did not result in any apparent phenotype (except for Δ*doaA*) and did not impair the faith of the heterologous protein. These observations pose the question, whether degradation of misfolded proteins is mediated by the ERAD pathway in *A. niger*, or whether this might be accomplished by an alternative, yet unknown mechanism, e.g., by the activity of ER-resident proteases as discussed for mammals (Evnouchidou et al. 2009, Carvalho et al. 2011). Taken together, the data from *A. niger* suggest that both the UPR and the ERAD pathways are cellular responses to forced protein overexpression and/or ER stress; however, induction of the UPR seems to be sufficient to maintain ER homeostasis, whereas the ERAD pathway is of minor importance for efficient protein production and secretion.

Improving secretory protein fluxes by targeting protein glycosylation

In eukaryotic organisms, most of the secretory proteins become glycosylated during their passage through the secretory pathway. Congurently, *Aspergilli* are able to perform both N- and O-glycosylation which are accomplished in the ER and the Golgi apparatus. In contrast to homologous proteins, heterologous proteins often fail to become properly N-glycosylated. Protein glycosylation, however, is an essential step for effective protein secretion because it increases the solubility of the yet unstructured polypeptides (Helenius and Aebi 2004). A detailed overview of the N-glycosylation pathway performed in *A. fumigatus* and a comparison to the pathways present in yeast and mammals were recently published (Jin 2012). In short, a lipid anchored $Glc_3Man_9GlcNAc_2$ oligosaccharide is synthesized at the ER lumen in several consecutive steps via a series of glycosyltransferases, which are either located in the cytoplasm or the ER (Fig. 2). The oligosaccharide is subsequently transferred to secretory proteins within the ER by the activity of an oligosaccharyltransferase complex (OST in *S. cerevisiae*). Thereafter, two of the three glucose molecules are removed by glucosidase I and II and the glycoprotein binds to the calcium-binding proteins calnexin/calreticulin. After correct folding, the last glucose residue is cleaved off by glucosidase II, causing the dissociation of the glycoprotein from calnexin/calreticulin. In case that the protein did not fold properly, an UDP-glucose: glycoprotein glucosyltransferase (UGT), a key enzyme of the ER quality control, reintegrates again one glucose molecule to the misfolded protein, thereby causing retention of the glycoprotein and thus preventing its premature exit from the ER (Ruddock and Molinari 2006). If the protein is correctly folded, the glycoprotein is transported to the Golgi, where additional modifications take place, including the removal of mannose residues and the addition of other monosaccharides (Jin 2012).

With the assumption that incorrect protein glycosylation decreases heterologous protein production, several studies have been conducted in *Aspergillus*. For example, overexpression of calnexin in *A. niger* resulted in a fourfold increase of the secretion of manganese peroxidase from *Phanerochaete chrysosporium* (Conesa et al. 2002). In another study, a poorly used glycosylation site within bovine chymosin was genetically modified, subsequently leading to a doubling of the protein secreted by *A. niger* (van den Brink et al. 2006). In conclusion, the introduction of an additional glycosylation site between the GlaA carrier and prochymosin considerably increased the amount of secreted fusion protein in *A. niger* (van den Brink et al. 2006).

Improving secretory capacities of Aspergillus *by establishing alternative secretion routes*

So far, this chapter has dealt with the production of secretory proteins channelled through the classical secretory pathway. However, many proteins and enzymes of industrial interest are intracellular proteins but their production is commercially not vital due to very high downstream processing cost. These enzymes would have, however, realistic potential for industrial applications. To address this limitation, an alternative, artificial secretion route, designated peroxicretion, was established and validated in *A. niger* (Sagt et al. 2009). The idea behind was that any intracellular (homologous or heterologous) protein is targeted to peroxisomes by genetically fusing its C-terminus to the peroxisome import signal (SKL). In addition, peroxisomes were decorated with Golgi-derived v-SNAREs, enabling peroxisomes to be transported to the hyphal tip, where they should fuse with the plasma membrane thus releasing their content into the culture broth. As a proof of concept, the fate of the green fluorescent protein (GFP) tagged with the SKL motif was studied in *A. niger* (Sagt et al. 2009). 55% of total GFP was found extracellularly suggesting that peroxicretion might indeed be a feasible alternative secretion route. This conclusion was corroborated by the observation that addition of C2-ceramide, causing enrichment of t-SNAREs, which are important for v-SNARE pin formation prior to membrane fusion (Bonifacino and Glick 2004), increased the amount of secreted protein even further. However, increasing the amount of peroxisomes did not increase the amount of protein released into the medium (Sagt et al. 2009). These observations imply that rerouting of proteins is in general possible in *Aspergillus* but his approach requires more understanding before becoming feasible for industrial production of intracellular heterologous proteins.

Improving recombinant protein expression by preventing product degradation

According to their saprophytic lifestyle, *Aspergilli* are capable of secreting vast amounts of different hydrolytic enzymes including proteases and glycan hydrolases required for the degradation of dead organic matter (Lopes et al. 2011, van den Hombergh et al. 1997). Although this high secretion capability makes *Aspergilli* as outstanding expression platforms, parallel secretion of proteases limits productivity as the protein of interest is likely to become degraded extracellularly. About 200 proteases have been predicted in the genome of *A. niger* CBS513.88, of which approximately 50 have a signal peptide sequence, thus being potentially secreted (Pel et al. 2007). Consequently, deactivation of single proteases is probably not sufficient to prevent proteolytic degradation as this will be bypassed by other secreted proteases. Hence, efforts have been undertaken to deactivate multiple proteases in *Aspergillus*. For example, double disruption of the proteases TppA and PepE increased the production of human lysozyme in *A. oryzae* (Jin et al. 2007). Through selection marker recycling, even quintuple (Yoon et al. 2009) and tenfold deletion (Yoon et al. 2011) of proteases have been achieved in *A. oryzae* leading to considerable further increase of the production yields for human lysozyme and bovine chymosin.

Already 20 years ago, a UV mutant of *A. niger* was isolated (AB1.13) which secreted about 80% less proteases than the wild type strain (Mattern et al. 1992). The underlying genotype remained unclear until complementation approaches identified the mutated gene to encode for a fungal Gal4-type Zn(2)-Cys(6) transcription factor designated PrtT (Punt et al. 2008). Most interestingly, PrtT orthologues are not present in any other non-*Aspergillus* (or related) species and also not in *A. nidulans*. In all *Aspergillus* species carrying a *prtT* orthologue, the gene is clustered with the AmyR amylolytic gene cluster (Punt et al. 2008). Whereas the point mutated version of the *prtT* gene of the *A. niger* AB1.13 mutant (causing a change of a conserved leucin at position 112 to prolin) considerably decreased protease expression, complete deletion of the *prtT* gene in *A. fumigatus* not only lowered protease activities but also adversely affected expression of genes involved in iron uptake, ergosterol synthesis and secondary metabolism, suggesting that PrtT fulfils multiple cellular functions for *Aspergillus* (Hagag et al. 2012).

Improving recombinant protein expression by optimising filamentous morphology

Polarised growth and secretion is a defining attribute of the lifestyle of filamentous fungi. Highly polarised hyphae send out branches from apical or lateral regions thereby forming a dense cellular meshwork, the mycelium (Fig. 3, A-B). Long-distance transport of RNA, proteins, vesicles and organelles along cytoskeletal tracks ensures that new cell material is exclusively added to the growth zone at the hyphal tip. These transport processes are also a prerequisite for polarised protein secretion (Harris 2008,

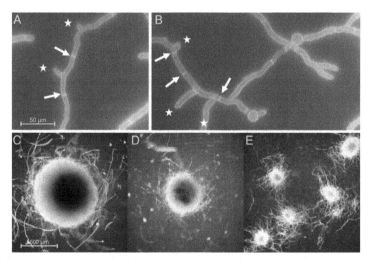

Fig. 3. Macromorphologies of *A. niger*. A, B: Young hyphae of *A. niger* cultivated in liquid medium. Hyphae were stained with calcofluor white to visualize sepate (arrows) and newly formed branches (stars). C-E: *A. niger* cultivated in shake flask cultures with increasing concentration of talc micro particles (C: 0 g/l, D: 2.5 g/l, E: 10 g/l).

Torralba et al. 1998, 1996, Harris 2006). Although a link between protein production and the abundance of actively growing hyphal tips has been proposed for a long time in *Aspergillus* (Wösten et al. 1991, Gordon et al. 2000), only contradictory results have been reported so far. An increase in the number of hyphal tips has been reported to improve protein secretion in some cases but not in all. Thus, no generally accepted model can be used as basis for rationally optimising the morphology of filamentous fungi with respect to protein secretion and their rheological behaviour in a bioreactor.

Basically, *Aspergilli* and all other filamentous fungi can form two different macromorphologies during submerged growth. They grow either as dense mycelial aggregates, so called pellets, which get formed when hyphae branch out at a high frequency (Fig. 3, C–E). Alternatively, they grow as freely dispersed mycelium, a result of low branching frequencies. Fungal macromorphologies affect the productivity of *Aspergillus* cultivation and the preferred morphology would consist of highly branched dispersed mycelia. Whereas the formation of pellets is less desirable because of the high proportion of biomass in a pellet that does not contribute to product formation, long, unbranched hyphae tend to entangle and are also sensitive to shear forces in the reactor. Lysis of hyphae and the subsequent release of intracellular proteases have thus a negative effect on protein production. Fungal macromorphologies can be controlled by environmental conditions including pH level, amount of spore inoculum, power input, osmolarity and the presence of shear stress provoking micro particles (Wucherpfennig et al. 2010, 2011). The addition of talc micro particles to the cultivation medium, for example, affects pellet size and densities allowing the identification of the best performing macromorphology for a certain process (Fig 3, C–E). Lowering pellet size has been shown to lead to a 4 and 9 fold increase of fructofuranosidase and glucoamylase production, respectively, in *A. niger* (Driouch et al. 2012a).

However, to systematically improve the morphological features of filamentous fungi in industrial processes, much more basic knowledge is required to obtain a deeper insight into the molecular networks regulating fungal morphology. To understand the connection between the processes of polarized growth and secretion in the industrially important fungus *A. niger*, genome-wide expression profiling studies were used to predict and identify signalling molecules and networks involved in these processes (Jørgensen et al. 2010, Jacobs et al. 2009, Meyer et al. 2009, 2007b). For example, the transcriptomic fingerprint of apically branched hyphae of *A. niger* uncovered that the stage for the formation of two new branches is set by increased activity of different signalling pathways including TORC2 signalling, phospholipid signalling, cell wall integrity signalling and calcium signalling (Meyer et al. 2009, 2010a). In addition, functional genomics approaches uncovered cellular protagonists coordinating these processes in *A. niger*, such as the Rho GTPase RacA, the polarisome componenent SpaA, the TORC2 component RmsA and the transcription factor RlmA (Meyer et al. 2008, 2009, 2010a, Damveld et al. 2005, Kwon et al. 2011). Future studies are necessary to inventory the full set of genes and proteins defining the morphology of *Aspergillus* and to disclose the key players and their embedment in different signalling pathways which drive growth, morphology and secretion of *Aspergillus*.

The Genomics Toolbox for *Aspergillus*

During the last decade, the genomes of many industrial or medical relevant *Aspergillus* species have been sequenced and published (Table 4). Numerous sequencing projects are on-going as well and are expected to extend the current collection of *Aspergillus* genomes (http: //www.fgsc. net/Aspergillus and (Andersen and Nielsen 2009). Genomic resources have initiated multiple new research activities in the *Aspergillus* research community including comparative genomics, functional genomics and systems biology approaches, aiming at a comprehensive understanding of *Aspergilli* and at a descriptive and predictive modelling of their growth and behaviour. Establishing such models follows four progressive steps, defining a cycle: (i) definition of an experimental set-up to investigate the organism of choice under specific conditions; (ii) application of different omics methods including genomics, transcriptomics, proteomics and/or metabolomics, producing large amounts of data; (iii) mining of these data by bioinformatics tools to gain information; and (iv) using the information obtained to set up a model. At this point, phase one starts again in order to validate or refine the model in terms of its predictive function (Aldridge et al. 2006). Although the field of systems biology for filamentous fungi and *Aspergilli* is developing fast and enormous progress has been made, *Aspergillus* systems biology is still in its infancy (Fig. 4).

Transcriptomics

In the field of transcriptomics, microarray technology has been extensively used for several *Aspergillus* species using Affymetrix GenChip, Agilent, Nimblegen or other array technologies (Andersen and Nielsen 2009). Only recently, first studies applying RNA sequencing (RNA-seq) have been published for this genus (Yu et al. 2011, Wang et al. 2010, Coradetti et al. 2012, Gibbons et al. 2012). In comparison to microarray analysis, RNA-seq offers higher reliability when analysing lowly expressed genes and the possibility to absolutely quantify transcripts (Wang et al. 2009). Most recently, de Bekker and co-workers published a protocol on how to

Table 4. Sequenced *Aspergillus* species.

Strain	Genome size (Mb)	Predicted genes	Genome database*
A. clavatus NRRL1	27.9	9,125	NCBI, AspGD, CADRE
A. flavus NRRL 3357	36.8	12,197	NCBI, ACD, CADRE
A. fumigatus A1163	29.2	9,906	NCBI, CADRE
A. fumigatus Af293	29.4	9,926	NCBI, AspGD, ACD, CADRE
A. kawachii IFO 4308	36.57	11,488	NCBI
A. nidulans FGSC A4	30	9,541	NCBI, AspGD, ACD, CADRE
A. niger CBS513.88	33.9	14,165	NCBI, AspGD
A. niger ATCC 1015	34.85	11,200	NCBI, ACD, CADRE
A. oryzae RIB40	37	14,063	NCBI, AspGD, CADRE
A. sojae NBRC4239	39.5	13,033	NCBI
A. terreus NIH 2624	29.33	10,406	NCBI, ACD, CADRE

*NCBI, National Centre for Biotechnology Information, www.ncbi.nlm.nih.gov; AspGD, *Aspergillus* Genome Database, www.aspgd.org; ACD, *Aspergillus* Comparative Database, www.broadinstitute.org/annotation/genome/aspergillus_group; CADRE, Central Aspergillus Data Repository, www.cadre-genomes.org.uk.

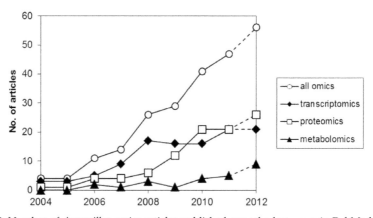

Fig. 4. Number of *Aspergillus*-omics articles published over the last years in PubMed. The number of articles for 2012 was extrapolated based on data obtained in July 2012.

use laser pressure catapulting to isolate RNA of single cells from *A. niger* hyphae (de Bekker et al. 2011a), opening up a new door towards single cell (transcript)omics analyses.

Transcriptomic data analysis requires several evaluation steps including background correction, normalization and identification of co-expressed or differentially expressed genes. Several commercial and open source programs like R/Bioconductor (Gentleman et al. 2004) are available for transcriptomic data analysis. After identification of subsets of co-expressed or differentially expressed genes, enrichment analysis of functional annotations, including Gene Ontology (GO), functional categories (FunCat), Pfam domain and KEGG pathway annotations, has been shown to strongly facilitate omics data analysis (Pel et al. 2007, Nitsche et al. 2011, 2012). Among the numerous tools that can be used for enrichment analysis of functional annotations are Blast2GO (http://www.blast2go.de/b2ghome) (Conesa et al. 2005), GSEA (http://www.broadinstitute.org/gsea/index. jsp) (Subramanian et al. 2005) and FetGOat (http://www.broadinstitute. org/fetgoat/index.html) (Nitsche et al. 2011) which was recently released as an online tool for enrichment analysis providing GO annotation for *A. nidulans, A. niger, A. fumigatus, A. terreus, A. flavus, A. oryzae, A. clavatus* and *N. fischeri*.

Proteomics

Proteomics research for filamentous fungi is a relatively new, though fast growing branch of fungal systems biology. Recently, a comprehensive review of the latest developments in the proteomics field of filamentous fungi has been released (de Oliveira and de Graaff 2011). Several intra- and extracellular proteomes have been published for different *Aspergillus* species based on two dimensional gel electrophoresis 2D-GE coupled with mass spectrometry (MS) or based on liquid chromatography (LC) coupled with mass spectrometry (LC-MS/MS) methods. Most of the MS based methods applied in proteomics make use of selective labelling of peptides and proteins with stable isotopes, e.g., ^{2}H, ^{13}C, ^{15}N or ^{18}O. The methods can be classified into two categories: (i) absolute quantification methods, where known amounts of labelled peptide or protein standards are mixed with the sample prior to MS analysis to estimate the protein concentrations or ii) relative quantification methods, where all proteins of a single biological sample are labelled and afterwards compared to the unlabelled proteins of a reference sample (de Oliveira and de Graaff 2011). For labelling of the whole proteome, different *in vivo* and *in vitro* methods have been developed. Cells can be grown in ^{15}N-medium (Krijgsveld et al. 2003) or in medium containing one or two labelled essential amino acids (stable isotope labelling by amino acids, SILAC) (Ong et al. 2002). *In vitro* methods cover the incorporation of heavy ^{18}O during tryptic digestion of proteins or linkage of isobaric tags for relative and absolute quantitation (iTRAQ) to all proteins (Adav et al. 2010). In addition to different labelling methods, advanced methods for protein extraction and purification have been developed, even allowing the purification of single organelles and determination of their proteome. One such example has been given for microbodies of *P. chrysogenum* (Kiel et al. 2009).

Metabolomics

Even fewer metabolomics studies have been published for *Aspergilli*. Metabolomics and flux analysis, describe the identification and quantification of all metabolites of a cell per time and the flux of metabolites through different catabolic and anabolic pathways. Metabolomics and flux analysis are suitable for identifying bottlenecks that can be targeted by genetic modification approaches or alternative feeding strategies to improve product formation. Classical chromatographic methods (gas chromatography, LC) coupled with MS are applied to detect a wide range of intracellular metabolites (Oldiges et al. 2007). For metabolomics analyses, short sampling times followed by an immediate stop of any cellular metabolic activity (quenching) are crucial. As the metabolome of a cell responds within subseconds to changes in the environment (de Koning and van Dam 1992), extremely fast sampling is mandatory which can usually be accomplished by automatic samplers. Cell samples become quenched by an immediate transfer to methanol (de Koning and van Dam 1992), liquid nitrogen (Hajjaj et al. 1998) or perchloric acid (Shryock et al. 1986). Subsequently, metabolites can be quantified by isotope dilution MS (IDMS) (Hintenberger et al. 1955), whereby cell extracts labelled with ^{13}C are widely applied as internal standards (Wu et al. 2005).

The first metabolic models for *A. niger* and *A. oryzae* have been established in 2008 by using bibliome and genomic information of both strains (Vongsangnak et al. 2008, Andersen et al. 2008). In case of *A. niger*, a network has been constructed that comprises 1,190 biochemical reactions from 52 enzyme complexes including a total of 1,045 metabolites distributed across the cytosol, mitochondria and the extracellular space. It covers the central carbon metabolism, catabolic pathways, 115 different carbon and 23 different nitrogen sources as well as anabolic pathways and *de novo* assembled pathways such as the ergosterol synthesis pathway. Additionally, secretion rates of the endogenous genes glucoamylase and alpha-amylase have been modelled to allow predictions of theoretical yields. The model was further validated by predicting yields of organics acids including oxalic acid, citric acid and gluconic acid.

Applying ^{13}C metabolic flux analysis, Driouch and co-workers compared the fluxes through the central carbon metabolism of an *A. niger* strain overproducing fructofuranosidase and its parental wild type strain (Driouch et al. 2012b). It was found that the carbon flux in the overproducing strain was redirected from the Krebs cycle towards the pentose phosphate pathway, probably to supply the cells with a surplus of NADPH important for anabolic pathways. By simulating the optimal fluxes *in silico* and comparing them to the *in vivo* fluxes, different candidate genes were identified (including enzymes of the Krebs cycle and gluconate synthesis), which might be promising target genes for rational metabolic engineering approaches. In addition, the metabolic model for *A. niger* (Andersen et al. 2008) was used and applied to correlate fluxes of independent pathways to biomass and product formation (flux mode analysis) and thereby the production of fructofuranosidase was predicted was predicted for *A. niger* cultivated on different carbon sources (Melzer et al. 2009).

Bioinformatics

Bioinformatic tools for the statistical analysis of omics data are a prerequisite for effective data interpretation and integration. The development of such tools for *Aspergillus* is an on-going process. Recently, a step-by-step protocol on how to dissect microarray data from *Aspergillus* by using open source bioinformatics was published (Nitsche et al. 2012). These instructions which combined with the *Aspergillus* FetGOat gene set enrichment tool, is freely accessible via the Broad Institute's website (http://www.broadinstitute. org/fetgoat/index.html) (Nitsche et al. 2011), gives guidance for systematic analyses of transcriptomic data obtained for *Aspergillus*. Moreover, the web-based toolbox "BiotMet" (www.sysbio.se/BioMet/) allows the integration of genomics, transcriptomics and metabolomics data from *A. niger* and *A. oryzae* with their genome scale metabolic models (Cvijovic et al. 2010). Another tool, MEMOSys (MEtabolic MOdel research and development System), a bioinformatics platform for genome-scale metabolic models (https://memosys.i-med.ac.at/MEMOSys/home.seam), has been launched recently. It allows the management, storage, and development of metabolic models for *A. niger, A. oryzae, A. nidulans* and other publically available models for non-*Aspergillus* species (Pabinger et al. 2011). Finally, the first web-based modelling tool FAME (Flux Analysis and Modeling Environment, http://f-a-m-e.org/) which combines the creation, edition, simulation and visualisation of stoichiometric models of *A. niger, A. oryzae* and *A. nidulans* in a single program, was recently published (Boele et al. 2012).

A major challenge in omics data analysis is the fact that the datasets can be error-prone due to technical irreproducibility (e.g., cultivations of *Aspergillus* in shake flasks, long sampling times, target degradation, inaccurate measurements) or due to biological variations caused by hyphal heterogeneity in *Aspergillus* populations (Levin et al. 2007, Vinck et al. 2005, de Bekker et al. 2011b, Vinck et al. 2011). Hence, statistical analyses for handling and modeling errors as well as internal controls are crucial to obtain legitimate interpretations from high-throughput omics data.

An omics view on **Aspergillus** *as acid and protein producer*

For *A. niger*, strains have evolved that are both excellent producers of organic acids and protein secretors. To understand the genotypic traits that are beneficial for protein secretion and/or organic acid production, the genomic landscape of two industrial predecessor strains of *A. niger* have been dissected (Andersen et al. 2011). The genome of the acidogenic strain ATCC 1015 was compared with the genome of the GlaA-overproducing strain CBS513.88. This comparison revealed a high amount of single nucleotide polymorphisms (SNPs), insertions/deletions of up to 200 kb, differences in transposon populations, a set of about 400–500 unique genes in both strains and an inversion of an entire chromosomal arm. Thirty-seven proteins involved in transcriptional regulation were predicted to be non-functional in CBS 315.88. All genes encoding for proteins that display differences in amino acid composition were mapped to the metabolic network of *A. niger*. Differences were found in biosynthetic pathways for proline, aspartate, asparagine, tryptophan and histidine as well as in the electron transport chain and the Krebs cycle. Transcriptomic comparison of both strains, cultivated under identical growth conditions, revealed that approximately 6,000 genes are differentially expressed in both strains. ATCC 1015 showed a higher expression of genes involved in the alternative oxidative pathway, whereas metabolic pathway genes involved in glycolysis and TCA cycle, tRNA synthases as well as genes of the entire biosynthetic pathways of threonine, serine, and tryptophan were induced in CBS513.88. These are interestingly exactly those amino acids which are required for efficient GlaA production as they are overrepresented in the GlaA protein sequence.

It has further been shown that the UPR is not exclusively activated in response to artificially induced protein overexpression in *Aspergillus*. When an *A. niger* wild type strain is cultivated at the same specific growth rate on two different carbon sources—xylose or maltose—it secretes about three times more (glyco)proteins on maltose compared to xylose. Transcriptomic analyses disclosed that the higher secretion level on maltose is permitted by up-regulation of more than 90 genes related to protein secretion including UPR target genes (Ferreira de Oliveira et al. 2011, Jørgensen et al. 2009). Furthermore, up-regulation of the UPR genes *hacA* and *pdiA* is also evident, when *A. niger* is cultivated at specific growth rates approaching zero (Jørgensen et al. 2010). These observations suggest that the UPR should be viewed as a general homeostatic control mechanism of *A. niger*, which not only gets activated under ER stress provoked by artificially forced protein overexpression, but also when *A. niger* has to modulate its secretory flux in order to flexibly adapt its cellular capacities and needs to changes in the environmental conditions.

An omics view on Aspergillus *as plant polysaccharide degrader*

The use of second-generation feedstocks, i.e., the use of non-starch and non-edible plant polysaccharides as carbon sources for microbial biotechnology, requires the availability of complex mixtures of enzymes for the complete conversion of plant polymers such as lignocelluloses, xylan, pectin, inulin and other non-starch polymers to fermentable monosaccharides. To evaluate the potential of *Aspergillus* for polysaccharide degradation, the starch-, pectin- and inulin-degrading enzyme network of *A. niger* was uncovered using genomics and transcriptomics approaches (Martens-Uzunova et al. 2006, Yuan et al. 2006, Martens-Uzunova and Schaap 2009). Using RNA-Seq, the degradation potential of *A. niger* was studied when cultivated on wheat straw as a model lignocellulosic substrate and during carbon starvation. The data obtained supports the very interesting model that a subset of polymer-degrading enzymes is expressed and secreted by *A. niger*, which acts as scout to test which polymers are available in the environment. The liberation of inducing sugars from actually present polymers subsequently induces the expression of hydrolases, whereby their respective mRNA species can account for up to 20% of the total mRNA (Delmas et al. 2012).

To obtain a comprehensive and even graphical overview on the complete carbohydrate-active enzyme network of *A. niger*, a bibliomic and genomic survey was recently undertaken, where data from 203 articles on the structure and degradation of 16 different types of plant polysaccharides was compiled and combined with a list of all 188 known or predicted carbohydrate-active enzymes from *A. niger* (Andersen et al. 2012). This data was combined with transcriptomic analyses on three monosaccharides and three complex carbohydrates, which identified enzymes being cross-induced and/or acting in a concerted manner on these carbon sources.

An omics view on Aspergillus *as secondary metabolite producer*

Not only do filamentous fungi have high secretion capacities making them outstanding industrial production hosts, but genome sequencing projects have revealed that they have versatile secondary metabolisms of which most products yet remained unexplored. As incidentally discovered by Alexander Fleming in 1928, fungal secondary metabolites constitute a valuable source for potent antibiotics. In light of increased microbial resistance to antibiotics, it is of utmost importance to continue exploring this natural treasure chest. Only recently, the World Health Organization has forecasted a disaster, should the current number of newly acquired antibiotic resistances continue to increase without a corresponding improvement in the exploration of new antibiotic drugs and their approval (Cooper and Shlaes 2011).

What are secondary metabolites? They have been generally described as structurally diverse natural products of low-molecular weight that often exert bioactivity (Keller et al. 2005). Their synthesis is restricted to specific tissues or phases during growth and development. Although secondary metabolites are extremely versatile, they are synthesized from a rather small pool of primary metabolic precursors including short-chain carboxylic acids, amino acids and terpenes (Keller et al. 2005). Secondary metabolites are produced by microorganisms as well as higher eukaryotes such as algae, plants and animals (Hoffmeister and Keller 2007). It has been estimated that more than 40% of the known microbial secondary metabolites originate from fungal species (Brakhage and Schroeckh 2011, Lazzarini et al. 2001). The roles of secondary metabolites in microbial habitats are not well understood and opposing views on their evolution and natural functions have emerged (Firn and Jones 2003). Among the fungal secondary metabolites identified to date are potent pharmaceutical agents that are clinically applied like the well-known beta-lactam antibiotics (penicillin and cephalosporin) as well as immunosuppressive (cyclosporine A), cholesterol-lowering (lovastatin), anticancer (taxol) and antifungal (griseofulvin) compounds (Meyer 2008). In contrast to many natural products with beneficial bioactivities, several fungal secondary metabolites are toxic and/or carcinogenic to humans and animals, e.g., aflatoxin, fumonisin B1 and ochratoxin A (Frisvad et al. 2007, Nielsen et al. 2009). In field infection and post-harvest contamination of food and feed with potent mycotoxin producing fungi such as *A. flavus* and *A. niger* have been suggested as the main risk (Keller et al. 2005), thus emphasizing the necessity of strict guidelines for monitoring of food and feed.

The availability of fungal genome sequences has made 'genome-mining' for new secondary metabolic genes by bioinformatics feasible and has revealed that the genes are physically clustered, which is in contrast to fungal primary metabolic genes and even secondary metabolite genes from plants (Sanchez et al. 2012). *Aspergilli* have been estimated to typically carry 30–40 secondary metabolite clusters (Brakhage and Schroeckh 2011). Besides large multidomain enzymes including polyketide synthases (PKSs) and nonribosomal peptide synthetases (NRPSs), secondary metabolite clusters often encode transporters, oxidases, hydroxylases and regulatory proteins that are generally thought to be cluster-specific (Brakhage and Schroeckh 2011). The large number of fungal secondary metabolite clusters has exceeded the number of known and expected secondary metabolites by far. However, genome-wide transcriptomic studies have shown that most secondary metabolite clusters are silent under standard laboratory conditions and consequently their corresponding putative secondary metabolites are unknown (Brakhage and Schroeckh 2011, Pel et al. 2007). These silent clusters are also referred to as 'orphan' or 'cryptic' and have initiated many approaches to explore this fungal treasure chest as resource for new potential pharmacologically or biologically active compounds.

Several excellent reviews have been published describing these approaches (Gross 2007, Hertweck 2009, Brakhage and Schroeckh 2011, Sanchez et al. 2012). The most straightforward and powerful approach is referred to as the "one strain many compounds" (OSMAC) strategy which, instead of genetically modifying organisms, simply aims at changing secondary metabolic profiles by altering the cultivation conditions, including C-, N-, P-sources and concentrations, pH, water activity, oxygen availability, light, temperature and specific growth rates (Bide 2002). The OSMAC strategy has for example been applied to demonstrate the toxigenic potential of *A. niger*. Despite the fact that *A. niger* has been considered to be nontoxic under industrial conditions (Schuster et al. 2002), genome sequencing revealed two secondary metabolite clusters sharing homology to those required for the synthesis of the mycotoxins fumonisin and ochratoxin. Both clusters have been shown to be silent under standard cultivation conditions (Pel et al. 2007), while several studies have confirmed the general toxigenic potential of *A. niger*, in particular as a contaminant of food and feed, by cultivation under non-standard conditions, e.g., on substrates with low water activity (Frisvad et al. 2007) or at specific growth rates near zero (Jørgensen et al. 2010, 2011). In another OSMAC approach, *A. nidulans* was cultivated in different single nutrient-limited (N, C and P) chemostat cultures at low specific growth rates to induce silent PKS genes. This approach resulted in an induction of the two PKSs encoding genes *orsA* and *mdpG* under N-limitation, the detection of several known secondary metabolites under N- and P-limitations as well as the identification of three novel secondary metabolites, namely sanghaspirodin A and B as well as pre-shamixanthone (Sarkar et al. 2012).

Although the biological functions of secondary metabolites are unknown and remain a matter of speculation, it has been generally believed that they play a role in intermicrobial communication and niche securement (Rohlfs et al. 2007, Kobayashi and Crouch 2009). This has motivated testing the hypothesis that co-cultivation between *Aspergillus* and soil bacteria—which share the same habitat in nature—might trigger the synthesis of certain secondary metabolites. Indeed, an early example of this OSMAC approach was the discovery of the antibiotic pestalone from co-cultivation of a marine fungus (*Pestaltia* sp.) with an unidentified Gram-negative bacterium (Cueto et al. 2001). This strategy has been recently refined by combining microarray analysis and systematic co-cultivation of *A. nidulans* with 58 actinomycetes (Schroeckh et al. 2009). Surprisingly, only one of the screened actinomycetes (*Sreptomyces hygroscopicus*) specifically induced secondary metabolism genes in *A. nidulans*. Subsequent transcriptome and metabolome analyses led to the identification of an induced PKS gene cluster required for orsellinic acid synthesis.

Alternative approaches for studying secondary metabolite clusters and biosynthetic pathways involve molecular genetic manipulation and subsequent comparative metabolic analysis of wild-type and mutant strains. One example for a successfully applied knock-out strategy is the targeted deletion of six randomly selected NRPS encoding genes in *A. nidulans* and comparative metabolic profiling. This approach led to the identification of five secondary metabolites, emericellamide A, C-F (Chiang et al. 2008). Interestingly, only one of the six gene deletions resulted in an altered metabolic profile under laboratory conditions, implying that the other five clusters were silent. Hence, a combinatorial approach of the OSMAC strategy to analyse the generated mutants is more promising. Another knock-out project has systematically targeted 32 genes putatively encoding PKSs in *A. nidulans* and identified two austinol meroterpenoids and improved the understanding of the biosynthetic routes for arugosins and violaceols (Nielsen et al. 2011).

The concept of OSMAC implies that environmental pH as well as carbon and nitrogen sources do influence fungal secondary metabolite production. Three wide-domain transcriptional regulators, PacC, CreA and AreA, are known to mediate the adaption of fungi to these environmental conditions (Peñalva et al. 2008, Etxebeste et al. 2010) and have indeed been proven to either induce or repress secondary metabolite production in a metabolite-dependent manner (Hoffmeister and Keller 2007). Another approach relates to the observation that many secondary metabolite clusters contain potential pathway-specific transcription factors. This was first discovered for the aflatoxin biosynthesis cluster in *A. nidulans* which contains the *aflR* gene encoding a cluster specific C(6) zinc finger transcription factor that specifically recognizes a palindromic motif found upstream of most aflatoxin biosynthesis genes. Deletion of *aflR* or additional *aflR* copies have been shown to abolish or induce the transcription of the aflatoxin cluster genes, respectively (Hoffmeister and Keller 2007). The existence of cluster-specific regulators was further demonstrated for the *apd* gene cluster of *A. nidulans* which was awakened from its silent transcriptomic state simply by conditional overexpression of its regulatory gene *apdR* leading to the production of the two new PKS-NRPS hybrids aspyridone A and B (Bergmann et al. 2007).

The analysis of classical developmental mutants of *A. nidulans* being defective in alleles of the fluffy genes *fluG*, *flbA* or *fadA* have shown that developmental regulation and secondary metabolism are closely linked via a shared G-protein signalling pathway (Calvo et al. 2002). Mutants affected in sterigmatocystin biosynthesis but with negligible developmental defects have been isolated and studied, whereby *laeA* (loss of *aflR* expression) has been identified to complement one of the mutants. LaeA acts as a global regulatory protein of secondary metabolism. Expression of secondary metabolic clusters including penicillin, lovastatin and stergimatocystin is blocked or induced by deletion or induction of *laeA*, respectively. The *laeA* gene encodes a nuclear protein that shares homology with methyltransferases suggesting that its global regulatory mechanism is mediated by chromatin remodelling (Bok and Keller 2004). LaeA has not been identified in yeast and the identification of the velvet complex in *A. nidulans* indicates a mechanistic link between light-induced fungal differentiation by VeA and secondary metabolism by LaeA (Bayram et al. 2008).

The discovery of LaeA as a global regulator of secondary metabolism suggests an upper-hierarchy, epigenetic level of regulation through chromatin remodelling. Two principally distinct chromatin forms exist for genomic DNA, i.e., a tightly packed and hence transcriptionally silenced chromatin (heterochromatin), and a less dense packed form, which allows higher transcriptional activities (euchromatin). The structure and packaging of chromatin is determined by histones and their post-translational modifications which comprises more than hundred distinct modifications including acetylation, methylation, phosphorylation and sumoylation (Brakhage and Schroeckh 2011, Rando 2012). It is generally thought that heterochromatin and gene silencing are associated with hypoacetylation. Accordingly, it has been shown in *A. nidulans* that deletion of the *hdaA* gene encoding the major histone deacetylase resulted in increased production of sterigmatocystin and penicillin. This induction was specific for subtelomeric secondary metabolite clusters. Through a chemical epigenetic approach with the histone deacetylase inhibitor TSA, this epigenetic regulatory mechanism has been demonstrated to be also present in other filamentous fungi including *Penicillium* and *Fusarium* (Shwab et al. 2007). Further evidence for a general regulatory mechanism of secondary metabolite cluster expression by histone (de-)acetylation in filamentous fungi was provided by another chemical epigenetic approach, where *A. niger* was treated with suberoylanilidehydrozamic acid, which lead to the discovery of the new fungal metabolite nygerone A (Henrikson et al. 2009).

Beside chemical approaches, molecular genetic approaches are also powerful tools to access the potential of fungi as secondary metabolite producers. Using such approaches, it was shown that besides acetylation, methylation and sumoylation of histones also affect the secondary metabolic landscape of *A. nidulans*. For example, deletion of the *cclA* genes, which encodes a subunit of the COMPASS complex that catalyses H3K4 methylation led to the identification of F9775A, F9775B, monodicyphenon, emodin and derivatives thereof. Deletion of the *sumO* gene encoding a small ubiquitin-like modifier (SUMO) protein in *A. nidulans* resulted in a strong induction of asperthecin production which allowed the identification of its biosynthetic genes. Furthermore, synthesis of austinol/dehydroaustinol and sterigmatocystin have been reported to be severely reduced in Δ*sumO* strains (Szewczyk et al. 2008). Taken together, different chemical, molecular and OSMAC approaches have shown that *Aspergilli* can be considered a valuable source for new medically interesting compounds—the discovery of many new secondary metabolites can be expected in the near future.

Conclusion: A Glimpse into the Future of *Aspergillus* as Cell Factory

Aspergilli are natural producers and secretors of enzymes, primary and secondary metabolites. All these products have an enormous commercial potential—a potential, which has, so far, only partly been tapped. On the one hand, new *Aspergillus* enzymes will play a key role for the implementation of secondary feedstocks in microbial biotechnology. On the other hand, interesting fungal secondary metabolites have the potential to become new commercial bestsellers: compounds including NRPS as new antibacterials to fight pathogenic bacteria but also isoprenoids, which are interesting as neutraceuticals or aroma compounds, lipopeptides and small proteins applicable as new antifungals and poly-unsaturated fatty acids and lipids with high application potential as food additives or fuel feedstocks.

Life sciences are currently moving from descriptive, qualitative sciences to predictive, quantitative sciences. For the *Aspergillus* research community, where systems and synthetic biology are still in their infancy, it will be a formidable task to analyse, understand, control and manipulate these hosts for targeted production of any compound of interest. The tight-linking of experiment and modelling requires a multi-disciplinary collaboration with biologists, mathematicians, physicians, chemists, engineers, and statisticians. Hence, the new generation of *Aspergillus* researchers should be well trained on fungal physiology and genetics, extended by training in genomics, transcriptomics, proteomics and metabolomics. These experimental expertise should be combined with mathematical modelling, fermentation skills and industrial know-how to address fundamental and applied questions.

References

Adav, S. S., A. A. Li, A. Manavalan, P. Punt and S. K. Sze. 2010. Quantitative iTRAQ Secretome Analysis of *Aspergillus niger* Reveals Novel Hydrolytic Enzymes research articles. Journal of Proteome Research. 9: 3932–3940.
Al-Sheikh, H., A. J. Watson, G. A. Lacey, P. J. Punt, D. A. MacKenzie, D. J. Jeenes, T. Pakula, M. Penttilä, M. J. C. Alcocer and D. B. Archer. 2004. Endoplasmic reticulum stress leads

to the selective transcriptional downregulation of the glucoamylase gene in *Aspergillus niger*. Molecular Microbiology. 53: 1731–42.

Aldridge, B. B., J. M. Burke, D. A. Lauffenburger and P. K. Sorger. 2006. Physicochemical modelling of cell signalling pathways. Nature Cell Biology. 8: 1195–203.

Aleksenko, A. and A. J. Clutterbuck. 1997. Autonomous plasmid replication in Aspergillus nidulans: AMA1 and MATE elements. Fungal Genetics and Biology. 21: 373–87.

Andersen, M. R., M. Giese, R. P. de Vries and J. Nielsen. 2012. Mapping the polysaccaride degradation potential of *Aspergillus niger*. BMC Genomics. 13: 313.

Andersen, M. R. and J. Nielsen. 2009. Current status of systems biology in Aspergilli. Fungal Genetics and Biology. 46: 180–90.

Andersen, M. R., M. L. Nielsen and J. Nielsen. 2008. Metabolic model integration of the bibliome, genome, metabolome and reactome of *Aspergillus niger*. Molecular Systems Biology. 4: 178.

Andersen, M. R., M. P. Salazar, P. J. Schaap, P. J. I. van de Vondervoort, D. Culley, J. Thykaer, J. C. Frisvad, K. F. Nielsen, R. Albang, K. Albermann, R. M. Berka, G. H. Braus, S. A. Braus-Stromeyer, L. M. Corrochano, Z. Dai, P. W. M. van Dijck, G. Hofmann, L. L. Lasure, J. K. Magnuson, H. Menke, M. Meijer, S. L. Meijer, J. B. Nielsen, M. L. Nielsen, A. J. J. van Ooyen, H. J. Pel, L. Poulsen, R. A. Samson, H. Stam, A. Tsang, J. M. van den Brink, A. Atkins, A. Aerts, H. Shapiro, J. Pangilinan, A. Salamov, Y. Lou, E. Lindquist, S. Lucas, J. Grimwood, I. V. Grigoriev, C. P. Kubicek, D. Martinez, N. N. M. E. van Peij, J. A. Roubos, J. Nielsen and S. E. Baker. 2011. Comparative genomics of citric-acid-producing *Aspergillus niger* ATCC 1015 versus enzyme-producing CBS 513. 88. Genome Research. 21: 885–97.

Arentshorst, M., A. F. J. Ram and V. Meyer. 2012. Using non-homologous end-joining-deficient strains for functional gene analyses in filamentous fungi. Methods in Molecular Biology (Clifton, N. J.). 835: 133–50.

Bando, H., H. Hisada, H. Ishida, Y. Hata, Y. Katakura and A. Kondo. 2011. Isolation of a novel promoter for efficient protein expression by Aspergillus oryzae in solid-state culture. Applied Microbiology and Biotechnology. 92: 561–9.

Bayram, O., S. Krappmann, M. Ni, J. W. Bok, K. Helmstaedt, O. Valerius, S. Braus-Stromeyer, N. -J. Kwon, N. P. Keller, J. -H. Yu and G. H. Braus. 2008. VelB/VeA/LaeA complex coordinates light signal with fungal development and secondary metabolism. Science (New York, N. Y.). 320: 1504–6.

Beadle, G. W. and E. L. Tatum. 1941. Genetic control of biochemical reactions in Neurospora. Proceedings of the National Academy of Sciences of the United States of America. 27: 499–506.

Bergmann, S., J. Schümann, K. Scherlach, C. Lange, A. a Brakhage and C. Hertweck. 2007. Genomics-driven discovery of PKS-NRPS hybrid metabolites from Aspergillus nidulans. Nature Chemical Biology. 3: 213–7.

Bermúdez-Humarán, L. G., N. G. Cortes-Perez, Y. Le Loir, A. Gruss, C. Rodriguez-Padilla, O. Saucedo-Cardenas, P. Langella and R. Montes de Oca-Luna. 2003. Fusion to a carrier protein and a synthetic propeptide enhances E7 HPV-16 production and secretion in Lactococcus lactis. Biotechnology Progress. 19: 1101–4.

Bide, H. B. 2002. Big effects from small changes: possible ways to explore nature's chemical diversity. Chembiochem. 3: 619–27.

Blumhoff, M., M. G. Steiger, H. Marx, D. Mattanovich and M. Sauer. 2012. Six novel constitutive promoters for metabolic engineering of *Aspergillus niger*. Applied Microbiology and Biotechnology.

Boele, J., B. G. Olivier and B. Teusink. 2012. FAME, the Flux Analysis and Modeling Environment. BMC Systems Biology. 6: 8.

Bok, J. W. and N. P. Keller. 2004. LaeA, a Regulator of Secondary Metabolism in Aspergillus spp. Eukaryotic Cell. 3: 527–535.

Bonifacino, J. S. and B. S. Glick. 2004. The mechanisms of vesicle budding and fusion. Cell. 116: 153–66.

Brakhage, A. a and V. Schroeckh. 2011. Fungal secondary metabolites - strategies to activate silent gene clusters. Fungal Genetics and Biology. 48: 15–22.

Calvo, A. M., R. A. Wilson, J. W. Bok, P. Nancy and N. P. Keller. 2002. Relationship between Secondary Metabolism and Fungal Development Relationship between Secondary Metabolism and Fungal Development. Microbiol. Mol. Biol. Rev. 66: 447–59.

Cardoza, R. E., S. Gutiérrez, N. Ortega, A. Colina, J. Casqueiro and J. F. Martín. 2003. Expression of a synthetic copy of the bovine chymosin gene in Aspergillus awamori from constitutive and pH-regulated promoters and secretion using two different pre-pro sequences. Biotechnology and Bioengineering. 83: 249–59.

Cardoza, R. E., F. J. Moralejo, S. Gutiérrez, J. Casqueiro, F. Fierro and J. F. Martín. 1998. Characterization and nitrogen-source regulation at the transcriptional level of the gdhA gene of Aspergillus awamori encoding an NADP-dependent glutamate dehydrogenase. Current Genetics. 34: 50–9.

Carvalho, N. D. S. P., T. R. Jørgensen, M. Arentshorst, B. M. Nitsche, C. A. van den Hondel, D. B. Archer and A. F. Ram. 2012. Genome-wide expression analysis upon constitutive activation of the HacA bZIP transcription factor in *Aspergillus niger* reveals a coordinated cellular response to counteract ER stress. BMC Genomics. 13: 350.

Carvalho, N. D. S. P., M. Arentshorst, M. J. Kwon, V. Meyer and A. F. J. Ram. 2010. Expanding the ku70 toolbox for filamentous fungi: establishment of complementation vectors and recipient strains for advanced gene analyses. Applied Microbiology. 87: 1463–1473.

Carvalho, N. D. S. P., M. Arentshorst, R. Kooistra, H. Stam, C. M. J. Sagt, C. a M. J. J. van den Hondel and A. F. J. Ram. 2011. Effects of a defective ERAD pathway on growth and heterologous protein production in *Aspergillus niger*. Applied Genetics and Molecular Biotechnology. 357–373.

Case, M. E., M. Schweizer, S. R. Kushner and N. H. Giles. 1979. Efficient transformation of Neurospora crassa by utilizing hybrid plasmid DNA. Proceedings of the National Academy of Sciences of the United States of America. 76: 5259–63.

Chen, W., T. Xie, Y. Shao and F. Chen. 2012. Genomic characteristics comparisons of 12 food-related filamentous fungi in tRNA gene set, codon usage and amino acid composition. Gene. 497: 116–24.

Chiang, Y., E. Szewczyk, T. Nayak and A. Davidson. 2008. Molecular Genetic Mining of the Aspergillus Secondary Metabolome: Discovery of the Emericellamide Biosynthetic Pathway. Chemistry Biology. 15: 527–532.

Conesa, a, P. J. Punt, N. van Luijk and C. a van den Hondel. 2001. The secretion pathway in filamentous fungi: a biotechnological view. Fungal Genetics and Biology. 33: 155–71.

Conesa, A., S. Götz, J. M. García-Gómez, J. Terol, M. Talón, M. Robles, S. Gotz, J. M. Garcia-Gomez and M. Talon. 2005. Blast2GO: a universal tool for annotation, visualization and analysis in functional genomics research. Bioinformatics (Oxford, England). 21: 3674–6.

Conesa, A., D. Jeenes, D. B. Archer, C. A. M. J. J. van den Hondel and P. J. Punt. 2002. Calnexin overexpression increases manganese peroxidase production in *Aspergillus niger*. Applied and Environmental Microbiology. 68: 846–51.

Contreras, R., D. Carrez, J. R. Kinghorn, C. A. van den Hondel and W. Fiers. 1991. Efficient KEX2-like processing of a glucoamylase-interleukin-6 fusion protein by Aspergillus nidulans and secretion of mature interleukin-6. Bio/technology (Nature Publishing Company). 9: 378–81.

Cooper, M. A. and D. Shlaes. 2011. Fix the antibiotics pipeline. Nature. 472: 32.

Coradetti, S. T., J. P. Craig, Y. Xiong, T. Shock, C. Tian and N. L. Glass. 2012. Conserved and essential transcription factors for cellulase gene expression in ascomycete fungi. Proceedings of the National Academy of Sciences. 109: 7397–402.

Cueto, M., P. Jensen, C. Kauffman, W. Fenical, E. Lobkovsky and J. Clardy. 2001. Pestalone, a new antibiotic produced by a marine fungus in response to bacterial challenge. J. Nat. Prod. 1: 1444–1446.

Cullen, D., G. L. Gray, L. J. Wilson, K. J. Hayenga, M. H. Lamsa, M. W. Rey, S. Norton and R. M. Berka. 1987. Controlled Expression and Secretion of Bovine Chymosin in Aspergillus Nidulans. Bio/Technology. 5: 369–376.

Cvijovic, M., R. Olivares-Hernández, R. Agren, N. Dahr, W. Vongsangnak, I. Nookaew, K. R. Patil and J. Nielsen. 2010. BioMet Toolbox: genome-wide analysis of metabolism. Nucleic Acids Research. 38: W144–9.

Damveld, R. A., M. Arentshorst, A. Franken, P. a vanKuyk, F. M. Klis, C. a M. J. J. van den Hondel and A. F. J. Ram. 2005. The *Aspergillus niger* MADS-box transcription factor RlmA is required for cell wall reinforcement in response to cell wall stress. Molecular microbiology. 58: 305–319.

Dave, K., M. Ahuja, T. N. Jayashri, R. B. Sirola and N. S. Punekar. 2012. A novel selectable marker based on *Aspergillus niger* arginase expression. Enzyme and Microbial Technology. 51: 53–8.

Dave, K. and N. S. Punekar. 2011. Utility of *Aspergillus niger* citrate synthase promoter for heterologous expression. Journal of Biotechnology. 155: 173–7.

de Bekker, C., O. Bruning, M. J. Jonker, T. M. Breit and H. a B. Wösten. 2011a. Single cell transcriptomics of neighboring hyphae of *Aspergillus niger*. Genome Biology. 12: R71.

de Bekker, C., G. J. van Veluw, A. Vinck, L. A. Wiebenga and H. A. B. Wösten. 2011b. Heterogeneity of *Aspergillus niger* microcolonies in liquid shaken cultures. Applied and Environmental Microbiology. 77: 1263–7.

de Bekker, C., A. Wiebenga, G. Aguilar and H. A. B. Wösten. 2009. An enzyme cocktail for efficient protoplast formation in *Aspergillus niger*. Journal of Microbiological Methods. 76: 305–6.

Delmas, S., S. T. Pullan, S. Gaddipati, M. Kokolski, S. Malla, M. J. Blythe, R. Ibbett, M. Campbell, S. Liddell, A. Aboobaker, G. a. Tucker and D. B. Archer. 2012. Uncovering the Genome-Wide Transcriptional Responses of the Filamentous Fungus *Aspergillus niger* to Lignocellulose Using RNA Sequencing. PLoS Genetics. 8: e1002875.

de Souza, W. R., P. F. de Gouvea, M. Savoldi, I. Malavazi, L. a de Souza Bernardes, M. H. S. Goldman, R. P. de Vries, J. V. de Castro Oliveira and G. H. Goldman. 2011. Transcriptome analysis of *Aspergillus niger* grown on sugarcane bagasse. Biotechnology for Biofuels. 4: 40.

de Oliveira, J. M. P. F. and L. H. de Graaff. 2011. Proteomics of industrial fungi: trends and insights for biotechnology. Applied Microbiology and Biotechnology. 89: 225–37.

Driouch, H., R. Hänsch, T. Wucherpfennig, R. Krull and C. Wittmann. 2012a. Improved enzyme production by bio-pellets of *Aspergillus niger*: targeted morphology engineering using titanate microparticles. Biotechnology and Bioengineering. 109: 462–71.

Driouch, H., G. Melzer and C. Wittmann. 2012b. Integration of in vivo and in silico metabolic fluxes for improvement of recombinant protein production. Metabolic Engineering. 14: 47–58.

Dudásová, Z., A. Dudás and M. Chovanec. 2004. Non-homologous end-joining factors of Saccharomyces cerevisiae. FEMS Microbiology Reviews. 28: 581–601.

Edelmann, S. E. and C. Staben. 1994. A Statistical Analysis of Sequence Features within Genes from Neurospora crassa. Experimental Mycology. 18: 70–81.

Elander, R. P. 2003. Industrial production of beta-lactam antibiotics. Applied Microbiology and Biotechnology. 61: 385–92.

Etxebeste, O., U. Ugalde and E. A. Espeso. 2010. Adaptive and developmental responses to stress in Aspergillus nidulans. Current protein & peptide science. 11: 704–18.

Evnouchidou, I., A. Papakyriakou and E. Stratikos. 2009. A new role for Zn(II) aminopeptidases: antigenic peptide generation and destruction. Current Pharmaceutical Design. 15: 3656–70.

Felenbok, B. 1991. The ethanol utilization regulon of Aspergillus nidulans: the alcA-alcR system as a tool for the expression of recombinant proteins. Journal of Biotechnology. 17: 11–7.

Ferreira de Oliveira, J. M. P., M. W. J. van Passel, P. J. Schaap and L. H. de Graaff. 2011. Proteomic Analysis of the Secretory Response of *Aspergillus niger* to D-Maltose and D-Xylose. PloS one. 6: e20865.

Fincham, J. R. 1989. Transformation in fungi. Microbiological Reviews. 53: 148–70.

Finkelstein, D. B. 1987. Improvement of enzyme production in Aspergillus. Antonie van Leeuwenhoek. 53: 349–352.

Firn, R. D. and C. G. Jones. 2003. Natural products? a simple model to explain chemical diversity. Natural Product Reports. 20: 382.

Fischer, M., W. Hilt, B. Richter-Ruoff, H. Gonen, A. Ciechanover and D. H. Wolf. 1994. The 26S proteasome of the yeast Saccharomyces cerevisiae. FEBS Letters. 355: 69–75.

Fleissner, A. and P. Dersch. 2010. Expression and export: recombinant protein production systems for Aspergillus. Applied Microbiology and Biotechnology. 87: 1255–70.

Forment, J. V., D. Ramón and A. P. MacCabe. 2006. Consecutive gene deletions in Aspergillus nidulans: application of the Cre/loxP system. Current Genetics. 50: 217–24.

Fowler, T., R. M. Berka, M. Ward, K. Dave and N. S. Punekar. 1990. Regulation of the glaA gene of *Aspergillus niger*. Current Genetics. 18: 537–545.

Frisvad, J. C., J. Smedsgaard, R. A. Samson, T. O. Larsen and U. Thrane. 2007. Fumonisin B2 production by *Aspergillus niger*. Journal of Agricultural and Food chemistry. 55: 9727–32.

Gems, D., I. L. Johnstone and a J. Clutterbuck. 1991. An autonomously replicating plasmid transforms Aspergillus nidulans at high frequency. Gene. 98: 61–7.

Gentleman, R. C., V. J. Carey, D. M. Bates, B. Bolstad, M. Dettling, S. Dudoit, B. Ellis, L. Gautier, Y. Ge, J. Gentry, K. Hornik, T. Hothorn, W. Huber, S. Iacus, R. A. Irizarry, F. Leisch, C. Li, M. Maechler, A. J. Rossini, G. Sawitzki, C. Smith, G. Smyth, L. Tierney, J. Y. Yang and J. Zhang. 2004. Bioconductor: open software development for computational biology and bioinformatics. Genome Biol. 5: R80.

Geysens, S., G. Whyteside and D. B. Archer. 2009. Genomics of protein folding in the endoplasmic reticulum, secretion stress and glycosylation in the Aspergilli. Fungal Genetics and Biology. 46: S121–S140.

Gibbons, J. G., A. Beauvais, R. Beau, K. L. McGary, J. -P. Latgé and A. Rokas. 2012. Global transcriptome changes underlying colony growth in the opportunistic human pathogen Aspergillus fumigatus. Eukaryotic Cell. 11: 68–78.

Goder, V., P. Carvalho and T. A. Rapoport. 2008. The ER-associated degradation component Der1p and its homolog Dfm1p are contained in complexes with distinct cofactors of the ATPase Cdc48p. FEBS Letters. 582: 1575–80.

Gonzalez, D. S., K. Karaveg, A. S. Vandersall-Nairn, A. Lal and K. W. Moremen. 1999. Identification, expression, and characterization of a cDNA encoding human endoplasmic reticulum mannosidase I, the enzyme that catalyzes the first mannose trimming step in mammalian Asn-linked oligosaccharide biosynthesis. The Journal of Biological Chemistry. 274: 21375–86.

Gordon, C. L., V. Khalaj, a F. Ram, D. B. Archer, J. L. Brookman, a P. Trinci, D. J. Jeenes, J. H. Doonan, B. Wells, P. J. Punt, C. a van den Hondel and G. D. Robson. 2000. Glucoamylase: green fluorescent protein fusions to monitor protein secretion in *Aspergillus niger*. Microbiology (Reading, England). 146: 415–26.

Gouka, R. J., J. G. Hessing, P. J. Punt, H. Stam, W. Musters and C. A. Van den Hondel. 1996. An expression system based on the promoter region of the Aspergillus awamori 1,4-beta-endoxylanase A gene. Applied Microbiology and Biotechnology. 46: 28–35.

Gouka, R. J., P. J. Punt and C. A. van den Hondel. 1997. Glucoamylase gene fusions alleviate limitations for protein production in Aspergillus awamori at the transcriptional and (post) translational levels. Applied and Environmental Microbiology. 63: 488–97.

Griffiths, a J. 1995. Natural plasmids of filamentous fungi. Microbiological Reviews. 59: 673–85.

de Groot, M. J., P. Bundock, P. J. Hooykaas and A. G. Beijersbergen. 1998. Agrobacterium tumefaciens-mediated transformation of filamentous fungi. Nature Biotechnology. 16: 839–42.

Gross, H. 2007. Strategies to unravel the function of orphan biosynthesis pathways: recent examples and future prospects. Applied Microbiology and Biotechnology. 75: 267–77.

Guillemette, T., N. N. M. E. van Peij, T. Goosen, K. Lanthaler, G. D. Robson, C. a M. J. J. van den Hondel, H. Stam and D. B. Archer. 2007. Genomic analysis of the secretion stress response in the enzyme-producing cell factory *Aspergillus niger*. BMC Genomics. 8: 158.

Gwynne, D. I., F. P. Buxton, S. A. Williams, A. M. Sills, J. A. Johnstone, J. K. Buch, Z. M. Guo, D. Drake, M. Westphal and R. W. Davies. 1989. Development of an expression system in Aspergillus nidulans. Biochemical Society transactions. 17: 338–40.

Hagag, S., P. Kubitschek-Barreira, G. W. P. Neves, D. Amar, W. Nierman, I. Shalit, R. Shamir, L. Lopes-Bezerra and N. Osherov. 2012. Transcriptional and Proteomic Analysis of the Aspergillus fumigatus ΔprtT Protease-Deficient Mutant. PLoS ONE. 7: e33604.

Hajjaj, H., P. Blanc, G. Goma and J. FranÃ§ois. 1998. Sampling techniques and comparative extraction procedures for quantitative determination of intra- and extracellular metabolites in filamentous fungi. FEMS Microbiology Letters. 164: 195–200.

Harris, S. D. 2008. Branching of fungal hyphae: regulation, mechanisms and comparison with other branching systems. Mycologia. 100: 823–832.

Harris, S. D. 2006. Cell polarity in filamentous fungi: shaping the mold. International Review of Cytology. 251: 41–77.

Hawksworth, D. L. 1991. The fungal dimension of biodiversity: magnitude, significance, and conservation. Mycological Research. 95: 641–655.

Helenius, A. and M. Aebi. 2004. Roles of N-linked glycans in the endoplasmic reticulum. Annual Review of Biochemistry. 73: 1019–49.

Henrikson, J. C., A. R. Hoover, P. M. Joyner and R. H. Cichewicz. 2009. A chemical epigenetics approach for engineering the in situ biosynthesis of a cryptic natural product from *Aspergillus niger*. Organic & Biomolecular Chemistry. 7: 435–8.

Hinterberger, H. 1956. *In:* Electromagnetically Enriched Isotopes and Mass Spectrometry (Smith, M. L., ed.) Butterworth, London. pp. 177–189.

Hertweck, C. 2009. Hidden biosynthetic treasures brought to light. Nature Chemical Biology. 5: 450–2.

Hoffmeister, D. and N. P. Keller. 2007. Natural products of filamentous fungi: enzymes, genes, and their regulation. Natural Product Reports. 24: 393–416.

Hu, W., S. Sillaots, S. Lemieux, J. Davison, S. Kauffman, A. Breton, A. Linteau, C. Xin, J. Bowman, J. Becker, B. Jiang and T. Roemer. 2007. Essential gene identification and drug target prioritization in Aspergillus fumigatus. PLoS Pathogens. 3: e24.

Ikemura, T. 1981. Correlation between the abundance of Escherichia coli transfer RNAs and the occurrence of the respective codons in its protein genes: a proposal for a synonymous codon choice that is optimal for the *E. coli* translational system. Journal of Molecular Biology. 151: 389–409.

Ikemura, T. 1982. Correlation between the abundance of yeast transfer RNAs and the occurrence of the respective codons in protein genes. Differences in synonymous codon choice patterns of yeast and Escherichia coli with reference to the abundance of isoaccepting transfer R. Journal of Molecular Biology. 158: 573–97.

Iriarte, A., M. Sanguinetti, T. Fernández-Calero, H. Naya, A. Ramón and H. Musto. 2012. Translational selection on codon usage in the genus Aspergillus. Gene. 506: 98–105.

Ishida, H., Y. Hata, A. Kawato, Y. Abe and Y. Kashiwagi. 2004. Isolation of a novel promoter for efficient protein production in Aspergillus oryzae. Bioscience, Biotechnology, and Biochemistry. 68: 1849–57.

Jacobs, D. I., M. M. A. Olsthoorn, I. Maillet, M. Akeroyd, S. Breestraat, S. Donkers, R. A. M. V. D. Hoeven, C. A. M. J. J. V. D. Hondel, R. Kooistra, T. Lapointe, H. Menke, R. Meulenberg, M. Misset, W. H. Müller, N. N. M. E. V. Peij, A. Ram, S. Rodriguez, M. S. Roelofs, J. A. Roubos, M. W. E. M. V. Tilborg, A. J. Verkleij, H. J. Pel, H. Stam and C. M. J. Sagt. 2009.

Effective lead selection for improved protein production in *Aspergillus niger* based on integrated genomics. Fungal Genetics and Biology. 46: S141–S152.

Jalving, R., P. J. van de Vondervoort, J. Visser and P. J. Schaap. 2000. Characterization of the kexin-like maturase of *Aspergillus niger*. Applied and Environmental Microbiology. 66: 363–8.

Jeenes, D. J., D. A. Mackenzie and D. B. Archer. 1994. Transcriptional and post-transcriptional events affect the production of secreted hen egg white lysozyme by *Aspergillus niger*. Transgenic Research. 3: 297–303.

Jin, C. 2012. Protein Glycosylation in Aspergillus fumigatus Is Essential for Cell Wall Synthesis and Serves as a Promising Model of Multicellular Eukaryotic Development. International Journal of Microbiology. 2012: 654251.

Jin, F. J., T. Watanabe, P. R. Juvvadi, J. -ichi Maruyama, M. Arioka and K. Kitamoto. 2007. Double disruption of the proteinase genes, tppA and pepE, increases the production level of human lysozyme by Aspergillus oryzae. Applied Microbiology and Biotechnology. 76: 1059–68.

Jørgensen, T. R., T. Goosen, C. a M. J. J. V. D. Hondel, A. F. J. Ram and J. J. L. Iversen. 2009. Transcriptomic comparison of *Aspergillus niger* growing on two different sugars reveals coordinated regulation of the secretory pathway. BMC Genomics. 10: 44.

Jørgensen, T. R., K. F. Nielsen, M. Arentshorst, J. Park, C. a van den Hondel, J. C. Frisvad and A. F. Ram. 2011. Submerged conidiation and product formation by *Aspergillus niger* at low specific growth rates are affected in aerial developmental mutants. Applied and Environmental Microbiology. 77: 5270–7.

Jørgensen, T. R., B. M. Nitsche, G. E. Lamers, M. Arentshorst, C. a van den Hondel and A. F. Ram. 2010. Transcriptomic insights into the physiology of *Aspergillus niger* approaching a specific growth rate of zero. Applied and Environmental Microbiology. 76: 5344–55.

Kahali, B., S. Ahmad and T. C. Ghosh. 2011. Selective constraints in yeast genes with differential expressivity: codon pair usage and mRNA stability perspectives. Gene. 481: 76–82.

Kaufman, R. J. 1999. Stress signaling from the lumen of the endoplasmic reticulum: coordination of gene transcriptional and translational controls. Genes & Development. 13: 1211–1233.

Kaufman, R. J., D. Scheuner, M. Schröder, X. Shen, K. Lee, C. Y. Liu and S. M. Arnold. 2002. The unfolded protein response in nutrient sensing and differentiation. Nature Reviews. Molecular Cell Biology. 3: 411–21.

Kawai, S., W. Hashimoto and K. Murata. 2010. Transformation of Saccharomyces cerevisiae and other fungi: methods and possible underlying mechanism. Bioengineered Bugs. 1: 395–403.

Keller, N. P., G. Turner and J. W. Bennett. 2005. Fungal secondary metabolism—from biochemistry to genomics. Nature reviews. Microbiology. 3: 937–47.

Kelly, J. M. and M. J. Hynes. 1987. Multiple copies of the amdS gene of Aspergillus nidulans cause titration of trans-acting regulatory proteins. Current Genetics. 12: 21–31.

Khalaj, V., H. Eslami, M. Azizi, N. Rovira-Graells and M. Bromley. 2007. Efficient downregulation of alb1 gene using an AMA1-based episomal expressionofRNAi construct in Aspergillus fumigatus. FEMS Microbiology Letters. 270: 250–254.

Kiel, J. A. K. W., M. A. van den Berg, F. Fusetti, B. Poolman, R. A. L. Bovenberg, M. Veenhuis and I. J. van der Klei. 2009. Matching the proteome to the genome: the microbody of penicillin-producing Penicillium chrysogenum cells. Functional & Integrative Genomics. 9: 167–84.

Kinnaird, J. H., P. A. Burns and J. R. Fincham. 1991. An apparent rare-codon effect on the rate of translation of a Neurospora gene. Journal of Molecular Biology. 221: 733–6.

Kitamoto, N., J. Matsui, Y. Kawai, A. Kato, S. Yoshino, K. Ohmiya and N. Tsukagoshi. 1998. Utilization of the TEF1-alpha gene (TEF1) promoter for expression of polygalacturonase genes, pgaA and pgaB, in Aspergillus oryzae. Applied Microbiology and Biotechnology. 50: 85–92.

Kito, H., T. Fujikawa, A. Moriwaki, A. Tomono, M. Izawa, T. Kamakura, M. Ohashi, H. Sato, K. Abe and M. Nishimura. 2008. MgLig4, a homolog of Neurospora crassa Mus-53 (DNA ligase IV), is involved in, but not essential for, non-homologous end-joining events in Magnaporthe grisea. Fungal Genetics and Biology. 45: 1543–51.

Kobayashi, D. Y. and J. A. Crouch. 2009. Bacterial/Fungal interactions: from pathogens to mutualistic endosymbionts. Annual Review of Phytopathology. 47: 63–82.

Koda, A., T. Bogaki, T. Minetoki and M. Hirotsune. 2005. High expression of a synthetic gene encoding potato alpha-glucan phosphorylase in *Aspergillus niger*. Journal of Bioscience and Bioengineering. 100: 531–7.

de Koning, W. and K. van Dam. 1992. A method for the determination of changes of glycolytic metabolites in yeast on a subsecond time scale using extraction at neutral pH. Analytical Biochemistry. 204: 118–23.

Korman, D. R., F. T. Bayliss, C. C. Barnett, C. L. Carmona, K. H. Kodama, T. J. Royer, S. A. Thompson, M. Ward, L. J. Wilson and R. M. Berka. 1990. Cloning, characterization, and expression of two alpha-amylase genes from *Aspergillus niger* var. awamori. Current Genetics. 17: 203–12.

Kostova, Z., Y. C. Tsai and A. M. Weissman. 2007. Ubiquitin ligases, critical mediators of endoplasmic reticulum-associated degradation. Seminars in Cell & Developmental biology. 18: 770–9.

Krappmann, S., O. Bayram and G. H. Braus. 2005. Deletion and allelic exchange of the Aspergillus fumigatus veA locus via a novel recyclable marker module. Eukaryotic cell. 4: 1298–307.

Krappmann, S., C. Sasse and G. H. Braus. 2006. Gene targeting in Aspergillus fumigatus by homologous recombination is facilitated in a nonhomologous end- joining-deficient genetic background. Eukaryotic Cell. 5: 212–5.

Krijgsveld, J., R. F. Ketting, T. Mahmoudi, J. Johansen, M. Artal-Sanz, C. P. Verrijzer, R. H. A. Plasterk and A. J. R. Heck. 2003. Metabolic labeling of C. elegans and D. melanogaster for quantitative proteomics. Nature Biotechnology. 21: 927–31.

Krogh, B. O. and L. S. Symington. 2004. Recombination proteins in yeast. Annual Review of Genetics. 38: 233–71.

Kwon, M. J., M. Arentshorst, E. D. Roos, C. a M. J. J. van den Hondel, V. Meyer and A. F. J. Ram. 2011. Functional characterization of Rho GTPases in *Aspergillus niger* uncovers conserved and diverged roles of Rho proteins within filamentous fungi. Molecular Microbiology. 79: 1151–67.

Kück, U. and B. Hoff. 2010. New tools for the genetic manipulation of filamentous fungi. Applied Microbiology and Biotechnology. 86: 51–62.

Lazzarini, A., L. Cavaletti, G. Toppo and F. Marinelli. 2001. Rare genera of actinomycetes as potential producers of new antibiotics. Antonie van Leeuwenhoek. 79: 399–405.

Levin, A. M., R. P. de Vries, A. Conesa, C. de Bekker, M. Talon, H. H. Menke, N. N. M. E. van Peij and H. a B. Wösten. 2007. Spatial differentiation in the vegetative mycelium of *Aspergillus niger*. Eukaryotic Cell. 6: 2311–22.

Lloyd, A. T. and P. M. Sharp. 1991. Codon usage in Aspergillus nidulans. Molecular & General Genetics. 230: 288–94.

Le Loir, Y., S. Nouaille, J. Commissaire, L. Brétigny, A. Gruss and P. Langella. 2001. Signal peptide and propeptide optimization for heterologous protein secretion in Lactococcus lactis. Applied and Environmental Microbiology. 67: 4119–27.

Lombraña, M., F. J. Moralejo, R. Pinto and J. F. Martín. 2004. Modulation of Aspergillus awamori thaumatin secretion by modification of bipA gene expression. Applied and Environmental Microbiology. 70: 5145–52.

Lopes, F. C., L. A. D. E. Silva, D. M. Tichota, D. J. Daroit, R. V. Velho, J. Q. Pereira, A. P. F. Corrêa and A. Brandelli. 2011. Production of Proteolytic Enzymes by a Keratin-Degrading *Aspergillus niger*. Enzyme Research. 487093.

Lubertozzi, D. and J. D. Keasling. 2009. Developing Aspergillus as a host for heterologous expression. Biotechnology Advances. 27: 53–75.

Malik, M., K. C. Nitiss, V. Enriquez-Rios and J. L. Nitiss. 2006. Roles of nonhomologous end-joining pathways in surviving topoisomerase II-mediated DNA damage. Molecular Cancer Therapeutics. 5: 1405–14.

Martens-Uzunova, E. S. and P. J. Schaap. 2009. Assessment of the pectin degrading enzyme network of *Aspergillus niger* by functional genomics. Fungal Genetics and Biology. 46 Suppl 1: S170–S179.

Martens-Uzunova, E. S., J. S. Zandleven, J. A. Benen, H. Awad, H. J. Kools, G. Beldman, A. G. Voragen, J. A. van den Berg and P. J. Schaap. 2006. A new group of exo-acting family 28 glycoside hydrolases of *Aspergillus niger* that are involved in pectin degradation. Biochem J. 400: 43–52.

Martin, J. A., R. A. Murphy and R. F. G. Power. 2003. Cloning and expression of fungal phytases in genetically modified strains of Aspergillus awamori. Journal of Industrial Microbiology & Biotechnology. 30: 568–76.

Mattern, I. E., J. M. van Noort, P. van den Berg, D. B. Archer, I. N. Roberts and C. A. van den Hondel. 1992. Isolation and characterization of mutants of *Aspergillus niger* deficient in extracellular proteases. Molecular & General Genetics. 234: 332–6.

Melzer, G., M. E. Esfandabadi, E. Franco-Lara and C. Wittmann. 2009. Flux Design: In silico design of cell factories based on correlation of pathway fluxes to desired properties. BMC Systems Biology. 3: 120.

Meyer, V. 2008. Genetic engineering of filamentous fungi—progress, obstacles and future trends. Biotechnology Advances. 26: 177–85.

Meyer, V., M. Arentshorst, A. El-Ghezal, A. -C. Drews, R. Kooistra, C. A. M. J. J. van den Hondel and A. F. J. Ram. 2007a. Highly efficient gene targeting in the *Aspergillus niger* kusA mutant. Journal of Biotechnology. 128: 770–5.

Meyer, V., M. Arentshorst, S. J. Flitter, B. M. Nitsche, M. J. Kwon, C. G. Reynaga-Peña, S. Bartnicki-Garcia, C. a M. J. J. van den Hondel and A. F. J. Ram. 2009. Reconstruction of signaling networks regulating fungal morphogenesis by transcriptomics. Eukaryotic cell. 8: 1677–91.

Meyer, V., M. Arentshorst, C. A. M. J. J. van den Hondel and A. F. J. Ram. 2008. The polarisome component SpaA localises to hyphal tips of *Aspergillus niger* and is important for polar growth. Fungal Genetics and Biology 45: 152–64.

Meyer, V., R. a Damveld, M. Arentshorst, U. Stahl, C. a M. J. J. van den Hondel and A. F. J. Ram. 2007b. Survival in the presence of antifungals: genome-wide expression profiling of *Aspergillus niger* in response to sublethal concentrations of caspofungin and fenpropimorph. The Journal of Biological Chemistry. 282: 32935–48.

Meyer, V., S. Minkwitz, T. Schütze, C. A. M. J. J. V. D. Hondel and A. F. J. Ram. 2010a. The *Aspergillus niger* RmsA protein. A node in a genetic network? Communicative & Integrative Biology. 3: 195–197.

Meyer, V., D. Mueller, T. Strowig and U. Stahl. 2003. Comparison of different transformation methods for Aspergillus giganteus. Current Genetics. 43: 371–7.

Meyer, V., A. F. J. Ram and P. J. Punt. 2010b. Genetics, Genetic Manipulation, and Approaches to Strain Improvement of Filamentous Fungi. In: Manual of Industrial Microbiology and Biotechnology, 3rd Editio. NY: Wiley. pp. 318–329.

Meyer, V., F. Wanka, J. van Gent, M. Arentshorst, C. A. M. J. J. van den Hondel and A. F. J. Ram. 2011a. Fungal gene expression on demand: an inducible, tunable, and metabolism-independent expression system for *Aspergillus niger*. Applied and Environmental Microbiology. 77: 2975–83.

Meyer, V., B. Wu and A. F. J. Ram. 2011b. Aspergillus as a multi-purpose cell factory: current status and perspectives. Biotechnology Letters. 33: 469–76.

Michielse, C. B., P. J. J. Hooykaas, C. A. M. J. J. van den Hondel and A. F. J. Ram. 2008. Agrobacterium-mediated transformation of the filamentous fungus Aspergillus awamori. Nature Protocols. 3: 1671–8.

Mizutani, O., K. Masaki, K. Gomi and H. Iefuji. 2012. Modified Cre-loxP Recombination in Aspergillus oryzae by Direct Introduction of Cre Recombinase for Marker Gene Rescue. Applied and Environmental Microbiology. 78: 4126–33.

Moralejo, F. J., R. E. Cardoza, S. Gutierrez and J. F. Martin. 1999. Thaumatin production in Aspergillus awamori by use of expression cassettes with strong fungal promoters and high gene dosage. Applied and Environmental Microbiology. 65: 1168–74.

Moralejo, F. J., A. J. Watson, D. J. Jeenes, D. B. Archer and J. F. Martín. 2001. A defined level of protein disulfide isomerase expression is required for optimal secretion of thaumatin by Aspegillus awamori. Molecular Genetics and Genomics. 266: 246–53.

Mulder, H. J., M. Saloheimo, M. Penttilä and S. M. Madrid. 2004. The transcription factor HACA mediates the unfolded protein response in *Aspergillus niger*, and up-regulates its own transcription. Molecular Genetics and Genomics. 271: 130–40.

Määttänen, P., K. Gehring, J. J. M. Bergeron and D. Y. Thomas. 2010. Protein quality control in the ER: the recognition of misfolded proteins. Seminars in Cell & Developmental Biology. 21: 500–11.

Nakajima, K. -ichiro, T. Asakura, J. -ichi Maruyama, Y. Morita, H. Oike, A. Shimizu-Ibuka, T. Misaka, H. Sorimachi, S. Arai, K. Kitamoto and K. Abe. 2006. Extracellular production of neoculin, a sweet-tasting heterodimeric protein with taste-modifying activity, by Aspergillus oryzae. Applied and Environmental Microbiology. 72: 3716–23.

Nayak, T., E. Szewczyk, C. E. Oakley, A. Osmani, L. Ukil, S. L. Murray, M. J. Hynes, S. A. Osmani and B. R. Oakley. 2006. A versatile and efficient gene-targeting system for Aspergillus nidulans. Genetics. 172: 1557–66.

Nelson, G., O. Kozlova-Zwinderman, A. J. Collis, M. R. Knight, J. R. S. Fincham, C. P. Stanger, A. Renwick, J. G. M. Hessing, P. J. Punt, C. A. M. J. J. van den Hondel and N. D. Read. 2004. Calcium measurement in living filamentous fungi expressing codon-optimized aequorin. Molecular Microbiology. 52: 1437–50.

Nemoto, T., J. -ichi Maruyama and K. Kitamoto. 2009. Improvement of heterologous protein production in Aspergillus oryzae by RNA interference with alpha-amylase genes. Bioscience, Biotechnology, and Biochemistry. 73: 2370–3.

Ngiam, C., D. J. Jeenes, P. J. Punt, C. A. Van Den Hondel and D. B. Archer. 2000. Characterization of a foldase, protein disulfide isomerase A, in the protein secretory pathway of *Aspergillus niger*. Applied and Environmental Microbiology. 66: 775–82.

Nielsen, J. B., M. L. Nielsen and U. H. Mortensen. 2008. Transient disruption of non-homologous end-joining facilitates targeted genome manipulations in the filamentous fungus Aspergillus nidulans. Fungal Genetics and Biology 45: 165–70.

Nielsen, K. F., J. M. Mogensen, M. Johansen, T. O. Larsen and J. C. Frisvad. 2009. Review of secondary metabolites and mycotoxins from the *Aspergillus niger* group. Analytical and Bioanalytical Chemistry. 395: 1225–42.

Nielsen, M. L., J. B. Nielsen, C. Rank, M. L. Klejnstrup, D. K. Holm, K. H. Brogaard, B. G. Hansen, J. C. Frisvad, T. O. Larsen and U. H. Mortensen. 2011. A genome-wide polyketide synthase deletion library uncovers novel genetic links to polyketides and meroterpenoids in Aspergillus nidulans. FEMS Microbiology Letters. 321: 157–66.

Nikolaev, I., M. Mathieu, P. van de Vondervoort, J. Visser and B. Felenbok. 2002. Heterologous expression of the Aspergillus nidulans alcR-alcA system in *Aspergillus niger*. Fungal Genetics and Biology. 37: 89–97.

Ninomiya, Y., K. Suzuki, C. Ishii and H. Inoue. 2004. Highly efficient gene replacements in Neurospora strains deficient for nonhomologous end-joining. Proceedings of the National Academy of Sciences of the United States of America. 101: 12248–53.

Nitsche, B. M., J. Crabtree, G. C. Cerqueira, V. Meyer, A. F. J. Ram, J. R. Wortman and A. Fj. 2011. New resources for functional analysis of omics data for the genus Aspergillus. BMC Genomics. 12: 486.

Nitsche, B. M., A. F. J. Ram and V. Meyer. 2012. The use of open source bioinformatics tools to dissect transcriptomic data. Methods in Molecular Biology (Clifton, N. J.). 835: 311–31.

Nyyssönen, E. and S. Keränen. 1995. Multiple roles of the cellulase CBHI in enhancing production of fusion antibodies by the filamentous fungus Trichoderma reesei. Current Genetics. 28: 71–9.

Oldiges, M., S. Lütz, S. Pflug, K. Schroer, N. Stein and C. Wiendahl. 2007. Metabolomics: current state and evolving methodologies and tools. Applied Microbiology and Biotechnology. 76: 495–511.

Ong, S. -E., B. Blagoev, I. Kratchmarova, D. B. Kristensen, H. Steen, A. Pandey and M. Mann. 2002. Stable isotope labeling by amino acids in cell culture, SILAC, as a simple and accurate approach to expression proteomics. Molecular & Cellular Proteomics. 1: 376–86.

Pabinger, S., R. Rader, R. Agren, J. Nielsen and Z. Trajanoski. 2011. MEMOSys: Bioinformatics platform for genome-scale metabolic models. BMC Systems Biology. 5: 20.

Pachlinger, R., R. Mitterbauer, G. Adam and J. Strauss. 2005. Metabolically independent and accurately adjustable Aspergillus sp. expression system. Applied and Environmental Microbiology. 71: 672–8.

Pakula, T. M., M. Laxell, A. Huuskonen, J. Uusitalo, M. Saloheimo and M. Penttilä. 2003. The effects of drugs inhibiting protein secretion in the filamentous fungus Trichoderma reesei. Evidence for down-regulation of genes that encode secreted proteins in the stressed cells. The Journal of Biological Chemistry. 278: 45011–20.

Panozzo, C., V. Capuano, S. Fillinger and B. Felenbok. 1997. The zinc binuclear cluster activator AlcR is able to bind to single sites but requires multiple repeated sites for synergistic activation of the alcA gene in Aspergillus nidulans. The Journal of Biological Chemistry. 272: 22859–65.

Pel, H. J., J. H. de Winde, D. B. Archer, P. S. Dyer, G. Hofmann, P. J. Schaap, G. Turner, R. P. de Vries, R. Albang, K. Albermann, M. R. Andersen, J. D. Bendtsen, J. A. E. Benen, M. van den Berg, S. Breestraat, M. X. Caddick, R. Contreras, M. Cornell, P. M. Coutinho, E. G. J. Danchin, A. J. M. Debets, P. Dekker, P. W. M. van Dijck, A. van Dijk, L. Dijkhuizen, A. J. M. Driessen, C. D'Enfert, S. Geysens, C. Goosen, G. S. P. Groot, P. W. J. de Groot, T. Guillemette, B. Henrissat, M. Herweijer, J. P. T. W. van den Hombergh, C. A. M. J. J. van den Hondel, R. T. J. M. van der Heijden, R. M. van der Kaaij, F. M. Klis, H. J. Kools, C. P. Kubicek, P. A. van Kuyk, J. Lauber, X. Lu, M. J. E. C. van der Maarel, R. Meulenberg, H. Menke, M. A. Mortimer, J. Nielsen, S. G. Oliver, M. Olsthoorn, K. Pal, N. N. M. E. van Peij, A. F. J. Ram, U. Rinas, J. A. Roubos, C. M. J. Sagt, M. Schmoll, J. Sun, D. Ussery, J. Varga, W. Vervecken, P. J. J. van de Vondervoort, H. Wedler, H. A. B. Wösten, A. -P. P. Zeng, A. J. J. van Ooyen, J. Visser, H. Stam, M. A. van den Berg, Hondel, P. A. VanKuyk and H. A. Wosten. 2007. Genome sequencing and analysis of the versatile cell factory *Aspergillus niger* CBS 513. 88. Nature Biotechnology. 25: 221–31.

Perrone, G., G. Stea, F. Epifani, J. Varga, J. C. Frisvad and R. a Samson. 2011. *Aspergillus niger* contains the cryptic phylogenetic species A. awamori. Fungal Biology. 115: 1138–50.

Peñalva, M. a, J. Tilburn, E. Bignell and H. N. Arst. 2008. Ambient pH gene regulation in fungi: making connections. Trends in Microbiology. 16: 291–300.

Prathumpai, W., M. McIntyre and J. Nielsen. 2004. The effect of CreA in glucose and xylose catabolism in Aspergillus nidulans. Applied Microbiology and Biotechnology. 63: 748–53.

Punt, P. J., M. A. Dingemanse, A. Kuyvenhoven, R. D. M. Seede, P. H. Peuwels and C. a M. J. J. van den Hondel. 1990. Functional elements in the promoter region of the Aspergillus nidulans gpdA gene encoding glyceraldehyde-3-phosphate dehydrogenase. Gene. 93: 101–109.

Punt, P. J., A. Drint-Kuijvenhoven, B. C. Lokman, J. A. Spencer, D. Jeenes, D. A. Archer and C. A. M. J. J. van den Hondel. 2003. The role of the *Aspergillus niger* furin-type protease gene in processing of fungal proproteins and fusion proteins. Evidence for alternative processing of recombinant (fusion-) proteins. Journal of Biotechnology. 106: 23–32.

Punt, P. J., I. A. van Gemeren, J. Drint-Kuijvenhoven, J. G. Hessing, G. M. van Muijlwijk-Harteveld, A. Beijersbergen, C. T. Verrips and C. A. van den Hondel. 1998. Analysis of the role of the gene bipA, encoding the major endoplasmic reticulum chaperone protein in the secretion of homologous and heterologous proteins in black Aspergilli. Applied Microbiology and Biotechnology. 50: 447–54.

Punt, P. J., F. H. J. Schuren, J. Lehmbeck, T. Christensen, C. Hjort and C. A. M. J. J. van den Hondel. 2008. Characterization of the *Aspergillus niger* prtT, a unique regulator of extracellular protease encoding genes. Fungal Genetics and Biology. 45: 1591–9.

Qin, L. -N., F. -R. Cai, X. -R. Dong, Z. -B. Huang, Y. Tao, J. -Z. Huang and Z. -Y. Dong. 2012. Improved production of heterologous lipase in Trichoderma reesei by RNAi mediated gene silencing of an endogenic highly expressed gene. Bioresource Technology. 109: 116–22.

Rando, O. J. 2012. Combinatorial Complexity in Chromatin Structure and Function: Revisiting the Histone Code. Current Opinion in Genetics & Development. 22: 148–155.

Rohlfs, M., M. Albert, N. P. Keller and F. Kempken. 2007. Secondary chemicals protect mould from fungivory. Biology Letters. 3: 523–5.

Roth, A. H. F. J. and P. Dersch. 2010. A novel expression system for intracellular production and purification of recombinant affinity-tagged proteins in *Aspergillus niger*. Applied Microbiology and Biotechnology. 86: 659–70.

Ruddock, L. W. and M. Molinari. 2006. N-glycan processing in ER quality control. Journal of Cell Science. 119: 4373–80.

Rumbold, K., H. J. J. van Buijsen, V. M. Gray, J. W. van Groenestijn, K. M. Overkamp, R. S. Slomp, M. J. van der Werf and P. J. Punt. 2010. Microbial renewable feedstock utilization: a substrate-oriented approach. Bioengineered Bugs. 1: 359–66.

Rumbold, K., H. J. J. van Buijsen, K. M. Overkamp, J. W. van Groenestijn, P. J. Punt and M. J. van der Werf. 2009. Microbial production host selection for converting second-generation feedstocks into bioproducts. Microbial Cell Factories. 8: 64.

Römisch, K. 1999. Surfing the Sec61 channel: bidirectional protein translocation across the ER membrane. Journal of Cell Science. 112 (Pt 2: 4185–91).

Sagt, C. M. J., P. J. ten Haaft, I. M. Minneboo, M. P. Hartog, R. a Damveld, J. M. van der Laan, M. Akeroyd, T. J. Wenzel, F. a Luesken, M. Veenhuis, I. van der Klei and J. H. de Winde. 2009. Peroxicretion: a novel secretion pathway in the eukaryotic cell. BMC Biotechnology. 9: 48.

Saloheimo, M. and T. M. Pakula. 2012. The cargo and the transport system: secreted proteins and protein secretion in Trichoderma reesei (Hypocrea jecorina). Microbiology (Reading, England). 158: 46–57.

Saloheimo, M., M. Valkonen and M. Penttilä. 2003. Activation mechanisms of the HAC1-mediated unfolded protein response in filamentous fungi. Molecular Microbiology. 47: 1149–61.

Sanchez, J. F., A. D. Somoza, N. P. Keller and C. C. C. Wang. 2012. Advances in Aspergillus secondary metabolite research in the post-genomic era. Natural Product Reports. 29: 351–71.

Sarkar, A., A. N. Funk, K. Scherlach, F. Horn, V. Schroeckh, P. Chankhamjon, M. Westermann, M. Roth, A. A. Brakhage, C. Hertweck and U. Horn. 2012. Differential expression of silent polyketide biosynthesis gene clusters in chemostat cultures of Aspergillus nidulans. Journal of Biotechnology. 160: 64–71.

Sasaguri, S., J. -ichi Maruyama, S. Moriya, T. Kudo, K. Kitamoto and M. Arioka. 2008. Codon optimization prevents premature polyadenylation of heterologously-expressed cellulases from termite-gut symbionts in Aspergillus oryzae. The Journal of General and applied Microbiology. 54: 343–51.

Schroeckh, V., K. Scherlach, H. -W. Nützmann, E. Shelest, W. Schmidt-Heck, J. Schuemann, K. Martin, C. Hertweck and A. a Brakhage. 2009. Intimate bacterial-fungal interaction triggers biosynthesis of archetypal polyketides in Aspergillus nidulans. Proceedings of the National Academy of Sciences of the United States of America. 106: 14558–63.

Schuster, E., N. Dunn-Coleman, J. C. Frisvad and P. W. M. Van Dijck. 2002. On the safety of *Aspergillus niger*-a review. Applied Microbiology and Biotechnology. 59: 426–35.

Schäfer, A. and D. H. Wolf. 2009. Sec61p is part of the endoplasmic reticulum-associated degradation machinery. The EMBO Journal. 28: 2874–84.

Sharma, R., M. Katoch, N. Govindappa, P. S. Srivastava, K. N. Sastry and G. N. Qazi. 2012. Evaluation of the catalase promoter for expressing the alkaline xylanase gene (alx) in *Aspergillus niger*. FEMS microbiology letters. 327: 33–40.

Shima, Y., Y. Ito, S. Kaneko, H. Hatabayashi, Y. Watanabe, Y. Adachi and K. Yabe. 2009. Identification of three mutant loci conferring carboxin-resistance and development of a novel transformation system in Aspergillus oryzae. Fungal Genetics and Biology. 46: 67–76.

Shoji, J. -Y., J. -ichi Maruyama, M. Arioka and K. Kitamoto. 2005. Development of Aspergillus oryzae thiA promoter as a tool for molecular biological studies. FEMS Microbiology Letters. 244: 41–6.

Shrivastav, M., L. P. De Haro and J. a Nickoloff. 2008. Regulation of DNA double-strand break repair pathway choice. Cell Research. 18: 134–47.

Shryock, J. C., R. Rubio and R. M. Berne. 1986. Extraction of adenine nucleotides from cultured endothelial cells. Analytical Biochemistry. 159: 73–81.

Shwab, E. K., J. W. Bok, M. Tribus, J. Galehr, S. Graessle and N. P. Keller. 2007. Histone deacetylase activity regulates chemical diversity in Aspergillus. Eukaryotic Cell. 6: 1656–64.

Snoek, I. S. I., Z. A. van der Krogt, H. Touw, R. Kerkman, J. T. Pronk, R. A. L. Bovenberg, M. A. van den Berg and J. M. Daran. 2009. Construction of an hdfA Penicillium chrysogenum strain impaired in non-homologous end-joining and analysis of its potential for functional analysis studies. Fungal Genetics and Biology. 46: 418–26.

Spencer, J. A., D. J. Jeenes, D. A. MacKenzie, D. T. Haynie and D. B. Archer. 1998. Determinants of the fidelity of processing glucoamylase-lysozyme fusions by *Aspergillus niger*. European Journal of Biochemistry / FEBS. 258: 107–12.

Storms, R., Y. Zheng, H. Li, S. Sillaots, A. Martinez-Perez and A. Tsang. 2005. Plasmid vectors for protein production, gene expression and molecular manipulations in *Aspergillus niger*. Plasmid. 53: 191–204.

Subramanian, A., P. Tamayo, V. K. Mootha, S. Mukherjee, B. L. Ebert, M. A. Gillette, A. Paulovich, S. L. Pomeroy, T. R. Golub, E. S. Lander and J. P. Mesirov. 2005. Gene set enrichment analysis: a knowledge-based approach for interpreting genome-wide expression profiles. Proc. Natl. Acad. Sci. USA. 102: 15545–15550.

Szewczyk, E., Y. -M. Chiang, C. E. Oakley, A. D. Davidson, C. C. C. Wang and B. R. Oakley. 2008. Identification and characterization of the asperthecin gene cluster of Aspergillus nidulans. Applied and Environmental Microbiology. 74: 7607–12.

Tada, S., K. Gomi, K. Kitamoto, K. Takahashi, G. Tamura and S. Hara. 1991. Construction of a fusion gene comprising the Taka-amylase A promoter and the Escherichia coli beta-glucuronidase gene and analysis of its expression in Aspergillus oryzae. Molecular & General Genetics. 229: 301–6.

Taheri-Talesh, N., T. Horio, L. Araujo-baza, X. Dou, E. A. Espeso, M. A. Pen, S. A. Osmani and B. R. Oakley. 2008. The Tip Growth Apparatus of Aspergillus nidulans. Molecular Biology of the Cell. 19: 1439–1449.

Takahashi, T., T. Masuda and Y. Koyama. 2006. Enhanced gene targeting frequency in ku70 and ku80 disruption mutants of Aspergillus sojae and Aspergillus oryzae. Molecular Genetics and Genomics. 275: 460–70.

Tanaka, M., M. Tokuoka, T. Shintani and K. Gomi. 2012. Transcripts of a heterologous gene encoding mite allergen Der f 7 are stabilized by codon optimization in Aspergillus oryzae. Applied Microbiology and Biotechnology.

Torralba, S., A. M. Pedregosa, J. R. De Lucas, M. S. Diaz, I. F. Monistrol and L. F. 1996. Effect of the microtubule inhibitor methyl benzimidazol-2-yl carbamate (MBC) on production and secretion of enzymes in Aspergillus nidulans. Mycological Research. 11: 1375–1382.

Torralba, S., M. Raudaskoski, a M. Pedregosa and F. Laborda. 1998. Effect of cytochalasin A on apical growth, actin cytoskeleton organization and enzyme secretion in Aspergillus nidulans. Microbiology (Reading, England). 144 (Pt 1: 45–53).

Tremblay, L. O. and A. Herscovics. 1999. Cloning and expression of a specific human alpha 1,2-mannosidase that trims Man9GlcNAc2 to Man8GlcNAc2 isomer B during N-glycan biosynthesis. Glycobiology. 9: 1073–8.

Upshall, A., A. A. Kumar, M. C. Bailey, M. D. Parker, M. A. Favreau, K. P. Lewison, M. L. Joseph, J. M. Maraganore and G. L. McKnight. 1987. Secretion of Active Human Tissue Plasminogen Activator from the Filamentous Fungus Aspergillus Nidulans. Bio/Technology. 5: 1301–1304.

Valkonen, M., M. Ward, H. Wang, M. Penttilä and M. Saloheimo. 2003. Improvement of foreign-protein production in *Aspergillus niger* var. awamori by constitutive induction of the unfolded-protein response. Applied and environmental microbiology. 69: 6979–86.

van den Hombergh, J. P., P. J. van de Vondervoort, L. Fraissinet-Tachet and J. Visser. 1997. Aspergillus as a host for heterologous protein production: the problem of proteases. Trends in Biotechnology. 15: 256–63.

van den Berg, B. A., M. J. T. Reinders, M. Hulsman, L. Wu, J. Herman, H. J. Pel, J. A. Roubos and D. de Ridder. 2012. Relating sequence characteristics to high-level protein production in *Aspergillus niger*. PloS one (in Press).

Varadarajalu, L. P. and N. S. Punekar. 2005. Cloning and use of sC as homologous marker for *Aspergillus niger* transformation. Journal of Microbiological Methods. 61: 219–24.

van Gemeren, I. A., A. Beijersbergen, C. A. van den Hondel and C. T. Verrips. 1998. Expression and secretion of defined cutinase variants by Aspergillus awamori. Applied and Environmental Microbiology. 64: 2794–9.

van Laar, T., A. J. van der Eb and C. Terleth. 2001. Mif1: a missing link between the unfolded protein response pathway and ER-associated protein degradation? Current Protein & Peptide Science. 2: 169–90.

van den Brink, H. J. M., S. G. Petersen, H. Rahbek-Nielsen, K. Hellmuth and M. Harboe. 2006. Increased production of chymosin by glycosylation. Journal of Biotechnology. 125: 304–10.

van Peij, N. N. M. E., A. Los, J. Bakhuis, L. Wu, J. -M. van der Laan, M. M. A. Olsthoorn, J. . Roubos and H. J. Pel. 2012. Novel approaches for solving bottlenecks and improving recombinant protein production by Aspergillus. In: The 9th International Aspergillus Meeting. Marburg.

Verdoes, J. C., A. D. van Diepeningen, P. J. Punt, A. J. Debets, A. H. Stouthamer and C. A. van den Hondel. 1994a. Evaluation of molecular and genetic approaches to generate glucoamylase overproducing strains of *Aspergillus niger*. Journal of Biotechnology. 36: 165–75.

Verdoes, J. C., P. J. Punt, P. van der Berg, F. Debets, A. H. Stouthamer and C. A. van den Hondel. 1994b. Characterization of an efficient gene cloning strategy for *Aspergillus niger* based on an autonomously replicating plasmid: cloning of the nicB gene of *A. niger*. Gene. 146: 159–65.

Verdoes, J. C., P. J. Punt, J. M. Schrickx, H. W. van Verseveld, A. H. Stouthamer and C. A. van den Hondel. 1993. Glucoamylase overexpression in *Aspergillus niger*: molecular genetic analysis of strains containing multiple copies of the glaA gene. Transgenic Research. 2: 84–92.

Verdoes, J. C., P. J. Punt, A. H. Stouthamer and C. A. van den Hondel. 1994c. The effect of multiple copies of the upstream region on expression of the *Aspergillus niger* glucoamylase-encoding gene. Gene. 145: 179–87.

Vinck, A., C. de Bekker, A. Ossin, R. A. Ohm, R. P. de Vries and H. A. B. Wösten. 2011. Heterogenic expression of genes encoding secreted proteins at the periphery of *Aspergillus niger* colonies. Environmental Microbiology. 13: 216–25.

Vinck, A., M. Terlou, W. R. Pestman, E. P. Martens, A. F. Ram, C. a M. J. J. van den Hondel and H. a B. Wösten. 2005. Hyphal differentiation in the exploring mycelium of *Aspergillus niger*. Molecular Microbiology. 58: 693–9.

Vogt, K., R. Bhabhra, J. Rhodes and D. S. Askew. 2005. Doxycycline-regulated gene expression in the opportunistic fungal pathogen Aspergillus fumigatus. BMC Microbiology. 11: 1–11.

Vongsangnak, W., P. Olsen, K. Hansen, S. Krogsgaard and J. Nielsen. 2008. Improved annotation through genome-scale metabolic modeling of Aspergillus oryzae. BMC genomics. 9: 245.

Vongsangnak, W., M. Salazar, K. Hansen and J. Nielsen. 2009. Genome-wide analysis of maltose utilization and regulation in aspergilli. Microbiology (Reading, England). 155: 3893–902.

Wang, B., G. Guo, C. Wang, Y. Lin, X. Wang, M. Zhao, Y. Guo, M. He, Y. Zhang and L. Pan. 2010. Survey of the transcriptome of Aspergillus oryzae via massively parallel mRNA sequencing. Nucleic Acids Research. 38: 5075–87.

Wang, H., J. Entwistle, E. Morlon, D. B. Archer, J. F. Peberdy, M. Ward and D. J. Jeenes. 2003. Isolation and characterisation of a calnexin homologue, clxA, from *Aspergillus niger*. Molecular Genetics and Genomics. 268: 684–91.

Wang, Z., M. Gerstein and M. Snyder. 2009. RNA-Seq: a revolutionary tool for transcriptomics. Nature Reviews. Genetics. 10: 57–63.

Ward, M., C. Lin, D. C. Victoria, B. P. Fox, J. A. Fox, D. L. Wong, H. J. Meerman, J. P. Pucci, R. B. Fong, M. H. Heng, N. Tsurushita, C. Gieswein, M. Park and H. Wang. 2004. Characterization of humanized antibodies secreted by *Aspergillus niger*. Applied and Environmental Microbiology. 70: 2567–76.

Ward, M., L. J. Wilson, K. H. Kodama, M. W. Rey and R. M. Berka. 1990. Improved Production of Chymosin in Aspergillus by Expression as a Glucoamylase-Chymosin Fusion. Bio/Technology. 8: 435–440.

Ward, O. P. 2011. Production of recombinant proteins by filamentous fungi. Biotechnology Advances. 30: 1119–1139.

Ward, P. P., C. S. Piddington, G. A. Cunningham, X. Zhou, R. D. Wyatt and O. M. Conneely. 1995. A system for production of commercial quantities of human lactoferrin: a broad spectrum natural antibiotic. Bio/technology (Nature Publishing Company). 13: 498–503.

Whittle, C. A., Y. Sun and H. Johannesson. 2012. Genome-Wide Selection on Codon Usage at the Population Level in the Fungal Model Organism Neurospora crassa. Molecular Biology and Evolution.

Wu, L., M. R. Mashego, J. C. van Dam, A. M. Proell, J. L. Vinke, C. Ras, W. A. van Winden, W. M. van Gulik and J. J. Heijnen. 2005. Quantitative analysis of the microbial metabolome by isotope dilution mass spectrometry using uniformly 13C-labeled cell extracts as internal standards. Analytical Biochemistry. 336: 164–71.

Wucherpfennig, T., T. Hestler and R. Krull. 2011. Morphology engineering—Osmolality and its effect on *Aspergillus niger* morphology and productivity. Microbial Cell Factories. 10: 58.

Wucherpfennig, T., K. A. Kiep, H. Driouch, C. Wittmann, R. Krull, C. Cordes, H. Horn, I. Kampen, A. Kwade, T. R. Neu and B. No. 2010. Morphology and rheology in filamentous cultivations. Advances in Applied Microbiology. 72: 89–136.

Wösten, H. a B., S. M. Moukha, J. H. Sietsma and J. G. Wessels. 1991. Localization of growth and secretion of proteins in *Aspergillus niger*. Journal of General Microbiology. 137: 2017–23.

Yoon, J., S. Kimura, J. -ichi Maruyama and K. Kitamoto. 2009. Construction of quintuple protease gene disruptant for heterologous protein production in Aspergillus oryzae. Applied Microbiology and Biotechnology. 82: 691–701.

Yoon, J., J. -ichi Maruyama and K. Kitamoto. 2011. Disruption of ten protease genes in the filamentous fungus Aspergillus oryzae highly improves production of heterologous proteins. Applied microbiology and biotechnology. 89: 747–59.

Yoshida, H., T. Matsui, a Yamamoto, T. Okada and K. Mori. 2001. XBP1 mRNA is induced by ATF6 and spliced by IRE1 in response to ER stress to produce a highly active transcription factor. Cell. 107: 881–91.

Yu, J., N. D. Fedorova, B. G. Montalbano, D. Bhatnagar, T. E. Cleveland, J. W. Bennett and W. C. Nierman. 2011. Tight control of mycotoxin biosynthesis gene expression in Aspergillus flavus by temperature as revealed by RNA-Seq. FEMS Microbiology Letters. 322: 145–9.

Yuan, X. -L., C. Goosen, H. Kools, M. J. E. C. van der Maarel, C. A. M. J. J. van den Hondel, L. Dijkhuizen and A. F. J. Ram. 2006. Database mining and transcriptional analysis of genes encoding inulin-modifying enzymes of *Aspergillus niger*. Microbiology (Reading, England). 152: 3061–73.

Yuan, X. -L., J. A. Roubos, C. A. M. J. J. van den Hondel and A. F. J. Ram. 2008. Identification of InuR, a new Zn(II)2Cys6 transcriptional activator involved in the regulation of inulinolytic genes in *Aspergillus niger*. Molecular Genetics and Genomics. 279: 11–26.

Zheng, H. -Z., H. -H. Liu, S. -X. Chen, Z. -X. Lu, Z. -L. Zhang, D. -W. Pang, Z. -X. Xie and P. Shen. 2005. Yeast transformation process studied by fluorescence labeling technique. Bioconjugate Chemistry. 16: 250–4.

http://bioinformatics. tudelft. nl/hipsec.

http://www.fgsc.net/Aspergillus

http://www.blast2go.de/b2ghome

http://www.broadinstitute.org/gsea/index.jsp

http://www.broadinstitute.org/fetgoat/index.html

www.sysbio.se/BioMet/

https://memosys.i-med.ac.at/MEMOSys/home.seam

http://f-a-m-e.org/

www.ncbi.nlm.nih.gov

www.aspgd.org

www.broadinstitute.org/annotation/genome/aspergillus_group

www.cadre-genomes.org.uk

2

Industrial Production of Organic Acids by Fungi
State of the Art and Opportunities

An Li and *Peter Punt**

ABSTRACT

The major commercially produced organic acid products are derivatives of components of the tricarboxylic acids cycle or simple oxidation products of sugars. In this chapter, biological production of industrially useful organic acids, especially di- and tri-carboxylic acids are reviewed. Bio-production processes have already been applied for several of these organic acids. Besides their biosynthetic pathways, other relevant aspects of organic acid transport are covered in the review. The commercial production of organic acids is mainly performed using fungi, in particular, members of the genus *Aspergillus* and *Rhizopus*. These species are natural producers of organic acids providing an ecological advantage in growing well at conditions of organic acid production, i.e., lower pH levels. Moreover, like many other fungi, these species are capable of utilizing renewable biomass resources.

Division of Microbiology & Systems Biology, TNO Innovation of Life, P.O. Box 360 I 3700 AJ
  Zeist I the Netherlands.
  Email: an.li@tno.nl
* Corresponding author: peter.punt@tno.nl

Introduction

Industrially useful organic acids especially di-, tricarboxylic acids have a broad range of applications in our daily life. Traditionally, many of them were isolated from natural plant resources like lemons and apples. Subsequently most of them have been produced chemically from fossil resources, and more recently several of them are being produced by microorganisms using carbohydrate substrates.

As one of the largest group of microorganisms, filamentous fungi are well-known for their ability in secreting organic acids. These acids are generated from glycolysis and the TCA (Tri-Carboxylic Acid) cycle. Many fungal species are capable of degrading plant biomass derived, so-called second generation feedstock materials (agriculture waste, etc.), into monosaccharide to serve as carbon source. Filamentous fungi such as *Aspergillus* are capable of growing at low pH levels, based on their ecological niche which makes them useful host strains for optimal production and secretion of organic acids.

In general, the organic acid production pathway in fungi takes place in the interplay between mitochondrial and cytosolic biochemical steps. Glycolysis glucose is degraded to pyruvate (Fig. 1). Pyruvate can be further hydrolyzed into lactate, or oxidized into acetyl-CoA. In the TCA cycle, acetyl-CoA together with oxaloacetate are converted into citrate which is the starting point for the TCA cycle. In this chapter, the production of 9 organic acids will be described: citric acid, succinic acid, fumaric acid, malic acid, lactic acid, gluconic acid, oxalic acid, itaconic acid and its derivative citraconic

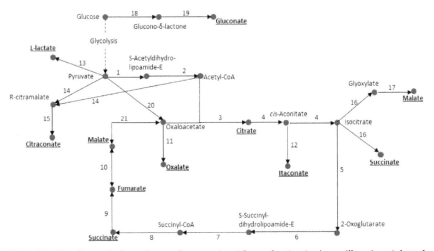

Fig. 1. (Predicted) metabolic pathways for organic acids production in *Aspergillus niger*. (plotted based on Kegg: http://www.genome.jp/kegg/pathway.html).

acid. Among them, citric-, gluconic-, itaconic- and fumaric acid have already been industrially produced by filamentous fungi *Aspergillus niger, Aspergillus terreus* and *Rhizopus oryzae* respectively. The other acids are currently not commercially produced by fungi. An overview of the characteristics and the application of each of these organic acids are presented in Table 1. In Table 2, the result of a genome mining effort in *Aspergillus niger* and the eukaryotic model species *Saccharomyces cerevisiae* is given, showing all genes potentially encoding proteins relevant for biochemical steps presented in Fig. 1. As shown from the predicted localization of the encoded proteins, a clear involvement of both cytosol and mitochondrion is suggested, given the involvement of the mitochondrial and cytosolic compartment in organic acid biosynthesis. In this review, organic acid transport, in particular the role of di, tri-carboxylic acid transporters in fungi, is discussed as well.

Citric Acid

By volume, citric acid, a C6 tricarboxylic acid is a major industrially produced organic acid. It is nowadays industrially produced by *Aspergillus niger (A. niger)* and has a large demand as acidulant in food/beverages, and further as preservative in pharmaceuticals and antioxidant in cosmetic manufactures.

Its production pathway in *A. niger* was first published in 1917 (Currie 1917). In the following years, Currie and his colleagues worked on production of citric acid from fermentation process instead of extraction from citrus fruits. A fermentation system was then developed which utilized a strain of the organism *Aspergillus niger*. The reason for involving this organism was the discovery of its defective citric acid cycle, which almost quantitatively converted glucose to citric acid. Although a large number of fungal species such as *Aspergillus niger, Aspergillus aculeatus, Aspergillus carbonarius, Aspergillus awamori, Aspergillus foetidus, Penicillium janthinellum, Candida tropicalis* and *Yarrowia lipolytica* were employed for citric acid production (Karaffa and Kubicek 2003, Grewal and Kalra 1995, Vandenberghe et al. 2000), *Aspergillus niger* has remained the organism of choice for commercial production as it still has the highest productivity of citric acid in product per time unit. Since the discovery of Penicillin and the concomitant submerged culture system for filamentous fungi, citric acid production process was adapted to large volume submerged cultures. By 2007, annual world citric acid production had increased to 1,600,000 tonnes (Berovic and Legisa 2007).

The metabolic pathway for citric acid explored in *A. niger* is as follows (see Fig. 1): After glucose is up taken by fungus, this six carbon compound directly enters glycolysis in the cytosol and is metabolized to pyruvate. Further in the mitochondria, pyruvate is converted into acetyl-CoA. Under

Table 1. Overview of the current status of several organic acids

Name	structure	current application	key organisms	current production method	annual production (tonnes)
citric acid		food, beverage, phamaceutical preservative, comsmatics antioxidant, etc.	*Aspergillus niger*	citrus fruit extraction, submerged fungal cultivation process	1,7 millon (Francielo et al. 2008)
succinic acid		surfactant, ion chelator, food, pharmaceuticals	rumen bacterial (*Corynebacterium glutamicum* and *Anaerobiospirillum*	Petrochemical process, anaerobic bacterial fed-batch or continuous cultivation system	30,000 (Natrass and Higson 2010)
fumaric acid		food and feed additive, synthetic resins, etc.	*Rhizopus* spp. (*Rhizopus oryzae*)	Chemical process, submerged fungal cultivation process	50,000 (Lee et al. 2009)
Malic acid		food acidulant, biodegradable polymers	*Aspergillus flavus*	Petrochemical process, anaerobic bacterial fed-batch or continuous cultivation system	60,000 (UNESCO 2012)
lactic acid		food, pharmaceuticals, cosmetics	lactic acid bacteria, *Rhizopus* spp. (*Rhizopus oryzae*)	bacterial fermentaion	300,000–400,000 (Natrass and Higson 2010)
oxalic acid		bleaching agent, chelating agent	saprophytic and phytopathogenic fungi	chemical process	120,000 (Retired and Tanifuji 2011)
gluconic acid		food and phamaceuticals	*Aspergillus niger*, *Penicillium* spp.	submerged fungal fermentation process	60,000 (Ramachandran et al. 2006)
itaconic acid		synthetic fibers, resins, rubbers, etc.	*Aspergillus terreus*	petrochemical process, submerged fungal fermentation process	80,000 (Okabe et al. 2009)
citraconic acid		Dimers, copolymers	*Methanocaldococcus jannaschii*	petrochemical process	unknown

Table 2. Genome mining for *Aspergillus niger* and *Saccharomyces cerevisiae*. Enzymes with potential relevance to organic acids production:

No.[a]	EC no.	(Putative) Enzyme function	*Aspergillus niger*[b] Accession no.	Wolf-Psort[c]	*Saccharomyces cerevisiae*[b] Accession no.	Wolf-Psort	Reference[d]
1	EC:1.2.4.1	pyruvate dehydrogenase E1 component subunit beta	ANI_1_1206064	M: 26.0	YER178W	M: 23.0	
		pyruvate dehydrogenase E1 component subunit alpha	ANI_1_12014	M: 26.5	YBR221C	M: 27.0	
2	EC:2.3.1.2	pyruvate dehydrogenase E2 component	ANI_1_274064	M: 24.0	YNL071W	M: 26.5	
3	EC:2.3.3.1	citrate synthase	ANI_1_876084	M: 19.0	YCR005C	P: 8.0	Kubicek and Röhr 1980, Kubicek et al. 2009
		citrate synthase	ANI_1_1226134	M: 22.0	YNR001C	M: 15.0	
		citrate synthase	ANI_1_1474074	-	-	-	
		citrate synthase	ANI_1_2950014	-	-	-	
4	EC:4.2.1.3	aconitate hydratase	ANI_1_470084	M: 26.5	YLR304C	-	
		aconitate hydratase			YJL200C	-	
5	EC:1.1.1.41	isocitrate dehydrogenase (NAD+) subunit	ANI_1_906164	M: 26.5	YOR136W	M: 27.0	
		isocitrate dehydrogenase [NAD] subunit 2	ANI_1_798074	M: 19.0	YNL037C	M: 23.0	
6	EC:1.2.4.2	2-oxoglutarate dehydrogenase	ANI_1_826184	M: 22.0	YIL125W	M: 21.0	
7	EC:2.3.1.61	dihydrolipoyllysine-residue succinyltransferase	ANI_1_1482094	M: 27.0	YDR148C	M: 26.0	
8	EC:6.2.1.4	succinyl-CoA ligase [GDP-forming] subunit alpha	ANI_1_230154	M: 22.5	YOR142W	M: 18.5	
		succinyl-CoA ligase [GDP-forming] subunit beta	ANI_1_58124	M: 21.5	YGR244C	M: 23.0	

No.	EC	Enzyme	Gene 1	Value 1	Gene 2	Value 2	Reference
9	EC:1.3.5.1	succinate dehydrogenase [ubiquinone] flavoprotein subunit	ANI_1_1750024	M: 27.0	YKL148C	M: 27.0	
		succinate dehydrogenase [ubiquinone] flavoprotein subunit	ANI_1_2706024	M: 11.0	YLL041C	M: 26.0	
10	EC:4.2.1.2	Fumarate hydratase	ANI_1_952104	M: 25.5	YPL262W	M: 21.5	
		Fumarase	ACY91850 (*Rhizopus oryzae*)[e]	M: 17.5	–	–	Song et al. 2011
		Fumarase	CAA55314 (*Rhizopus oryzae*)[e]	M: 23.5	–	–	Friedberg et al. 1995
11	EC:3.7.1.1	oxaloacetate acetylhydrolase	ANI_1_92174	–	–	–	Pedersen et al. 2000
		oxaloacetate acetylhydrolase	ANI_1_2054064	M: 13.0	–	–	
12	EC:4.1.1.6	cis-aconitate decarboxylase	ATEG_09971 (*Aspergillus terreus*)[e]	–	–	–	Kanamasa et al. 2008, Li et al. 2011
13	EC:1.1.1.27	L-lactate dehydrogenase	ANI_1_2114184	M: 11.0	–	–	
14	EC:2.3.1.182	citramalate synthase	MJ_1392 (*Methanocaldococcus jannaschii*)[e]	–	–	–	Howell et al. 1999
			ANI_1_1952184	–	–	–	
15	EC:4.2.1.33	isopropylmalate isomerase (IPMI)	MJ_0499 (*Methanocaldococcus jannaschii*)[e]	–	YDR234W	M: 10.0	
			ANI_1_440024	–	–	–	
16	EC:4.1.3.1	isocitrate lyase	ANI_1_1256014		YER065C		
		isocitrate lyase			YPR006C	M: 17.5	
17	EC:2.3.3.9	malate synthase	ANI_1_320134	P:12.5	YIR031C	P: 12.3	
		malate synthase			YNL117W	P: 8.5	

Table 2. contd.....

Table 2. contd.

No.[a]	EC no.	(Putative) Enzyme function	Aspergillus Niger[b] Accession no.	Wolf-Psort[c]	Saccharomyces cerevisiae[b] Accession no.	Wolf-Psort[c]	Reference[d]
18	EC:1.1.3.4	glucose oxidase	ANI_1_1678104	E: 18	-	-	
		glucose oxidase	ANI_1_748094	E: 24	-	-	
		glucose oxidase	ANI_1_1398064	E: 24	-	-	
		glucose oxidase	ANI_1_1992014	E: 27	-	-	
19	EC:3.1.1.17	gluconolactonase	ANI_1_106174	E: 6.0	-	-	
20	EC:6.4.1.1	Pyruvate carboxylase	ANI_1_440184	-	YBR218C	-	
		Pyruvate carboxylase	-	-	YGL062W	-	
21	EC: 1.1.1.37	Malate dehydrogenase	ANI_1_12134	-	YOL126C	-	Brown et al. 2011, Minard and Mcalister-henn. 1991
			AOR_1_26114 (*Aspergillus oryzae*)[e]		YDL078C	-	
					YKL085W	-	
		Malate dehydrogenase	ANI_1_268064	M: 17.5			Brown et al. 2011
			AOR_1_766174 (*Aspergillus oryzae*)[e]	M: 23			

[a]These numbers refer to the numbering in Fig. 1.

[b]Indicate orthologous gene of *Saccharomyces cerevisiae* and *Aspergillus niger* which are included in the KEGG orthology (KO) data base http://www.genome.jp/kegg/ko.html. NCBI accession numbers are used.

[c]Wolf-Psort prediction (http://wolfpsort.org/). M: mitochondria, E: extracellular, P: peroxisome. ·: no specific localization signal predicted. Note that these classifications are only based on prediction and not on experimental validation.

[d]References describe functionally validated fungal enzymes.

[e]These genes represent fungal but non-A. niger enzymes which have been studied in more detail (see indicated references). For the citraconate pathway, only results from a non-fungal archaeal *Methanocaldococcus jannaschii* pathway are available.

the catalysis of citrate synthase, acetyl-CoA and oxaloacetate, which is derived by cytosolic pyruvate carboxylation and mitochondrial import are converted into citrate entering into the TCA cycle (Verhoff and Spradlin 1976) (Fig.1). As indicated in Table 2, there are several homologues for genes encoding putative citrate synthases (EC: 2.3.3.1) in *A. niger* (Pel et al. 2007, Sun et al. 2007). Two of them are predicted by 'WoLF PSORT' (Horton et al. 2007) to be localized in the mitochondrion of which one is also functionally characterized (Kirimura et al. 1999) (Table 2).

As already stated, current industrial production for citric acid is mainly carried out in submerged fermentations. Accumulation of citric acid is strongly influenced by the composition of the fermentation medium. The main factors affecting citric acid fermentation are: type and concentration of the carbon source, pH, aeration, temperature, concentration of trace elements and the morphology of the producer organism. Certain nutrients need to be in excess such as sugar, oxygen, etc., while others have to be within limit (e.g., nitrogen and phosphate source) and some components had to remain below defined limits (e.g., trace metals, especially manganese) (Schreferl et al. 1986). Similar to most filamentous fungi, *A. niger* can grow in a wide variety of morphologies depending on the strains, the initial cell density, the nutrient conditions and the type of fermenter. As indicated in many studies, mycelia with shortly branched hyphae are important for efficient production of citric acid in a stirred tank fermenter (Ju and Wang 1986, Aden et al. 2004, Tsao et al. 1999). This limited mycelial growth in combination with high concentrations of a carbohydrate source allow a high final yield of citric acid (Mattey et al. 1992). Morphology is strongly influenced by the amount of Mn^{2+} (Ju and Wang 1986, Tsao et al. 1999), where levels of around 10 ppb are found to be most suitable for citric acid production (Dai et al. 2004). Also a low cultivation pH (pH2-3), limiting extensive biomass formation is vital for high yield of citric acid. For optimal citric acid accumulation, the pH should fall below 3 within a few hours after initiation of the spore inoculation. Temperatures between 28 and 30°C are the normal range for obtain high rates of accumulation and high yields of citric acid. Other temperatures result in either increased by-product levels or reduced rates of accumulation.

Citric acid fermentation is an aerobic process. An increase in the oxygen supply results in an increase in citric acid yields during submerged fermentation. However, the aeration should not reduce carbon dioxide levels which are important as a substrate for pyruvate carboxylase replenishing the supply of oxaloacetate for citrate synthase. Trace elements are considered to be the main factor influencing the success of submerged citric acid production. *A. niger* requires certain trace metals for growth, a limitation in other trace metals is necessary for production of citric acid especially in submerged fermentation. Besides manganese, others including Zn, Fe, Cu should be limited as well (Papagianni et al. 2007).

Succinic Acid

Succinic acid (Butanedioic acid) is a C4 dicarboxylic acid, an intermediate of the TCA cycle in fungus (see Fig. 1). In industry, succinic acid is mainly used in four areas, as detergent/surfactant, ion chelator, food acidulant and anti-oxidant in the pharmaceutical industry (Beauprez et al. 2010). This acid is produced naturally by several microorganisms as a fermentation end-product. The rumen bacteria *Actinobacilus succinogenes* and *Mannheimia succiniporducens*, recombinant *E. coli*, *Corynebacterium glutamicum* and *Anaerobiospirillum succiniproducens* are among the best studied microorganisms for succinic acid production (Raab et al. 2010). Besides, succinic acid production is also established in yeast species *Saccharomyces cerevisiae* and *Yarrowia lipolytica* (Raab et al. 2010, Yuzbashev et al. 2010). Several filamentous fungi which are known to produce and secrete this acid are *Fusarium* spp. (Foster 1949), *Aspergillus* spp. (Bercovitz et al. 1990) and *Peniciluium simplicissimum* (Gallmetzer et al. 2002).

In fungi, the metabolic pathway of succinic acid may occur via two routes (see Fig. 1): 1) deriving from isocitric acid via the TCA cycle, 2) from isocitric acid via the glyoxylate bypass-pathway (Raab et al. 2010, Beauprez et al. 2010, Gallmetzer et al. 2002). In the TCA cycle, isocitric acid is converted to 2-oxoglutarate and further via succinyl-CoA to achieve succinate by succinyl-CoA ligase. In the glyoxylate route, isocitric acid is converted one step into succinic acid by iso-citrate lyase.

It is still unknown which one of these actually may lead to a high succinic acid accumulation in fungi. Although currently not employed, the production of succinate by filamentous fungi is suggested as a promising field (Magnuson and Lasure 2004). The enzyme iso-citrate lyase in *A. niger* is encoded by a unique gene ANI_1_1256014. Interestingly, this enzyme is predicted by 'WoLF PSORT' not to be localized in the mitochondrial compartment. No further functional analysis of this enzyme is yet reported to verify its actual localization.

Current industrial succinate production processes mainly use bacterial strains in anaerobic fed-batch or continuous cultivation (Song and Lee 2006). Mostly glucose is supplied as a carbon source and cultivation is carried out at neutral pH. The most commonly performed downstream processing (DSP) approaches are liquid-liquid extraction, ultrafiltration and precipitation (Beauprez et al. 2010). The production titers of succinic acid in the fermentation broth are reported to be up to 146 g/L (Okino et al. 2008).

Although fungi especially *Aspergillus* spp. have not been applied by industry for producing succinic acid as its titer and yields are not comparable with bacterial strains, the potential biological pathway and the availability of the required systems biology tools illustrate options for future direction of employing fungal strains for succinic acid production. In particular their natural acid tolerance provides an additional advantage over current production hosts for growth and downstream processing options.

Fumaric Acid

Fumaric acid is a C4 white crystalline di-carboxylic acid, an intermediate from the TCA (tri-carboxylic acid) cycle (Fig. 1). In pharmacy, the ester form of this acid is used to treat psoriasis and is considered as possible treatment of multiple sclerosis(Moharregh-Khiabani et al. 2009, Anstey 2010). In food and feed industry, fumaric acid is used as additives in the feed of broiler (Pirgozliev et al. 2008). Beyond this, this acid is also applied in the field of chemical engineering and as additive in paint, resin and plasticizer, etc. (Li et al. 2005, Roa Engel et al. 2008).

Production of fumaric acid by microorganisms started in 1940s in United States (Forster and Waksman 1939) following the earlier discovery of fumaric acid in *Rhizopus nigricans* by Felix Ehrlich in 1911 (Ehrlich 1911), leading to a patented fermentation process (Ling and Ng 1989). The most efficient producing strains for this acid are various *Rhizopus* species such as *R. nigiricans* 45, *R. arrhizus* NRRL 1526, *R. formosa* MUCL 28422, *R. arhizus* NRRL 2582, *R. oryzae* ATCC 20344, among which the latter two have the highest volumetric productivity, product titer and product yield values (Roa Engel et al. 2008).

The production pathway of fumaric acid in *Rhizopus oryzae* had been reviewed by Goldberg (Goldberg et al. 2006). Friedberg et al. suggested that, *Rhizopus oryzae* harbors the genetic information encoding two different fumarase, one localizing in mitochondria, which catalyzes the conversion of fumaric to L-malic acid, and a cytosolic enzyme, which catalyzes the conversion of L-malic to fumaric acid (Friedberg et al. 1995). However, more recent research suggests that both are probably encoded by the same gene, one having a truncation of 15 amino acids in the N-terminal region in comparison to the other (Song et al. 2011), while subcellular prediction suggests that both are located in mitochondria (Table 2). However, in the TCA cycle, fumarate can also be converted from succinate via succinate dehydrogenase.

Industrial fumaric acid production is carried out mainly in stirred tank fermentation, with glucose as a carbon source and a relatively neutral pH around 6.0 using *Rhizopus* (Roa Engel et al. 2008). Several reactor designs such as traditional stirred-tank, bubble fermentor, air-lift fermentor, or airlift loop fermentor have been used but stirred-tanks are the most commonly practiced reactors. Although the submerged fermentation process using *Rhizopus* species achieved economically attractive yields, these species are non-acid resistant and the fact of adding neutralizing agents inhibit the production of fumaric acid (Roa Engel et al. 2008). Thus, *A. niger* as an acid tolerant fungus is a potentially attractive new production host for future fumaric acid production improvement, given also its amenability for custom metabolic engineering, which is still lacking for *Rhizopus* species.

Malic Acid

Malic acid is a C4 dicarboxylic acid mainly used as an acidulant and taste enhancer in the beverage and food industry. In addition, it has therapeutic value for the treatment of hyperammonemia and liver dysfunction and as a component for amino acid infusion (Gong et al. 1996). Because of its increased application in the food industry as a citric acid replacement and its potential use as a raw material for the manufacturing of biodegradable polymers, it is included by the U.S. Department of Energy in the top 12 most interesting chemical building blocks that can be derived from biomass (Aden et al. 2004). The malic acid producing fungal species are *Aspergillus flavus, Rhizopus arrhizus and Paecilomyces varioti, Monascus araneosus, Schizophyllum commune, Zygosaccharomyces rouxii* and *Saccharomyces cerevisiae* (Zelle et al. 2008). The metabolic pathway for the production of malic acid from glucose is possible via three routes: 1) direct reduction of oxaloacetate, 2) oxidation of citrate via the TCA cycle, 3) formation from acetyl-CoA via the cyclic glyoxylate route (Fig. 1). As predicted by Zelle (Zelle et al. 2008), a most promising pathway (route 1) for malic acid production from glucose proceeds via carboxylation of pyruvate, followed by reduction of oxaloacetate to malate. The key enzyme involved in converting oxaloacetate to malate is the cytosolic isoenyzme of malate dehydrogenase (Table 2).

Malic acid production is preferably carried out in the 18–25°C degrees, high pH and low concentrations of sugar. The production of malic acid from an engineered *S. cerevisiae* strain achieved up to 59g/L using glucose as substrate (Zelle et al. 2008). A most recent research from West (West 2011) discovered the potential of *Aspergillus species* for malic acid production using thin stillage among which *A. niger* strains (ATCC 9142 and ATCC 10577) have highest production of 17–19 g/L. Thin stillage is a low value co-product of many ethanol plants containing a high percentage of glycerol and around 1% nitrogen (Kim et al. 2008). West concluded that the accumulation of malic acid instead of citric acid in *A. niger* is likely due to the composition of thin stillage, which make *A. niger* a potential commercial production host of malic acid. Following the same reasoning as followed for *S. cerevisiae*, overexpression of malate dehydrogenase has also been successfully explored for malic acid production in *Aspergillus oryzae* (Brown et al. 2011).

Lactic Acid

Lactic acid is a C3 carboxylic acid considered to be the mostly widely used organic acid in food, pharmaceutical, cosmetics and chemical industry (Zhang et al. 2007, Wee et al. 2006). Besides production by lactic acid bacteria, this acid can also be biologically produced by several *Rhizopus* spp. (Litchfield 1996). Among them, the fungus *Rhizopus oryzae* is the most suitable host for lactic acid production in fungi (Soccol et al. 1994). The production pathway of lactic acid was illustrated by Wright (Wright et al. 1996) and Longacre (Longacre et al. 1997) using a flux analysis of glucose metabolism of *R. oryzae*. In the model developed from MCT (metabolic control theory) (Westerhoff et al. 1984), *R. oryzae* can secrete amylase to digest starch and glucose from this hydrolysis enters glycolysis. In addition, *R. oryzae* can utilize xylose via the pentose phosphate pathway. The final product of the glycolysis and the pentose phosphate pathway is pyruvate. Pyruvate is converted to lactate by an NAD+ dependent lactate dehydrogenase (LDH) (Pritchard 1973).

Semi-continuous fermentation using pellets of *Rhizopus oryzae* has been shown to be a promising approach for L-lactic acid production and the titer reaches 103.7 g/L (Wu et al. 2011). The optimal conditions for L-lactic acid production is around 32–34°C, using high concentration of glucose as a carbon source, together with $(NH_4)_2SO_4$, KH_2PO_4, $ZnSO_4 \cdot 7H_2O$, $MgSO_4 \cdot 7H_2O$, and $CaCO_3$ in the production medium (Wu et al. 2011).

Given the high titers produced by *Rhizopus* and in particular lactic acid bacteria, development of other filamentous fungi for lactic acid production seems not to be very attractive.

Gluconic Acid

Gluconic acid is produced from glucose through a simple dehydrogenation reaction catalyzed by glucose oxidase (Roehr et al. 1992). This acid and its derivatives have wide applications in the food and pharmaceutical industry (Roehr et al. 1992, Milsom and Meers 1985, Miall 1978, Das and Kundu 1987). Calcium gluconate finds use as a readily available and widely used material for calcium therapy, when calcium deficiency occurs, or when allergy is severe. Sodium gluconate is widely used as a sequestering agent to prevent the precipitation of lime soap scums on cleaned products. Free gluconic acid is used as mild acidulant in metal processing, in tanning processes, and in food application. Although its production has long been known in some of the acetic acid bacteria (Ramachandran et al. 2006), gluconic acid was first reported as a product of fungal metabolism by Milliard (1922). Among several fungal species, *Aspergillus niger* is able to produce high yields of gluconic acid, gluconate salts, gluconolactone through the action of glucose dehydrogenase. Solid-state fermentation was reported to produce gluconic acid using cheaper raw materials as substrates (Singh et al. 2005, Singh et al. 2003, Roukas 2000). The fermentation from Singh using pretreated sugarcane bagasse achieved a high yield of 95% (Singh et al. 2003).

Gluconic acid is industrially produced by *Aspergillus niger*, and *Penicillium* spp. in submerged fermentation process (Singh and Kumar 2007). The production condition requires high glucose concentration at 30–37°C, and pH higher than 5. A typical medium for this fermentation includes $MgSO_4.7H_2O$, KH_2PO_4 $(NH_4)_2HPO_4$ and $CaCO_3$.

Gluconic acid production is an oxygen-consuming process with a high oxygen demand for the bioconversion reaction which is strongly influenced by the dissolved oxygen concentration (Sakurai et al. 1989, Kapat et al. 2001).

The formation of gluconic acid is different from the pathway of synthesis of the other acids. The process is extracellular and is a two step conversion of glucose. The first step is an oxidation of glucose to glucono-δ-lactone by glucose oxidase (encoded by gene *goxC*) and the subsequent step is hydrolysis of the lactone to gluconic acid, which can occur spontaneously or be catalyzed by a lactonase (Roehr et al. 1992, Witteveen et al. 1993) (Fig. 1).

Oxalic Acid

Oxalic acid is a C2 organic acid which has the simplest structure among dicarboxylic acids. Oxalic acid and its conjugate base oxalate can be used as bleaching agent in textile and wood industry. Besides, this acid causes leaching and transformation of insoluble inorganic metal compounds in nature through acidification and via its chelating property (Gadd 1999, Pernet 1991, Gadd and Raven 2010, Burgstaller and Schinner 1992, Drever and Stillings 1997). Therefore, related studies on utilization of oxalic acid in hydrometallurgy for extracting iron and some heavy metals to improve clays, quartz, etc. (Groudev and Groudeva 1986, Mandal et al. 2002, Mandal and Banerjee 2004, Vegliò et al. 1999) have been initiated.

In nature, it is produced by many saprophytic and phytopathogenic fungi (Dutton and Evans 1996, Sayer et al. 1999). Immobilized *A. niger* could efficiently produce oxalic acid from glucose (Mandal and Banerjee 2005), which is under usual submerged fermentation converted to gluconic acid. The production of oxalic acid by *A. niger* is highly regulated depending on the factors like carbon, nitrogen and the cultivation pH (Strasser et al.

1994). Most of the production processes use glucose or lactose as carbon source, and nitrate as nitrogen source at 30°C with controlled pH around 6.5. The production pathway of oxalic acid in *A. niger* involves hydrolytic cleavage of oxaloacetate catalyzed by oxaloacetate acetylhydrolase (OAH) (Han et al. 2007).

Oxalic acid can efficiently be produced by chemical processes. However, the increasing demand in hydrometallurgy for boarder applications, make production of oxalic acid from a cheap carbon source by fermentation more attractive and eco-friendly (Mandal and Banerjee 2005).

Itaconic Acid and its Derivatives

Itaconic acid is a C5 substituted acrylic acid synthesized from carbohydrates by the fungus *Aspergillus itaconicus* and *Aspergillus terreus*. Its primary use is as a copolymer with synthetic resins. Itaconic acid is used in synthetic fiber manufacture as part of an acrylonitrile copolymer. As part of styrene butadiene copolymers, uses are found in carpet backing and paper coating (Miall 1978, Willke and Vorlop 2001, Okabe et al. 2009).

The biosynthesis of itaconic acid in *A. terreus* occurs via glycolysis and through the TCA cycle. Citric acid is generally considered to be precursor of itaconic acid (Bentley and Thiessen 1957).

The itaconic acid biosynthesis route in *A. terreus* has not yet been fully established. A complicating factor in this respect is that the pathway towards itaconic acid occurs in two compartments, in the cytosol and in the mitochondria, similar to that for citric acid. The proposed itaconic acid biosynthesis pathway starts with glycolysis in the cytosol. Glucose is metabolized to pyruvate. Pyruvate is then dehydrogenated into acetyl-CoA in the mitochondria. In mitochondria, citrate synthase converts acetyl-CoA and oxaloacetate to citric acid. Further on in the TCA cycle, citrate is converted into *cis*-aconitate which is transported back into cytosol and then decarboxylated to itaconate.

The complete manufacturing process of itaconic acid using *A. terreus* strain derived from NRRL 1960 was established by Kobayashi and Nakamura in 1964. Since then, *A. terreus* became the main itaconic acid producing fungus used in industry. The important environmental conditions for high yield using *A. terreus* include: high percentage of sugar in the production medium, low nitrogen and phosphate source, low pH of 2 to 3, high oxygen supply, and limited but adequate trace elements (Willke and Vorlop 2001).

Only recently in a transcriptomics study in *A. terreus*, several itaconic acid-related genes were identified (Li et al. 2011). Among them, *cadA* encoding *cis*-aconitate decarboxylase (CAD) which is the key enzyme for itaconic acid production from citric acid in the tri-carboxylic acid

(TCA) cycle. This enzyme was also identified using a classical reversed genetics approach (Kanamasa et al. 2008). Based on its high citric acid producing capability and broad applicability in industry, *Aspergillus niger* was selected as a novel itaconic acid production host strain in recent work of Li. Expression of *cadA* in *A. niger* leads to itaconic acid production in *A. niger* (Li et al. 2011).

Itaconic Acid Derivatives

Some derivatives of itaconic acid are also used in industrial applications as building blocks for forming dimers, copolymers in multiple applications. Among them, the most commonly used are *cis*-aconitic acid and citraconic acid. Similarly as itaconic acid, production of *cis*-aconitic acid shares the same metabolic pathway which is derived one step from precursor citric acid. Different from *cis*-aconitic acid, a natural route of citraconic acid production does not exist in fungi. In the bacterial species *Methanocaldococcus jannaschii*, citraconate can be produced in a two step conversion from pyruvate. The citramalate synthase enzyme (CMS) catalyzes the first step in this pathway, condensation of acetyl-CoA with pyruvate to form R-citramalate (Drevland et al. 2007). In the following step, R-citramalate is converted to citraconate by an isopropylmalate isomerase (IPMI). In *Methanocaldococcus jannaschii*, IPMI functions in both leucine and isoleucine pathways.

The citramalate synthase enzyme catalyzing the conversion of pyruvate and acetyl-CoA into R-citramalate has only been found in *Methanococcus jannaschii, Leptospira interrogans* and a few other microorganisms (Ma et al. 2008). The orthologs in filamentous fungi include *A. niger* ANI_1_1952184, *A. terreus* ATEG_00906, *A. clavatus* ACLA_049660, *A. flavus* AFLA_026530. The fungal IPMI orthologue proteins include *Saccharomyces cerevisiae* NP_010520 and *A. niger* ANI_1_440024. The potential citraconate synthesis route in *A. niger* is presented in Fig. 1. However, no functional validation of the enzyme activity has been performed.

Organic Acids Transporters in Fungi

In this chapter, the production or potential production pathway of several organic acids have been described. As discussed, for several of these acids the pathway includes cytosolic and mitochondrial steps. Moreover, as these acids are finally ending up in the culture fluid, transport steps are relevant for efficient production as well. In this section, we focus on di- and tri-carboxylic acid transporters.

Although studies on fungal organic acid transporters are limited, this is an inevitable and important research area as most of the organic acids are produced intracellular and may require transporters to exit the cells.

Organic acid transporters can be divided into two groups based on their localization and related function in fungi: one group is responsible of transportation across subcellular compartments (e.g., between cytosol and mitochondria), the other group functions as exporters for organic acids to exit fungal cells. In fungi, the main tri-carboxylic acid transporters are responsible for citrate, *cis*-aconitate, iso-citrate transport and the di-carboxylic acid transporters are for malate, succinate, fumarate, etc. (Aliverdieva and Mamaev 2009).

In yeast *S. cerevisiae*, the transporters for citric acid in between mitochondria and cytosol are identified to be Ctp1p (P38152) (Kaplan et al. 1995) and Yhm2p (Castegna et al. 2010). The former displays a high degree of similarity to a mitochondrial carrier protein (CMC) (XP_001395080.1) in *A. niger* (Pel et al. 2007), with 48% identity in a 282 amino acids overlap. In *A. niger*, CMC is suggested to be a citrate-malate mitochondrial transporter with the same function as Ctp1p in *S. cerevisiae*. The other mitochondrial carrier Yhm2p works as a $NADP^+/NADPH$ dependent anti-porter for citrate and oxoglutarate (Fig. 3). Several proteins encoded by fungal genes are found to be very likely orthologs of Yhm2p (NP_013968.1). These include XP_001216612.1 (74%) from *Aspergillus terreus*, XP_001822773.1 (73%) from *Aspergillus oryzae RIB40*, XP_001393982.1 (72%) from *Aspergillus niger*, XP_681543.1 (71%) from *Aspergillus nidulans FGSC A4*, XP_002557384.1 (71%) from *Penicillium chrysogenum*, etc. Beside these two citrate mitochondrial carrier proteins, the *S.cerevisiae* mitochondrial dicarboxylate transporter (Dic1p) (Q06143) (Palmieri et al. 1996) is 49% identical with 277 amino acids coverage to a putative dicarboxylate transporter (XP_001399314.2) in *A. niger*.

The mitochondrial transporters mentioned here will be part of a complex system which is partly elucidated in the cells of yeast, and may function similarly in filamentous fungi. Although the exact properties of this system are not known, various models have been described. Palmieri suggested the existence of a system transporting citrate and malate such as citrate-malate or malate pyruvate shuttle (Palmieri 2004) (Fig. 3A/B) whereas recently a citrate-oxoglutarate shuttle has also been suggested (Fig. 3C) (Castegna et al. 2010).

In respect to the other class of transporters in organic acid export, the research of Camarasa describe increased malate export in *S. cerevisiae* by expression of a permease gene *MAE1* (Grobler et al. 1995) from *Schizosaccharomyces pombe* (Camarasa et al. 2001). This protein is a plasma membrane localized dicarboxylate transporter also established in *S. cerevisiae* to catalyze the transport of succinate, possible L-malate and malonate (Aliverdieva et al. 2006). Although no potential homologs of Mae1p is presented in *S. cerevisiae*, a putative C4-dicarboxylate transporter/malic

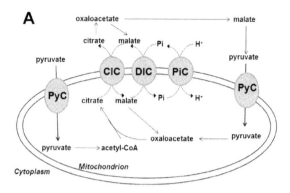

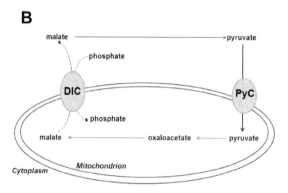

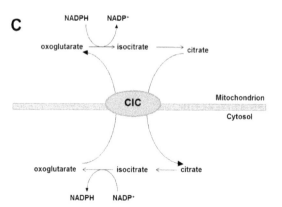

Fig. 3. Models represent the citrate and malate transporters (CIC citrate carrier, DIC dicarboxylate carrier, PyC pyruvate carrier, PiC phosphate carrier) **A** the citrate-malate-pyruvate shuttle, **B** the pyruvate-malate shuttle, and **C** the citrate-oxoglutarate shuttle. Images are modified from Palmieri (Palmieri 2004) (A.B.) and Castegna (Castegna et al. 2010) (C).

acid transport protein (XP_001398131.1) from *A. niger* has been indicated with an identity of 37% (Pel et al. 2007).

In *Aspergillus terreus*, two putative organic acid transporters MTT (a putative mitochondrial carrier protein) and MFS (a putative dicarboxylate carrier protein belongs to the 'Major Facilitator Superfamily' transporters) are found to be expressed in a hypothetical itaconic acid gene cluster (Li et al. 2011). MTT is considered to function as a tricarboxylic acid transporter shuttling citrate and/or *cis*-aconitate from mitochondria to cytosol to provide substrate for producing itaconic acid, while MFS works potentially as a dicarboxylic acid exporter for itaconate to exit the cell. The MFS exporter has homologs from *Aspergillus flavus* XP_002382138.1 (71%), *Aspergillus oryzae* XP_001819137.1 (71%), etc. BlastP shows no highly identical hits (> 70%) of MTT or MFS transporters in *A. niger*. *Aspergillus niger* contains two other putative mitochondrial tricarboxylate carrier proteins, a putative citrate-malate transporter CMC (An11g11230/XP_001395080.1) as described above and a putative tricarboxylic acid transporter (An18g00070/XP_001398478.1) (Pel et al. 2007) which has a homology of 42% to CMC. Besides, *A. niger* also has 3 putative MFS transporters involved in carboxylic acid transport: XP_001394176.1, XP_001391338.1 and XP_001397361.1 (Pel et al. 2007). BlastP search shows that two of these (XP_001391338.1 and XP_001397361.1) are related to a MFS carboxylic acid transporter Jen1p (NP_012705.1) from *S. cerevisiae* (Casal et al. 2008), with 39% and 28% homology.

Although a lot of research has been performed on some of the yeast transporters (Palmieri et al. 2004, Casal et al. 2008) and the system by which they operate, in depth knowledge on the transport mechanism is still limited. In filamentous fungi, especially, very little research has been conducted to verify whether the relevant transporters might function under the same circumstances as present in the models. For industrial scale organic acid fermentation, these transporters might be crucial factors to improve the productivity. As shown by Kaplan, modification of single amino acid residues can have both positive and detrimental effects on carrier-substrate specificity as well as protein kinetics (Kaplan et al. 1995, Kaplan et al. 2000). Therefore, improving industrial organic acids production using related transporters is recommended, which can be achieved by, e.g., site-directed mutagenesis

Conclusion

Fungal production of industrial organic acids especially tri- and dicarboxylic acids as chemical building blocks has a great potential for the future biobased chemical industry. Currently, production of organic acids such as succinic acid, fumaric acid, L-malic acid is still limited in fungi. Exploration of the biochemical pathway of organic acids by modification of key pathway enzymes connects the path to break through the current limitations in productivity. Studies on organic acid transporters provide knowledge on shuttling pathway intermediates between subcellular compartments. Although coordination of metabolic pathway modification and metabolite transport may be complicated, combination of these two is crucial for efficient fungal organic acids production in the future.

References

Aden, A., J. Bozell, J. Holladay, J. White and A. Manheim. 2004. Top Value Added Chemicals from Biomass http://www1. eere. energy. gov/biomass/pdfs/35523. pdf. Eds LLSJaGP D. Elliot and LLaSKN K. Ibsen. Results of Screening for Potential Candidates from Sugars and Synthesis Gas. 1.

Aliverdieva, D. A. and D. V. Mamaev. 2009. Molecular characteristics of transporters of C_4-dicarboxylates and mechanism of translocation. Journal of Evolutionary Biochemistry and Physiology. 45: 323–339.

Aliverdieva, D. A., D. V. Mamaev, D. I. Bondarenko and K. F. Sholtz. 2006. Properties of yeast Saccharomyces cerevisiae plasma membrane dicarboxylate transporter. Biochemistry. 71: 1161–1169.

Anstey, A. V. 2010. Fumaric acid esters in the treatment of psoriasis. British Journal of Dermatology. 162: 237–238.

Beauprez, J. J., M. De Mey and W. K. Soetaert. 2010. Microbial succinic acid production: Natural versus metabolic engineered producers. Process Biochemistry. 45: 1103–1114.

Bercovitz, A., Y. Peleg, E. Battat, J. S. Rokem and I. Goldberg. 1990. Localization of pyruvate carboxylase in organic acid-producing *Aspergillus* strains. Applied and Environmental Microbiology. 56: 1594–1597.

Berovic, M. and M. Legisa. 2007. Citric acid production. Biotechnology annual review 13: 303–343.

Brown, S., S. Lutteringer, D. Yaver and A. Berry. 2011. Methods for improving malic acid production in filamentous fungi, WO 2011/028643.

Burgstaller, W. and F. Schinner. 1992. Leaching of metals with fungi. Journal of Biotechnology. 27: 91–116.

Camarasa, C., F. Bidard, M. Bony, P. Barre and S. Dequin. 2001. Characterization of *Schizosaccharomyces pombe* Malate Permease by Expression in *Saccharomyces cerevisiae*. Appl. Environ. Microbiol. 67: 4144–4151.

Casal, M., S. Paiva, O. Queirós and I. Soares-Silva. 2008. Transport of carboxylic acids in yeasts. FEMS Microbiol. Rev. 32: 974–994.

Castegna, A., P. Scarcia, G. Agrimi, L. Palmieri, H. Rottensteiner, I. Spera, L. Germinario and F. Palmieri. 2010. Identification and functional characterization of a novel mitochondrial carrier for citrate and oxoglutarate in *Saccharomyces cerevisiae*. Journal of Biological Chemistry. 285: 17359–17370.

Currie, J. N. 1917. The citric acid fermentation of Aspergills niger. Journal of Biological Chemistry. 31: 15–37.

Dai, Z., X. Mao, J. K. Magnuson and L. L. Lasure. 2004. Identification of Genes Associated with Morphology in *Aspergillus niger* by Using Suppression Subtractive Hybridization. Applied and Environmental Microbiology. 70: 2474–2485.

Das, A. and P. N. Kundu. 1987. Microbial production of gluconic acid. Journal of Scientific and Industrial Research. 46: 307–311.

Drever, J. I. and L. L. Stillings. 1997. The role of organic acids in mineral weathering. Colloids and Surfaces A: Physicochemical and Engineering Aspects. 120: 167–181.

Drevland, R. M., A. Waheed and D. E. Graham. 2007. Enzymology and evolution of the pyruvate pathway to 2-oxobutyrate in Methanocaldococcus jannaschii. Journal of Bacteriology. 189: 4391–4400.

Dutton, M. V. and C. S. Evans. 1996. Oxalate production by fungi: Its role in pathogenicity and ecology in the soil environment. Canadian Journal of Microbiology. 42: 881–895.

Ehrlich, F. 1911. Formation of fumaric acid by means of molds. Ber. Dtsch. Chem. Ges. 44: 3737–3742.

Francielo, V., M. Patricia and S. A. Fernanda. 2008. Apple pomace: a versatile substrate for biotechnological applications. Critical Reviews in Biotechnology. 28: 1–12.

Forster, J. W. and S. A. Waksman. 1939. The production of fumaric acid by molds belonging to the genus *Rhizopus*. J. Am. Chem. Soc. 61: 127–135.

Foster, J. W. 1949. chemical activities of fungi. Academic Press. New York.

Gadd, G. M. 1999. Fungal production of citric and oxalic acid: Importance in metal speciation, physiology and biogeochemical processes. 41: 47–92.

Gadd, G. M. and J. A. Raven. 2010. Geomicrobiology of eukaryotic microorganisms. Geomicrobiology Journal 27: 491–519.

Gallmetzer, M., J. Meraner and W. Burgstaller. 2002. Succinate synthesis and excretion by Penicillium simplicissimum under aerobic and anaerobic conditions. FEMS Microbiology Letters. 210: 221–225.

Goldberg, I., J. S. Rokem and O. Pines. 2006. Organic acids: Old metabolites, new themes. Journal of Chemical Technology and Biotechnology. 81: 1601–1611.

Gong, C. S., N. Cao, Y. Sun and G. T. Tsao. 1996. Production of L-malic acid from fumaric acid by resting cells of brevibacterium sp. Applied Biochemistry and Biotechnology. 57–58: 481–487.

Grewal, H. S. and K. L. Kalra. 1995. Fungal production of citric acid. Biotechnology Advances. 13: 209–234.

Grobler, J., F. Bauer, R. E. Subden and H. J. J. van Vuuren. 1995. The mae1 gene of *Schizosaccharomyces pombe* encodes a permease for malate and other C4 dicarboxylic acids. Yeast. 11: 1485–1491.

Groudev, S. N. and V. I. Groudeva. 1986. Biological leaching of aluminium from clays. Biotechnol. Bioeng. Symp. 16: 91–99.

Han, Y., H. J. Joosten, W. Niu, Z. Zhao, P. S. Mariano, M. McCalman, J. Van Kan, P. J. Schaap and D. Dunaway-Mariano. 2007. Oxaloacetate hydrolase, the C-C bond lyase of oxalate secreting fungi. Journal of Biological Chemistry. 282: 9581–9590.

Horton, P., K. -J. Park, T. Obayashi, N. Fujita, H. Harada, C. J. Adams-Collier and K. Nakai. 2007. WoLF PSORT: Protein localization predictor. Nucleic acids research. 35: 585–587

Howell, D. M., H. Xu and R. H. White. 1999. (R)-Citramalate Synthase in Methanogenic Archaea. J. Bacteriol. 181: 331.

Ju, N. and S. S. Wang. 1986. Continuous production of itaconic acid by *Aspergillus terreus* immobilized in a porous disk bioreactor. Applied Microbiology and Biotechnology. 23: 311–314.

Kanamasa, S., L. Dwiarti, M. Okabe and E. Y. Park. 2008. Cloning and functional characterization of the cis-aconitic acid decarboxylase (CAD) gene from *Aspergillus terreus*. Applied Microbiology and Biotechnology. 80: 223–229.

Kapat, A., J. K. Jung and Y. H. Park. 2001. Enhancement of glucose oxidase production in batch cultivation of recombinant Saccharomyces cerevisiae: Optimization of oxygen transfer condition. Journal of Applied Microbiology. 90: 216–222.

Kaplan, R. S., J. A. Mayor, D. A. Gremse and D. O. Wood. 1995. High level expression and characterization of the mitochondrial citrate transport protein from the yeast Saccharomyces cerevisiae. Journal of Biological Chemistry. 270: 4108–4114.

Kaplan, R. S., J. A. Mayor, R. Kotaria, D. E. Walters and H. S. Mchaourab. 2000. The yeast mitochondrial citrate transport protein: Determination of secondary structure and solvent accessibility of transmembrane domain IV using site-directed spin labeling. Biochemistry. 39: 9157–9163.

Karaffa, L. and C. P. Kubicek. 2003. *Aspergillus niger* citric acid accumulation: Do we understand this well working black box? Applied Microbiology and Biotechnology. 61: 189–196.

Kim, Y., N. S. Mosier, R. Hendrickson, T. Ezeji, H. Blaschek, B. Dien, M. Cotta, B. Dale and M. R. Ladisch. 2008. Composition of corn dry-grind ethanol by-products: DDGS, wet cake, and thin stillage. Bioresour Technol. 99(12): 5165–5176.

Kubicek, C. P., P. Punt and J. Visser. 2010. Production of organic acids by filamentous fungi. The Mycota X: Industrial Applications. 10: 215–234.

Lee, S. Y., Y. K. Jung, H. song, J. M. Kim and J. H. Park. 2009. Fermentative production of organic acids for polymer synthesis. *In:* B. Rehm [ed.]. Microbial Production of Biopolymers and Polymer Precursors: Applications and Perspectives. Caister Academic Press, Norfolk, UK. pp. 181–196.

Li, A. N. van Luijk, M. ter Beek, M. Caspers, P. Punt and M. van der Werf. 2011. A clone-based transcriptomics approach for the identification of genes relevant for itaconic acid production in *Aspergillus*. Fungal Genet. Biol. 48: 602–611.

Li, X. K., K. Zhang, Z. Gao, N. Liu and H. Huang. 2005. Progress in synthesis and application of fumaric acid. Xiandai Huagong/Modern Chemical Industry. 25: 81–83+88.

Litchfield, J. H. 1996. Microbiological production of lactic acid. Adv. Appl. Microbiol. 42: 45–95.

Ling, L. B. and T. K. Ng. 1989. Fermentation process for carboxylic acids. US4877731.

Longacre, A., J. M. Reimers, J. E. Gannon and B. E. Wright. 1997. Flux analysis of glucose metabolism in rhizopus oryzae for the purpose of increasing lactate yields. Fungal Genetics and Biology. 21: 30–39.

Ma, J., P. Zhang, Z. Zhang, M. Zha, H. Xu, G. Zhao and J. Ding. 2008. Molecular basis of the substrate specificity and the catalytic mechanism of citramalate synthase from Leptospira interrogans. Biochemical Journal. 415: 45–56.

Magnuson, J. K. and L. L. Lasure. 2004. Organic acid produciton by filamentous fungi. Advances in Fungal Biotechnology for Industry Agriculture and Medicine. *In:* J. S. Tkacz and L. Lange [eds.]. Kluwer Academic, Plenum Publishers, New York, USA. pp. 307–340.

Mandal, S. K. and P. C. Banerjee. 2004. Iron leaching from China clay with oxalic acid: Effect of different physico-chemical parameters. *International* Journal of Mineral Processing 74: 263–270.

Mandal, S. K. and P. C. Banerjee. 2005. Submerged production of oxalic acid from glucose by immobilized *Aspergillus niger*. Process Biochemistry. 40: 1605–1610.

Mandal, S. K., A. Roy and P. C. Banerjee. 2002. Iron leaching from China clay by fungal strains. Transactions of the Indian Institute of Metals. 55: 1–7.

Mattey, M. 1992. The production of organic acids. Critic Review Biotechnology. 12: 87–132.

Miall, L. M. 1978. Organic acids. Econ Microbiol. 2: 48–119.

Milsom, P. E. and J. L. Meers. 1985. Gluconic acid and itaconic acids. Compr. Biotechnol. 3: 681–700.

Minard, K. I. and L. Mcalister-henn. 1991. Isolation, nucleotide sequence analysis, and disruption of the MDH2 gene from *Saccharomyces cerevisiae*: evidence for three isozymes of yeast malate dehydrogenase Molecular and cellular biology. 11: 370–380.

Moharregh-Khiabani, D., R. A. Linker, R. Gold and M. Stangel. 2009. Fumaric acid and its esters: An emerging treatment for multiple sclerosis. Current Neuropharmacology. 7: 60–64.

Nattrass, L. and A. Higson. 2010. NNFCC Renewable chemicals factsheet: Succinic acid, http://www. nnfcc. co. uk/publications/nnfcc-renewable-chemicals-factsheet-succinic-acid.

Nattrass, L. and A. Higson. 2010. NNFCC Renewable chemicals factsheet: Lactic acid, http://www. nnfcc. co. uk/publications/nnfcc-renewable-chemicals-factsheet-lactic-acid.

Okabe, M., D. Lies, S. Kanamasa and E. Y. Park. 2009. Biotechnological production of itaconic acid and its biosynthesis in *Aspergillus terreus*. Applied Microbiology and Biotechnology. 84: 597–606.

Okino, S., R. Noburyu, M. Suda, T. Jojima, M. Inui and H. Yukawa. 2008. An efficient succinic acid production process in a metabolically engineered Corynebacterium glutamicum strain. Applied Microbiology and Biotechnology. 81: 495–464.

Palmieri, F. 2004. The mitochondrial transporter family (SLC25): Physiological and pathological implications. Pflugers Archiv European Journal of Physiology. 447: 689–709.

Palmieri, L., F. Palmieri, M. J. Runswick and J. E. Walker. 1996. Identification by bacterial expression and functional reconstitution of the yeast genomic sequence encoding the mitochondrial dicarboxylate carrier protein. FEBS Lett. 399: 299–302.

Papagianni, M. 2007. Advances in citric acid fermentation by *Aspergillus niger*: Biochemical aspects, membrane transport and modeling. Biotechnology Advances. 25: 244–263.

Pedersen, H., C. Hjort and J. Nielsen. 2000. Cloning and characterization of oah, the gene encoding oxaloacetate hydrolase in *Aspergillus niger*. Mol. Gen. Genet. 263: 281–286.

Pel, H. J., J. H. De Winde, D. B. Archer, P. S. Dyer, G. Hofmann, P. J. Schaap, G. Turner, R. P. De Vries, R. Albang, K. Albermann, M. R. Andersen, J. D. Bendtsen, J. A. E. Benen, M. Van Den Berg, S. Breestraat, M. X. Caddick, R. Contreras, M. Cornell, P. M. Coutinho, E. G. J. Danchin, A. J. M. Debets, P. Dekker, P. W. M. Van Dijck, A. Van Dijk, L. Dijkhuizen, A. J. M. Driessen, C. D'Enfert, S. Geysens, C. Goosen, G. S. P. Groot, P. W. J. De Groot, T. Guillemette, B. Henrissat, M. Herweijer, J. P. T. W. Van Den Hombergh, C. A. M. J. Van Den Hondel, R. T. J. M. Van Der Heijden, R. M. Van Der Kaaij, F. M. Klis, H. J. Kools, C. P. Kubicek, P. A. Van Kuyk, J. Lauber, X. Lu, M. J. E. C. Van Der Maarel, R. Meulenberg, H. Menke, M. A. Mortimer, J. Nielsen, S. G. Oliver, M. Olsthoorn, K. Pal, N. N. M. E. van Peij, A. F. J. Ram, U. Rinas, J. A. Roubos, C. M. J. Sagt, M. Schmoll, J. Sun, D. Ussery, J. Varga, W. Vervecken,, P. J. J. Van De Vondervoort, H. Wedler, H. A. B. Sten, A. P. Zeng, A. J. J. Van Ooyen, J. Visser and H. Stam. 2007. Genome sequencing and analysis of the versatile cell factory *Aspergillus niger* CBS 513. 88. Nature Biotechnology. 25: 221–231.

Pernet, J. C. 1991. Oxalic acid. *In*: Encyclopedia of Chemical Techonology,. R. E. Kirk and D. F. Othmer [eds.]. Interscience Publishers Inc, New York, USA. pp. 661–674.

Pirgozliev, V., T. C. Murphy, B. Owens, J. George and M. E. E. McCann. 2008. Fumaric and sorbic acid as additives in broiler feed. Research in Veterinary Science. 84: 387–394.

Pritchard, G. G. 1973. Factors affecting the activity and synthesis of NAD dependent lactate dehydrogenase in Rhizopus oryzae. Journal of General Microbiology. 78: 125–137.

Raab, A. M., G. Gebhardt, N. Bolotina, D. Weuster-Botz and C. Lang. 2010. Metabolic engineering of *Saccharomyces cerevisiae* for the biotechnological production of succinic acid. Metabolic Engineering 12: 518–525.

Ramachandran, S., P. Fontanille, A. Pandey and C. Larroche. 2006. Gluconic Acid: A Review, Food Technol. Biotechnol. 44: 185–195.

Retired, W. R. and M. Tanifuji. 2011. Oxalic acid. Ullmann's Encyclopedia of Industrial Chemistry. 2002. Wiley-VCH, Verlag GmbH & Co. KGaA.

Roa Engel, C. A., A. J. J. Straathof, T. W. Zijlmans, W. M. Van Gulik and L. A. M. Van Der Wielen. 2008. Fumaric acid production by fermentation. Applied Microbiology and Biotechnology. 78: 379–389.

Roehr, M., C. P. Kubicek and J. Kominek. 1992. Industrial acids and other small molecules. Biotechnology Reading, Mass. 23: 91–131.

Roukas, T. 2000. Citric and gluconic acid production from fig by *Aspergillus niger* using solid-state fermentation. Journal of Industrial Microbiology and Biotechnology. 25: 298–304.

Sakurai, H., W. L. Hang, S. Sato, S. Mukataka and J. Takahashi. 1989. Gluconic acid production at high concentrations by *Aspergillus niger* immobilized on a nonwoven fabric. Journal of Fermentation and Bioengineering. 67: 404–408.

Sayer, J. A., J. D. Cotter-Howells, C. Watson, S. Millier and G. M. Gadd. 1999. Lead mineral transformation by fungi. Current Biology. 9: 691–694.

Schreferl, G., C. P. Kubicek and M. Rohr. 1986. Inhibition of citric acid accumulation by manganese ions in *Aspergillus niger* mutants with reduced citrate control of phosphofructokinase. Journal of Bacteriology. 165: 1019–1022.

Singh, O. V., R. K. Jain and R. P. Singh. 2003. Gluconic acid production under varying fermentation conditions by *Aspergillus niger*. Journal of Chemical Technology and Biotechnology. 78: 208–212.

Singh, O. V., N. Kapur and R. P. Singh. 2005. Evaluation of agro-food byproducts for gluconic acid production by *Aspergillus niger* ORS-4. 410. World Journal of Microbiology and Biotechnology. 21: 519–524.

Singh, O. V. and R. Kumar. 2007. Biotechnological production of gluconic acid: Future implications. Applied Microbiology and Biotechnology. 75: 713–722.

Soccol, C. R., B. Marin, M. Raimbault and J. M. Lebeault. 1994. Potential of solid state fermentation for production of L(+)-lactic acid by *Rhizopus oryzae*. Applied Microbiology and Biotechnology. 41: 286–290.

Song, H. and S. Y. Lee. 2006. Production of succinic acid by bacterial fermentation. Enzyme and Microbial Technology. 39: 352–361.

Strasser, H., W. Burgstaller and F. Schinner. 1994. High-yield production of oxalic acid for metal leaching processes by *Aspergillus niger*. FEMS Microbiology Letters. 119: 365–370.

Sun, J., X. Lu, U. Rinas and P. Z. An. 2007. Metabolic peculiarities of *Aspergillus niger* disclosed by comparative metabolic genomics. Genome Biology. 8: R182.

Tsao, G. T., N. J. Cao, J. Du and C. S. Gong. 1999. Production of multifunctional organic acids from renewable resources. Advances in biochemical engineering/biotechnology. 65: 243–280.

UNESCO. 2012. Novozymes develops fungus to produce biochemicals. http://www.greencarcongress. com/2012/08/novozymes-20120816. html.

Vandenberghe, L. P. S., C. R. Soccol, A. Pandey and J. M. Lebeault. 2000. Solid-state fermentation for the synthesis of citric acid by *Aspergillus niger*. Bioresource Technology. 74: 175–178.

Vegliò, F., B. Passariello and C. Abbruzzese. 1999. Iron removal process for high-purity silica sands production by oxalic acid leaching. Industrial and Engineering Chemistry Research. 38: 4443–4448.

Verhoff, F. H. and J. E. Spradlin. 1976. Mass and energy balance analysis of metabolic pathways applied to citric acid production by *Aspergillus niger*. Biotechnology and Bioengineering. 18: 425–432.

Wee, Y. J., J. N. Kim and H. W. Ryu. 2006. Biotechnological production of lactic acid and its recent applications. Food Technology and Biotechnology. 44: 163–172.

West, T. P. 2011. Malic acid production from thin stillage by *Aspergillus* species. Biotechnology Letters. 33: 2463–2467.

Westerhoff, H. V. and Y. -D. Chen. 1984. How do enzyme activities control metabolite concentrations? Eur. J. Biochem. 142: 425–430.

Willke, T. and K. D. Vorlop. 2001. Biotechnological production of itaconic acid. Applied Microbiology and Biotechnology. 56: 289–295.

Witteveen, C. F. B., P. J. I. van de Vondervoort, H. C. Van Den Broeck, F. A. C. Van Engelenburg, L. H. De Graaff, M. H. B. C. Hillebrand, P. J. Schaap and J. Visser. 1993. Induction of glucose oxidase, catalase, and lactonase in *Aspergillus niger*. Current Genetics. 24: 408–416.

Wright, B. E., A. Longacre and J. Reimers. 1996. Models of metabolism in Rhizopus oryzae. Journal of Theoretical Biology. 182: 453–457.

Wu, X., S. Jiang, M. Liu, L. Pan, Z. Zheng and S. Luo. 2011. Production of L-lactic acid by Rhizopus oryzae using semicontinuous fermentation in bioreactor. Journal of Industrial Microbiology and Biotechnology. 38: 565–571.

Yuzbashev, T. V., E. Y. Yuzbasheva, T. I. Sobolevskaya, I. A. Laptev, T. V. Vybornaya, A. S. Larina, K. Matsui, K. Fukui and S. P. Sineoky. 2010. Production of succinic acid at low pH by a recombinant strain of the aerobic yeast Yarrowia lipolytica. Biotechnology and Bioengineering. 107: 673–682.

Zelle, R. M., E. De Hulster, W. A. Van Winden, P. De Waard, C. Dijkema, A. A. Winkler, J. M. A. Geertman, J. P. Van Dijken, J. T. Pronk and A. J. A. Van Maris. 2008. Malic acid production by *Saccharomyces cerevisiae*: Engineering of pyruvate carboxylation, oxaloacetate reduction, and malate export. Applied and Environmental Microbiology. 74: 2766–2777.

Zhang, Z. Y., B. Jin and J. M. Kelly. 2007. Production of lactic acid from renewable materials by Rhizopus fungi. Biochemical Engineering Journal. 35: 251–263.

3

Microbial Enzyme Technology in Baked Cereal Foods

Deborah Waters

ABSTRACT

In recent years, there has been greater emphasis on standardized, safe and affordable alternative to chemical additives, preservatives and sensory improvers. This has been particularly noticeable in cereal-based foods. It results from government directives, consumer preferences and food industry economics. Bread and other cereal baked goods represent a staple food in many countries. Traditional lactic acid bacteria (LAB) microbial fermentation of sourdough has been reinvented for application in many modern cereal products. Other advancements in bread-making technology include fungal enzyme application obtained through natural fermentation technology, specific enzyme over-expression or *in situ* enzymatic modification, all of which are now commonplace industrially. These technologies may include the application of purified highly specific enzymes, genetic modification of microbial cell factories for enzyme production, and application of crude enzyme cocktails. The overall aim of these methods is to provide a cheap alternative for the cereal food industry to modify existing products and develop novel functional foods and raw materials. However, product safety, nutritional attributes, and 'green label' status must be maintained

School of Food and Nutritional Sciences, University College Cork, College Road, Cork, Ireland.
Email: d.waters@ucc.ie

or preferably ameliorated whilst also fulfilling an obligation to adhere to increasingly strict EFSA (European Food Science Authority) and FDA (Food and Drug Administration) guidelines. Natural microbial enzyme technology provides many potential benefits for the standardization of foods. This review aims to discuss the implementation of these technologies in an industrial production setting for the betterment of baked cereal goods with references to relevant developments in the last decade.

Introduction

For many centuries, baked cereal foods have been a staple part of the human diet. Baked cereal goods primarily include breads, cakes, biscuits and pastries. They are generated by combining milled cereal flour with water and other ingredients such as fat, salt, sugar, eggs, etc., to make dough or batter formulations which are baked before consumption. Countless traditional processes have led to the production of numerous baked cereal goods of various forms throughout the continents. These products can be classified based on, flavor (sweet, savory or bland), unleavened or leavened (biological, chemical, physical), textural profile (crumb, softness, brittleness, etc.), specific volume (high—pan bread, medium—rye bread, or flat breads —typically Asian) and technological attributes (pH, moisture, etc.).

In recent years, there has been an increased pressure on food producers inclined towards a standardized, safe and affordable alternative to chemical additives, preservatives and sensory improvers. EU and USA directives, local government regulations, industrial economics, and consumer preferences all influence the trend towards 'clean-labeled' natural food ingredients. Bread making aids, specifically additives, must be disclosed on labels (CFR 2011, EC and DG SANCO 2003, USFDA 2012). This aim can be realized through exogenous (microbial) enzyme application and the incorporation of age old traditional lactic acid bacteria (LAB) fermentation technology, which can also include yeasts or fungi. Changes have been particularly noticeable in cereal-based food processing and these alterations results from government directives, consumer preferences and food industry economics.

To comprehend the roles of LAB and exogenous enzyme application in cereal foods, we must first understand the raw materials which will subsequently act as substrates. Economically and industrially, the most significant cereals in decreasing order of global annual production for food use are, wheat, rice and maize, with others such as barley, sorghum, millet, oat, triticale and rye also contributing significantly to the world supply (FAOSTAT 2012). These grains are quite diverse, however all have certain common chemical/nutritional components which serve as enzyme substrates including starch, fibers, proteins, lipids, micronutrients (vitamins and minerals) and others (anti-nutrients, ash etc.). The ratios and combinations of these constituents in the grain or cereal fraction being used in a particular food, for example whole meal flour versus refined flours or starch, etc., will determine the effectiveness of adding particular exogenous enzymes or efficiency of the fermentation process.

Using LAB fermentation technology can impart many different attributes to a cereal product. The exact effects of the fermentation process are dependent on many factors including, starter culture properties (inoculum or natural starter), fermentation time, temperature, cereal substrate being used, etc. There are a wide range of commercially available starter cultures currently available for industrial use in cereal processing (Böcker 2012, Clerici Sacco 2012, Ed Wood 2012, Lallemand 1998, Puratos 2012), some of which are listed in Table 2.

Additionally, exogenous enzyme incorporation into cereal-based food formulations can also cause a range of effects, as shown in Table 1. These changes depend heavily on the type of enzyme and substrate availability. For instance, an endo-acting amylolytic enzyme will readily hydrolyze starch, but only if the substrate is accessible, i.e., after processing steps such as extraction or milling, which increases the proportion of damaged starch or through gelatinization (i.e., cooking or heating in the presence of moisture). Furthermore, even if the starch is a heavily damaged, heat-moisture-treated raw material, the enzyme will only cause hydrolysis if is used within its range of functional physicochemical parameters, such as, temperature, pH, dosage levels, maltose concentrations present (end-product inhibition of certain enzymes is also an issue) (Kirk et al. 2002), etc. In spite of these restrictions, enzymes have proven to be highly specific and potent utensils in modifying food formulations. In particular, part of their perceived effectiveness is due to their ability to fulfill the necessity for environmentally clean, non-chemical-based, natural and sustainable sources of processing aids in the food chain. Moreover, enzymes with broad or very restricted specificities which function at high temperatures and low dosages are particularly useful in many food applications. Like LAB fermentation, exogenous (primarily fungal or bacterial) enzymes can also enhance textural (Barrett et al. 2005, Błaszczak et al. 2004) and flavor (Moayedallaie et al. 2010) attributes of foods, as well as extend shelf-life (Caballero et al. 2007, Moayedallaie et al. 2010), improving processability (Caballero et al. 2007, Gujral and Rosell 2004, Waters et al. 2011) and providing a consumer-friendly and clean-label (Simpson et al. 2012) approach to modern day high-throughput food processing chains. Many enzyme-producing companies currently market food-safe enzymes specifically engineered for cereal-based food processes, examples of which are listed in Table 1.

Table 1. Examples of and information about some commercially available starter cultures/sourdough (also called cultured flour, levain or sour preferment) for industrial or home purchase.

Company products	Product form	Dosage*	Food uses	LAB role in this food	Reference
Böcker (industry supplier)					
Florapan®	Freeze-dried	0.1%	Ferments wheat, rye, spelt, buckwheat, barley, oat, whole grain, organic flours, etc.	• Preparation of LAB and aromatic dry yeast • They are stable, high-concentration live cultures • Very controlled and reproducible sourdough	(Böcker 2012)
Ready-to-use (RTU) cultured flours	Powder, paste or liquid	2–15%	Suitable for use during milling, blending or directly added into dough formulations	• Fully fermented products added to dough • Stabilized it by drying, salting or acidifying • No capacity for further fermentation	(Böcker 2012)
Böcker F	Dried sourdough	1–3%	Mixed wheat-rye (up to 5:1) bread, or wheat bread	• Increased crispiness • Long-lasting crispiness	(Böcker 2012)
Böcker Germe 80	Dried wheat germ sourdough	1–3%	Wheat bread	• Nutty flavor with yellowish crust and crumb • Long-lasting crispiness	(Böcker 2012)
Böcker M	RTU	1–3%	Wheat bread and pastries	• Flavor and texture improver	(Böcker 2012)
Böcker M	Dried sourdough (RTU)	1–3%	Baguettes, white bread and pastries	• Mild, wheat-like flavor • Sweet flavor	(Böcker 2012)
Böcker 350	Dried sourdough	1–3%	Rustic mixed rye or wheat breads	• Rich taste • Concentrated aroma precursors	(Böcker 2012)
Böcker Poolish	Dried	At own discretion	Production of bright products: white bread, rolls, baguettes, pizza, biscuit	• Improved aroma • Better taste	(Böcker 2012)

Böcker Direkt 25	Liquid wheat pre-dough	5–10%	Mixed wheat bread, wheat bread and pastries	• Dry dough, high proofing tolerance • Long lasting crispiness, freshness, juiciness • Mild wheat-like flavor	(Böcker 2012)
Böcker Wheat Sprouts	Paste-like wheat sprout sourdough	30%	Mixed wheat bread and wheat bread	• Increased nutrition and taste • Juicy, fresh and moist crumb • Increased shelf life and anti-staling	(Böcker 2012)
Clerici Sacco (industry supplier)					
Lyoflora B 2	*Lactobacillus plantarum*	10^7–10^8 CFU/g	Bread starter cultures	• Controlled indigenous bacteria • Minimize batch inter-variation • Improved texture and flavor of breads	(Clerici Sacco 2012)
Lyoflora B 4	*Lactobacillus brevis*	10^7–10^8 CFU/g	Bread starter cultures	• Controlled indigenous bacteria • Minimizes batch inter-variation • Improved texture and flavor of breads	(Clerici Sacco 2012)
Lyoflora B 7	Eight LAB blend	10^7–10^8 CFU/g	Bread starter cultures	• Controlled indigenous bacteria • Minimizes batch inter-variation • Improve texture and flavor of breads	(Clerici Sacco 2012)
Lyoflora B 8	LAB and yeast blend	10^7–10^8 CFU/g	Bread starter cultures	• Controlled indigenous bacteria • Minimizes batch inter-variation • Improved texture and flavor of breads	(Clerici Sacco 2012)
Puratos (industry supplier)					
Sapore Tosca	RTU powder form durum wheat sourdough	1–3%	Italian breads, French baguettes, crusty bread, crusty rolls, pizza, dough balls, breadsticks, grissini, ciabatta, focaccia, sourdough, whole meal bread, par-baked, frozen goods	• Imparts unique Italian flavor • Optimal flavor development • Clean label • Improved crustiness • Yellow crumb color	(Puratos 2012)

Table 1. contd....

Table 1. contd.

Company products	Product form	Dosage*	Food uses	LAB role in this food	Reference
Puratos (industry supplier)					
Sapore Traviata PR 80	Powder from rye flour based mild LAB fermentation	1–3%	Wheat breads, rye breads, frozen dough and par-baked bakery products	• Natural improved taste and flavor • Consistent results and prolonged freshness • Better volume, crumb texture, elasticity and slicing properties	(Puratos 2012)
Sapore Fidelio S500	Liquid San Francisco sourdough	10–12%	Spelt loaf, Italian cheese and herb bread, San Francisco sourdough, cheese pillow, sourdough bread	• Improved taste and flavor with consistent quality in straight dough process • Desired volume, crumb texture, elasticity • Improved freshness and sliceability	(Puratos 2012)
Sapore Panarome LW	Liquid	1–3%	Toast breads, buns, frozen dough, American sourdough and black olive bread	• Reproduces fresh flavor of sponge and dough method • Clean label and convenient	(Puratos 2012)
Sapore Rigoletto	RTU sponge powder concentrate	1–3%	White tin breads, buns, toast and frozen dough	• Optimal flavor development • Replaces sponge and dough process	(Puratos 2012)
Orfeo PR 200°	RTU powder rye sourdough concentrate	1–3%	Sourdough, brown bread, whole meal bread, seeded bread and rye bread	• Clean label • Improved flavor • Better crumb elasticity and sliceability	(Puratos 2012)
Trovatore	RTU powder wheat flour sponge concentrate	1–3%	Toast bread, burger buns, soft rolls, fingers, English muffins, tear & share, bun goods, Panini, scotch rolls, wheat bread and bagels	• Reproduces the fresh method associated with the sponge or dough • Obtains optimal flavor development • Reduced fermentation time	(Puratos 2012)

Traviata	Powder form durum wheat sourdough	1–3%	Italian style breads, French sticks, baguettes, ciabatta, focaccia, bread, sourdough, par-baked, frozen goods	• Natural sourdough flavor and taste • Straight dough process with consistency • Desired volume, crumb texture, crustiness and prolonged freshness • Yellow crumb color with better elasticity and sliceability	(Puratos 2012)
Norma PW55	Powder form wheat flour based sourdough	3–5%	White breads, Italian breads (Toscana, pugliese, ciabatta,....), French baguettes, frozen dough, par baked goods	• Straight dough process • Can be tailored to impart various flavors • Gives a "rural touch" consistent product • Compensates for flavor loss (par-baked/frozen)	(Puratos 2012)
O-tentic	Dried durum wheat sourdough with yeast	4% (with addition of water and salt)	French baguettes, crusty bread/rolls, pizza, dough balls, cheese bread, breadsticks, grissini, ciabatta, focaccia, sourdough, Italian style, seeded bread, fruited bread, tiger bread, par-baked dough types and frozen dough	• Mediterranean taste profile • Crustiness is increased	(Puratos 2012)
Othello PR 200	Powder form rye flour based caramelized fermentation	0.5–2/4%	White or whole grain loaves up to full-flavor, no-time rye bread production	• Clean label, straight dough process • Flavor can be tailored to lift/round (0.5–1%) • Full flavor with some acidity (1.5–2%) • Can compensate for flavor loss in par-baked and frozen goods	(Puratos 2012)
Sourdoughs International (home baking supplier)					
Australia (Tasmanian Devil)	Dried wild yeast and LAB spelt flour fermentate	–	Works well with spelt and kamut	• Distinctive flavor and texture	(Ed Wood 2012)
Austria Culture	Dried	–	–	• Strongly sour flavor • Slow activity culture	(Ed Wood 2012)

Table 1. contd.....

Table 1. contd.

Company products	Product form	Dosage*	Food uses	LAB role in this food	Reference
Sourdoughs International (home baking supplier)					
Bahrain Culture	Dried	–	–	• Rises well • Very sour dough	(Ed Wood 2012)
Egypt: The Giza culture	Dried from wheat flour fermentate	–	–	• Likely the first culture for leavened bread	(Ed Wood 2012)
Egypt: The Red Sea culture	Dried	–	–	• Mild flavor • Works well in home bread machines	(Ed Wood 2012)
Finland Culture	Dried	–	–	• Distinctive flavor and aroma • It rises well	(Ed Wood 2012)
France Culture	Dried	–	–	• Mild sourdough flavor • It rises very well	(Ed Wood 2012)
Italian Cultures (two)	Dried	–	Pizza and Italian country bread	• Italian bread flavors and texture	(Ed Wood 2012)
New Zealand Cultures	Dried (two)	–	Rye breads	• Special flavor	(Ed Wood 2012)
Original San Francisco Culture	Dried *Candida humilis and L. sanfrancisco*	–	San Francisco sourdough bread	• Original San Francisco sourdough culture imparting the flavor and textures associated with these traditional products	(Ed Wood 2012)
Poland Culture	Dried (rye/ pumpernickel)	–	Rye and pumpernickel breads	• Traditional Polish pumpernickel bread flavor	(Ed Wood 2012)
Russia culture	Dried	–	Wheat breads	• Fast leavening culture • Suitable for home baking machines	(Ed Wood 2012)
Saudi Arabia Culture	Dried	–	–	• Rises moderately well • Distinctive flavor	(Ed Wood 2012)

South African Culture	Dried	—	Ferments whole wheat, spelt and kamut flours	• Leavens whole wheat flour best • Unique texture, nutty flavor and sourness	(Ed Wood 2012)
Yukon Culture	Dried	—	—	• Unique flavors and other properties	(Ed Wood 2012)
Lallemand Baking Solutions (industry supplier)					
DV 1-10	Paste (5% moisture) concentrated	*Yeast and bacteria*	Brioche and panettone	• Starter for traditional sourdough breads • Adds mild acidity and fresh aromatic flavor • Compatible with sponge and dough or flour brew methods with textural improvements • 1 day consistent fermentations	(Lallemand 1998)
DV 1-11	Paste (5% moisture)	*Yeast and bacteria*	Pain au Levain and ciabatta	• Complex fermentation and aromatic notes • 1 day consistent fermentations, moderate acidity	(Lallemand 1998)
DV 1-12	Paste (5% moisture)	*Yeast and bacteria*	Pain Campagne, Pain Rustique and Pain au Levain	• Produces levain with good acidity • Spicy fermentation notes • 1 day consistent fermentations	(Lallemand 1998)

*Dosage given in % is based on flour weight.

Table 2. Exogenous enzymes which function in industrial baking applications, their producer companies, enzyme sources, properties and applications.

Company	Commercial name	Enzyme name	Applications	References
Amylolytic enzymes				
Amano Enzyme	Kleistase E5S	Amylase	• Starch industry (glucose/fructose/maltose syrup, dextrin, etc.) • Improves quality of biscuits, crackers, etc.	(Amano Enzyme 2007)
AB Enzymes	Veron® Amylofresh	*Bacillus subtilis* amylase	• Flour treatment • Extension of baked goods' shelf life	(AB Enzymes 2007)
AB Enzymes	Veron ® xTender	Maltogenic amylase	• Increases crumb softness over time, improved elasticity and freshness • More tender texture, improved flavor and taste	(AB Enzymes 2012)
AB Enzymes	Veron® 2000	Fungal amylase	• Has xylanolytic side activities • Stable fluffy dough • Increases loaf volume and gives soft elastic crumb	(AB Enzymes 2012)
AB Enzymes	Veron® AX	Fungal amylase	• Fungal xylanolytic side activities • Dry and stable dough properties with increased loaf volume	(AB Enzymes 2012)
AB Enzymes	Veron® SX	Fungal amylase	• Additional bacterial xylanase side activity • Increased volume and softer crumb structure	(AB Enzymes 2012)
AB Enzymes	Veron® M4	Fungal α-amylase	• Regulation of gassing and faster fermentation dextrin/sugar creation • Improves dough processing, baking volume and crust browning	(AB Enzymes 2012)
Novozymes	Novamyl® 10000 GB	Maltogenic amylase	• Extension of shelf life • Increases softness and cohesiveness • Decreases chewiness and firmness	(Bollaín et al. 2005, Gámbaro et al. 2006, Giménez et al. 2007)
Novozymes	Fungamyl 2500 BG	Fungal amylase	• Increases baking volume	
Novozymes	Cold Crust	Glucoamylase	• Increased rate of fermentation by increasing glucose content	(Selinheimo et al. 2007)

Company	Product	Enzyme	Properties	Reference
Genencor/ Danisco	G4-amylase	Exo-amylase	• Produces maltotetraose • Anti-staling	(Calabria et al. 2009)
Leveking	ABK-I80	*Aspergillus oryzae* α-amylase	• Improve pliable dough handling properties • Improve bread elasticity and volume • Softer finer crumb structure with good bread color (low sugar levels)	(Leveking 2012)
Proteolytic enzymes				
AB Enzymes	Veron® PS	*Aspergillus oryzae* protease enzyme preparation	• Improves gluten elasticity and dough relaxation • Contributes to ease of kneading and reduces time • Decreases dough fermentation time and increases gas retention • Flour standardization	(AB Enzymes 2007, 2012)
AB Enzymes	Veron® P	*Bacillus subtilis* protease enzyme preparation	• Flour standardization • Allows short dough resting times for biscuit/crackers • Pleasant brown color, smooth surface and round edges	(AB Enzymes 2007, 2012)
AB Enzymes	VERON® HPP	Bacterial protease	• Makes smoother dough for cookies and crackers • Improves dough processability and extensibility, reduces resting time • Brown, smooth, rounded final product	(AB Enzymes 2008)
AB Enzymes	VERON® L10	Papain based proteinase	• Liquid form • Hydrolyses gluten to give loss of elasticity • Easier product lamination and no product shrinkage	(AB Enzymes 2008)
AB Enzymes	VERON® S50	Protease	• Improves dough properties • Improves friability, browning and texture	(AB Enzymes 2012)
Xylamolytic/ Cellulolytic enzymes				
AB Enzymes	Veron® 191	Fungal xylanase (highly concentrated)	• High baking volume and improved dough properties • Works well with fungal amylase (Veron® M4)	(AB Enzymes 2012)

Table 2. contd....

Table 2. contd.

Company	Commercial name	Enzyme name	Applications	References
AB Enzymes	Veron® 292	*Aspergillus niger* xylanase plus bacterial xylanases (hemicellulolytic)	• Improves dough extensibility and high baking volume • Improves break-and-shred with a soft, fine crumb structure	(AB Enzymes 2007)
AB Enzymes	Veron® 393	Xylanase	• Dry stable dough and high baking volume • Improves retarded and frozen dough	(AB Enzymes 2012)
AB Enzymes	Veron® Special	Bacterial xylanase (concentrated)	• Works well with fungal amylase (Veron® M4) • Excellent baking performance, good dough properties	(AB Enzymes 2012)
AB Enzymes	Veron® RL	Bacterial xylanase	• Performs like Veron® Special	(AB Enzymes 2012)
Novozymes	Pentopan® Mono BG	Xylanase from *Thermomyces lanuginosus*	• Improves gluten elasticity—easy dough handling • Better crumb structure and increased volume	(Novozymes 2010)
Novozymes	Celluclast 1.5 L	Cellulase from *Trichoderma reesei* ATCC 26921	• Increases volume and decreases firmness-reduces starch retrogradation • Less gumminess and chewiness	(Haros et al. 2002)
Novozymes	Pentopan Plus BG	Endo-1,4-xylanase	• Increases loaf specific volume • Works well synergistically	(Selinheimo et al. 2007)
Novozymes	Panzea (10 XBG concentrated form)	*Bacillus Licheniformis* multifunctional xylanase	• Increases volume • Desired texture • Dry, balanced dough • Low dosage • Uninhibited by wheat flour inhibitors • Redistributes water thus improves gluten	(Culliney 2012)
Leveking	XBK-B300	Xylanase	• Enhances dough stability and increases oven-spring and loaf volume • Better crumb softness, fineness and shelf life and improves flavor	(Leveking 2012)
Leveking	LVK-HC100	Xylanase	• Improves bread loaf volume, crumb structure, flavor and taste • Enhances dough stability and replaced chemical additives	(Leveking 2012)

Lipolytic enzymes

Novozymes	Lipopan F™	Lipase	• Increases softness	(Novozymes 2010)
Leveking	LBK-B400	Lipase	• Enhances dough stability and increases oven-spring and loaf volume • Whiter crumb and replaces chemical additives as a natural emulsifier	(Leveking 2012)
Leveking	LBK-BX300	Lipase	• Enhances dough stability and increases oven-spring and loaf volume • Better crumb softness and fineness with improved flavor and taste • Extends shelf life	(Leveking 2012)
Leveking	LBK-S200	Lipase	• Improves bread volume and internal structure • Better flavor, taste and extended shelf life	(Leveking 2012)

Other enzymes and mixtures

AB Enzymes	Veron® Oxibake	Glucose oxidase from *Aspergillus* sp.	• Natural oxidizing system cross-linking the protein network by creation of additional disulfur-links and can reduce ascorbic acid addition • Increases dough elasticity and fermentation tolerance • Gives surface cut products such as French Baguettes or German rolls a nice burst and open cut	(AB Enzymes 2012)
AB Enzymes	Veron® TG	Transglutaminase	• Irreversible strong protein cross-linking effect • Increases dough elasticity and thus fermentation tolerance	(AB Enzymes 2012)
AB Enzymes	Veron® GMS+	Amylolytic/ Lipolytic enzyme mix	• Replaces monoglycerides thus decreases cost of bread production • Clean label breads	(AB Enzymes 2006)

Table 2. contd.....

Table 2. contd.

Company	Commercial name	Enzyme name	Applications	References
AB Enzymes	Veron® HF	Xylanase/ Transglutaminase *Aspergillus niger* enzyme mix (no α-amylase)	• Flour treatment and bread/roll production to stabilize rye systems • Improves dough and fermentation stability with stable and dry dough • Increases volume and bread and shred properties • Improves bread shape and can replace gluten	(AB Enzymes 2008)
AB Enzymes	Veron® W	Proteinase/ Xylanase	• Reduces batter viscosity • Reduces water content—saving energy costs	(AB Enzymes 2008)
Novozymes	Fungamyl® Super MA	α-Amylase/ xylanase-Blend of Pentopan Mono and Fungamyl	• Improves mixing tolerance • Improves machinability, stability and oven-spring • Improves crumb, volume and smoother grain	(Novozymes 2010)
Novozymes	Gluzyme® Mono	Glucose oxidase	• Strengthens gluten network and improves gas retention	(Novozymes 2010)
Novozymes	Acrylaway	*Aspergillus oryzae* asparaginase	• Reduces the suspected carcinogen, acrylamide, in starchy baked goods by up to 90% converting it to harmless aspartic acid • Desirable Maillard products are unaffected	(Novozymes 2012)
Novozymes	Panzea DualBG	Xylanase (Panzea 10XBG) and amylase	• Increases volume and reduced staling • Improves product standardization	(Culliney 2012)

Wheat

In food manufacturing in general, and particularly in the baking industry, wheat is the most abundantly processed cereal and raw material. Other grains, whether cereal or pseudocereal, are also becoming more popular and some, such as rice and corn, due to their gluten-free statuses. However, this review concentrates on wheat-based products with occasional reference to others.

Wheat has been consumed as food for around 19,000 years, having been domesticated 10,000 years ago (Kislev et al. 1992). This crop is a member of the grass family, *Triticeae*. Common bread wheat, *Triticum aestivum*, is hexaploid, unlike emmer or durum wheat species which are tetraploid. However, it is the unique elastic nature of baking wheat gluten which, when combined with water, forms an extensible viscoelastic dough allowing it to expand during yeast or chemical leavening whilst retaining gas to produce an even and open structure when baked, which we associate with typical white pan loaf (Cornell 2012). On a dry weight basis, wheat flour which is primarily milled endosperm, contains approximately 75% starch (7–10% of which is damaged starch), 10–15% protein, 2% lipids and 1% ash (Hager et al. 2012). These proportions are roughly retained in standard baking flour. However they may differ due to seasonal variations such as heavy rainfall, excessively dry conditions, crop infection, etc. Details of the structure and physiochemical nature of baking wheat (herein referred to simply as wheat) and its flour has been well described in many reviews (Abdel-Aal et al. 2002, Evers and Bechtel 1988, Hager et al. 2012). In particular, wheat components and their functionality in bread making has been detailed in an extensive review (Goesaert et al. 2005).

Annual wheat consumption globally averages in excess of 108 kg/person, most of which is consumed as baked goods, with wheat beers and porridges also accounting for part of this intake. Before wheat is suitable for food uses, it is cleaned, tempered, milled, and sifted to divide according to particle size. The resulting flour is then graded based on its ability to satisfy the bakers' requirements for a standardized and consistent performing raw material.

Typical bread making processes

Bread has been a staple food for many generations across the globe, probably due to its ease of production, dietary energy contribution and the feeling of satiety associated with its consumption. The most lean bread formulation consists of flour, water, yeast and salt. When these components are mixed, a strong gluten network is formed and during the subsequent proving step, the yeast fermentation gases are captured in the viscoelastic dough matrix. The water is responsible for hydrating the gluten, however, it concurrently solubilizes other critical flour constituents such as starch (Dreese et al. 1988). Salt is also involved in ameliorating the gluten rheology as well as contributing to overall product taste and shelf life (Belz et al. 2012). After proving, the leavened dough is baked and expanding gases (CO_2 and ethanol primarily) cause oven-rise resulting in a large light loaf with a brown crust and pale crumb matrix punctuated by various sized holes (Goesaert et al. 2005).

The over-simplified description of bread making referred to above is not reflective of the current situation in industrial bakeries where various baking aids, preservatives, and taste enhancers are added to the recipe. This ensures that natural seasonal variations in the raw ingredients do not result in an unpredictable end product which causes problems regarding processability and end product consumer acceptance.

Dough making is the most important step in the bread making process and this can be done in a number of ways; straight-dough process, sponge-dough process, liquid-sponge process, Chorleywood process, or sourdough process.

- The straight-dough process simply involves mixing the ingredients into dough and fermenting for 1–4 hours followed by cutting and molding. Then the dough pieces are fermented for a further period, to reach the desired size, followed by baking which typically results in bread with a coarse crumb and relatively bland flavor (Yang 2007).
- The sponge-dough process involves separation of the dough with one part being fermented for a prolonged period and then re-combined with the remainder of the dough followed by a second intermediate proof to allow dough relaxation. The dough is then divided, molded and baked. The resulting bread has a fine cell structure and well-developed flavor (Yang 2007).
- The liquid-sponge process involves fermentation in the liquid state with the fermentate being pumped into the mixer allowing a shorter fermentation time and thus, is favorable in an industrial high throughput bakery (Fowler and Priestley 1980).
- The Chorleywood process is used to produce bread from weaker flour (low protein content). It typically has a dramatic reduction in proofing time due to the lower restrictions of a weak gluten network. In this process, mixing is done under pressure to incorporate an optimal volume of air (Martin et al. 2004).
- The sourdough process is quite a complex procedure but has been optimized through generations of traditional fermentations involving indigenous LAB and yeasts. However, industrial processes generally use standardized commercially available starter cultures (Table 2). The resulting breads are rich in flavor, texture and often

have increased nutritional and shelf-life profiles. Typical sourdough is a liquid/paste mixture of flour, water and fermenting LAB/yeasts with specific species of the latter combination dictating the acid, aroma and leavening characteristics of the final breads (Hammes and Gänzle 1998). The sourdough ecosystem favors LAB which are most suited to a particular flour, normally wheat or rye, with others such as barley sourdough also being reported (Zannini et al. 2009) in the fermentation mixture and thus, the system dictates specific characteristics of each sourdough type (De Vuyst and Vancanneyt 2007). There are three sourdough classifications: Types I, II and III (Böcker et al. 1995). Type I needs daily propagation/back-slopping to keep the ecosystem active. Type II is common industrially and is a semi-liquid sourdough fermented at > 30°C for 48–120h. Type III is also common industrially and is ready-to-use (RTU) dried sourdough. Table 2 has a list of commercially available starter cultures which is indicative of the types available, but not fully inclusive.

Exogenous enzymes in baking

Due to the low or variable contents of various endogenous (wheat) flour enzymes, including β-amylase and protease, exogenous microbial enzymes are frequently supplemented into flour and baking formulations. Positive effects of enzyme addition to cereal-based formulations include, improved fermentation, increased dough machinability and stability, as well as ameliorated crumb brightness, fineness and structure, enhanced shelf life, and more developed crust color of the baked product. These enzymes are frequently naturally produced by microbial fermentation, derived from over-expression in genetically modified microorganisms, or produced in the natural host but as genetically engineered proteins. However, because these biocatalysts are typically destroyed during the baking process, the resultant products are 'clean-label' and thus, exogenous enzymes can partially or completely replace the chemicals frequently incorporated into baking formulations in a consumer friendly way.

Enzymes generally have very specific activity towards their substrate, however, a large range of biocatalyst activity types have been reported. Furthermore, other environmental factors such as substrate availability, end-products, wheat enzyme inhibitors, co-enzymes, co-factors, pH, water activity, temperature, etc., can influence the enzymatic reactions in the dough (Fig. 1). Enzyme reactions can comprise catalysis of oxidation/reduction reactions, functional group transfer, specific bond hydrolysis, specific bond cleavage, molecular isomerization, or molecular ligation via covalent bond formation (Moss 2012). Generally, enzymes which are hydrolytic or oxidative find functionality in the bread making industry.

Given the relative abundance of starch in wheat flour and the importance of gluten proteins in dough extensibility and baking performance, it is predictable that exogenous amylolytic and protein-active (proteolytic or oxidative) enzymes play a significant role in modulating the dough and the final baked product's characteristics. Additionally, it is widely reported that the xylanolytic hydrolases also have substantial effects on the rheological and handling properties of dough as well as the texture and final volume of breads (Butt et al. 2008). Furthermore, certain lipolytic and oxidative enzymes can modify the properties of dough and bread lipid or lipid/gluten components, respectively (Moayedallaie et al. 2010, Nicolas and Potus 2000). The most important and commonly used baking enzymes are discussed below. Additionally, Table 1 lists a range of industrially available enzymes (not all-inclusive) which are currently used to ameliorate flours, dough and baked goods with standardized and predictable results regardless of natural or seasonal variation in the raw materials, particularly wheat flour.

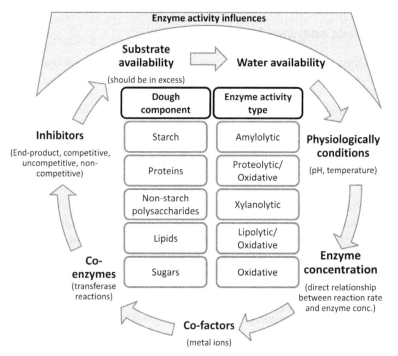

Fig. 1. Baking relevant enzymes, the dough components upon which they act, and the environmental influences on the enzyme-mediated hydrolysis or oxidative activity (NSP, non-starch polysaccharides).

Commercial Enzyme Production

Baking enzymes for industrial use are generally derived from microbial sources. However, due to the typically low enzyme levels necessary for host organism functionality, this is reflected in their enzyme production levels. Furthermore, fermenting these microbes in substrates aimed to produce higher levels of the desired enzyme often results in a heterogeneous enzyme cocktail with potentially undesired side-activities, metabolites, or other by-products. As such, it is more common industrially to reduce production and purification costs by using genetic modification as a means of generating large volumes of relatively pure biocatalysts. These enzymes are produced from the natural gene sequence in a heterologous host or by means of an engineered gene product which better suits the end application. In these cases, the production strain must have GRAS (Generally Regarded As Safe) status, readily secrete the recombinant enzyme as the dominant component in the fermentate, and be capable of optimal growth and enzyme production on a relatively cheap and abundantly available carbohydrate source.

The primary enzyme production processes are either solid-state fermentation or, more commonly, submerged fermentation due to its ease of control (homogenization of pH, temperature, nutrients and aeration, as well as easy scale-up, decreased product recovery costs, etc.) and automation (Couto and Sanromán 2006). The fermentation can be continuous or in a batch format, but must be sterile in either system. After fermentation, the enzyme broth is refined, concentrated and purified where necessary. Downstream processing involves separation of the biomass from the broth (filtration/centrifugation), enzyme concentration (ultra-filtration) and lyophilization/spray drying/agglomeration/encapsulation within a carrier, if the final product is to be packaged for sale as a powder. Quality control steps are implemented to ensure safety and functionality of the enzyme product in the baking industry (Kornbrust et al. 2007).

Amylolytic enzymes

Starch is only available as a substrate for amylolysis when the granules are damaged during processing or gelatinized by heat-treatment in the presence of moisture. A certain proportion of normal baker's flour (~ 8%) (Hager et al. 2012) is damaged as a result of the milling process, and is susceptible to enzymatic attack. Amylase, and other starch-active hydrolases, are frequently used in bread formulations to positively influence dough consistency, to increase simple sugars for amelioration of fermentation activity, and modify amylopectin chains to retard staling (Barrett et al. 2005, Błaszczak et al. 2004, Morgan et al. 1997). Due to Maillard activity and other chemical reactions, the amylolytic enzyme products can also influence bread color and flavor. Indigenous α-amylase is either negligble or too unpredictable in normal baking flours (wheat/rye), thus exogenous activity (typically microbial or cereal) is added as liquid/powder or malt preparations, respectively. In general, the amylolytic enzymes associated with baking include α-amylase, maltogenic α-amylase and amyloglucosidase.

The use of α-amylase (EC 3.2.1.1, endo-α-amylase) is associated with improved gas production during fermentation (likely due to the release of maltose through exo-activity on oligosaccharides), improved dough stability and crumb structure (decreased water absorption by starch, increased availability for gluten), increased bread volume (due to better fermentation and subsequently oven-spring), better crust color development (resulting from Maillard reaction products), and an extension of shelf life (through retardation of amylopectin retrogradation), industrially used examples of which are listed in Table 1. This enzyme shows random endo-1,4-glycosidic link cleavage of starch polymers into dextrins, which subsequently act as a substrate for flour derived β-amylase. Maltogenic α-amylase (EC 3.2.1.33, amylo-α-1,6-glucosidase) is associated with extended shelf life and functions by hydrolyzing un-substituted α-1,6-glucose residues from a branching point on the α-1,4-glucose backbone of a starch chain (ExPASY 2012). Fungal amyloglucosidase (EC 3.2.1.3, glucoamylase or glucan 1,4-alpha-glucosidase) causes rapid hydrolysis of 1,6-α-D-glucosidic bonds beside a 1,4-α-D-glucosidic bond and terminal 1,4-α-D-glucose residues form the non-terminal end of a starch polymer (ExPASY 2012) leading to improved fermentation gas production, increased bread volume and good crust color development.

In addition to enzyme activity type, the origin of the enzyme (i.e., fungal or bacterial) also has a strong impact on its activity. Fungal and bacterial α-amylase is endo-acting. However, the fungal dextrins produced are generally smaller than their bacterial counterpart's products. Additionally, fungal α-amylase optimal activity is at 55°C, with only 65% of this activity remaining at 70°C. Conversely, the bacterial enzymes have temperature optima between 70–80°C, with activity still remaining at 90°C allowing them to act on gelatinized starch (Błaszczak et al. 2004) as well as the damaged starch fractions. Bacterial amylase over-dosing may be a problem as the large dextrins produced result in a sticky dough and bread crumb, as such maltogenic amylases are currently preferred to increase shelf life. It has also been noted that fungal and bacterial α-amylases contributed different characteristics to bread (Błaszczak et al. 2004) such as: specific volume—the fungal enzyme gave a higher specific volume, crumb porosity—the bacterial enzyme-treated bread had higher Dallmann's porosity values, and texture

—the bacterial enzyme-treated bread had decreased hardness and staling with similar elasticity and increased cohesiveness compared to the fungal enzyme.

Amyloglucosidase (EC 3.2.1.3, glucan 1,4-alpha-glucosidase) is an exo-acting amylase which breaks down starch and starch-derived polymers by cleaving single α-1,4-glucose-linked (and at a slower rate α-1,6-glucose-linked) molecules off the non-reducing ends of a starch polymer in a stepwise manner (ExPASY 2012). This enzyme increases rate of fermentation by elevating glucose content (Table 1).

Maltogenic α-amylase works on both amylopectin and amylose during starch gelatinization (Hedeba et al. 1990). It does not disrupt the amylopectin backbone, instead hydrolyses the ends of starch polymers at branching points, thus decreasing the maltooligosaccharide lengths and releasing large amounts of maltose and other small sugar molecules. This slows amylopectin retrogradation and allows bread to remain softer and resilient for longer during storage (Derde et al. 2012) without causing excessive weakening of the amylose network (Goesaert et al. 2009). Many studies have been completed which explore the role of α-amylases as anti-staling agents. One particular study investigated the functionality of maltogenic α-amylase and concluded that the starch bread studied had reduced staling due to a decrease in starch retrogradation (Morgan et al. 1997). The authors measured staling over 6 days and found that the enzyme-treated starch bread increased in hardness by approximately 11% over the time of the study, compared to over 200% for the untreated control bread (Morgan et al. 1997).

Previous research compares *Bacillus stearothermophilus* maltogenic α-amlyase (BStA) and *B. subtilis* α-amylase (BSuA) during the straight dough making procedure (Lagrain et al. 2008). BSuA activity continued throughout bread making whilst BStA activity was more active at the end of baking and during cooling resulting in a higher initial firmness than BSuA treated breads, leading the authors to conclude that the mode of action and temperature optima of the enzymes are critical parameters in the applicability of enzymes for bread making purposes (Lagrain et al. 2008).

A more recent study compared BStA α-amylase with a recombinant *Pseudomonas saccharophila* maltotetraogenic (PSA) α-amylase (both optimally active at approximately 60°C) (Derde et al. 2012). This research group found that the PSA was able to reduce starch polymers' molecular weights faster than BStA (based on iodine staining capacity) and that BStA produced a greater proportion of reducing sugars indicating that the former is most likely endo-acting or has multiple attack sites. PSA hydrolysis products were also comparatively smaller. The authors concluded that due to the increased relative endo-activity of PSA on amylose compared to soluble starch, and lack of variation in its activity as temperature varied in contrast with BStA (higher activity >70°C than at lower temperatures), may have significant implications for PSA in bread making as the temperature gradually increases during the process (Goesaert et al. 2009).

In addition to bread making applications, amylolytic enzymes also function in flour standardization. Supplementation of flour, which has variable β-amylase contents (depending on wheat growth environmental conditions and variety), with fungal α-amylase has been commonplace for decades, resulting in breads with a higher specific volume and more reproducible dough and end-product characteristics (Drapron and Godon 1987).

Proteolytic and network-forming enzymes

Proteases and transglutaminase enzymes are both active on the protein (gluten) fraction of the dough. Protease enzymes exhibit hydrolytic activity towards peptide bonds and thus are responsible for degradation of proteins into smaller peptides, dipeptides or amino acids. Conversely, transglutaminase (EC 2.3.2.13, protein-glutamine gamma-glutamyltransferase) catalyzes the formation of links between the acyl donor γ-carboxymide groups of a glutamine residue and the 6-amino-group acceptor of a lysine residue in proteins (ExPASY 2012), thus making it suitable to increase the strength and resistance of dough formed using a weak (low protein) flour. Research into the role of transglutaminase in wheat bread and whole meal wheat bread production revealed that this enzyme causes aggregation and polymerization of proteins by disulfide interchange reactions. This leads to improved viscoelasticity in the dough resulting in a final product with better elasticity, water-holding capacity, crumb strength, firmness and heat stability (Collar and Bollaín 2004).

Proteases are hydrolyzing enzymes, as previously stated, and can act in an endo-manner (randomly attacking the protein backbone resulting in smaller peptides) or in an exo-fashion (cleaving single amino acids off the ends of peptide or protein chains). Endoproteases are roughly divided into four groups based on their amino acid specificities: serine proteases (EC 3.4.21, serine endopeptidases), cysteine proteases (EC 3.4.22, cysteine endopeptidases), aspartic proteases (EC 3.4.22, aspartic endopeptidases) and metalloproteases (EC 3.4.24, metalloendopeptidases). Exoproteases are either carboxypeptidases (EC 3.4.16) or aminopeptidases (EC 3.4.11) (ExPASY 2012). Protease/peptidase enzymes added to dough are typically of fungal origin (*Aspergillus oryzae* or *A. niger*), and are incorporated to modulate the dough properties and bread quality (Table 1). However, the release of peptides and amino acids also means that protease hydrolysis products can contribute to the Maillard reaction and thus influence aroma and color properties of the final baked product. Additionally, the produced compounds are precursors for the production of volatiles or themselves may contribute flavor attributes to the breads (Martínez-Anaya 1996).

Proteases from bacterial sources are also used in baking and these are generally more aggressive in their attack than their fungal counterparts. Either source of microbial enzyme is primarily used to weaken the gluten network in very strong dough thereby reducing mixing time, ameliorating pan bread shape, preventing shrinkage of sheeted dough and improving dough extensibility during fermentation whilst also replacing chemical reducing agents (Table 1).

Xylanolytic and hemicellulolytic enzymes

Non-starch polysaccharides constitute 1–3% of wheat flour dry weight and can be grouped into water extractable or unextractable depending on their solubility, a factor which can be modified using xylanolytic enzymes. Water unextractable arabinoxylanases are estimated to hold up to 25% of the total dough water content (Goesaert et al. 2005). Whether water extractable or unextractable, arabinoxylans are believed to be part of the gluten network in wheat bread and may interfere with the viscoelasticity by increasing its hardness (Kornbrust et al. 2007). However, using endoxylanases (EC 3.2.1.8) which specifically target water unextractable arabinoxylans and convert then to water extractable arabinoxylans, through cleavage of the backbone, will allow water to be released thus increasing gluten hydration and contributing to softer dough (Kornbrust et al. 2007).

Xylanases are normally confined to glycosyl hydrolase families (GH) 10 and 11 with some others being classified into families 5, 7, 8, and 43 additionally (CAZy 2012). However, those from GH 11 are primarily associated with improvements in dough and bread making attributes due to their endo-cleavage of unsubstituted regions of the water unextractable arabinoxylans thereby leading to increased viscosity and lack of specificity for water extractable arabinoxylans giving good baking results (Butt et al. 2008, Courtin and Delcour 2001). These include increases in oven rise and final loaf volume as well as a fine, soft crumb and increased shelf life (Courtin and Delcour 2002).

Investigations performed into the effect of an endoxylanase on wheat dough pentosans (primarily arabinoxylan) revealing that the enzyme could transform water unextractable hemicellulose into water extractable hemicellulose thereby allowing increased water binding capacity and improving the dough and bread characteristics (Rouau 1993). Xylanase has been shown to work well in synergy with other baking enzymes to give an improved dough or final product, as observed when combined with peroxidase (Hilhorst et al. 2002). Additionally, xylanase combined with protease, lipase, or α-amylase, resulted in an improved oven-spring and texture (Mathewson 2000). Various combinations of xylanase with transglutaminase, protease and laccase enzymes revealed that the former enzyme contributed to a better shape and specific volume of the final loaves (Caballero et al. 2007).

In spite of the positive baking results associated with xylanase addition in the bread formulation, there have also been continued reports of inhibition of these enzymes by indigenous cereal inhibitors including XIP (xylanase inhibitor protein) and TAXI (Triticum aestivum xylanase inhibitor) (Gebruers et al. 2004, Juge et al. 2004). However, wheat inhibitor insensitive enzymes occur naturally or can be engineered (Tison et al. 2009).

Lipolytic and lypoxygenic enzymes

Normal wheat flour has 1–2% lipid, a large proportion of which is intimately associated with the starch granules and so are unavailable for enzymatic modification. However, modulation of the remaining wheat lipid or any fat added to the formulation can dramatically affect the dough and final baked goods in various ways depending on enzyme specificity and the fats used, polar lipids (phospholipids or galactolipids) or non-polar lipids (mostly triglycerides). Generally, polar lipids are associated with stabilization of the air bubbles in the continuous gluten-starch matrix. Bread making formulations are frequently supplemented with emulsifiers to aid in air bubble stabilization thus ensuring that a there is a more stable dough and uniform dispersion of holes in the final product. However, lipase enzymes can replace these chemical emulsifiers, leading to a clean label product with the same functionality, which is highly desirable in the modern food industry.

Lipase (EC 3.1.1.3, Triacylglycerol lipase) acts on ester-water/air interfaces. In dough this is the bubble-dough interaction point with preferential hydrolysis of the outer ester links, which are the fatty acids located on position 1 or 3 of the non-polar lipid glycerol backbone (ExPASY 2012). More specific phospholipase (EC 3.1.1.32, Phospholipase A1) and phosphatidase (EC 3.1.1.4, Phospholipase A2) enzymes are currently used in bread making, with the former having a broader specificity (ExPASY 2012). Both phospholipase types mentioned can modify polar/non-polar lipids.

There have been numerous studies into the effects of lipase on wheat flour, dough and bread quality. Wheat flour farinograph properties were improved by lipase addition resulting in better baked, steamed and boiled Chinese wheat dough recipes (Huijing et al. 2008). It has been reported that lipase addition increased the amount of free fatty acid, thus imparting an emulsifying property to the dough, as well as improving the formation of the starch-lipid complex (Siswoyo et al. 1999). Additionally, lipase reduced starch retrogradation over 5 days and increased the volume of the bread, but not to the same extent as α-amylase and lipase combined (with the combination being superior to α-amylase only treated breads) (Siswoyo et al. 1999).

Treatment with lipase enzymes leads to easy dough handling properties, improved machinability, increased dough strength, stabilized air bubbles in the dough matrix during mixing, resting and fermentation, as well as contributing to a better oven spring and a finer more even crumb with a whiter appearance. However, using a lipase enzyme with non-polar lipid specificity in recipes containing a lot of diary fats may have poor results from a flavor/aroma perspective (Kornbrust et al. 2007).

Oxidases, including lipoxygenase (EC 1.13.11.12), which are incorporated into a bread formulation for gluten-strengthening purposes, may lead to negative lipid-derived aroma/flavor compounds. Lipoxygenase is known to oxidize polyunsaturated fatty acids containing cis-cis pentadiene into their corresponding hydroperoxides (ExPASY 2012). In one baking study, based on this reaction, linolenic and linoleic acids were lost in this way as oxygen was introduced to the system during mixing which, when combined with the release of bound lipids, lead to an increase in dough mixing tolerance and relaxation time with enhanced volume in the final bread (Nicolas et al. 2000). The lipoxygenase used in bread making is often sourced from soybean (non-microbial) and leads to an increased loaf volume and improved organoleptic qualities (Permyakova and Trufanov 2011).

Other Enzymes

Various other enzymes, such as oxidases/oxidoreductases, are incorporated into bakery formulations where the gluten network is too weak to exert its optimum functionality. Additionally, dough stickiness is reduced and bread volume is increased in weak flours but decreased in strong flours (Decamps et al. 2012). These enzymes can catalyze reduction-oxidation reactions and are used to oxidize sulfhydryl groups between gluten chains to disulfide bonds which bind the gluten proteins together into a strong network. These enzymes can replace dough conditioners such as ascorbic acid and potassium bromate thus providing a clean label alternative to chemical dough improvers, many of which have been banned for safety reasons. Lipoxygenase, which was briefly discussed in the previous section, is an example of an oxidoreductase, as well as glucose oxidase and pyranose oxidase (Decamps et al. 2012). Examples of commonly used bread making enzymes available industrially are listed in Table 2 below.

Lactic acid bacteria fermentation

LAB are abundant in natural unprocessed foods and using sourdough starter cultures takes advantage of this microflora through traditional fermentations. Typical LAB species isolated from wheat and rye sourdough primarily include Lactobacilli species. Traditionally, fermentation was used as a tool to leaven breads and acidify dough for anti-microbial and taste purposes (Gänzle et al. 2007). Nowadays, the impetus for natural microbial fermentation of dough relates to government regulations which limit chemical additives, a consumer driven desire for clean-label products, and industry's necessity to fulfill these aims whilst maintaining low costs and a technologically viable product. In summary, LAB fermentation products are necessary to address similar shortcomings in cereal food technology as previously discussed for exogenous enzyme processing.

Shelf-life effect of LAB fermentation

LAB have been widely reported to produce antimicrobial compounds during fermentation of cereal (typically wheat and rye) flours. Certain anti-fungal or anti-microbial strains are capable of inhibiting growth of fungal indicator species (commonly found in bakery or cereal environments), such as those from *Aspergillus, Eurotium, Penicillium,* and *Fusarium*. Some studies have investigated the ability of LAB to produce anti-fungal metabolites *in vitro*, however the *in situ* bakery system is of more interest in this review. One study in wheat bread revealed that sourdough fermented by antifungal *Lactobacillus plantarum* FST 1.7 was able to replace the chemical preservative calcium propionate, and it actually extended the shelf life by over 10 days (as opposed to 2 days for CAP at 0.03% addition level) against *A. niger, F. culmorum* or *P. expansum* (Ryan et al. 2008). Another study showed that *L. plantarum* CRL 778, *L. reuteri* CRL 1100, *L. brevis* CRL 772 and *L. brevis* CRL 796 showed antifungal activity against *Aspergillus, Fusarium,* and *Penicillium* species, primarily through production of acetic and phenyllactic acids (Gerez et al. 2009). Sourdough with these strains resulted in a 50 % reduction in the concentration of calcium propionate (to 0.2%) needed for the same preservation effect in wheat bread (Gerez et al. 2009). Additionally, given the higher water content of most gluten-free bread recipes (hydrocolloids need water to exert a structural effect which can replace gluten), it was impressive to see, in a study by Moore et al. (2008), that *L. plantarum* FST 1.7 sourdough could soften bread, reduce staling and extend bread shelf life by 3 days. More recently, it has been reported that *Lactobacillus amylovorous* DSM 19280 sourdough could prolong reduced NaCl bread shelf life by up to 14 days which compared favorably with 0.3% calcium propionate addition which decreased shelf life for up to 12 days (Belz et al. 2012). LAB has also been reviewed to discuss its potential as a type of biopreservative in food and feed applications (Schnürer and Magnusson 2005).

LAB-mediated nutrition, texture, flavor and aroma amelioration

Tasty, wholesome and nutritious foods are a highly desired commodity in current developed marketplaces and as such, the role of LAB sourdough in increasing the nutrient profile of baked cereal goods is of interest. Sourdough can improve the sensory appeal of high fiber, wholegrain and gluten free cereal-based products whilst also reducing the anti-nutrient components (for example phytate) and increasing the free amino acids and minerals (Waters et al. 2012, Zannini et al. 2012). These aspects of LAB sourdough have been comprehensively reviewed recently (Chavan and Chavan 2011, Poutanen et al. 2009, Zannini et al. 2012).

Conclusions

The bread producer wants clean label functional baking aids without increasing the cost of their product. Additionally, specialized breads or novel cereal-based baked goods improvers must increase the safety of bread or add a health benefit such as lower glycemic index, increased fiber content, reduced salt, mineral fortification, and gluten-free etc. The bread industry needs to find solutions which maintain/decrease costs whilst addressing these market needs. Given the broad range of consumer requirements and the necessity to move away from unnatural or chemical additives, the specificity of natural enzymes and traditional LAB fermentation can be exploited to perform precise roles in the bread making process, depending on the qualities needed in the final product.

High-throughput microbiological and enzyme screening methods in combination with molecular biology ensure that enzyme activities of interest can be monitored in a range of fungal/yeast or bacterial hosts. Modern biotechnological developments mean that genes can be mined from the ever growing databases and libraries of enzyme products. These techniques allow enzyme producers to specifically target certain gene/protein sequences for a functional motif such as the lipase catalytic triad or carbohydrate/starch binding domains, etc., to find the biocatalyst of interest. Additionally, protein engineering allows replacement of single amino acids to modulate enzyme characteristics such as specificity (give a broader activity spectrum by relieving steric inhibition, etc.), temperature/pH optima (add stabilizing disulfide bonds, etc.), alleviation of inhibitor problems (for example wheat XIP fungal xylanase inhibition, etc.).

Genetically modified and/or over-expressed natural enzymes which are eliminated during processing and so absent in the final product, can be easily incorporated into bread making without having negative impacts on labeling or marketing of the product. As such, it seems an obvious solution to incorporate biocatalysts into bread making formulations to perform specific roles whilst allowing exclusion of chemicals, which are increasingly unaccepted by consumers and may go against government regulations.

Additionally, LAB fermentation is a completely natural way to ameliorate cereal products. LAB can also be characterized using the high-throughput microbiological and enzyme/metabolite screening methods mentioned for enzyme research. This allows the use of a sourdough which imparts specific properties to baked goods, including a longer shelf life (anti-fungal LAB metabolites), acidic and other flavor compounds, texture modulation, and varied aroma profiles, etc.

Future prospects for enzyme and fermentation technology in the bread making industry are immensely promising. The market is awash with many baked products characterized by poor flavor, bad shelf life properties, as well as high salt and/or low nutrition levels. There is a growing necessity to produce natural products which adhere to stricter FDA and EFSA regulations whilst maintaining acceptability in the marketplace. Huge market opportunities exist for food manufacturers who can supply economically priced natural clean-label baked cereal goods and I propose that enzyme and fermentation technology represent the best solutions to do so.

References

AB Enzymes, I. 2006. AB Enzymes launches VERON® GMS+, an enzymatic monoglyceride replacer., from http://www.abenzymes.com/news/ab-enzymes-launches-veron%C2%AE-gms-an-enzymatic-monoglyceride-replacer.

AB Enzymes, I. 2007. Veron P Description and Specification. from http://www.ifi-sa.com/pdfs/P%20-e%20Rev4_mL.pdf.

AB Enzymes, I. 2007. Veron PS Description and Specification. from http://www.ifi-sa.com/pdfs/PS%20-e%20Rev4_mL.pdf.

AB Enzymes, I. 2007. Veron® 292 Description and Specification. from http://www.ifi-sa.com/pdfs/292-e%20Rev4_mL.pdf.

AB Enzymes, I. 2007. Veron® Amylofresh description and specification. from http://www.ifi-sa.com/pdfs/Amylofresh%20-e%20Rev4_mL.pdf.

AB Enzymes, I. 2008. Veron HF. from http://www.ifi-sa.com/pdfs/HF%20-e%20Rev7_mL.pdf.

AB Enzymes, I. 2008. Wafers, biscuits and crackers. from http://www.abenzymes.com/products/baking/wafers-biscuits-and-crackers.

AB Enzymes, I. 2012. AB Enzymes announce expansion of VERON® product range—AB Enzymes. from http://www.abenzymes.com/news/ab-enzymes-announce-expansion-of-veron-product-range.

AB Enzymes, I. 2012. Baking. 2012, from http://www.abenzymes.com/products/baking.
AB Enzymes, I. 2012. Dough relaxation. from http://www.abenzymes.com/products/baking/bromate-replacement.
AB Enzymes, I. 2012. Dough Stability. from http://www.abenzymes.com/products/baking/dough-stability-2.
AB Enzymes, I. 2012. Machinability and Volume. from http://www.abenzymes.com/products/baking/dough-and-volume.
AB Enzymes, I. 2012. Our range of enzymes. from http://www.ifi-sa.com/enzymesrange.
Abdel-Aal, E. S. M., P. Hucl, R. N. Chibbar, H. L. Han and T. Demeke. 2002. Hysicochemical and Structural Characteristics of Flours and Starches from Waxy and Nonwaxy Wheats1. Cereal Chem 79: 458–464.
Amano Enzyme, I. 2007. Product Listing Food Processing Use. 2012. from http://www.amano-enzyme.co.jp/aee/product/foodprocessing.html.
Arendt, E. K., L. A. M. Ryan and F. Dal Bello. 2007. Impact of sourdough on the texture of bread. Food Microbiol. 24: 165–174.
Barrett, A. H., G. Marando, H. Leung and G. Kaletunç. 2005. Effect of Different Enzymes on the Textural Stability of Shelf-Stable Bread. Cereal Chem. 82: 152–157.
Belz, M. C. E., R. Mairinger, E. Zannini, L. A. M. Ryan, K. D. Cashman and E. Arendt. 2012. The effect of sourdough and calcium propionate on the microbial shelf-life of salt reduced bread. Appl. Microbiol. Biot.
Błaszczak, W., J. Sadowska, C. M. Rosell and J. Fornal. 2004. Structural changes in the wheat dough and bread with the addition of alpha-amylases. Eur Food Res. Technol. 219: 348–354.
Böcker, E. 2012. Product line summary. from http://www.sauerteig.de/site/index_en.htm.
Böcker, G., P. Stolz and W. P. Hammes. 1995. Neue Erkenntnisse zum Ökosystem Sauerteig und zur Physiologie der sauerteigtypischen Stämme Lactobacillus sanfrancisco und Lactobacillus pontis. Getreide Mehl. Brot. 49: 370–374.
Bollaín, C., A. Angioloni and C. Collar. 2005. Bread staling assessment of enzyme-supplemented pan breads by dynamic and static deformation measurements. Eur Food Res. Technol. 220: 83–89.
Butt, M. S., M. Tahir-Nadeem, Z. Ahmad and M. T. Sultan. 2008. Xylanases and their applications in baking industry. Food Technol. Biotech. 46: 22–31.
Caballero, P. A., M. Gómez and C. M. Rosell. 2007. Improvement of dough rheology, bread quality and bread shelf-life by enzymes combination. J. Food Eng. 81: 42–53.
Calabria, A., C. Peres and J. Van Langeraert. 2009. Transferring fermentation process methods from R&D to the manufacturing scale. Recent Advances in Fermentation Technology VIII, Marriott Mission Valley, San Diego, CA.
CAZy, C. A. e. 2012. Glycoside Hydrolase family classification. Retrieved 25 June 2012. from http://www.cazy.org/Glycoside-Hydrolases.html.
CFR 2011. Title 21-Foods and Drugs. Department of Health and Human Services. Foods and Drugs Administration, Code of Federal Regulations.
Chavan, R. S. and S. R. Chavan. 2011. Sourdough Technology—A Traditional Way for Wholesome Foods: A Review. Compr. Rev. Food Sci. F. 10: 169–182.
Clarke, C. I. and E. K. Arendt. 2005. A Review of the Application of Sourdough Technology to Wheat Breads. Adv. Food Nutr. Res., Academic Press. Volume. 49: 137–161.
Clerici Sacco, I. 2012. Bread cultures. from http://www.saccosrl.it/catalog/?category=15.
Collar, C. and C. Bollaín. 2004. Impact of microbial transglutaminase on the viscoelastic profile of formulated bread doughs. Eur. Food Res. Technol. 218: 139–146.
Cornell, H. J. 2012. The chemistry and biochemistry of wheat. Breadmaking. S. P. Cauvain. Cambrigde, UK, Woodhead Publishing Limited. 35–76.
Courtin, C. M. and J. A. Delcour. 2001. Relative Activity of Endoxylanases Towards Water-extractable and Water-unextractable Arabinoxylan. J. Cereal. Sci. 33: 301–312.
Courtin, C. M. and J. A. Delcour. 2002. Arabinoxylans and Endoxylanases in Wheat Flour Bread-making. J. Cereal Sci. 35: 225–243.

Couto, S. R. and M. Á. Sanromán. 2006. Application of solid-state fermentation to food industry—A review. J. Food Eng. 76: 291–302.

Culliney, K. 2012. No need to mix xylanases anymore, says Novozymes. Bakery and Snacks. from http://www.bakeryandsnacks.com/Formulation/No-need-to-mix-xylanases-anymore-says-Novozymes.

De Vuyst, L. and M. Vancanneyt. 2007. Biodiversity and identification of sourdough lactic acid bacteria. Food Microbiol. 24: 120–127.

Decamps, K., I. J. Joye, C. M. Courtin and J. A. Delcour. 2012. Glucose and pyranose oxidase improve bread dough stability. J. Cereal Sci. 55: 380–384.

Derde, L. J., S. V. Gomand, C. M. Courtin and J. A. Delcour. 2012. Hydrolysis of β-limit dextrins by α-amylases from porcine pancreas, Bacillus subtilis, Pseudomonas saccharophila and Bacillus stearothermophilus. Food Hydrocolloid. 26: 231–239.

Drapron, R. and B. Godon. 1987. Role of enzymes in baking. Enzymes and their Role in Cereal Technology. J. E. Kruger, D. Lineback and C. E. Stauffer. St. Paul, Minnesota, American Association of Cereal Chemists. 281–304.

Dreese, P., J. Faubion and R. Hoseney. 1988. Dynamic rheological properties of flour, gluten, and gluten-starch doughs. I. Temperature-dependant changes during heating. Cereal Chem. 65: 348–353.

EC and DG Sanco. 2003. Evaluation of the food labelling legislation. T. E. Partnership. Middlesex, UK, The European Commission and The Directorate-General for Health and Consumer Protection.

Ed Wood, I. S. 2012. Sourdoughs International from http://www.sourdo.com/home/category/cultures/.

Evers, A. and B. Bechtel. 1988. Microscopic structure of the wheat grain". Wheat: chemistry and technology. Y. Pomeranz. St. Paul, Mn, American Association of Cereal Chemists, Inc. 1: 17–95.

ExPASY. 2012. Enzyme class: EC 3.1.1. Bioinformatice Resource Portal, from http://enzyme.expasy.org/EC/3.1.1.

ExPASY. 2012. Enzyme entry: EC 1.13.11.12. Bioinformatice Resource Portal, from http://enzyme.expasy.org/EC/1.13.11.12.

ExPASY. 2012. Enzyme entry: EC 2.3.2.13. Bioinformatice Resource Portal, from http://enzyme.expasy.org/EC/2.3.2.13.

ExPASY. 2012. Enzyme entry: EC 3.1.1.3. Bioinformatice Resource Portal, from http://enzyme.expasy.org/EC/3.1.1.3.

ExPASY. 2012. Enzyme entry: EC 3.2.1.3. Bioinformatice Resource Portal, from http://enzyme.expasy.org/EC/3.2.1.3.

ExPASY. 2012. Enzyme entry: EC 3.2.1.33. Bioinformatice Resource Portal, from http://enzyme.expasy.org/EC/3.2.1.33.

ExPASY. 2012. Enzyme entry: EC 3.4.11. Bioinformatice Resource Portal, from http://enzyme.expasy.org/EC/3.4.11.-.

ExPASY. 2012. Enzyme entry: EC 3.4.16. Bioinformatice Resource Portal, from http://enzyme.expasy.org/EC/3.4.16.-

FAOSTAT. 2012. Food and Agricultural commodities production. Food and Agricultural commodities production Retrieved 09062012, 2012, from http://faostat3.fao.org/home/index.html#DOWNLOAD.

Fowler, A. A. and R. J. Priestley. 1980. The evolution of panary fermentation and dough development—A review. Food Chem. 5: 283–301.

Galle, S., C. Schwab, E. Arendt and M. Ganzle. 2010. Exopolysaccharide-Forming Weissella Strains as Starter Cultures for Sorghum and Wheat Sourdoughs. J. Agr. Food Chem. 58: 5834–5841.

Gámbaro, A., A. N. A. Giménez, G. Ares and V. Gilardi. 2006. Influence of enzymes on the texture of brown pan bread. J. Texture Stud. 37: 300–314.

Gänzle, M. G., N. Vermeulen and R. F. Vogel. 2007. Carbohydrate, peptide and lipid metabolism of lactic acid bacteria in sourdough. Food Microbiol. 24: 128–138.

Gebruers, K., K. Brijs, C. M. Courtin, K. Fierens, H. Goesaert, A. Rabijns, G. Raedschelders, J. Robben, S. Sansen, J. F. Sørensen, S. Van Campenhout and J. A. Delcour. 2004. Properties of TAXI-type endoxylanase inhibitors. Biochimica Biophysica Acta Prot. 1696: 213–221.

Gerez, C. L., M. I. Torino, G. Rollán and G. Font de Valdez. 2009. Prevention of bread mould spoilage by using lactic acid bacteria with antifungal properties. Food Control. 20: 144–148.

Giménez, A., P. Varela, A. Salvador, G. Ares, S. Fiszman and L. Garitta. 2007. Shelf life estimation of brown pan bread: A consumer approach. Food Qual Pref. 18: 196–204.

Goesaert, H., K. Brijs, W. S. Veraverbeke, C. M. Courtin, K. Gebruers and J. A. Delcour. 2005. Wheat flour constituents: how they impact bread quality, and how to impact their functionality. Trends Food Sci. Tech. 16: 12–30.

Goesaert, H., L. Slade, H. Levine and J. A. Delcour. 2009. Amylases and bread firming—an integrated view. J. Cereal Sci. 50: 345–352.

Gujral, H. S. and C. M. Rosell. 2004. Functionality of rice flour modified with a microbial transglutaminase. J. Cereal Sci 39: 225–230.

Hager, A. S., A. Wolter, E. Zannini, M. Czerny, J. Bez, E. K. Arendt and M. Czerny. 2012. Investigation of product quality, sensory profile and ultra-structure of breads made from a range of commercial gluten free flours compared to their wheat counterparts. European Food Research and Technology Accepted.

Hammes, W. P. and M. Gänzle. 1998. Sourdough and related products. Microbiology of fermented foods. Woods, B. J. B. 199–216.

Haros, M., C. Rosell and C. Benedito. 2002. Effect of different carbohydrases on fresh bread texture and bread staling. Eur. Food Res. Technol. 215: 425–430.

Hedeba, R., L. Bowles and W. Teague. 1990. Developments in enzymes for retarding staling of baked goods. Cer Food World. 35: 453–457.

Hilhorst, R., H. Gruppen, R. Orsel, C. Laane, H. A. Schols and A. G. J. Voragen. 2002. Effects of Xylanase and Peroxidase on Soluble and Insoluble Arabinoxylans in Wheat Bread Dough. J. Food Sci. 67: 497–506.

Huijing, L., J. Yingmin, T. Yiling and X. Liqiang. 2008. Effects of Lipase on Wheat Flour Quality. J. Chin. Cer Oil Assoc: 1–8.

Jensen, H., S. Grimmer, K. Naterstad and L. Axelsson. 2012. *In vitro* testing of commercial and potential probiotic lactic acid bacteria. Int. J. Food Microbiol. 153: 216–222.

Juge, N., F. Payan and G. Williamson. 2004. XIP-I, a xylanase inhibitor protein from wheat: a novel protein function. Biochimica Biophysica Acta Prot. 1696: 203–211.

Katina, K., E. Arendt, K. H. Liukkonen, K. Autio, L. Flander and K. Poutanen. 2005. Potential of sourdough for healthier cereal products. Trends Food Sci. Technol. 16: 104–112.

Kirk, O., T. V. Borchert and C. C. Fuglsang. 2002. Industrial enzyme applications. Curr. Opin. Biotechnol. 13: 345–351.

Kislev, M. E., D. Nadel and I. Carmi. 1992. Epipalaeolithic (19,000 BP) cereal and fruit diet at Ohalo II, Sea of Galilee, Israel. Rev. Palaeobot Palyno. 73: 161–166.

Kornbrust, B., T. Forman and I. Matveeva. 2007. Applications of enzymes in breadmaking. Baked Products. S. P. Cauvain. New Delhi, India, Woodhead publishing Ltd. 1: 470–498.

Lagrain, B., P. Leman, H. Goesaert and J. A. Delcour. 2008. Impact of thermostable amylases during bread making on wheat bread crumb structure and texture. Food Research International. 41: 819–827.

Lahtinen, S., A. C. Outwehand, S. Salminen and A. Von Wright. 2011. Lactic Acid Bacteria Microbiological and Functional Aspects. Bosa Roca, Taylor & Francis Inc. 798.

Lallemand. 1998. Baking Update Pain au Levain. Lallemand Baking Update. Montréal, QC, Canada, Lallemand Inc. 2: 1–3.

Leveking. 2012. Food Enzymes. from http://www.leveking.com/en/product-default-74.html.

Martin, P. J., N. L. Chin, G. M. Campbell and C. J. Morrant. 2004. Aeration During Bread Dough Mixing: III. Effect of Scale-up. Food Bioproduct Process. 82: 282–290.

Martínez-Anaya, M. A. 1996. Enzymes and Bread Flavor†. J. Agr. Food Chem. 44: 2469–2480.

Mathewson, P. R. 2000. Enzymatic activity during bread baking. Cer Food World. 45: 98–101.

Moayedallaie, S., M. Mirzaei and J. Paterson. 2010. Bread improvers: Comparison of a range of lipases with a traditional emulsifier. Food Chem. 122: 495–499.

Moore, M., F. Bello and E. Arendt. 2008. Sourdough fermented by Lactobacillus plantarum FST 1.7 improves the quality and shelf life of gluten-free bread. Eur. Food Res. Technol. 226: 1309–1316.

Morgan, K. R., L. Hutt, J. Gerrard, D. Every, M. Ross and M. Gilpin. 1997. Staling in Starch Breads: The Effect of Antistaling α-Amylase. Starch. 49: 54–59.

Moss, G. 2012. IUBMB Nomenclature Committee Recommendations. from http://www.chem.qmul.ac.uk/iubmb/.

Nicolas, J. and J. Potus. 2000. Interactions between lipoxygenase and other oxidoreductases in baking. 2nd European Symposium on Enzymes in Grain Processing, Helsinki, Finland, Technical Research Centre of Finland (VTT).

Novozymes. 2010. Enzymes produced by genetically modified microorganisms. 2012. from http://www.novozymes.com/en/about-us/vision-and-values/positions/Pages/Enzymes-produced-by-GMMs.aspx.

Novozymes. 2012. Applications for Acrylaway. from http://www.acrylaway.novozymes.com/en/acrylaway-applications/Pages/default.aspx.

Permyakova, M. and V. Trufanov. 2011. Effect of soybean lipoxygenase on baking properties of wheat flour. Appl. Biochem. Microbiol. 47: 315–320.

Poutanen, K., L. Flander and K. Katina. 2009. Sourdough and cereal fermentation in a nutritional perspective. Food Microbiol. 26: 693–699.

Puratos. 2012. Flavour and Sourdough. 2012. from http://www.puratos.co.uk/products_solutions/bakery/flavours_sourdoughs/default.aspx.

Rouau, X. 1993. Investigations into the Effects of an Enzyme Preparation for Baking on Wheat Flour Dough Pentosans. J. Cereal Sci. 18: 145–157.

Ryan, L. A. M., F. Dal Bello and E. K. Arendt. 2008. The use of sourdough fermented by antifungal LAB to reduce the amount of calcium propionate in bread. Int. J. Food Microbiol. 125: 274–278.

Schnürer, J. and J. Magnusson. 2005. Antifungal lactic acid bacteria as biopreservatives. Trends Food Sci. Tech. 16: 70–78.

Selinheimo, E., K. Autio, K. Kruus and J. Buchert. 2007. Elucidating the Mechanism of Laccase and Tyrosinase in Wheat Bread Making. J. Agr. Food Chem. 55: 6357–6365.

Simpson, B. K., X. Rui and J. XiuJie. 2012. Enzyme-assisted food processing: Green Technologies in Food Production and Processing. J. I. Boye and Y. Arcand, Springer New York. 327–361.

Siswoyo, T. A., N. Tanaka and N. Morita. 1999. Effect of Lipase Combined with ALPHA.-Amylase of Retrogradation of Bread. Food Sci. Technol. Res. 5: 356–361.

Thiele, C., M. G. Gänzle and R. F. Vogel. 2002. Contribution of Sourdough Lactobacilli, Yeast, and Cereal Enzymes to the Generation of Amino Acids in Dough Relevant for Bread Flavor. Cer Chem. J. 79: 45–51.

Tison, M., A. L. Gwénaëlle, M. Lafond, J. Georis, N. Juge and J. G. Berrin. 2009. Molecular determinants of substrate and inhibitor specificities of the Penicillium griseofulvum family 11 xylanases. Biochimica Biophysica Acta Prot. 1794: 438–445.

USFDA. 2012. U.S. FDA Food, Beverage, and Supplement Labeling Requirements. F. Labeling.

Waters, D., F. Jacob, J. Titze, E. Arendt and E. Zannini. 2012. Fibre, protein and mineral fortification of wheat bread through milled and fermented brewer's spent grain enrichment. Eur. Food Res. Technol. Submitted.

Waters, D. M., L. A. M. Ryan, P. G. Murray, E. K. Arendt and M. G. Tuohy. 2011. Characterisation of a Talaromyces emersonii thermostable enzyme cocktail with applications in wheat dough rheology. Enzyme Microb. Technol. 49: 229–236.

Yang, C. H. 2007. Fermentation. Bakery Products, Blackwell Publishing. 261–272.

Zannini, E., C. Garofalo, L. Aquilanti, S. Santarelli, G. Silvestri and F. Clementi. 2009. Microbiological and technological characterization of sourdoughs destined for bread-making with barley flour. Food. Micro. 26: 744–753.

Zannini, E., M. Paoloni, R. Papa and F. Clementi. 2006. Use of selected sourdoughs for bread-making with barley flour. Tecnica Molitoria. 57: 650–658.

Zannini, E., E. Pontonio, D. Waters and E. Arendt. 2012. Applications of microbial fermentations for production of gluten-free products and perspectives. Appl. Microbiol. Biot. 93: 473–485.

4

Recent Developments on Amylase Production by Fermentation using Agroindustrial Residues as Substrates

Juliana M. Gasparotto, Rodrigo Klaic, Jéssica M. Moscon,
Bruno C. Aita, Daniel P. Chielle, Fabiane M. Stringhini,
Gabrielly V. Ribeiro, Luiz J. Visioli, Paulo R. S. Salbego
*and Marcio A. Mazutti**

ABSTRACT

Amylases are starch degrading enzymes that have great significance in present day biotechnology. These enzymes have a wide area of potential applications including food processing, animal nutrition, beverages production, pharmaceuticals, textiles, detergents, paper and pulp, biofuel, etc. and their applicability has been expanding especially due to the increasing interest in using agroindustrial residues as substrates associated with the development of solid state fermentation technology. In this context, this chapter brings a review of the most recent developments in the fermentative processes of amylases production from agroindustrial residues.

Department of Chemical Engineering, Federal University of Santa Maria, Av. Roraima, 1000, Santa Maria 97105-900, Brazil.
* Corresponding author: mazutti@ufsm.br

Introduction

Amylases are one of the most important enzymes in present-day biotechnology due to their wide range of applications in numerous industrial processes such as in the food, textiles, paper and bioethanol industries. They are used for starch hydrolysis in the starch liquefaction process that converts starch into fructose and glucose syrups. They are also used as a partial replacement for the expensive malt in the brewing industry, to improve flour in the baking industry, and to produce modified starches for the paper industry. In addition to this, they are used to remove starch in the manufacture of textiles (desizing) and as additives to detergents for both washing machines and automated dishwashers (Pandey et al. 2000a, Gangadharan et al. 2006).

It should be emphasized that the application of amylases in different industrial processes demands specific characteristics including thermostability, particular pH activity profiles, pH stability, and Ca-independency. Consequently, demand for novel amylases is increasing worldwide, as the variety of enzyme applications is increasing in various industrial sectors (Hashemi et al. 2010, Pandey et al. 2000b). Although amylases can be obtained from several sources, such as plants and animals, the enzymes from microbial sources generally meet industrial demand. Currently, a large number of microbial amylases are available commercially and they have almost completely replaced the chemical hydrolysis of starch in the starch processing industry, especially in food applications, where there is great interest in developing and using natural food and additives, since they are more desirable than the synthetic ones produced by chemical processes. The enzymatic hydrolysis is preferred to acid hydrolysis in starch processing industry due to a number of advantages such as specificity of the reaction, stability of the generated products, lower energy requirements and elimination of neutralization steps (Alva et al. 2007, Couto and Sanromán 2006, Sivaramakrishnan et al. 2006).

Amylases can be classified into three major groups: endoamylases, exoamylases and debranching amylases. Endoamylases (α-amylase) catalyze hydrolysis in a random manner in the interior of the starch molecule. This action causes the formation of linear and branched oligosaccharides of various chain lengths. Enzymes from this group present, therefore, a crucial role in starch liquefaction. Exoamylases (β-amylases, α-glucosidade and glucoamilase) hydrolyze from the non-reducing end, successively resulting in low molar mass product, mainly glucose and maltose. Debranching amylases (pullulanase and isoamylase) catalyze the hydrolysis of amylopectin. Today, a large number of enzymes are known which hydrolyse starch molecules into different products and a combined action of various enzymes is required to hydrolyse starch completely (Castro et al. 2011, Gupta et al. 2003).

Traditionally, amylases have been produced by submerged fermentation (SmF) and used in a one-way process in solution. In recent years, however, solid state fermentation (SSF) processes have been increasingly utilized for the production of this enzyme. In fact, SSF has gained renewed interest from researchers for the production of these enzymes in view of its economic and engineering advantages such as improved yields and lower operation costs, especially because inexpensive agriculture and agro-industrial residues are generally considered the best substrates for SSF processes and for enzyme production in these systems (Kunamneni et al. 2005).

In this chapter we have made an overview discussing the recent developments in amylase production by both submerged and solid-state fermentative processes from agriculture and agroindustrial residues. First we collected the most recent scientific publications in the area and disposed the information in two different tables: one for studies using solid state fermentation and the other for those using submerged fermentation. Thereafter, the major trends in microorganisms, substrates, some operation conditions and types of bioreactors are analyzed and discussed. Some of the most relevant works are discussed individually.

Substrates

For decades, agroindustrial materials have been used for the production of amylolytic enzymes, as well as feedstock for bioethanol production, e.g., corn kernel, potato, wheat, rye, sorghum, cassava, rice, barley, castor bean and babassu cake. However, the use of materials that are also used for food was making the final product very expensive (Castro and Castro 2012). In the last years, there has been an increasing trend to convert industrial and agro-industrial residues into valuable products, which can result in potential reductions in the residue volume and the disposal cost of the remaining fraction. Regarding their composition, feedstocks for amylase production must present considerable contents of starch, in order to contribute to the cost-effectiveness of the processes (Castro et al. 2011a).

Different solid substrates that are rich in starch like rice bran, wheat bran, groundnut oil cake, cassava fibrous residue, coconut oil cake, rice bran, castor seed residue and others, have started being used as feedstock for the enzyme production. These agro industrial residues are cheap raw materials for amylase production. From Tables 1 and 2, it is possible to note that there is a slight trend to use wheat bran in solid state fermentation due to its substrate structure, whereas there is a preference for using cassava flour and cassava fibrous residue as substrate in submerged fermentation.

Table 1. Production of amylolytic enzymes from agroindustrial materials by SSF.

Enzyme Type	Raw material	Microorganism	Ferm. time (h); Temp. (°C)	Max. activity (U/g)	Reactor type	Reference
Alpha amylases	Wheat bran and groundnut oil cake	*Bacillus amyloliquefaciens* ATCC 23842	72; 37	62 470*	Erlenmeyer flask	Gangadharan et al. (2006)
Alpha amylases	Cassava fibrous residue	*Bacillus brevis* MTCC 7521	50; 36	23 050**	Roux bottles	Ray et al. (2008)
Alpha amylases	Wheat bran	*Bacillus* sp. KR-8104	48; 37	140*	Erlenmeyer flasks	Hashemi et al. (2010)
Amylases (NS)	Black gram bran	*Aspergillus niger* BAN3E	144; 37	86* U/mg protein	NS	Suganthi et al. (2011)
Alpha amylases, glucoamylases	Coconut oil cake, groundnut oil cake, rice bran	*Aspergillus* sp. JGI 12	96; 25	11 000; 16 420*	Erlenmeyer flasks	Alva et al. (2007)
Alpha amylases	Wheat bran	*Aspergillus niger*	120; 28	1 140*	Erlenmeyer flasks	Khan and Yadav (2011)
Alpha amylases	Wheat bran	*Aspergillus niger* JGI 24	96; 30	74* U/mg protein	Erlenmeyer flasks	Varalakshmi et al. (2009)
Alpha amylases	Wheat bran	*Aspergillus oryzae* IFO-30103	66; 30	15 460*	Conventional tray reactor	Bhanja et al. (2007)
Alpha amylases	Wheat bran	*Aspergillus oryzae* IFO-30103	54; 30	19 655*	GROWTEK bioreactor	Bhanja et al. (2007)
Alpha amylases; glucoamylases	Babassu cake	*Aspergillus awamori* IOC-3914	48; 37	64,3; 45,7***	Lab scale tray bioreactors	Castro et al. (2010a)
Glucoamylases; Alpha amylases	Castor seed residue	*Aspergillus awamori* IOC-3914; *Aspergillus wentii*	96; 30	29,8; 47,8***	Lab scale tray bioreactors	Castro et al. (2011)

Table 1. contd....

Table 1. contd.

Enzyme Type	Raw material	Microorganism	Ferm. time (h); Temp. (°C)	Max. activity (U/g)	Reactor type	Reference
Alpha amylases; glucoamylases	Babassu cake	Aspergillus awamori IOC-3914	144; NS	40,5; 42,7	Lab scale tray bioreactors	Castro et al. (2010b)
Alpha amylases	Wheat bran	Aspergillus oryzae IFO 30103	48; 32	22 317*	2-cm bed height with air flow rate of 0,1L/(min.g wheat bran)	Dey and Banerjee (2011)
endoamylases, exoamylases and proteases	Babassu cake	Aspergillus awamori I OC-3914	120; 30	55,4; 104,3; 17,0***	Lab scale tray bioreactors	Castro et al. (2011)
Alpha amylases	Wheat bran	Aspergillus oryzae	72; 30	15095*	Erlenmeyer flasks	Sivaramakrishnan et al. (2007)
Alpha amylases	Wheat bran	Bacillus sp. KR-8104	48; 40 (first 20h) 32 (last 28h)	95	Erlenmeyer flasks	Hashemi et al. (2012)

*One unit of enzyme activity is defined as the amount of enzyme that liberateds 1 µmol of reducing sugar as glucose equivalent in 1 min under the assay conditions.

**One unit of enzyme activity is defined as the quantity of enzyme that causes 0.01% reduction of blue colour intensity of starch iodine solution at 50°C per minute per liter.

***One unit of enzyme activity is defined as the amount of enzyme that liquefies 1 mg of starch per minute, under the assay conditions.

NS—not specified.

Table 2. Production of amylolytic enzymes from agroindustrial materials by SmF.

Enzyme Type	Raw material	Microorganism	Ferm. time (h); Temp. (°C)	Max. activity (U/mL)	Reactor type	Reference
Alpha amylases	Soluble starch, cassava starch, cassava flour	*Bacillus brevis* MTCC 7521	36; 50	2611** U/mg protein	Erlenmeyer flasks	Ray et al. (2008)
Alpha amylases	Potato starch	*Geobacillus thermodenitrificans* HRO10	28; 49	30,20*	1,2L bioreactor	Ezeji and Bahl (2007)
Alpha amylases	Sugarcane bagasse hydrolysate	*Bacillus subtilis* KCC103	18; 37	67,4*	NS	Rajagopalan and Krishnan (2008)
Amylase (NS)	Potato dextrose	*Aspergillus niger*	96; RT	52* U/mg protein	NS	Suganthi et al. (2011)
Alpha amylases	Wheat bran; sunflower oil cake	*Bacillus subtilis* KCC103	10 a 12; 37	1258*	3,5L laboratory-scale stirred-tank type	Rajagopalan and Krishnan (2009)
Alpha amylases	Cassava peels waste	*Aspergillus niger*	120; 35	54,08*	Erlenmeyer flasks	Cruz et al. (2011)
Alpha amylases	Wheat bran	*Aspergillus niger* JGI 24	96; 30	58* U/mg protein	Erlenmeyer flasks	Varalakshmi et al. (2009)
Amylases (NS)	Soluble Starch, potato dextrose broth	fungus *Piriformospora indica*	42; 30	39,84*	500 mL Erlenmeyer flasks and 14 L bioreactor	Kumar et al. (2012)
Alpha amylases	Wheat bran; groundnut oil cake	*Bacillus amyloliquefaciens*	42; 27	965,9*	Erlenmeyer flask	Gangadharan et al. (2008)

Table 2. contd....

Table 2. contd.

Enzyme Type	Raw material	Microorganism	Ferm. time (h); Temp. (°C)	Max. activity (U/mL)	Reactor type	Reference
Alpha amylases	Brewer's spent grain	*Bacillus* sp. KR8104	31; 37	0,0236*	Erlenmeyer flask	Maryam et al. (2011)
Alpha amylases; pullulanases	Cassava bagasse starch	*Streptomyces erumpens* MTCC 7317	48; 50	222,5; 69,5*	Erlenmeyer flask	Kar et al. (2010)
Alpha amylases; glucoamylases	Cassava peels	*Aspergillus niveus*	48; 35	60	Erlenmeyer flask	Silva et al. (2009)
Alpha amylases; glucoamylases	65% Polished barley	*Aspergillus kawachii* NBRC4308	48; 37	7,7; 150,8	Não especificado	Shoji et al. (2007)
Glucoamylases and amylases	Indonesian cassava flour	*Aspergillus kawachii* FS005	72; 37	NS	Erlenmeyer flasks	Sugimoto et al. (2011)

*One unit of enzyme activity was defined as the amount of enzyme that liberated 1 μmol of reducing sugar as glucose equivalent in 1 min under the assay conditions.

**One unit of enzyme activity is defined as the quantity of enzyme that causes 0.01% reduction of blue colour intensity of starch iodine solution at 50°C per minute mg of protein.

NS—not specified.

RT—room temperature.

Microbial Source

Amylases can be derived from several sources, including plants, animals and microorganisms. Microbial enzymes generally meet industrial demands (Pandey et al. 2000a). Among bacteria, *Bacillus* sp. is widely used for thermostable α-amylase production to meet industrial needs. *B. subtilis, B. stearothermophilus, B. licheniformis* and *B. amyloliquefaciens* are known to be good producers of α-amylase and these have been widely used for commercial enzyme production for various applications. Similarly, filamentous fungi have been widely used for the production of amylases for centuries. As these moulds are known to be prolific producers of extracellular proteins, they are widely exploited for the production of different enzymes including α-amylase. Fungi belonging to the genus *Aspergillus* have been most commonly employed for the production of α-amylase. Production of enzymes by solid-state fermentation (SSF) using these moulds has turned out to be a cost-effective production technique (Sivaramakrishnan et al. 2006).

As shown in Fig. 1, it is noticeable that there is a preference for fungi in solid state fermentation, principally the *Aspergillus* sp. genus. This happens because, in SSF, the reduced amount of water on the substrate greatly limits the number of microorganisms that are able to adapt to this process. Fungi, however, appear quite tolerant to this environment. Regarding submerged fermentation, there is an increase in the use of bacteria as these are better suited to more moist environments, but the use of fungi still prevails.

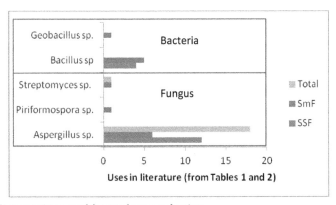

Fig. 1. Microorganisms used for amylases production.
Color image of this figure appears in the color plate section at the end of the book.

SSF vs. SmF

Amylase has been traditionally produced by submerged fermentation due to its facility to control process parameters. However, solid-state fermentation processes have been increasingly utilized for the production of this enzyme lately because it has a number of advantages over SmF, including higher product concentration, lower catabolite repression, and lower operational costs, although, SSF is still not the most widely used large-scale fermentation technology, as shown in reactor types in Tables 1 and 2. This can be attributed to engineering bottlenecks for its scale up. For this reason, several works have been developed to make SSF an economically feasible technology for bioprocessing (Kunamneni et al. 2005, Holker and Lenz 2005, Das et al. 2011).

Based on a comparative study between solid-state fermentation and submerged fermentation conducted on the same substrate (wheat bran), using the same microorganism (*Aspergillus niger*), Varalakshmi et al. (2009) had found better activity values for SSF. It probably happened because wheat bran is a dry substrate and for this reason it is more propitious to be used for SSF. The same author had also compared five different fungi and the higher amylase activity values for both SSF and SmF was achieved for *Aspergilus niger*. Suganthi et al. (2011) had compared SSF and SmF using the same microorganisms (*Aspergillus niger*), but they used potato dextrose as substrate for SmF and different substrates like rice bran, wheat bran, black gram bran, coconut oil cake, groundnut oil cake, and gingely oil cake were used as solid substrates for SSF. This work also showed higher amylase activity for SSF.

Operational Conditions

Regarding fermentation time, it was observed in most of the reviewed papers that it is longer than 72 hours for bacteria and may reach values of up to 144 hours for fungi (Tables 1 and 2). This is mainly due to the fact that fungi requires a longer time than bacteria for multiplying and consequently further development of the fermentation process. From Tables 1 and 2 it was also found that the temperature values required for fermentation were higher for the bacteria than for the fungi, as the fungi have optimum temperature in the range of 25–30°C, whereas the bacteria require temperatures higher than these values.

Hashemi et al. (2012) investigated the effects of temperature shift in α-amylase production by the bacterium *Bacillus* sp. KR-8104 in a SSF system, and found that the biosynthesis of amylase can be improved through a change in the fermentation temperature of 40°C in the initial 20 hours to 32°C for the remaining period of 20–48 hours. This reduction in temperature is particularly attractive from an economic point of view for large scale industrial fermentation.

Enzymatic Activities

Regarding enzymatic activities for both types of fermentation, it can be seen that there is a large difference in the order of magnitude among the values obtained in the different studies (Tables 1 and 2). This is due to the fact that there are different methods for determination of enzymatic activity and, hence, a simple comparison of the numerical values of enzyme activity can only be made among those studies that used the same method. Comparing studies that employed the method based on the hydrolysis of starch to reducing sugars, the highest enzymatic activity obtained in SSF was reported by Gangadharan et al. (2006), wherein after testing fourteen different substrates on different ratios, as well as making an optimization of fermentation time and supplements addition, the author found a value of enzymatic activity of 62,470 U/g, using wheat bran and groundnut oil cake as substrate in 1:1 ratio and using *Bacillus amyloliquefaciens* as the microorganism. On the other hand, employing SmF, the highest enzyme activity was obtained by Rajagopalan and Krishnan (2009), using sugarcane bagasse as substrate and *Bacillus subtilis* as the microorganism in a stirred tank reactor (KLF 2000 R4) gave a value of 1258 U/mL.

Bioreactors

The analysis of bioreactors employed in the great majority of studies published in recent years for solid state fermentation mainly shows the use of Erlenmeyer flasks as reactor, since the control of process parameters is simpler and facilitates process optimization. However, few studies have been developed to make this production feasible on a larger scale. Bhanja et al. (2007) made a comparative study on the solid state fermentation of wheat bran, using microorganisms such as *Aspergillus oryzae*, in two types of bioreactors: the conventional tray reactor and the reactor GROWTEK. Comparing the results, the author concluded that using the reactor GROWTEK, besides increasing the enzymatic activity from 15,460 U/g (found in the conventional tray reactor) to 19,655 U/g, it was possible to decrease the fermentation time from 66 to 54 hours. In addition, Dey and Banerjee (2011) proposed a new type of bioreactor for producing amylases, using the same substrate and microorganisms [Bhanja et al. (2007)]. The process parameters were optimized, and for 2 cm height bed bioreactor with an air flow of 0.1 L/(min.g) operated at 32°C, the best activity obtained was 22,316 U/g. The increase in the enzymatic activity found in this study may possibly be attributed to the high rate of heat transfer due to air flow present in the bioreactor during fermentation.

For submerged fermentation, flasks also have been widely used as reactors. However, it can be noted that some studies used bioreactors with larger workloads. Rajagopalan and Krishnan (2009) conducted a comparison study between data obtained from a 3.5L bioreactor (KLF 2000 R4) and data obtained using flasks. Comparing the results, the authors concluded that the use of the bioreactor provided a 14 times increase (90 U/ml to 1258 U/ml) in the production of α-amylase and a 12 hours reduction in the fermentation time (24 h to 12 h). The bioreactor optimum operation conditions were observed at 37°C temperature, pH 7 and a 700 rpm of agitation. Another work to be highlighted is the one done by Ezeji and Bahl (2007), who performed tests with a 1.2L bioreactor, achieving a maximum activity of 30.19 U/mL during 28 hours of fermentation under optimal conditions (49°C and pH 7.1).

Analysis of Process Feasibility from Screening to Production Cost

Castro et al. (2010a) have evaluated the use of eight different filamentous fungi of *Aspergillus* and *Penicillium* genus on the amylases production by solid-state fermentation using babassu cake as substrate. Amongst all tested fungi, *Aspergillus awamori* IOC-3914 presented the highest endoamylases and exoamylases production, as well as the highest yields on babassu cake hydrolysis, both at high temperatures and warmer temperatures, indicating its potential to be used in the cold hydrolysis of starch. The same authors have also studied the enzyme production by solid-state fermentation using different fungi and four different industrial residues as substrate (babassu, canola, sunflower cake and castor seed residue). The babassu cake was considered as the more appropriate substrate for not only endoamylases production, but also for exoamylases and proteases production. The highest endoamylase activity was also reached using the same microorganism (*A. awamori* IOC-3914), emphasizing the importance of this microorganism in the hydrolysis of starch (Castro et al. 2011a).

In addition to these studies, the authors also carried out an economic analysis of amylases and other hydrolases production by solid-state fermentation using the software SuperProDesigner. They verified that the fermentation time and the propagation of inoculums were the limiting steps in the process. Using the elimination of bottlenecks tools available in the software, the optimal medium of propagation and configuration process was defined. The simulations using the optimized process indicated the need to consider the fermented cake after enzymatic extraction, as co-product for commercialization on animal feed marketing. A competitive unit cost of 10.40 USD kg^{-1} for producing the final product was thus obtained. This is mainly due to the advantages of the presence of accessory hydrolytic enzymes and the ability to hydrolyze starch in mild conditions and this is consistent with cold hydrolysis processes (Castro et al. 2010b).

Another study used a multiresponse approach based on the statistical function of 'desirability' to determine the optimum conditions of inoculum for simultaneous production of amylolytic and proteolytic enzymes by solid-state fermentation. As a result of the statistical analysis, the optimum values for age of the inoculum (28.4 h), C/N ratio of the propagation medium (25.8) and inoculum concentration to be added in the fermentation medium (9, 1 mg g^{-1}) were obtained. These optimal conditions were experimentally validated, resulting in a high enzymatic activity and then confirming the use of the statistical function of 'desirability' as a suitable tool for optimization of enzyme production (Castro et al. 2011b).

Conclusion

In this chapter, we presented and discussed the most recent studies for production of amylases. Generally there is vast scope of applicability of amylases in different areas of industry not only in food and starch industries, in which demand for these enzymes is increasing continuously, but also in various other industries such as paper and pulp, textile, etc. Also, it should be emphasized that amylases production occurs preferentially through the use of filamentous fungi by submerged fermentation at industrial scale. However, solid-state fermentation is being looked at as a potential tool for its production, specially applying agroindustrial starchy residues as substrates, which can soften the residues disposal problem as well as decrease operational costs in the amylases production process. Finally, trend towards a continuous search for technical improvements in amylases production especially by solid-state fermentation is evident, as it presents numerous advantages over submerged fermentation and has been emerging as the future technology in the bioprocesses development.

References

Alva, S., J. Anupama, J. Savla, Y. Y. Chiu, P. Vyshali, M. Shruti, B. S. Yogeetha, D. Bhavya, J. Purvi, K. Ruchi, B. S. Kumudini and K. N. Varalakshmi. 2007. Production and characterization of fungal amylase enzyme isolated from *Aspergillus* sp. JGI 12 in solid state culture. Afr. J. Biotechnol. 6(5): 576–581.

Bhanja, T., A. S. Rout, R. Banerjee and B. C. Bhattacharyya. 2007. Comparative profiles of α-amylase production in conventional tray reactor and GROWTEK bioreactor. Bioproc. Biosys. Eng. 30: 369–376.

Castro, A. M, T. V. Andréa, D. F. Carvalho, M. M. P. Teixeira, L. R. Castilho and D. M. G. Freire. 2011a. Valorization of Residual Agroindustrial Cakes by Fungal Production of Multienzyme Complexes and Their Use in Cold Hydrolysis of Raw Starch. Waste Biomass. Valor. 2: 291–302.

Castro, A. M., T. V. Andréa, L. R. Castilho and D. M. G. Freire. 2010a. Use of Mesophilic Fungal Amylases Produced by Solid-State Fermentation in the Cold Hydrolysis of Raw Babassu Cake Starch. Appl. Biochem. Biotech. 162: 1612–1625.

Castro, A. M., D. F. Carvalho, L. R. Castilho and D. M. G. Freire. 2010b. Economic Analysis of the Production of Amylases and Other Hydrolases by *Aspergillus awamori* in Solid-State Fermentation of Babassu Cake. Enzyme Res. 2010: 1–9.

Castro, A. M., L. R. Castilho, and D. M. G. Freire. 2011a. An overview on advances of amylases production and their use in the production of bioethanol by conventional and non-conventional processes. Biomass Conv. Bioref. 1: 245–255.

Castro, A. M., M. M. P. Teixeira, D. F. Carvalho, D. M. G Freire and L. R. Castilho. 2011b. Multiresponse Optimization of Inoculum Conditions for Amylases and Proteases by *Aspergillus awamori* in Solid-State Fermentation of Babassu Cake. Enzyme Res. 2011: 1–9.

Castro, S. M. and A. M. Castro. 2012. Assessment of the Brazilian potential for the production of enzymes for biofuels from agroindustrial materials. Biomass Conv. Bioref. 2: 87–107.

Couto, S. R. and M. A. Sanromán. 2006. Application of solid-state fermentation to food industry—A review. J. Food Eng. 76: 291–302.

Cruz, E. A, M. C. Melo, N. B. Santana, M. Franco, R. S. M. Santana, L. S. Santos and Z. S. Gonçalves. 2011. Alpha-Amylase Production by *Aspergillus niger* in Cassava Peels Waste. UNOPAR Cient. Ciênc. Biol. Saúde. 13(4): 245–9.

Das, S., S. Singh, V. Sharma and M. L. Soni. 2011. Biotechnological applications of industrially important amylase enzyme. Int. J. Pharm. Bio. Sci. 2(1): 486–496.

Dey, T. B. and R. Banerjee. 2011. Hyperactive a-amylase production by *Aspergillus oryzae* IFO 30103 in a new bioreactor. Lett. Appl. Microbiol. 54: 102–107.

Ezeji, T. C and H. Bahl. 2007. Production of raw-starch-hydrolysing α-amylase from the newly isolated Geobacillus thermodenitrificans HRO10. World J. Microb. Biot. 23: 1311–1315.

Gangadharan, D., S. Sivaramakrishnan, K. M. Nampoothiri and A. Pandey. 2006. Solid Culturing of Bacillus amyloliquefaciens for Alpha Amylase Production. Food Technol. Biotech. 44(2): 269–274.

Gangadharan, D., S. Sivaramakrishnan, K. M. Nampoothiri, R. K. Sukumaran and A. Pandey. 2008. Response surface methodology for the optimization of alpha amylase production by Bacillus amyloliquefaciens. Bioresource Technol. 99: 4597–4602.

Gupta, R., P. Gigras, H. Mohapatra, V. K. Goswami and B. Chauhan. 2003. Microbial α-amylases: a biotechnological perspective. Process Biochem. 38: 1599–1616.

Hashemi, M., S. H. Razavi, S. A. Shojaosadati, S. M. Mousavi, K. Khajeh and M. Safari. 2010. Development of a solid-state fermentation process for production of an alpha amylase with potentially interesting properties. J. Biosci. Bioeng. 110(3): 333–337.

Hashemi, M., S. A. Shojaosadati, S. H. Razavi, S. M. Mousavi, K. Khajeh and M. Safari. 2012. The Efficiency of Temperature-Shift Strategy to Improve the Production of α-Amylase by *Bacillus* sp. in a Solid-State Fermentation System. Food Bioprocess Tech. 5: 1093–1099.

Holker, U. and J. Lenz. 2005. Solid-state fermentation—are there any biotechnological advantages? Curr. Opin. Microbiol. 8(3): 301–306.

Kar, S., R. C. Ray and U. B. Mohapatra. 2010. Purification, characterization and application of thermostable amylopullulanase from *Streptomyces erumpens* MTCC 7317 under submerged fermentation. Ann. Microbiol. 62: 931–937.

Khan, J. A. and S. K. Yadav. 2011. Production of alpha amylases by *Aspergillus niger* using cheaper substrates employing solid state fermentation. Int. J. Plant Anim. Environ. Sci. 1(3): 100–108.

Kumar, V., V. Sahai and V. S. Bisaria. 2012. Production of amylase and chlamydospores by *Piriformospora indica*, a root endophytic fungus. Biocatal. Agric. Biotechnol. 1: 124–128.

Kunamneni, A., K. Permaul and S. Singh. 2005. Amylase Production in Solid-State Fermentation by the Thermophilic Fungus Thermomyces lanuginosus. J. Biosci. Bioeng. 100(2): 168–171.

Maryam, H., S. H. Razavi, S. A. Shojaosadati and S. M. Mousavi. 2011. The potential of brewer's spent grain to improve the production of a-amylase by *Bacillus* sp. KR-8104 in submerged fermentation system. New Biotechnol. 28(2): 165–172.

Pandey, A., P. Nigam, C. R. Soccol, V. T. Soccol, D. Singh and R. Mohan. 2000a. Review: Advances in microbial amylases. Biotechnol. Appl. Bioc. 31: 135–152.

Pandey, A., C. R. Soccol and D. Mitchell. 2000b. New developments in solid state fermentation: I-bioprocesses and products. Process Biochem. 35: 1153–1169.

Rajagopalan, G. and C. Krishnan. 2008. α-Amylase production from catabolite derepressed *Bacillus subtilis* KCC103 utilizing sugarcane bagasse hydrolysate. Bioresource Technol. 99: 3044–3050.

Rajagopalan, G. and C. Krishnan. 2009. Optimization of agro-residual medium for α-amylase production from a hyper-producing *Bacillus subtilis* KCC103 in submerged fermentation. J. Chem. Technol. Biotechnol. 84: 618–625.

Ray, R. C., S. Kar, S. Nayak, and M. R. Swain. 2008. Extracellular α-Amylase Production by *Bacillus brevis* MTCC 7521. Food Biotechnol. 22: 234–246.

Ray, R. C, S. Mohapatra., S. Panda and S. Kar. 2008. Solid substrate fermentation of cassava fibrous residue for production of α-amylase, lactic acid and ethanol. J. Environ. Biol. 29(1): 111–115.

Shoji, H., T. Sugimoto, K. Hosoi, K. Shibata, M. Tanabe and K. Kawatsura. 2007. Simultaneous production of glucoamylase and acid-stable α-amylase using novel submerged culture of *Aspergillus kawachii* NBRC4308. J. Biosci. Bioeng. 103(2): 203–205.

Silva, T. M., R. F. Alarcon, A. R. L. Damasio, M. Michelin, A. Maller, D. C. Masui, H. F. Terenzi, J. A. Jorge and M. L. T. M. Polizeli. 2009. Use of Cassava Peel as Carbon Source for Production of Amylolytic Enzymes by *Aspergillus niveus*. Int. J. Food Eng. 5(5): 1–11.

Sivaramakrishnan, S., D. Gangadharan, K. M. Nampoothiri, C. R. Soccol and A. Pandey. 2006. α-Amylases from Microbial Sources—An Overview on Recent Developments. Food Technol. Biotech. 44(2): 173–184.

Sivaramakrishnan, S., D. Gangadharan, K. M. Nampoothiri, C. R. Soccol and A. Pandey. 2007. Alpha-amylase production by *Aspergillus oryzae* employing Solid-State Fermentation. J. Sci. Ind. Res. India. 66: 621–626.

Suganthi, R., J. F. Benazir, R. Santhi, V. Ramesh Kumar, A. Hari, N. Meenakshi, K. A. Nidhiya, G. Kavitha and R. Lakshmi. 2011. Amylase Production by *Aspergillus niger* under solid state fermentation using agroindustrial wastes. Int. J. Eng. Sci. Technol. 3(2): 1756–1763.

Sugimoto, T., T. Makita, K. Wantanabe and H. Shoji. 2011. Production of multiple extracellular enzyme activities by novel submerged culture of *Aspergillus kawachii* for ethanol production from raw cassava flour. J. Ind. Microbiol. Biotechnol. 39: 605–612.

Varalakshmi, K. N., B. S. Kumudini, B. N. Nandini, J. Solomon, R. Suhas, B. Mahesh and A. P. Kavitha. 2009. Production and characterization of α-amylase from *Aspergillus niger* JGI 24 isolated in Bangalore. Pol. J. Microbiol. 58(1): 29–36.

5

Recent Developments and Industrial Perspectives in the Microbial Production of Bioflavors

Gustavo Molina,[1], Juliano L. Bicas,[2] Érica A. Moraes,[3] Mário R. Maróstica-JR[3] and Gláucia M. Pastore[1]*

ABSTRACT

Nowadays the use of microbial and enzymatic processes for the production of food ingredients is increasingly being developed and biotechnology offers unique opportunities to produce natural food ingredients. Thus, in this chapter the main aspects involving the properties and biotechnological production methods of food additives, focusing on bioflavors, have been discussed. This chapter is also intended to cover all aspects of flavor compounds formation by microorganisms and plant cells focusing on recent progress and highlighting the most relevant developments in this area. Representative

[1] Laboratory of Bioflavors and Bioactive Compounds, Department of Food Science, Faculty of Food Engineering, State University of Campinas, R. Monteiro Lobato, 80, CEP: 13083-862, Campinas–São Paulo, Brazil.
[2] Department of Chemistry, Biotechnology and Bioprocess Engineering, Campus Alto Paraopeba, University of São João Del-Rei. Cx. Postal 131, CEP: 36420-000, Ouro Branco–Minas Gerais, Brazil.
[3] Laboratory of Nutrition and Metabolism, Department of Food and Nutrition, Faculty of Food Engineering, State University of Campinas, R. Monteiro Lobato, 80, CEP: 13083-862, Campinas–São Paulo, Brazil.
* Corresponding author: gustavomolinagm@gmail. com

data are presented to demonstrate that a wide variety of compounds possessing unique and potent flavor properties have been isolated as naturally occurring constituents of biocatalysts.

Introduction

Flavors and fragrances have a wide application in the food, feed, cosmetic, chemical and pharmaceutical sectors (Vandamme 2003). These compounds greatly influence the flavor of food products and govern their acceptance by consumers and their market success. The increasing consumer preference for natural products has encouraged remarkable efforts towards the development of biotechnological processes for the production of flavor compounds (Bicas et al. 2009).

These compounds present a wide variety of chemical functions: there are hydrocarbons, alcohols, aldehydes, ketones, acids, esters or lactones that are present in low concentrations in the composition of food products (Bicas et al. 2010a). This directly reflects the great diversity of possible applications of such compounds as food additives.

The methods for obtaining aroma compounds include the direct extraction from nature, chemical transformations and biotechnological transformations (which include microbial and enzymatic biotransformations, *de novo* synthesis and the use of genetic engineering tools) (Franco 2004).

Most of the natural compounds currently processed by the flavor and fragrance industry are obtained by extraction or distillation of parts of field-grown plants (Berger et al. 2010), although many disadvantages are encountered, such as (i) low concentrations of the product of interest, which increases the extraction and purification procedures, (ii) dependency on seasonal, climatic and political features, and (iii) possible ecological problems involved with the extraction (Bicas et al. 2009).

Despite the great industrial application of aroma compounds produced *via* chemical synthesis (still responsible for a large portion of the market due to the satisfactory yields), this strategy is also associated with a number of environmental challenges and hardly presents adequate regio- and enantio-selectivity to the substrate, resulting in a mixture of molecules. In addition, the increasing interest for "natural" labeled products has led to intense research on the microbial production of the so-called 'bioflavors' (Krings and Berger 1998, Berger 1995).

In this sense, there is a growing demand for biotechnology to provide alternatives for the production of natural flavorings and fragrances (Margetts 2005). Thus, biotechnology offers the potential to produce natural aroma compounds in a commercial scale with many advantages over the chemical processes: the reactions occur at mild conditions, present high regio- and enantio-selectivity and do not generate toxic wastes. Additionally, there are some compounds that may be produced exclusively *via* biotechnology (Demyttenaere 2001).

Although microorganisms have been used to produce flavors for a long time, especially when considering the preparation of traditional fermented foods and beverages, only recently the relationship between microbial development and the typical desirable flavor of fermented foodstuff was recognized. Further analysis and optimization of such food fermentations led to the investigation of pure microbial strains and their capacity to produce specific single flavor molecules, either by *de novo* synthesis or by converting an added substrate/precursor molecule (Vandamme 2003). Thus, this chapter is intended to cover all aspects of flavor compounds formation from fungi and bacteria, focusing on recent progress and highlighting the most relevant developments in this area. This chapter is intended to demonstrate important advances with micro-organisms and plant cell that are able to transform flavor precursors directly into flavor molecules (biotransformation) or produce flavors along multi-step processes (bioconversion and *de novo* synthesis). Representative data are presented to demonstrate that a wide variety of compounds possessing unique and potent flavor properties have been isolated as naturally occurring constituents of biocatalysts, such as fungi and bacteria.

Bioflavors: Microorganisms, Process and Production

It has been known for a long time that some microorganisms can generate pleasant odors (Feron and Waché 2006). The use of biocatalysts such as microorganisms has led to a great advance in the industrial research of flavor compounds and a wide variety of compounds obtained, thus the biotechnological generation of natural aroma compounds is rapidly expanding. The biotechnological production of these compounds can be achieved by single-step biotransformations, bioconversions and *de novo* synthesis using microorganisms, plant cells or isolated enzymes (Krings and Berger 1998).

De novo synthesis comprises the production of complex substances from simple molecules through complex metabolic pathways (Bicas et al. 2010a) and the production of chemically quite different volatile flavors, such as short-chain alcohols, esters aldehydes, ketones, methylketones and acids as well as pyrazines and lactones that could be formed concurrently (Krings and Berger 1998). Some examples of bioflavors obtained by *de novo* synthesis are summarized in Table 1 and their structure presented in Fig. 1.

Biotransformations or bioconversions consist of, respectively, a single or few reactions catalyzed enzymatically (which includes the use of whole cells) to result in a product structurally similar to the substrate molecule (Feron and Waché 2006). In some cases, the precursors employed can be

Table 1. Example of the production of bioflavors obtained by *de novo* synthesis.

Microorganism	Culture medium	Products	References
Fungi			
Neurospora sp.	Malt extract	Ethyl hexanoate	Pastore et al. (1994)
Saccharomyces cerevisiae	Malt extract	Isoamyl acetate	Kobayashi et al. (2008)
Ceratocystis fimbriata	Coffee husk	Ethyl acetate	Soares et al. (2000)
Geotrichum fragrans	Cassava liquid	2-Phenylethanol	Damasceno et al. (2003)
Kluyveromyces marxianus	Cassava bagasse	Butyl acetate	Medeiros et al. (2000)
Bacteria			
Bacillus cereus	Plate count agar	2,5-Dimethylpyrazine, Trimethylpyrazine	Adams and De Kimpe (2007)
Bacillus sp. RX3-17	Glucose	2,3,5,6-Tetramethylpyrazine	Xiao et al. (2006)
Corynebacterium glutamicum	Amino acids	Tetramethylpyrazine	Smith et al. (2010)
Pseudomonas fragi	Skim milk	Ethyl butyrate	Cormier et al. (1991)
Lactococcus lactis	Starch	Butyric acid, Diacetyl	Escamilla-Hurtado et al. (2000)

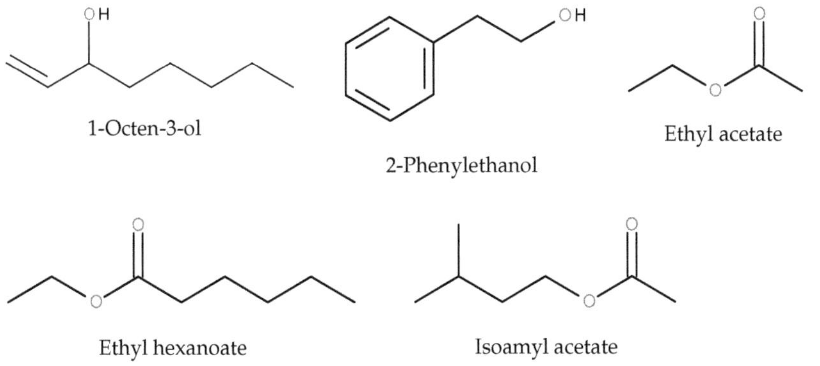

1-Octen-3-ol

2-Phenylethanol

Ethyl acetate

Ethyl hexanoate

Isoamyl acetate

Fig. 1. Structure of some bioflavors obtained by *de novo* synthesis.

considered inexpensive and readily available, such as fatty or amino acids, and can be converted to more highly valued flavors (Krings and Berger 1998). Thus, these approaches may offer more economic advantages, since high yields may be achieved (Feron and Waché 2006).

Although for a multitude of microorganisms the metabolic potential for *de novo* flavor biosynthesis is immense and a wide variety of valuable products can be detected in microbial culture media or their headspaces, the concentrations found in nature are usually too low for commercial applications. Therefore, the biocatalytic conversion of a structurally related precursor molecule is often a more adequate strategy which allows significantly enhanced accumulation of a desired flavor product. As a prerequisite for this strategy, the precursor must be present in nature and its isolation in sufficient amounts from the natural source must be easily feasible in an economically viable fashion (e.g., the monoterpenes limonene and α-pinene).

Among the most targeted substrates for biotransformation/ bioconversion approaches are the terpenes. These compounds are the most abundant group in nature responsible for the characteristic odors of essential oils and are a preferable substrate for bioconversion studies (Janssens et al. 1992). The biotransformation of terpenes is of interest because it allows the production of enentiomerically pure flavors and fragrances under mild reaction conditions. Products produced by biotransformation processes may be considered as 'natural' (De Carvalho and Da Fonseca 2006). The use of microorganisms in monoterpene biotransformation is relatively recent, dating from the late 1950s and mid 1960s (Bicas et al. 2009) and since then the biotransformation of these compounds have been published over the years using different biocatalysts and substrates, and some examples include enzymes, cell extracts and whole cells of bacteria, cyanobacteria, yeasts, microalgae, fungi and plants (De Carvalho and Da Fonseca 2006). Some examples of substrates and products obtained by microbial biotransformation are summarized in Table 2 and Fig. 2.

This chapter attempts to review some recent developments and industrial perspectives in the microbial production of bioflavors, including the most recognized microorganisms used for the production of natural flavors by *de novo* synthesis or by biotransformations.

Fungal production of natural flavor compounds

Several fungal strains are related to the production of natural flavor compounds, mainly due to the large biodiversity that occurs especially in the ascomycetes and basidiomycetes orders (Feron and Waché 2006).

Several filamentous fungi are able to produce novel odorous compounds by *de novo* synthesis such as floral flavors by *Ceratocystis* sp. and *Trichoderma viride* (Vandamme 2003). Different aroma products have also been produced by biotransformation using fungi as biocatalysts, such as vanillin, important terpenes-derivatives and others. Also, precursors as fatty acids and PUFA's can be converted by fungi (such as *Penicillum* sp. and *Botryodiplodia* sp.) to provide flavor compounds with "green notes", mushroom flavor, fruity lactones and cheese-flavored methylketones.

Yeasts, unicellular non-filamentous fungi, are eukaryotic cells that show a broad range of volatile metabolites, such as esters, lactones, aldehydes and

Table 2. Example of microbial biotransformations for the production of bioflavors.

Microorganism	Substrate	Products	References
Fungi			
Penicillium digitatum	R-(+)-Limonene	R-(+)-α-Terpineol	Adams et al. (2003)
Penicillium solitum	α-Pinene	Verbenone	Pescheck et al. (2009)
Aspergillus niger	Citronellol	Rose oxide	Demyttenaere et al. (2004)
Fusarium oxysporum 152B	R-(+)-Limonene	R-(+)-α-Terpineol	Bicas et al. (2008a)
Pleurotus sapidus	Valencene	Nootkatone	Fraatz et al. (2009)
Bacteria			
Bacillus pumilus	Isoeugenol	Vanillin	Hua et al. (2007)
Pseudomonas fluorescens	Limonene	α-Terpineol	Bicas et al. (2010c)
Pseudomonas putida	α-Pinene	Verbenol	Divyashree et al. (2006)
Rhodococcus erythropolis MLT1	β-Myrcene	Geraniol	Thompson et al. (2010)
Rhodococcus opacus	Limonene	Carvone	De Carvalho and Da Fonseca (2006)
Microalgae			
Chlorella sp.	Limonene	Carveol	Rasoul-Amini et al. (2011)
Oocystis pusilla	Limonene	Carveol, Carvone	Ghasemi et al. (2009)
Synechococcus sp. PCC 7942	S-(–)-Limonene	Carveol	Hamada et al. (2003)
Chlorella sp.	Valencene	Nootkatone	Furusawa et al. (2005b)

phenolics (Debourg 2000). Volatile flavors from all chemical classes were also found in basidiomycete fruit bodies and cell cultures, with a particular emphasis on volatile phenylpropanoic and phenolic compounds (Berger and Zorn 2004, Wu et al. 2007, Berger et al. 2010). Moreover, the variety of products obtained can be further increased by the modification of simple constituents of the culture medium as carbon or nitrogen source (Feron and Waché 2006).

Thus, these microorganisms represent interesting biocatalysts for the microbial production of flavor compounds. Below, there are some examples described in detail covering the most important genus of fungi for the production of bioflavors through *de novo* synthesis or biotransformation/bioconversion processes.

Penicillium sp.

Most of the studies involving the production of aroma compounds by genus are associated with biotransformation processes. Nevertheless, the production of interesting flavors also occurs *via de novo* synthesis. One example involves the strain *Penicillium camemberti*, which could grow

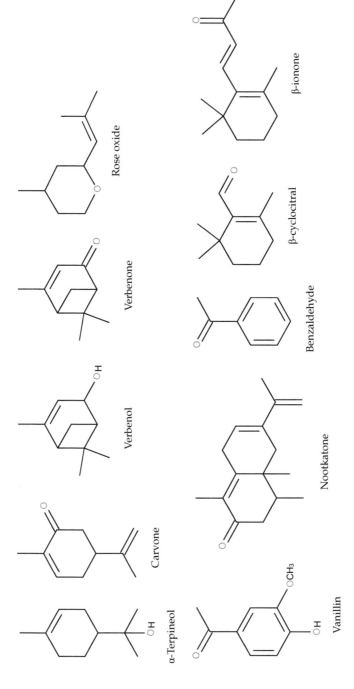

Fig. 2. Example of bioflavors obtained by microbial biotransformation processes.

on medium containing glucose, some salts and linoleic acid, producing 1-octen-3-ol after 5 days of fermentation. This compound is used in lavender compositions and in mushroom aromas due to its powerful, sweet, earthy odor with a strong and herbaceous note (Bauer et al. 2001). Another well known example occurs during blue cheese preparation, whose typical flavor is attributed to methyl ketones formed by the degradation of fatty acids by *Penicillium roqueforti* (Demyttenaere et al. 1996).

Meanwhile, many studies involving the biotransformation of terpenes by *Penicillium* sp. were published. Some examples include the use of five strains of *Penicillium digitatum* for the bioconversion of R-(+)-limonene in R-(+)-α-terpineol with a high enantioselectivity (ee>99%) (Adams et al. 2003). The product obtained is one of the most frequently used and inexpensive fragrance sustances, commonly applied in cosmetics and household products (Bicas et al. 2009).

Another *Penicillium* sp. strain was applied to biotransform citronellol in an unusual culture medium, cassava wastewater (manipueira), for the production of *cis*- and *trans*-rose oxides (Maróstica Jr. and Pastore 2006). Fungal spores of *P. digitatum* were able to biotransform geraniol, nerol, citrol (mixture of the alcohols nerol and geraniol) and citral (mixture of the aldehydes neral and geranial) to 6-methyl-5-hepten-2-one (Demyttenaere et al. 2000). Another study concerning the biotransformation of nerol by *Penicillium* sp. and microbial transformation of citral by sporulated surface cultures method (SSCM) of *P. digitatum* has also been reported (Esmaeili and Tavassoli 2010).

Recently, Pescheck et al. (2009) studied improved fungal biotransformation of monoterpenes avoiding the volatilization of the precursor in a closed gas loop bioreactor. *Penicillium solitum*, isolated from kiwi, was capable to convert α-pinene into a range of products including verbenone, a valuable aroma compound, reaching 35 mg.L^{-1}. Furthermore, the group evaluated the capacity of the strain *P. digitatum* DSM 62840 that converted (+)-limonene into the aroma compound α-terpineol and due to the beneficial effects of the system used, the concentration of the product reached 1,009 mg.L^{-1} and an average productivity of 8–9 mg.L^{-1}.h^{-1}, which represents a doubling of the respective values reported in a previous work (Pescheck et al. 2009).

Aspergillus sp.

To our knowledge and through the review performed, it seems that no recent work in the area deals with this genus in production of natural flavors through *de novo* synthesis, however these microorganisms are involved in many studies of biotransformation of terpenes leading to the production of flavor compounds and currently *A. niger* is one of the most extensively studied fungal species involved in monoterpene biotransformation (Bicas et al. 2009).

Some microorganisms identified as *Aspergillus niger* were used in hydroxylation reactions with terpene substrates (De Oliveira et al. 1999, Hashimoto et al. 2001), including the production of natural carveol (Divyashree et al. 2006) and perillyl alcohol (Menéndez et al. 2002) as the major products using limonene as the substrate.

In biotransformation processes of α-pinene, another important monoterpene due to its abundance and availability, verbenone is formed as a major product when using *A. niger* (Agrawal and Joseph 2000). Another *A. niger* strain performed the biotransformation of citronellol into rose oxide (Demyttenaere et al. 2004).

In addition to the most commonly studied monoterpenes, the oxifunctionalization of α-farnesene was studied for the production of natural flavor compounds by different isolates. The major oxidation product in all transforming cultures, 3,7,11-trimethyldodeca-1,3(E),5(E)10-tetraen-7-ol, showed a pleasant citrus-like odor and peak concentrations of 170 mg.L^{-1}. An *A. niger* isolated from mango could generate natural terpene alcohol with an apricot-like odor (Krings et al. 2006).

Specifically in the research field of biotransformation of terpenes, the importance of these two genera presented so far, *Aspergillus* sp. and *Penicillium* sp., can be noticed by studies that show the use of both for the production of natural flavor compounds, such as those presented by Demyttenaere et al. (2000) and Agrawal et al. (1999) when using these strains.

In addition, Rottava et al. (2010) conducted a screening of 405 microorganisms and tested their ability to bioconvert the substrates R-(+)-limonene and (–)-β-pinene for bioflavor production. Both precursors were biotransformed into α-terpineol. The highest concentration of α-terpineol, about 3,450 mg.L^{-1}, was obtained for the biotransformation of R-(+)-limonene by a *Penicillium* sp. isolated from eucalyptus steam. When using (–)-β-pinene as substrate, the highest product concentration, 675.5 mg.L^{-1}, was achieved by an *Aspergillus* sp. isolated from orange tree stem. Furthermore, the last process was optimized using the methodology of experimental design and the production achieved 770 mg.L^{-1} when (–)-β-pinene was used as substrate (Rottava et al. 2011).

Biotransformation of menthol by sporulated surface culture of *A. niger* and *Penicillium* sp. was studied. The main bioconversion product obtained from menthol of *A. niger* was cis-p-menthan-7-ol and the main products obtained by surface *Penicillium* sp. were limonene, p-cymene and γ-terpinene using sporulated surface culture (Esmaeili et al. 2009).

Microbial biotransformations of α-pinene into verbenol as the major product have also been reported. The microorganisms capable of doing such conversion were a hybrid (protoplast fusion) from *A. niger* and *P. digitatum* (Rao et al. 2003).

Fusarium sp.

The ability of the *Fusarium oxysporum* 152B strain to bioconvert R-(+)-limonene has been previously elucidated by some researchers. Initially, the production of α-terpineol from limonene was studied using cassava wastewater as culture media for the strain growth and orange essential oil as the sole carbon and energy source. The authors found that the biotransformation of R-(+)-limonene resulted in 450 mg.L⁻¹ of R-(+)-α-terpineol after 3 days of reaction (Maróstica Jr. and Pastore 2007). In the sequence, Bicas et al. (2008a) optimized the process conditions for the biotransformation of R-(+)-limonene to R-(+)-α-terpineol by this strain using response surface methodology. After extensive study to evaluate process parameters, the authors found a significant increase in the production of R-(+)-α-terpineol reaching up to 2.4 g.L⁻¹ after 72 h cultivation at 26°C/240 rpm. The last process also occurred for a biomass resulting from a lipase production process, which showed that the co-production of this enzyme and R-(+)-α-terpineol was feasible (Bicas et al. 2010b).

Others

The screening of several white-rot fungi has shown that numerous species are able to synthesize benzaldehyde through *de novo*, such as *Pleurotus sapidus*, *Polyporus* sp., and others (Lomascolo et al. 1999). Berger et al. (1987) studied the formation of methoxy benzaldehyde in *Ischnoderma benzoicum*. Natural benzaldehyde is used as an ingredient in cherry and other natural fruit flavors and has a market of approximately 20 tons per year and a price of approximately 240 €/kg. Benzaldehyde obtained from natural cinnamaldehyde can be purchased for 100 €/kg with an estimated market of more than 100 tons per year (Feron and Waché 2006).

The basidiomycete *Nidula niveo-tomentosa* synthesized traces of raspberry ketone and its corresponding *de novo* alcohol starting from simple nutrients, such as glucose and amino acids. A systematic attempt was made to improve the productivity of this fungus in submerged culture (Böker et al. 2001). Raspberry ketone, 4-(4-hydroxyphenyl)butan-2-one, is one of the character-impact components of raspberry flavor.

Several fungal strains belonging to *Neurospora* genus have been described for the production of flavor compounds through *de novo* synthesis. Brigido (2000) investigated the production of ethyl acetate and ethyl butyrate, depending on the culture medium, by a *Neurospora* sp. Pastore and co-workers (1994) obtained 59 mg.L^{-1} of ethyl hexanoate using malt extract broth as culture medium after 3–4 days of fermentation while Yamauchi et al. (1989) using pregelatinized rice impregnated with 5% malt broth as solid culture medium, obtained approximately 180 mg.L^{-1} of the same ester. The *de novo* bioproduction of isoamyl alcohol was also reported by this fungi (Kobayashi et al. 2008, Brown and Hammond 2003).

The microorganisms *Ceratocystis fimbriata* (Soares et al. 2000) and *C. moniliformis* (Bluemke and Schrader 2001) were described as producers of ethyl acetate by *de novo* synthesis. This ester has a fruity smelling liquid with a brandy note, and is the most common ester in fruits (Bauer et al. 2001). Also, different strains of *Geotrichum fragans* are related to the production of ethyl acetate by the *de novo* synthesis, an important ester for the organoleptic characteristics of distillates (Damasceno et al. 2003). Moreover, *Geotrichum fragans* is also responsible for the production of 2-phenylethanol (Damasceno et al. 2003).

Aleu and Collado (2001) reviewed the biotransformations carried out by *Botrytis* sp. Some of these transformations use grape terpenoids as substrate to produce volatile substances important as a distinctive aroma in wine.

The synthesis of nootkatone is performed by oxidation of valencene (Surburg and Panten 2006, Burdock and Fenaroli 2010). This might be carried out by microorganisms or their enzymes in biotransformation processes, but one of the highest yields was described for the basidiomycete *Pleurotus sapidus* (Fraatz et al. 2009). The biotransformation of valencene has also been reported for microorganisms such as *Botryosphaeria dothidea* and *Fusarium culmorum* to afford structurally interesting metabolites (Furusawa et al. 2005a).

Corynespora cassiicola DSM 62485 was identified as a novel highly stereoselective linalool transforming biocatalyst showing the highest productivity reported so far, reaching 120 mg.L^{-1} of linalool oxides with a conversion yield close to 100%. In the same screening, among 19 fungi isolated, *Aspergillus niger* DSM 821, *Botrytis cinerea* 5901/02, and *B. cinerea* 02/FBII/2.1 produced different isomers of lilac aldehyde and lilac alcohol from linalool (Mirata et al. 2008).

Submerged cultures of *Pleurotus ostreatus*, when supplemented with β-myrcene produced perilene, a furanoid monoterpene with a unique flowery citrus-like flavor. Also, myrcene diols were formed from the cleavage of several myrcene epoxides identified as the immediate reaction products (Krings et al. 2008).

The biotransformation of R-(+)-limonene was also investigated by Trytek et al. (2009) with the psychrotrophic fungus *Mortierella minutissima*. Perillyl alcohol and perillyl aldehyde were the main products obtained and the authors improved the biotransformation rate in H$_2$O$_2$-oxygenated culture.

An interesting alternative to generate flavor compounds was via the fungal conversion of larger terpene molecules into volatile breakdown products. In this sense, the bio-production of volatile compounds derived from carotenoids is gaining more attention from the flavor and fragrance industries, as it represents a feasible alternative to the chemical synthesis and offers the production of enantiomerically pure molecules. To date, biotransformation of carotenoids has been reported using integer cells (plant-cultured cells, fungi, bacteria and yeasts) and pure enzymes (see Bicas et al. 2009 for references). Moreover, the use of integer cells were evaluated by Zorn et al. (2003), using filamentous fungal to cleave β-carotene into flavor compounds. The products obtained were β-ionone, β-cyclocitral, dihydroactinidiolide and 2-hydroxy-2,6,6-trimethylcyclohexanone. Several of these fungal enzymes, all belonging to the large peroxidase family, were isolated from the basidiomycete source, sequenced, cloned and one of them is commercially available (De Boer et al. 2005).

Another example includes a mixed culture formed by *Bacillus* sp. and *Geotrichum* sp. that produced tobacco aroma compounds from lutein after formation of the intermediate β-ionone (Sanchez-Contreras et al. 2000, Maldonado-Robledo et al. 2003).

Yeasts

Yeasts play an important role in the formation of flavors of several fermented beverages. For example, over recent decades research into the role of yeast in the development of wine flavor has revealed complex interactions between this microbe and grape compounds, wherein these interactions contribute to the aroma and flavor of wine (Ugliano and Henschke 2009).

Besides the wide range of applications of yeasts, some have been used for the production of natural flavors through *de novo* synthesis, such as *Kluyveromyces marxianus* and *Saccharomyces cerevisiae*. Several strains of *Saccharomyces cerevisiae* are related to the production of esters, such as ethyl acetate (Brown and Hammond 2003, Carrau et al. 2010, Kobayashi et al. 2008, Sumby et al. 2010), ethyl hexanoate (Kobayashi et al. 2008, Sumby et al. 2010) and isoamyl acetate (Brown and Hammond 2003, Kobayashi et al. 2008) by *de novo* synthesis.

The biotechnological production of sulfur compounds, such as methional and 2-methyltetrahydrothiophen-3-one, 2-mercaptoethanol, *cis*-2-methyltetrahydro-thiophen-3-ol, *trans*-2-methyltetrahydro-thiophen-3-ol and 3-mercapto-1-propanol, was also observed for *S. cerevisiae* (Moreira et al. 2008). At optimal concentrations in wine, these compounds impart flavors of passion fruit, grapefruit, gooseberry, blackcurrant, lychee, guava and box hedge (Swiegers et al. 2008).

Kluyveromyces marxianus was reported as a producer of butyl acetate, which has a strong fruity odor and taste reminiscent of pineapple. This compound occurs in many fruits and is a constituent of apple (Bauer et al. 2001) but could be produced by *K. marxianus* in cassava bagasse (Medeiros et al. 2000). The production of ethyl butyrate and ethyl acetate by this microorganism was also evidenced depending on the culture medium (Dragone et al. 2009, Medeiros et al. 2000, Medeiros et al. 2001).

Another yeast, *Sporidiobolus salmonicolor*, was capable of producing an intense peach and fruity odor on standard medium through *de novo* synthesis (Vandamme 2003). Meanwhile, the same yeast and *Yarrowia lipolytica* are related to a well-established industrial process for the production of bioflavor, where yields over 10 g.L^{-1} of γ-decalactone can be obtained by the bioconversion of ricinoleic acid (12-hydroxy-C18:1), using castor oil as a source of that fatty acid (Vandamme 2003).

Many microorganisms, especially yeasts, are capable of producing 2-phenylethanol by normal metabolism (*de novo* synthesis), but the final concentration of the flavor in the culture broth generally remains very low (Etschmann et al. 2002). However, a concentration as high as 453.1 mg.L^{-1} after 16 hours of fermentation by *Pichia fermentans* was reported by Huang et al. (2001). Phenylethyl alcohol (2-phenylethanol) has a characteristic rose-like odor and an initially slightly bitter taste, then sweet and reminiscent of peach and it is the main component of rose oils obtained from rose blossoms (Burdock 2010). It is used frequently and in large amounts in perfumes, cosmetics and food (Bauer et al. 2001, Etschmann et al. 2002).

Interestingly, there are only a few descriptions of yeast-mediated terpene biotransformation processes. Van Dyk et al. (1998) obtained pinocamphone from (−)-β-pinene using *Hormonema* sp., while the yeast *Candida tropicalis* MTCC 230 has shown its capacity to oxidize α-pinene to α-terpineol with an overall yield of 77% after 96 h at 30°C, when 0.5 g.L^{-1} of substrate was used. The product concentration remained stable up to 120 h of reaction time (Chatterjee and Bhattacharyya 1999).

Recently, Ponzoni et al. (2008) screened yeasts of the genera *Debaryomyces*, *Kluyveromyces* and *Pichia*, for their ability to biotransform the acyclic monoterpenes geraniol and nerol. Some of the results showed that linalool was the main product for the biotransformation of geraniol, whereas linalool and α-terpineol were the main products obtained by the conversion of nerol.

Bacteria as Biocatalysts for the Production of Natural Flavor Compounds

Bacteria have historically played an important role in the elaboration of flavor components of many different foods (e.g., wine, vinegar, beer and many others). Bacterial pathways to volatile flavors are often based on strong hydrolytic properties. Through *de novo* synthesis bacteria can provide a wide and important variety of volatile to the profile of many products. Also, the microbial transformation of terpenes has received considerable attention and many bacteria strains are able to break down these substrates or carry out specific conversions, thus creating products with an added value (Longo and Sanromán 2006). This can be demonstrated since nearly two-thirds of the manuscripts published on production and/or biotransformation of terpenes in the last decade used either bacteria or fungi as biocatalysts (De Carvalho and Da Fonseca 2006).

Bacillus sp.

Through *de novo* synthesis, few but important flavor compounds were achieved using *Bacillus* sp. strains, such as pyrazines. These compounds are related to musty, fermented, coffee odor and are present in several cheeses (swiss, camembert, gruyère), roasted peanuts, soy products, beans, mushroom, cocoa and coffee products. Pyrazines have aroma detection varying from 1 to 10 ppm and present great industrial importance in beverages, baked goods, meat products and frozen dairy products (Burdock 2010). Several pyrazines such as 2,5-dimethylpyrazine (characteristic of earthy and potato-like odor), 2,6-dimethylpyrazine (nutty and coffee-like odor) or trimethylpyrazine were produced by strains of *Bacillus cereus* (Adams and De Kimpe 2007). The latter has a baked potato or roasted nut aroma and a great commercial importance (Burdock 2010). The concentrations of these pyrazines did not reach more than 4 mg.L^{-1} and the production depended on the temperature and the culture medium (Demyttenaere et al. 2001, Feron and Waché 2006).

Another pyrazine, 2,3,5,6-Tetramethylpyrazine, a flavor compound which presents a musty, fermented, coffee odor, was produced from glucose by a new isolated *Bacillus* mutant designated as *Bacillus* sp. RX3-17 (Burdock 2010, Xiao et al. 2006). The production of 2,3,5,6-tetramethylpyrazine achieved 4.33 g.L^{-1} after 64.6 h of cultivation in a 5-L fermenter, using the optimized medium which contained 20% glucose, 5% soytone, 3% $(NH_4)_2HPO_4$, and vitamin supplements. Fermentations were carried out with stirring at 700 rpm, air flow at 1.0 vvm, pH at 7.0, and 37°C. The product purity was 99.88% and the main impurities were 2,3,5-trimethylpyrazine (0.09%) and 2-ethyl-3,5,6-trimethylpyrazine (0.02%). The authors affirmed that this natural and high purity of 2,3,5,6-tetramethylpyrazine by using cheap green renewable materials makes this process promising to compete with chemical synthetic methods (Xiao et al. 2006).

Several *Bacillus* sp. strains have been reported as vanillin producers by bioconversion of phenyl propanoids. Isoeugenol is the main substrate of bioconversion of *Bacillus pumilus* (Hua et al. 2007), *Bacillus fusiformis* (Zhao et al. 2006) and *Bacillus subtilis* (Shimoni et al. 2000). There are reports on the bioconversion of ferulic acid by *Bacillus coagulans* to produce vanillin (Karmakar et al. 2000).

Hua et al. (2007) observed, using *Bacillus pumilus* strain and 10 g.L^{-1} of substrate (isoeugenol), a vanillin production of 3.75 g.L^{-1} after 150 h, with a molar yield of 40.5%. The authors highlighted that this yield was obtained in a batch process, indicating a great potential for much higher vanillin production if a fed-batch method is used.

A strain of *Bacillus fusiformis* was used by Zhao et al. (2006) to obtain a concentration of vanillin ranging from 7.56 to 8.10 g.L^{-1}. This production could be obtained under the optimum conditions, which consisted in the use of resin HD-8 (Cation exchange resin), which adsorb the vanillin and isoeugenol in a reaction liquid, and 50 g.L^{-1} isoeugenol, 72 h bioconversion at pH 7.0, 37 °C and 180 rpm.

Bacillus subtilis strains are reported as poorer producers of vanillin than other *Bacillus*. In the presence of isoeugenol at concentration of approximately 1% (v:v), cultures produced 0.61 g.L^{-1} of vanillin, with a molar yield of 12.4% (Shimoni et al. 2000). The vanillin production was demonstrated from ferulic acid bioconversion by *Bacillus coagulans*, using M9-yeast extract medium without glucose. This microorganism was able to convert more than 95% substrate after 7 h of growth with accumulation of 4-vinylguaiacol (908 mg.L^{-1}) as a major degradation product and very small amount of vanillic acid (10.5 mg.L^{-1}) and vanillin (15.6 mg.L^{-1}) as quantified by HPLC in the culture supernatant (Karmakar et al. 2000).

Besides *Bacillus subtilis* produces vanillin, these strains are involved in acetoin production. Acetoin (3-hydroxy-2-butanone) is an important flavor compound, widely present in dairy products and some fruits. A study using a *Bacillus subtilis* mutant strain produced a rate that reached 43.8 and 46.9 g.L^{-1} in flask and 10-L fermenter fermentations, respectively (Xu et al. 2011).

Another *Bacillus* species, *B. megaterium*, was isolated in a screening process from topsoil. It was used for bioconversion of N-demethylate N-methyl methyl anthranilate into *methyl anthranilate*. Methyl anthranilate is the characteristic flavor compound of concord grapes and also appears in several essential oils such as neroli and bergamot oils. Maximal productivity of 70 mg.L^{-1}. day^{-1} was achieved under laboratory-scale conditions without further optimization. The authors suggested that methyl anthranilate production by *B. megaterium* bioconversion was fast-growing and easy to handle bacterium when it was compared to production by fungal microorganisms (Taupp et al. 2005).

Corynebacterium sp.

Few studies have reported the use of strains from *Corynebacterium* sp. as flavor producers (Demain et al. 1967, Dickschat et al. 2010, Smith et al. 2010). A genetically uncharacterized *C. glutamicum* mutant was isolated, and it showed the ability to produce large amounts of tetramethylpyrazine, 3 g.L⁻¹ in a 5-day fermentation, besides being only able to grow with the supply of these amino acids.

A more recent study reported the biosynthesis of several pyrazines and acyloin derivatives, such as trimethylpyrazine, tetramethylpyrazine, and acetoin by *C. glutamicum* (Dickschat et al. 2010). The authors demonstrated that deletion of the ketol-acid reductoisomerase of the biosynthetic pathway resulted in a mutant that required addition of valine, leucine, and isoleucine to the medium to produce significantly higher amounts and more different compounds of the pyrazine and acyloin classes. Similarly, they reported that it was essential to supplement branched chain amino acids for the microorganism whose gene for acetolactate synthase was inactivated, which had produced only minor amounts of the pyrazines tetramethylpyrazine, 2,3-dimethylpyrazine and trimethylpyrazine. The authors explain these results on the basis of the use of the valine biosynthetic pathway operating in reverse order. The *ΔASΔKR* (double mutant of deletion of the acetolactate synthase) showed impaired production of pyrazines and acyloins, but the authors reported that the production of tetramethylpyrazine and trimethylpyrazine could be restored by the addition of acetoin to the culture medium (Dickschat et al. 2010).

The production of isobutanol by *Corynebacterium glutamicum* was described through 2-keto acid pathways. Overexpression of *alsS* of *Bacillus subtilis*, *ilvC* and *ilvD* of *C. glutamicum*, *kivd* of *Lactococcus lactis*, and a native alcohol dehydrogenase, *adhA*, led to the production of 2.6 g.L⁻¹ isobutanol and 0.4 g.L⁻¹ 3-methyl-1-butanol in 48 h. Using longer term batch cultures, isobutanol reached 4.0 g.L⁻¹ after 96 h. However, after the inactivation of several genes to direct more carbon through the isobutanol pathway, the production was increased by approximately 25% for 4.9 g.L⁻¹ isobutanol in a *ΔpycΔldh* background (Smith et al. 2010).

Pseudomonas sp.

This class of versatile biocatalysts is mainly involved in biotransformation works for the production of natural flavors, including processes with high yields, involvement of genetic engineering and also patented processes. In fact, this genus was applied in the pioneer works of terpene biotransformation for the production of flavor compounds.

Just a few reports describe the production of such compounds by *de novo* synthesis, as an example *Pseudomonas fragi* produced ethyl butyrate when grown in skim milk at 15°C (Medeiros et al. 2000) and other research describe the production of ethyl hexanoate (Cormier et al. 1991). These esters are related mainly with a fruity flavor, whose industrial interests are aroused due to their pleasant fruity notes, e.g., ethyl hexanoate, which are highly desirable in fruit-flavored and dairy products (Bauer et al. 2001).

Some strains of *Pseudomonas* sp. are involved in bioconversions to produce vanillin. Ashengroph et al. (2011) recently described a new strain of *Pseudomonas*, i.e., *P. resinovorans*, capable of bioconverting/transforming eugenol into vanillin and vanillic acid. Using 2.5 g.L^{-1} of eugenol substrate, the maximal vanillin concentration was 0.24 g.L^{-1}, molar yield of 10% and vanillic acid was 1.1 g.L^{-1}, molar yield of 44%. This concentration was achieved after 30 and 60 h of incubation, respectively. The authors affirmed that *P. resinovorans* strain produced vanillin in good molar yield despite eugenol bioconversion into vanillin often resulting in low concentrations.

In order to achieve high yields and productivity of vanillin, Di Gioia et al. (2011) inactivated the vanillin dehydrogenase (*vdh*)-encoding gene of *Pseudomonas fluorescens* BF13 via targeted mutagenesis. The results demonstrated that engineered derivatives of strain BF13 accumulate vanillin if inactivation of *vdh* is associated with concurrent expression of structural genes for feruloyl-CoA synthetase (*fcs*) and hydratase/aldolase (*ech*) from a low-copy plasmid. Optimization of culture conditions and bioconversion parameters at both small and larger scale results in vanillin production up to 8. 41 mM (1.28 g L^{-1}). The authors described this productivity as the highest level found in the literature for recombinant *Pseudomonas* strains. Other pseudomonads, such as *Pseudomonas putida* KT2440 (Plaggenborg et al. 2003), *Pseudomonas nitroreducens* Jin1 (Unno et al. 2007), *Pseudomonas* sp. ISPC2 (Ashengroph et al. 2008) and *Pseudomonas* sp. KOB10 (Ashengroph et al. 2010) have been described as vanillin producers.

Pseudomonas strains are well-known biocatalysts for the biotransformation of terpene substrates. In a study of metabolic profile of two *Pseudomonas* strains, *P. rhodesiae* was the most suitable biocatalyst for the production of isonovalal from α-pinene oxide and did not metabolize limonene. Differently, *P. fluorescens*, besides degrading α- and β-pinene in the same way as *P. rhodesiae*, also metabolized limonene through two pathways, the most efficient being the synthesis of α-terpineol (Bicas et al. 2008b). A conversion of R-(+)-α-terpineol from R-(+)-limonene by *P. fluorescens* (reidentified as *Sphingobium* sp.) was also reported. The study carried out by Bicas et al. (2010c) described a high conversion rate of α-terpineol from limonene using a biphasic medium in which the aqueous phase contained concentrated resting cells of *P. fluorescens* and the organic phase was sunflower oil. This product could be obtained in concentrations up to 130 g.L^{-1} in biphasic medium using vegetable oils as organic phase and it was by far the highest ever described for the bioproduction of α-terpineol (Bicas et al. 2010c). Linares et al. (2009) observed that novalic acid was not an end product for the metabolism of α-pinene by *P. rhodesiae*, since it could be transformed into 3,4-dimethylpentanoic acid by fresh (non permeabilized) cells with a production yield of 40% with a production of 16 g.L^{-1}, while isonovalic acid was not consumed.

Microbial biotransformations of α-pinene into verbenol as the major product have also been reported using *Pseudomonas putida* (Divyashree et al. 2006). More recently, literature demonstrated biotransformation of myrcene by *Pseudomonas aeruginosa*: using 4.47 g.L^{-1} of substrate with 0.1 g.L^{-1} methanol as co-solvent and an agitation speed of 150 rpm it was possible to obtain dihydrolinalool (79.5%) and 2,6-dimethyloctane (9.3%) as the main compounds after 1. 5 day of fermentation. After three days of incubation, the major products were 2,6-dimethyldoctane (90.0%) and α-terpineol (7.7%). The authors affirmed that the major products of this biotransformation (dihydrolinalool, 2,6-dimethyloctane, and α-terpineol) have many applications in flavoring, extracts, food and drug manufacturing industries, and as a fragrance ingredient used in decorative cosmetics, as well as in non-cosmetic products. Their use worldwide is in the region of greater than 1,000 metric tons per annum (Esmaeili and Hashemi 2011). A similar experiment was designed to convert myrcene by *P. putida*, however the main products formed were dihydrolinalool (59.5%), cis-β-dihydroterpineol (25.0%) and hexadecanoic acid (12.5%), with a yield of 97.0% in a period of 30 h (Esmaeili et al. 2011).

Rhodococcus sp.

Rhodococcus erythropolis DSM 44534 grown on ethanol, (R)- and (S)-1,2-propanediol medium was used for biotransformation of racemic 1,4-alkanediols into γ-lactones. The strain oxidized 1,4-decanediol and 1,4-nonanediol into the corresponding γ-lactones 5-hexyldihydro-2(3H)-furanone (γ-decalactone) and 5-pentyl-dihydro-2(3H)-furanone (γ-nonalactone), respectively, with an enantiomeric excess (R) of 40–75%. The yield of lactones formed was higher in slightly alkaline pH values (pH=8.8); however, the enatiomeric excesses were improved at pH of 5.2 (Moreno-Horn et al. 2007).

More recently, the *R. erythropolis* MLT1 was used to bioconvert β-myrcene into monoterpene alcohol geraniol. The biotransformation took place in 50 mM potassium phosphate buffer (pH 7.0) and 7.4 mM myrcene at 30°C for 1 h. This resulted in a conversion of approximately 2% (Thompson et al. 2010). This same strain was also reported by Morrish and Daugulis (2008) for the oxidation of (–)-trans-carveol into R-(–)-carvone with a volumetric productivity of 31 mg.L^{-1}h^{-1}.

The biotransformation of limonene to carvone has already been described for *Rhodococcus opacus* as biocatalysts (De Carvalho and Da Fonseca 2006). Carveol has also been used as a substrate using the bacteria *Rhodococcus globerus* or *R. erythropolis* (De Carvalho et al. 2005) as biocatalysts. Concentrations up to 150 g.L^{-1} have been achieved in small scale column reactors.

Bioconversion of geraniol to geranic acid was studied by Chatterjee (2004) using *Rhodococcus* sp. strain GR3 isolated from soil. The optimum temperature for this process was found to be 30°C and a reaction time lower than 12.5 h. However, no appreciable concentration of geranic acid was obtained. Geranic acid has been used as orange, tea, mint, ripe fruit and melon flavors (Burdock 2010).

Rhocococcus rhodochrous MTCC 265 was reported as a vanillin producer. The culture was grown on minimal salt medium (containing glucose and peptone) and when supplemented with 29.02 mg.L^{-1} of natural curcumin, could produce 3.56 mg.L^{-1} of vanillin (13.05% yield). Although the amount produced is very low, the authors highlighted that the process could be improved optimizing the medium composition and using higher cell density and immobilization of cell culture (Bharti and Gupta 2011).

Streptomyces sp.

4-Vinyl guaiacol (3-methoxy 4-hydroxystyrene) is a flavor compound with a particular interest for the food industry, which has a powerful, spicy, apple, rum, roasted peanut aroma (Burdock 2010). Its threshold is reported to be 0.3 mg.L^{-1}, whereas ferulic acid precursor has a much higher threshold, 600 mg.L^{-1} (Coghe et al. 2004). 4-Vinyl guaiacol was obtained by the bioconversion of ferulic acid by *Streptomyces setonii*. The formation of this metabolite was favored by microaerobic conditions, increasing progressively the product concentration from 543.3 to 885 mg.L^{-1} when aeration level was diminished, reaching a highest volumetric productivity of 70.4 mg.L^{-1} h^{-1} and a product yield of 1.11 mol. mol^{-1} (Max et al. 2012).

Gunnarsson and Palmqvist (2006), using the same strain, studied the influence of pH and carbon source in the bioconversion of ferulic acid to vanillin. They found that, at pH 8.2, the production of vanillin was the highest and conversion rates were several fold higher when arabinose was used as the carbon source. However, at pH 7.2, low vanillin yields were obtained due to reduction of vanillin to vanillyl alcohol. This was more pronounced when arabinose was used as the carbon source.

Streptomyces sannanensis when cultured using ferulic acid as sole carbon source, produced vanillic acid. The highest amount of product (400 mg.L^{-1}) occurred when cultures were grown on 5 mM ferulic acid at 28°C. Purification of vanillic acid was achieved by gel filtration chromatography using Sephadex™ LH-20 matrix (Ghosh et al. 2007).

In fact, *Streptomyces* is one of the most well explored microorganisms for the production of vanillin. Indeed, some processes were already patented and one of this reports the production of vanillin in high yields, using precursors such as ferulic acid and eugenol. The bioconversion of ferulic acid into vanillin by strains of *Amycolatopsis* sp. or *Streptomyces setonii* in a 10-L bioreactor could reach final yields of 11.5 g.L^{-1} (Rabenhorst and Hopp 2000) and 13.9 g.L^{-1}, respectively (Muheim and Lerch 1999).

Other bacteria

Some studies described the production of natural flavor compounds by *de novo* synthesis by naturally present bacteria strains in several foods, such as the lactic acid bacteria (e.g., *Lactococcus lactis*, *Lactobacillus* sp., *Streptococcus thermophilus*, *Leuconostoc mesenteroides*) that are responsible for the production of dairy flavor compounds. Mixed cultures of lactic acid bacteria growing in starch-based media, for example, produced butyric acid, lactic acid and diacetyl (Escamilla-Hurtado et al. 2000). The last product is mainly related to butter flavor and therefore it is extensively used in food and beverages (Bauer et al. 2001). Some other bacteria are involved in biotransformation of several substrates to produce flavor compounds. Guaiacol, for example, is a flavor characterized for sweet odor (Burdock 2010).

Recently, Ashengroph et al. (2012) described the first evidence for biotransformation of isoeugenol into vanillin in the genus *Psychrobacter*. Growing cells were able to produce vanillin resulting in a maximal concentration of 88.18 mg.L^{-1} after 24 h of reaction time using 1 g.L^{-1} isoeugenol as precursor (molar yield = 10.2%). When resting cells was used, the optimal yield of vanillin (141.45 mg.L^{-1}, 16.4%) was obtained after 18 h reaction using 1 g.L^{-1} isoeugenol and 3.1 g of dry weight of cells per liter harvested at the end of the exponential growth phase. Substrate concentration also has been analyzed for the use of resting cell. The highest vanillin concentration (1.28 g.L^{-1}) was obtained using 10 g.L^{-1} isoeugenol after a 48 h reaction, without further optimization.

Several reports on the literature describe biotransformation of ferulic acid to 4-vinyl guaiacol by *Lactobacillus farciminis* (Adamu et al. 2012). The 4-vinyl guaiacol known as 2-methoxy-4-vinylphenol has a powerful spicy, apple, rum, roasted peanut aroma (Burdock 2010). The *Lactobacillus farciminis* bioconversion was obtained after 5 h of incubation with the substrate in Man Regosa and Sharpe Broth at 37°C under 5% CO_2. Productivity was significantly affected by initial concentration of ferulic acid. Concentration of 1, 15 and 50 mg.L^{-1} of ferulic acid yielded 0, 3.34 and 10.26 mg.L^{-1} of 4-vinyl guaiacol, respectively (Adamu et al. 2012).

Cellulosimicrobium cellulans EB-8-4 is a strain able to bioconvert terpenes (Wang et al. 2009). This microorganism demonstrated to be a powerful catalyst for the regio- and stereoselective allylic hydroxylation of R-(+)-limonene to (+)-*trans*-carveol. Resting cells of *C. cellulans* EB-8-4 resulted in a maximum product concentration of 13. 4 mM of (+)-*trans*-carveol. The authors suggested that such hydroxylation was possibly catalyzed by a nicotinamide adenine dinucleotide (NADH)-dependent oxygenase involved in the degradation of aromatic ring during cell growth. Besides, EB-8-4 strain obtained more than 99% regio- and stereoselectivity, showed a specific hydroxylation activity of 4.0 U.g^{-1} cell dry weights, and accepted 62 mM of R-(+)-limonene without inhibition.

Microalgae as Biocatalysts for the Production of Natural Flavor Compounds

The term microalgae has been used to refer to a vast group of microscopic photosynthetic organisms, which includes prokaryotic (e.g., cyanobacteria [Cyanophyceae], formerly known as blue-green algae) and eukaryotic (e.g., green algae [Chlorophyta], red algae [Rhodophyta], diatoms [Bacillariophyta]) microbes, with a unicellular or simple multicellular structure. This is a very robust and versatile group of microorganisms, able to grow efficiently even in harsh conditions. Therefore, they are present in a wide range of environments, from aquatic (saline or freshwater) to terrestrial ecosystems. In terms of energy requirement, microalgae can assume many types of metabolism. Depending on the environmental conditions, the organisms can grow, for example, photoautotrophically (light as sole energy source, CO$_2$ as sole carbon source), heterotrophically (only uses organic compounds as source of energy and carbon), mixotrophically (photosynthesis as main energy source, but organic compounds may also be used) or photoheterotrophically (light is required to use organic compounds as carbon source, which is not very well distinguished from mixotrophy) (Brennan and Owende 2010, Mata et al. 2010, Singh and Sharma 2012).

In the last decades, considerable scientific and commercial interest has been paid to the biotechnological use of microalgae, particularly for the production of renewable energy and biofuels (Li et al. 2008, Pramar et al. 2011). But some authors (Foley et al. 2011, Wijffels et al. 2010) have indicated that the sole production of energy/biofuels from microalgae will hardly lead to viable processes and, thus, the algae biorefinery concept through which biomass is processed for the integrated (co)generation of energy (fuels, power, heat) and a wide range of other commercially interesting products (for examples see Milledge 2011) must be developed. One of such potential microalgae-derived high-value products are aroma compounds.

The production of aroma compounds by *de novo* synthesis from microalgal cultivation is not well documented. In fact, there are a considerable number of articles dealing with volatile production by cyanobacteria (Smith et al. 2008), but the compounds described (e.g., geosmin, 2-methylisoborneol) are usually those related to off-flavors (e.g., earthy-musty notes) in drinking water or aquaculture products, what can impair their commercialization. Naturally, some of these compounds may be desirable in other applications (e.g., 1-octen-3-ol, β-cyclocitral, β-ionone) (Höckelmann and Jüttner 2005). Apart from such substances, some microalgae may produce interesting aroma compounds through terpene biotransformation.

In comparison to the well-established biocatalysts (fungi and bacteria) described above, only a modest investigation on microalgae biotransformation processes has been observed: from all articles covering the biotransformation of terpenes that were published from 1996 to 2006, only 6% of them were related to microalgal or cyanobacterial processes (De Carvalho and Da Fonseca 2006). Therefore, novel descriptions of such processes will have great scientific value. Some of the few processes described so far, most of them involving regio- and stereoselective reduction reactions, will be presented in the following paragraphs.

It has been observed that a range of aromatic aldehydes could be biotransformed into their corresponding primary alcohols by five microalgal species (*Chlorella minutissima, Nannochloris atomus, Dunaliella parva, Porphyridium purpureum* and *Isochrysis galbana*) (Hook et al. 1999), researchers have used these strains in biotransformation of monoterpene ketones, by adding 100 ppm of substrate in conical flasks containing 100mL medium. After 5 days of incubation, all microalgae could reduce (4S)-(+)-carvone to (1S,4S)-(−)-dihydrocarvone, (1S,2R,4S)-(−)-neodihydrocarveol, (1R,4S)-(+)-isodihydrocarvone and (1R,2R,4S)-(−)-isodihydrocarveol, while (4R)-(−)-carvone was transformed into a lesser extent by all species yielding (1R,4R)-(+)-dihydrocarvone and (1R,2S,4R)-(+)-neodihydrocarveol. *P. purpureum* was the most effective strain for both biotransformations. The five microalgae tested could also perform diastereoselective oxidation of (−)-*trans*-carveol into (4R)-(−)-carvone (Hook et al. 2003).

Analogous behavior was evidenced for *Oocystis pusilla*, which was inoculated (ca. 2.8–3.0·10⁶ cells/mL) in 250-ml conical flasks containing mineral medium and 100ppm of substrate. After incubation, the samples were extracted with dichloromethane prior to TLC and GC/MS analysis. Many reduction reactions were observed (L-citronellal into citronellol, (−)-menthone and (+)-pulegone into menthol), including the conversion of (+)-carvone into *trans*-dihydrocarvone. Oxidative biotransformations ((+)-limonene into *trans*-carveol, carvone and *trans*-limonene oxide;

(+)-β-pinene into *trans*-pinocarveol; (–)-linalool into furanoid and pyranoid linalool oxides; thymol into thymoquinone) were also found, including the formation of carvone from (–)-carveol (Ghasemi et al. 2009). Following a similar procedure, another microalgal strain, *Chlorella vulgaris* MCCS 012, has also shown the ability to perform the reduction of C=C double bond of (+)-carvone to yield *trans*-dihydrocarvone and *cis*-dihydrocarvone. Other reduction reactions were observed (both (–)-menthone and (+)-pulegone could be converted into menthol) but no oxidation reaction was described for this strain (Ghasemi et al. 2010). Finally, Rasoul-Amini et al. (2011) reported the biotransformation of five terpene subtrates ((R)-(+)-carvone, (R)-(+)-limonene, (+)-β-pinene, thymol (12), and (+)-linalool) by immobilized cells of ten microalgae strains (four *Chlorella* sp., two *Oocystis* sp., two *Synechococcus* sp. and two *Chlamydomonas* sp.) applying 100ppm of substrate and incubation at 25°C/70rpm/24h. Again, data similar to that cited above were found, although different strains were used. The reduction of endocyclic C=C double bonds of carvone to yield *cis*- and (mainly) *trans*-dihydrocarvone was observed for nine strains, at different proportions. Also, a nonselective biotransformation of limonene into *cis*- and *trans*-carveol, carvone and *cis*- and *trans*-limonene oxide was detected for three microalgal strains (two *Chlorella* sp. and one *Synechococcus* sp.). When linalool was used as substrate, three strains (two *Chlorella* sp. and one *Chlamydomonas* sp.) reduced it to dihydro linalool and one of them (*Chlorella* sp.) was also able to produce *cis*- and *trans*-furanoid linalool oxide in low yields. Thymol was converted only by a strain of *Synechococcus* sp., yielding low amounts of thymoquinone, while β-pinene was not converted by any of the strains tested (Rasoul-Amini et al. 2011).

Other authors reported exclusively oxidative biotransformations by microalgae, most of them different from those already described for the microalgae strains mentioned so far. One example is the biotransformation of limonene and limonene oxide by the cyanobacterium *Synechococcus* sp. PCC 7942. The process consisted of incubating 100 ml pre-grown culture (containing ca. 0.1 g of wet cells) with 10 mg of substrate at 25°C with shaking and illumination by white fluorescent light for 6 h, followed by extraction with diethyl ether. The results showed that the biotransformation of limonene was enantiospecific, with a regioselective hydroxylation of the S-isomer in the 6-position: S-(–)-limonene was converted into *cis*- and *trans*-carveol (yields close to 10%) while the biotransformation of R-(+)-limonene hardly occurred. These biotransformations were also feasible with immobilized cells, which remained with almost the same reaction rate even after 7 batches. This microorganism was also able to stereospecifically convert limonene-oxide: (1S,2R,4R)-limonene oxide was biotransformed into (1S,2S,4R)-limonene-1,2-diol and (1S,4R)-limonene-1-ol-2-one, while the diasteroisomers (1R,2S,4R)-, (1S,2R,4S)- and (1R,2S,4S)-limonene oxide were not attacked by this strain (Hamada et al. 2003). In another study, different *Chlorella* species were tested for the conversion of (+)-valencene, a sesquiterpene which can be cheaply obtained from Valencia oranges. After seven days of growth at 25°C under illumination and stationary conditions, 400 mg.L^{-1} of substrate were added to the medium, and the biotransformation took place in the following days (up to 20 days). GC-MS analysis showed that conversion ratios of 89 to 100% of valencene into nootkatone could be obtained. This is a very interesting report showing how nootkatone, an expensive grapefruit aroma, can be produced from microalgae biotransformation of (+)-valencene (Furusawa et al. 2005b).

Genetic Engineering: General Aspects and Process Improvement

Genetic engineering provides tools such as simple gene deletions or amplifications, DNA rearrangements in a species, trans-species gene transfer and bioanalytical screening and monitoring techniques, and others that could benefit the biocatalysts and the biotechnological production of natural flavors (Pridmore et al. 2000). From the technical point of view, genetic engineering applied to food products has reached an interesting status, however, legal and public concern are the major drawbacks for its application. Furthermore, there are few, but important reports on genetic engineering to produce flavor compounds, specially reporting the overexpression of some genes in *E. coli* (Savithiry et al. 1998).

Some researchers have described *Escherichia coli* utilization to obtain vanillin by bioconversion (Lee et al. 2009, Torres et al. 2009). *Escherichia coli* JM109/pBB1 has been studied to produce vanillin from hydrolyzed corn cob. This by-product of corn industry was hydrolyzed for 6 h with 0.5N NaOH at solid/liquid ratio of 0.084 g.g^{-1} allowed obtaining a hydrolyzate containing 1171±34 mg.L^{-1} ferulic acid and 2156±63 mg.L^{-1} *p*-coumaric acid that was used as a medium for vanillin bioproduction. Biomass pre-cultivated once in unsterilized hydrolyzate was able to effectively convert ferulic and p-coumaric acids to a mixture of vanillin, vanillic acid and vanillyl alcohol provided with the typical vanilla flavor. At initial biomass concentration of 0.5 gL^{-1} (dry matter), a maximum value of vanillin concentration was 239±15 mg.L^{-1}. The authors suggested that the by-product of corn be used as an interesting raw material for vanillin bioproduction in bench-scale bioreactor (Torres et al. 2009).

Additionally, some other studies report on the production of modified crops with improved flavor, color and resistance to pests. Some examples are 'golden rice' (rich in carotenoids) and transgenic tomatoes with improved concentrations of (Z)-3-hexenal and (Z)-3-hexen-1-ol, characteristic compounds of tomato flavor, which are derivatives from lipid metabolism (Ye et al. 2000, Wang et al. 1996).

Gounaris et al. (2010) presents a very detailed review reporting the recent advances on the field. Improvements of flavor production after overexpressed enzymes involved on the biochemical routes were already reported. Large yields were also reported, making it possible to get commercial scale processes.

Classical reports, like the production of flavor compounds by genetic modified micro-organisms, the expression of the enoyl-SCoA hydratase/lyase enzyme from *P. fluorescens* in *E. coli* should be cited. The micro-organism proved to be capable of converting ferulic acid into vanillin (Barghini et al. 2007). A Research group from Washington State University has isolated cDNAs from mint species coding for limonene hydroxylases (Lupien et al. 1999, Haudenschild et al. 2000). The cDNAs were overexpressed in *E. coli* and *S. cerevisiae* and resulted in the production of (–)-*trans*-carveol and (-)-*trans*-isopiperitenol. The authors confirmed that the enzyme sequences were very similar to the ones from the original plant. Other reports about the overexpression of genes involved on the flavor biotransformation are already described (Van Beilen et al. 2005, Gounaris et al. 2010).

There are few, but important reports about the genetic engineering applied to biotransformation processes to obtain flavor compounds. It has been shown to be a promising topic to be addressed, as after the advent of DNA recombinant techniques, direct genetic approaches for increasing biotransformation rates and simplifying the process have been driving studies in this area.

Perspectives for the Industrial Production of Microbial Bioflavors

From an economical point of view, the production of many bioflavors is feasible. Examples of commercial natural aroma chemicals with probable biotechnological origin may be found, e.g., ethyl butanoate, 2-heptanone, β-ionone, nootkatone, 1-octen-3-ol, 4-undecalactone, vanillin, among others (Berger 2009). Additionally, a number of recent patents or patent applications may be found for this subject (Caputi and Aprea 2011, Zorn et al. 2012)—particularly for those processes involving vanillin production by the biotransformation of ferulic acid (Cheetham et al. 2005, Heald et al. 2011, Torres et al. 2011, Xu et al. 2009)—which is a reflex of the development of advanced and (possibly) more viable procedures.

The main motivation for the microbial production of aroma compounds is the market value of the so-called "biotechnological aroma", which is commonly far above their synthetic counterparts, although usually lower than those extracted from nature. For example: synthetic vanillin has a price of approximately US$11/Kg, natural vanilla flavor extracted from fermented pods of *Vanilla* orchids costs U$1,200–4000/Kg, while "biotech vanillin" is sold for a price of approximately US$1,000/Kg (Scharader et al. 2004, Xu et al. 2007). Equivalent data are also observed for other aroma compounds, as γ-decalactone (synthetic = US$150/Kg; natural = US$6,000/Kg; "biotec" = US$300/Kg) and ethyl butyrate (synthetic = US$4/Kg; natural = US$5,000/Kg; "biotec" = US$180/Kg), for example (Dubal et al. 2008). Additionally, according to Berger (2009), the biotechnological production of aroma compounds has also been stimulated by the increasing consumers' preference for "natural"-labeled products and also for the vivid discussion on functional foods and healthy ingredients. As a consequence, many biotechnological aroma compounds are already available in a commercial scale (Berger 2009).

In this context, the biotransformation of terpenes seems to be a very promising approach for the industrial production of aroma compounds, due to two main reasons: considerable productivities may be obtained and some terpene substrates are available in high amounts and low costs (e.g., R-(+)-limonene, which corresponds to >90% of orange peel oil; α-pinene, major component of turpentine; and β-farnesene, that will be produced by Amyris in large amounts in Brazil to be applied as biofuel, called Biofene™). Although there is much progress in this field (Bicas et al. 2009, Bicas et al. 2010c), there are yet some challenges related to the biotransformation of terpenes: (i) the chemical instability of both substrate and product, (ii) the low water solubility of the substrate, (iii) the high volatility of both substrate and product, (iv) the high toxicity of both substrate and product, (v) the low yields (Krings and Berger 1998) and (vi) the high costs related to fermentation processes (Bicas et al. 2010c). Many strategies have been applied in order to overcome these problems, as follows:

- A good strain selection can avoid problems associated to substrate toxicity and may also favor the discovery of microorganisms (or their enzymes) that are highly efficient in terms of terpene biotransformation, what may be a solution for the low yields traditionally associated with such processes. Example of procedures for the isolation and screening of potentially interesting strains may be found (Bicas and Pastore 2007, Bier et al. 2011, Rottava 2010).
- Biphasic systems consisting of an aqueous phase, which contains the biocatalyst, and an organic phase, which contains the hydrophobic substrate and product, are considered a promising strategy to overcome the low solubility, high volatility and high toxicity of both substrate and product. Additionally, these systems may also ease downstreaming process and enhance yields by shifting the chemical equilibrium

(Cabral 2001, Larroche et al. 2007, León et al. 1998). The choice of the solvent is crucial for the success of the process: organic compounds with log $P_{O/W} > 5$ are usually biocompatible to whole-cell biocatalysis in biphasic systems (León et al. 1998). Biphasic systems have been employed, for example, for the biotransformation of α-pinene oxide into isonovalal by *Pseudomonas rhodesiae* (Fontanille and Larroche 2003) or biotransformation of (*R*)-(+)-limonene into α-terpineol by *Sphingobium* sp. (Bicas et al. 2010c). In the last case, ca. 130g of product per liter of organic phase could be obtained, what is one of the highest yields already obtained for a microbial bioflavor.

- In most cases, simple changes in the basic parameters of a bioprocess (temperature, pressure, pH, substrate concentration, inoculums size, etc.) can significantly improve the productivity of a biotransformation. In this case, the use of Response Surface Methodology for optimizations may be a valuable tool (Berger 2009). Recently, some experiments for the production of aroma compounds using this technique were reported (Bicas et al. 2008a, Rottava et al. 2011).
- The use of agroindustrial by-products and residues is also an emergent trend for the reduction of costs inherent to bioprocess, including those related to microbial bioflavor production (Bicas et al. 2010c). In this sense, an interesting report showed that cassava wastewater could be used as culture medium for *Fusarium oxysporum* growth, and the resulting biomass could be used for the biotransformation of orange peel oil (source of *R*-(+)-limonene) into α-terpineol (Maróstica Jr. and Pastore 2007). The production of α-terpineol using orange peel oil was also viable for *Penicillium digitatum* NRRL 1202 (Badee et al. 2011).
- Other strategies, such as (i) *in situ* product recovery, (ii) use of fed-batch systems, (iii) use of specific bioreactors (e.g., membrane reactors), among others, may also be very useful approaches (Berger 2009).

Additionally, the use of recombinant DNA technology and protein engineering can significantly improve yields and increase the biocatalyst stability in bioprocess' conditions. Therefore, as presented earlier in this text, researches have been investigating novel engineered biocatalysts, which can efficiently be used for the production of bioflavors. However, the success of innovative bioflavor production does not depend exclusively on genetically modifying biocatalysts, but also on process engineering (Schrader et al. 2004, Berger 2009). In this sense, as emphasized by other authors, genetic engineering is expected to be a general solution in the future, but, until then, a careful selection of strain associated with appropriate process bioengineering will remain essential to obtain high-yield processes (Kaspera et al. 2006). So far, public prejudgment about food application of genetic engineering is a considerable barrier to be considered (Berger 2009).

Conclusion

The biotechnological production of natural flavor compounds is of great importance in view of the increasing demand for these compounds. Although some yields obtained are high enough for industrial application and many microbial processes are already operating in commercial scale, much is yet to be developed in order to meet industrial demand, including the increase in product concentration and the discovery of novel desired products. The low transformation rates and high production costs are still obstructing their wide-scale adoption. More progress in this field is required and biocatalytical systems will seriously compete on a broad industrial scale with the conventional sources. In this sense, recent advances in genetic engineering techniques, detailed pathway study and regulation, bioprocess engineering and use of by-products as an alternative substrate can stimulate the researches in biotechnological process to produce aroma compounds and will lead to new and successful commercial technologies.

References

Adams, A., J. C. R. Demyttenaere and N. De Kimpe. 2003. Biotransformation of (*R*)-(+)- and (*S*)-(–)-limonene to alpha-terpineol by *Penicillium digitatum*—Investigation of the culture conditions. Food Chem. 80: 525–534.

Adams, A. and N. De Kimpe. 2007. Formation of pyrazines and 2-acetyl-1-pyrroline by *Bacillus cereus*. Food Chem. 101: 1230–1238.

Adamu, H. A., S. Iqbal, K. W. Chan and I. Maznah. 2012. Biotransformation of ferulic acid to 4-vinyl guaiacol by *Lactobacillus farciminis*. Afr. J. Biotechnol. 11: 1177–1184.

Agrawal, R., N. U. Deepika and R. Joseph. 1999. Improvement in the biotransformation of a-pinene to verbenol in *Aspergillus* sp. and *Penicillium* sp. by induced mutation using chemicals and UV irradiation. Biotechnol. Bioeng. 63: 249–252.

Agrawal, R. and R. Joseph. 2000. Bioconversion of alpha-pinene to verbenone by resting cells of *Aspergillus niger*. Appl. Microbiol. Biotechnol. 53: 335–337.

Aleu, J. and I. G. Collado. 2001. Biotransformations by *Botrytis* species. J. Mol. Catal. B Enzyme 13: 77–93.

Ashengroph, M., I. Nahvi and H. Zarkesh-Esfahani. 2008. A bioconversion process using a novel isolated strain of *Pseudomonas* sp. ISPC2 to produce natural vanillin from isoeugenol. Res. Pharm. Sci. 3: 41–47.

Ashengroph, M., I. Nahvi, H. Zarkesh-Esfahani and F. Momenbeik. 2010. Optimization of media composition for improving conversion of isoeugenol into vanillin with *Pseudomonas* sp. strain KOB10 using the Taguchi method. Biocat. Biotrans. 28: 339–347.

Ashengroph, M., I. Nahvi, H. Zarkesh-Esfahani and F. Momenbeik. 2011. *Pseudomonas resinovorans* SPR1, a newly isolated strain with potential of transforming eugenol to vanillin and vanillic acid. New Biotechnol. 28: 656–664.

Ashengroph, M., I. Nahvi, H. Zarkesh-Esfahani and F. Momenbeik. 2012. Conversion of isoeugenol to vanillin by *Psychrobacter* sp. strain CSW4. Appl. Biochem. Biotechnol. 166: 1–12.

Badee, A. Z. M., S. A. Helmy and N. F. S. Morsy. 2011. Utilization of orange peel in the production of α-terpineol by *Penicillium digitatum* (NRRL 1202). Food Chem. 126: 849–854.

Barghini, P., D. Di Gioia,, F. Fava and M. Ruzzi. 2007. Vanillin production using metabolically engineered *Escherichia coli* under non-growing conditions. Microb. Cell Fact. **6**, 13; doi: 10.1186/1475-2859-6-14.

Bauer, K., D. Garbe and H. Surburg. 2001. Common fragance and flavor materials. Preparation, Properties and Uses, 4 ed. Wiley-VCH, Weinheim, Germany.

Berger, R. G., U. Krings and H. Zorn. 2010. Biotechnological flavour generation. *In:* A. J. Taylor and R. S. T. Linforth [eds.]. Food Flavour Technology. Blackwell Publishing Ltd, United Kingdom, UK. pp. 89–115.

Berger, R. G. 1995. Aroma Biotechnology. Springer, Berlin, Germany.

Berger, R. G. 2009. Biotechnology of flavours—the next generation. Biotechnol. Letters. **31**: 1651–1659.

Berger, R. G. and H. Zorn. 2004. Flavors and fragrances. *In:* J. S. Tkacz and L. Lange. [eds.]. Advances in Fungal Biotechnology for Industry, Agriculture, and Medicine. Kluwer Academic/Plenum Publishers, New York, USA. pp. 341–358.

Berger, R. G., K. Neuhaeuser and F. Drawert. 1987. Biotechnological production of flavour compounds: III. High productivity fermentation of volatile flavours using a strain of *Ischnoderma benzoinum*. Biotechnol. Bioeng. **30**: 987–990.

Bharti, A., A. L. Nagpure and R. K. Gupta. 2011. Biotransformation of curcumin to vanillin. Indian J. Chem. **50B**: 1119–1122.

Bicas, J. L. and G. M. Pastore. 2007. Isolation and screening of D-limonene-resistant microorganisms. Braz. J. Microbiol. **38**: 563–567.

Bicas, J. L., F. F. C. Barros, R. Wagner, H. T. Godoy and G. M. Pastore. 2008a. Optimization of R-(+)-α-terpineol production by the biotransformation of R-(+)-limonene. J. Ind. Microbiol. Biotechnol. **35**: 1061–1070.

Bicas, J. L., P. Fontanille, G. M. Pastore and C. Larroche. 2008b. Characterization of monoterpene biotransformation in two pseudomonads. J. Appl. Microbiol. **105**: 1991–2001.

Bicas, J. L., A. P. Dionísio and G. M. Pastore. 2009. Bio-oxidation of terpenes: an approach for the flavor industry. Chem. Rev. **109**: 4518–4531.

Bicas, J. L., J. C. Silva, A. P. Dionísio and G. M. Pastore. 2010a. Biotechnological production of bioflavors and functional sugars. Ciênc. Tecnol. Aliment. **30**: 7–18.

Bicas, J. L., C. P. de Quadros, I. A. Neri-Numa and G. M. Pastore. 2010b. Integrated process for co-production of alkaline lipase and R-(+)-α-terpineol by *Fusarium oxysporum*. Food Chem. **120**: 452–456.

Bicas, J. L., P. Fontanille, G. M. Pastore and C. Larroche. 2010c. A bioprocess for the production of high concentrations of R-(+)-α-terpineol from R-(+)-limonene. Proc. Biochem. **45**: 481–486.

Bier, M. C. J., S. Poletto, V. T. Soccol, C. R. Soccol and A. B. P. Medeiros. 2011. Isolation and screening of microorganisms with potential for biotransformation of terpenic substrates. Braz. Arch. Biol. Technol. **54**: 1019–1026.

Bluemke, W. and B. J. Schrader. 2001. Integrated bioprocess for enhanced production of natural flavors and fragrances by *Ceratocystis moniliformis*. Biomol. Eng. **17**: 137–142.

Böker, A., M. Fischer and R. G. Berger. 2001. Raspberry ketone from submerged cultured cells of the basidiomycete *Nidula niveo-tomentosa*. Biotechnol. Prog. **17**: 568–572.

Brennan, L. and P. Owende. 2010. Biofuels from microalgae—A review of technologies for production, processing, and extractions of biofuels and co-products. Renew. Sustain. Energy Rev. **14**: 557–577.

Brigido, B. M. 2000. Produção de compostos voláteis por novas linhagens de *Neurospora*. PhD Thesis. Universidade Estadual de Campinas, Brazil. 148p.

Brown, A. K. and J. R. M. Hammond. 2003. Flavor control in small-scale beer fermentations. Food Bioprod. Process. **81**: 40–49.

Burdock, G. A. 2010. Fenaroli's Handbook of Flavor Ingredients, Sixth Edition. CRC Press, Boca Raton, USA.

Cabral, J. M. S. 2001. Biotransformations. *In:* C. Ratledge and B. Kristiansen. [eds]. Basic Biotechnology, 2nd. Cambridge University Press, UK.

Caputi, L. and E. Aprea. 2011. Use of terpenoids as natural flavouring compounds in food industry. Rec. Pat. Food Nutrit. Agric. 3: 9–16.

Carrau, F., K. Medina, L. Fariña, E. Boido and E. Dellacassa. 2010. Effect of *Saccharomyces cerevisiae* inoculum size on wine fermentation aroma compounds and its relation with assimilable nitrogen content. Int. J. Food Microbiol. 143: 81–85.

Chatterjee, T. 2004. Biotransformation of geraniol by *Rhodococcus* sp. strain GR3. Biotechnol. Appl. Biochem. 39: 303–306.

Chatterjee, T., B. K. De and D. K. Bhattacharyya. 1999. Microbial oxidation of α-pinene to (+)-α-terpineol by *Candida tropicalis*. Indian J. Chem. B 38B: 515–517.

Cheetham, P. S. J., M. L. Gradley and J. T. Sime. 2005. Flavor/aroma materials and their preparation. US Patent n° 6,844,019 B1, Jan. 18, 2005.

Coghe, S., K. Benoot, F. Delvaux, B. Vanderhaegen and F. R. Delvaux. 2004. Ferulic acid release and 4-vinylguaiacol formation during brewing and fermentation: Indications for feruloyl esterase activity in *Saccharomyces cerevisiae*. J. Agric. Food Chem. 52: 602–608.

Cormier, F., Y. Raymond, C. P. Champagne and A. Morin. 1991. Analysis of odor-active volatiles from *Pseudomonas fragi* grown in milk. J. Agric. Food Chem. 39: 159–161.

Damasceno, S., M. P. Cereda, G. M. Pastore and J. G. Oliveira. 2003. Production of volatile compounds by *Geotrichum fragrans* using cassava wastewater as substrate. Proc. Biochem. 39: 411–414.

De Boer, L., R. B. Meima, H. Zorn, M. Scheibner, B. Hülsdau and R. G. Berger. 2005. Novel enzymes for use in enzymatic bleaching of food products, PTC/EPA2005/053329, PCT/EP2006064132.

De Carvalho, C. C. C. R. and M. M. R. Da Fonseca. 2006. Biotransformation of terpenes. Biotechnol Adv. 24: 134–142.

De Carvalho, C. C. C. R. and A. Poretti and M. M. R. Da Fonseca. 2005. Cell adaptation to solvent, substrate and product: a successful strategy to overcome product inhibition in a bioconversion system. Appl. Microbiol. Biotechnol. 69: 268–275.

De Oliveira, B. H., M. C. Dos Santos and P. C. Leal. 1999. Biotransformation of the diterpenoid isosteviol by *Aspergillus niger*, *Penicillium chrysogenum* and *Rhizopus arrhizus*. Phytochem. 51: 737– 41.

Debourg, A. 2000. Yeast flavor metabolites. Eur. Brew. Conv. Monogr. 28: 60–73.

Demain, A. L., M. Jackson and N. R. Trenner. 1967. Thiamine-dependent accumulation of tetramethylpyrazine accompanying a mutation in the isoleucine-valine pathway. J. Bacteriol. 94: 323–326.

Demyttenaere, J. C. R., M. C. Herrera and N. De Kimpe. 2000. Biotransformation of geraniol, nerol and citral by sporulated surface cultures of *Aspergillus niger* and *Penicillium* sp. Phytochem. 55: 363–373.

Demyttenaere, J. C. R. 2001. Biotransformation of Terpenoids by Microorganisms. *In:* A. Rahman. [ed.]. Studies in Natural Products Chemistry. Elsevier, London, UK. pp. 125–178.

Demyttenaere, J. C. R., I. E. I. Koninckx and A. Meersman. *In:* A. J. Taylor and D. S. Mottram. [eds.]. 1996. Flavour Science: Recent Developments. The Royal Society of Chemistry, Cambridge, UK. pp. 105–117.

Demyttenaere, J. C. R., J. Vanoverschelde and N. De Kimpe. 2004. Biotransformation of (*R*)-(+)- and (*S*)-(–)-citronellol by *Aspergillus* sp. and *Penicillium* sp., and the use of solid-phase microextraction for screening. J. Chromat. A 1027: 137–146.

Di Gioia, D., F. Luziatelli, A. Negroni, A. G. Ficca, F. Fava and M. Ruzzi. 2011. Metabolic engineering of *Pseudomonas fluorescens* for the production of vanillin from ferulic acid. J. Biotechnol. 156: 309–316.

Dickschat, J. S., S. Wickel, C. J. Bolten, T. Nawrath, S. Schulz and C. Wittmann. 2010. Pyrazine biosynthesis in *Corynebacterium glutamicum*. Eur. J. Org. Chem. 2010: 2687–2695.

Divyashree, M. S., J. George and R. Agrawal. 2006. Biotransformation of terpenic substrates by resting cells of *Aspergillus niger* and *Pseudomonas putida* isolates, J. Food Sci. Technol. 43: 73–76.

Dragone, G., S. I. Mussatto, J. M. Oliveira and J. A. Teixeira. 2009. Characterisation of volatile compounds in an alcoholic beverage produced by whey fermentation. Food Chem. 112: 29–935.

Dubal, S. A., Y. P. Tilkari, S. A. Momin and I. V. Borkar. 2008. Biotechnological routes in flavour industry. Adv. Biotechnol. 6: 20–31.

Escamilla-Hurtado, M. L., S. Valdes-Martinez, J. Soriano-Santos and A. Tomasini-Campocosio. 2000. Effect of some nutritional and environmental parameters on the production of diacetyl and on starch consumption by *Pediococcus pentosaceus* and *Lactobacillus acidophilus* in submerged cultures. J. Appl. Microbiol. 88: 142–153.

Esmaeili, A., S. Sharafiana, S. Safaiyanb, S. Rezazadehc and A. Rustaivan. 2009. Biotransformation of one monoterpene by sporulated surface cultures of *Aspergillus niger* and *Penicillium* sp. Nat. Prod. Res. 23: 1058–1061.

Esmaeili, A., A. Tavassoli. 2010. Microbial transformation of citral by *Penicillium* sp. Acta Biochim. Pol. 57: 265–268.

Esmaeili, A. and E. Hashemi. 2011. Biotransformation of myrcene by *Pseudomonas aeruginosa*. Chem. Central J., 5: 26–33.

Esmaeili, A., E. Hashemi, S. H. Safaiyan and A. Rustaiyan. 2011. Biotransformation of myrcene by *Pseudomonas putida* PTCC 1694. Herba Pol. 57: 51–58.

Etschmann, M. M. W., W. Bluemke, D. Sell and J. Schrader. 2002. Biotechnological production of 2-phenylethanol. Appl. Microbiol. Biotechnol. 59: 1–8.

Feron, G. and Y. Waché. 2006. Microbial biotechnology of food flavor production. *In:* K. Shetty, G. Paliyath, A. Pometto and R. Levin [eds.]. Food Biotechnology. CRC Press Taylor & Francis, New York, USA. pp. 407–442.

Foley, P. M., E. S. Beach and J. B. Zimmerman. 2011. Algae as a source of renewable chemicals: opportunities and challenges. Green Chem. 13: 1399–1405.

Fontanille, P. and C. Larroche. 2003. Optimization of isonovalal production from α-pinene oxide using permeabilized cells of *Pseudomonas rhodesiae* CIP107491. Appl. Microbiol. Biotechnol. 60: 534–540.

Fraatz, A., S. J. L. Riemer, R. Ströber, R. Kaspera, M. Nimtz, R. G. Berger and H. Zorn. 2009. A novel oxygenase from *Pleurotus sapidus* transforms valencene to nootkatone. J. Mol. Catal. B-Enzymatic. 61: 202–207.

Franco, M. R. B. 2004. Aroma e Sabor dos Alimentos: Temas Atuais. Livraria Varela, São Paulo, Brazil.

Furusawa, M., T. Hashimoto, Y. Noma and Y. Asakawa. 2005a. Biotransformation of citrus aromatics nootkatone and valencene by microorganisms. Chem. Pharm. Bull. 53: 1423–1429.

Furusawa, M., T. Hashimoto, Y. Noma and Y. Asakawa. 2005b. Highly efficient production of nootkatone, the grapefruit aroma from valencene, by biotransformation. Chem. Pharm. Bull. 53: 1513–1514.

Ghasemi, Y., A. Mohagheghzadeh, M. Moshavash, Z. Ostovan, S. Rasoul-Amini, M. H. Morowvat, M. B Ghoshoon, M. J. Raee and S. B. Mosavi-Azam. 2009. Biotransformation of monoterpenes by *Oocystis pusilla*. World J. Microbiol. Biotechnol. 25: 1301–1304.

Ghasemi, Y., A. Mohagheghzadeh, Z. Ostovan, M. Moshavash, S. Rasoul-Amini and M. H. Morowvat. 2010. Biotransformation of some monoterpenoid ketones by *Chlorella vulgaris* MCCS 012. Chem. Nat. Comp. 46: 734–737.

Ghosh, S., A. Sachan, S. Sen and A. Mitra. 2007. Microbial transformation of ferulic acid to vanillic acid by *Streptomyces sannanensis* MTCC 6637. J. Ind. Microbiol. Biotechnol. 34: 131–138.

Gounaris, Y. 2010. Biotechnology for the production of essential oils, flavours and volatile isolates. A review. Flavour Fragr. J. 25: 367–386.

Gunnarsson, N. and E. A. Palmqvist. 2006. Influence of pH and carbon source on the production of vanillin from ferulic acid by *Streptomyces setonii* ATCC 39116. *In:* L. P. B. Wender and P. M. Agerlin [eds.]. Developments in Food Science, vol. 43. Elsevier. pp. 73–76.

Hamada, H., Y. Kondo, K. Ishihara, N. Nakajima, H. Hamada, R. Kurihara and T. Hirata. 2003. Stereoselective biotransformation of limonene and limonene oxide by cyanobacterium, *Synechococcus* sp. PCC 7942. J. Biosci. Bioeng. 96: 581–584.

Hashimoto, T., Y. Noma and Y. Asakawa Y. 2001. Biotransformation of terpenoids from the crude drugs and animal origin by microorganisms. Heterocycles. 54: 529–59.

Haudenschild, C., M. Schalk, F. Karp and R. Croteau. 2000. Functional expression of regiospecific cytochrome P450 limonene hydroxylases from mint (*Mentha* ssp.) in *Escherichia coli* and *Saccharomyces cerevisiae*. Arch. Biochem. Biophys. 379: 127–136.

Heald, S., S. Mayers, T. Walford, K. Robbins and C. Hill. 2011. Preparation of vanillin from microbial transformation media by extraction by means supercritical fluids or gases. US Patent Application n° US 2011/0268858.

Höckelmann, C. and F. Jüttner. 2005. Off-flavours in water: hydroxyketones and β-ionine derivatives as new odour compounds of freshwater cyanobacteria. Flavour Frag. J. 20: 387–394.

Hook, I. L., S. Ryan and H. Sheridan. 1999. Biotransformation of aromatic aldehydes by five species of marine microalgae. Phytochem. 51: 621–627.

Hook, I. L., S. Ryan and H. Sheridan. 2003. Biotransformation of aliphatic and aromatic ketones, including several monoterpenoid ketones and their derivatives by five species of marine microalgae. Phytochem. 63: 31–36.

Hua, D., C. Ma, S. Lin, L. Song, Z. Deng, Z. Maomy, Z. Zhang, B. Yu and P. Xu. 2007. Biotransformation of isoeugenol to vanillin by a newly isolated *Bacillus pumilus* strain: Identification of major metabolites. J. Biotechnol. 130: 463–470.

Huang, C. Jr., S. L. Lu and C. C. Chou. 2001. Production of 2-phenylethanol, a flavor ingredient, by *Pichia fermentans* L-5 under various culture conditions. Food Res. Int. 34: 277–282.

Janssens, L., H. L. Depooter, N. M. Schamp and E. J. Vandamme. 1992. Production of flavors by microorganisms. Proc. Biochem. 27: 195–215.

Karmakar, B., R. M. Vohra, H. Nandanwar, P. Sharma, K. G. Gupta and R. C. Sobti. 2000. Rapid degradation of ferulic acid via 4-vinylguaiacol and vanillin by a newly isolated strain of *Bacillus coagulans*. J. Biotechnol. 80: 195–202.

Kaspera, R., U. Krings and R. G. Berger. 2006. Microbial terpene biotransformation. *In:* C. Larroche, A. Pandey and C. G. Dussap [eds.]. Current Topics on Bioprocesses in Food Industry. Asiatech Publisher, New Delhi. pp. 54–69.

Kobayashi, M., H. Shimizu and S. Shioya. 2008. Beer volatile compounds and their application to low-malt beer fermentation. J. Biosci. Bioeng. 106: 317–323.

Krings U. and R. G. Berger. 1998. Biotechnological production of flavors and fragrances. Appl. Microbiol. Biotechnol. 49: 1–8.

Krings, U., D. Hapetta and R. G. Berger. 2008. A labeling study on the formation of perillene by submerged cultured oyster mushroom, *Pleurotus ostreatus*. Appl. Microbiol. Biotechnol. 78: 533–541.

Krings, U., B. Hardebusch, D. Albert, R. G. Berger, M. R. Marostica Jr. and G. M. Pastore. 2006. Odor-active alcohols from the fungal transformation of α-farnesene. J. Agric. Food Chem. 54: 9079–9084.

Larroche, C., J. B. Gros and P. Fontanille. 2007. Microbial Processes. *In:* R. G. Berger [ed.]. Flavour and Fragrances: Chemistry, Bioprocessing and Sustainability. Springer-Verlag, Berlin, Germany. pp. 575–597.

Lee, E. G., S. H. Yoon, A. Das, S. H. Lee, C. Li, J. Y. Kim, M. S. Choi, D. K. Oh and S. W. Kim. 2009. Directing vanillin production from ferulic acid by increased acetyl-CoA consumption in recombinant *Escherichia coli*. Biotechnol. Bioeng. 102: 200–208.

León, R., P. Fernandes, H. M. Pinheiro and J. M. S. Cabral. 1998. Whole-cell biocatalysis in organic media. Enzyme Microb. Tech. 23: 483–500.

Li, Y., M. Horsman, N. Wu, C. Q. Lan and N. Dubois-Calero. 2008. Biofuels from microalgae. Biotechnol. Prog. 24: 815–820.

Linares, D., P. Fontanille and C. Larroche. 2009. Exploration of α-pinene degradation pathway of *Pseudomonas rhodesiae* CIP 107491. Application to novalic acid production in a bioreactor. Food Res. Int. 42: 461–469.

Lomascolo, A., C. Stentelaire, M. Asther and L. Lesage-Meessena. 1999. Basidiomycetes as new biotechnological tools to generate natural aromatic flavors for the food industry. Trends in Biotechnol. 17: 282–289.

Longo, M. A. and M. A. Sanromán. 2006. Production of food aroma compounds: microbial and enzymatic methodologies. Food Technol. Biotechnol. 44: 335–353.

Lupien, S., F. Karp, M. Wildung and R. Croteau. 1999. Regiospecific cytochrome P450 limonene hydroxylases from mint species: cDNA isolation, characterization, and functional expression of (–)-4S-limonene-3-hydroxylase and (–)-4S-limonene-6-hydroxylase. Arch. Biochem. Biophys. 368: 181–192.

Maldonado-Robledo, G., E. Rodriguez-Bustamante, A. Sanchez-Contreras, R. Rodriguez-Sanoja and S. Sanchez. 2003. Production of tobacco aroma from lutein. Specific role of the microorganisms involved in the process. Appl. Microbiol. Biotechnol. 62: 484–488.

Margetts, J. 2005. Aroma Chemicals V: Natural Aroma Chemicals. *In:* D. J. Rowe [ed.]. Chemistry and Technology of Flavors and Fragrances. Blackwell Publishing, UK. pp. 169–198.

Maróstica, Jr., M. R. and G. M. Pastore. 2007. Production of R-(+)-α-terpineol by the biotransformation of limonene from orange essential oil, using cassava waste water as medium. Food Chem. 101: 345–350.

Maróstica, Jr., M. R. and G. M. Pastore. 2006. Biotransformation of citronellol in rose-oxide using cassava wastewater as a medium. Ciênc. Tecnol. Aliment. 26: 690–696.

Mata, T. M., A. A. Martins and N. S. Caetano. 2010. Microalgae for biodiesel production and other applications: A review. Renew. Sust. Energ. Rev. 14: 217–232.

Max, B., J. Carballo, S. Cortés and J. Domínguez. 2012. Decarboxylation of ferulic acid to 4-vinyl guaiacol by *Streptomyces setonii*. Appl. Biochem. Biotechnol. 166: 289–299.

Medeiros, A. B. P., A. Pandey, P. Christen, P. S. G. Fontoura, R. J. S. Freitas and C. R. Soccol. 2001. Aroma compounds produced by *Kluyveromyces marxianus* in solid-state fermentation on packed bed column bioreactor. World J. Microbiol. Biotechnol. 17: 767–771.

Medeiros, A. B. P., A. Pandey, R. J. S Freitas, P. Christen and C. R. Soccol. 2000. Optimization of the production of aroma compounds by *Kluyveromyces marxianus* in solid-state fermentation using factorial design and response surface methodology. Biochem. Eng. J. 6: 33–39.

Menéndez, P., C. García, P. Rodríguez, P. Moyna and H. Heinzen. 2002. Enzymatic systems involved in d-limonene bio-oxidation. Braz. Arch. Biol. Technol. 45: 111–114.

Milledge, J. J. 2011. Commercial application of microalgae other than as biofuels: a brief review. Rev. Env. Sci. Biotechnol. 10: 31–41.

Mirata, M. A., M. Wüst, A. Mosandl and J. Schrader. 2008. Fungal Biotransformation of (±)-linalool. J. Agric. Food Chem. 56: 3287–3296.

Moreira, N., F. Mendes, P. G. Pinho, T. Hogg and I. Vasconcelos. 2008. Heavy sulphur compounds, higher alcohols and esters production profile of *Hanseniaspora uvarum* and *Hanseniaspora guilliermondii* grown as pure and mixed cultures in grape must. Int. J. Food Microbiol. 124: 231–238.

Moreno-Horn, M., E. Martinez-Rojas, H. Görisch, R. Tressl and L. A. Garbe. 2007. Oxidation of 1,4-alkanediols into γ-lactones via γ-lactols using *Rhodococcus erythropolis* as biocatalyst. J. Mol. Cat. B-Enzymatic. 49: 24–27.

Morrish, J. and A. Daugulis. 2008. Inhibitory effects of substrate and product on the carvone biotransformation activity of *Rhodococcus erythropolis*. Biotechnol. Lett. 30: 1245–1250.

Muheim, A. and K. Lerch. 1999. Towards a high-yield conversion of ferulic acid to vanillin. Appl. Microbiol. Biotechnol. 51: 456–461.

Pastore, G. M., Y. K. Park and D. B. Min. 1994. Production of fruity aroma by *Neurospora* from Beiju. Mycol. Res. 98: 1300–1302.

Pescheck, M., M. A. Mirata, B. Brauer, U. Krings, R. G. Berger and J. Schrader. 2009. Improved monoterpene biotransformation with *Penicillium* sp. by use of a closed gas loop bioreactor. J. Ind. Microbiol. Biotechnol. 36: 827–836.

Plaggenborg, R., J. Overhage, A. Steinbüchel and H. Priefert. 2003. Functional analyses of genes involved in the metabolism of ferulic acid in *Pseudomonas putida* KT2440. Appl. Microbiol. Biotechnol. 61: 528–535.

Ponzoni, C., C. Gasparetti, M. Goretti, B. Turchetti, U. G. Pagnoni, M. R. Cramarossa, L. Forti and P. Buzzini. 2008. Biotransformation of acyclic monoterpenoids by *Debaryomyces* sp., *Kluyveromyces* sp., and *Pichia* sp. strains of environmental origin. Chem. Biod. 5: 471–483.

Pramar, A., N. K. Singh, A. Pandey, E. Gnansounou and D. Madamwar. 2011. Cyanobacteria and microalgae: A positive prospect for biofuels. Bioresource Technol. 102: 10163–10172.

Pridmore, R. D., D. Crouzillat, C. Walker, S. Foley, R. Zink, M. C. Zwahlen, H. Brussow, V. Petiard and B. Mollet. 2000. Genomics, molecular genetics and the food industry. J. Biotechnol. 78: 251–258.

Rabenhorst, J. and R. Hopp. 2000. Process for the preparation of vanillin and suitable microorganisms. *DE patent*. 19532317.

Rao, S. C. V., R. Rao and R. Agrawal. 2003. Enhanced production of verbenol, a highly valued food flavourant, by an intergeneric fusant strain of *Aspergillus niger* and *Penicillium digitatum*. Biotechnol. Appl. Biochem. 37: 145–147.

Rasoul-Amini, S., E. Fotooh-Abadi and Y. Ghasemi. 2011. Biotransformation of monoterpenes by immobilized microalgae. J. Appl. Phycol. 23: 975–981.

Rottava, I., P. F. Cortina, C. E. Grando, A. R. S. Colla, E. Martello, R. L. Cansian, G. Toniazzo, H. Treichel, A. O. C. Antunes, E. G. Oestreicher and D. De Oliveira. 2010. Isolation and screening of microorganisms for *R*-(+)-Limonene and (–)-β-Pinene biotransformation. Appl. Biochem. Biotechnol. 162: 719–732.

Rottava, I., P. F. Cortina, E. Martello, R. L. Cansian, G. Toniazzo, A. O. C Antunes, E. G. Oestreicher, H. Treichel and D. De Oliveira. 2011. Optimization of α-terpineol production by the biotransformation of *R*-(+)-Limonene and (–)-β-pinene. Appl. Biochem. Biotechnol. 164: 514–523.

Sanchez-Contreras, A., M. Jiménez and S. Sanchez. 2000. Bioconversion of lutein to products with aroma. Appl. Microbiol. Biotechnol. 54: 528–534.

Savithiry, N., D. Gage, W. Fu and P. Oriel. 1998. Degradation of pinene by *Bacillus pallidus* BR425. Biodegradation. 9: 337–341.

Schrader, J., M. M. W. Etschmann, D. Sell, J. -M. Hilmer and J. Rabenhorst. 2004. Applied biocatalysis for the synthesis of natural flavour compounds—current industrial processes and future prospects. Biotechnol. Lett. 26: 463–472.

Shimoni, E., U. Ravid and Y. Shoham. 2000. Isolation of a *Bacillus* sp. capable of transforming isoeugenol to vanillin. J. Biotechnol. 78: 1–9.

Singh, R. N. and S. Sharma. 2012. Development of suitable photobioreactor for algae production —A review. Renew. Sust. Energ. Rev. 16: 2347–2353.

Smith, J. L., G. L. Boyer and P. V. Zimba. 2008. A review of cyanobacterial odorous and bioactive metabolites: Impacts and management alternatives in aquaculture. Aquaculture. 280: 5–20.

Smith, K., K. -M. Cho and J. Liao. 2010. Engineering *Corynebacterium glutamicum* for isobutanol production. Appl. Microbiol. Biotechnol. 87: 1045–1055.

Soares, M., P. Christen, A. Pandey and C. R. Soccol. 2000. Fruity flavor production by *Ceratocystis fimbriata* grown on coffee husk in solid-state fermentation. Process Biochem. 35: 857–861.

Sumby, K. M., P. R. Grbin and V. Jiranek. 2010. Microbial modulation of aromatic esters in wine: current knowledge and future prospects. Food Chem. 121: 1–16.

Surburg, H. and J. Panten. 2006. Common Fragrance and Flavor Materials. Preparation, Properties and Uses. 5th. Wiley-VCH Verlag, Weinheim.

Swiegers, J. H., S. M. G. Saerens and I. S. Pretorius. 2008. The development of yeast strains as tools for adjusting the flavor for fermented beverages to market specifications. *In:* D. Havkin-Frenkel and F. C. Belanger [ed.]. Biotechnology in Flavor Production. Blackwell Publishing., Oxford, UK. pp. 1–55

Taupp, M., D. Harmsen, F. Heckel, P. Schreier. 2005. Production of natural methyl anthranilate by microbial N-demethylation of N-methyl methyl anthranilate by the topsoil-isolated bacterium *Bacillus megaterium*. J. Agric. Food Chem. 53: 9586–9589.

Thompson, M., R. Marriott, A. Dowle and G. Grogan. 2010. Biotransformation of β-myrcene to geraniol by a strain of *Rhodococcus erythropolis* isolated by selective enrichment from hop plants. Appl. Microbiol. Biotechnol. 85: 721–730.

Torres, A. A., M. T. Martinez, A. B. Mir and R. M. M. Ochoa. 2011. Process of production of vanillin with immobilized micro-organisms. US Patent Application n° US 2011/0065156 A1.

Torres, B. R., B. Aliakbarian, P. Torre, P. Perego, J. M. Domínguez, M. Zilli and A. Converti. 2009. Vanillin bioproduction from alkaline hydrolyzate of corn cob by *Escherichia coli* JM109/pBB1. Enzyme Microb. Tech. 44: 154–158.

Trytek, M., J. Fiedurek and M. Skowronek. 2009. Biotransformation of (R)-(+)-limonene by the psychrotrophic fungus *Mortierella minutissima* in H_2O_2-oxygenated culture. Food Technol. Biotechnol. 47: 131–136.

Ugliano, M. and P. A. Henschke. 2009. Yeasts and wine flavour. *In:* M. V. Moreno-Arribas and M. C. Polo. [eds.]. Wine Chemistry and Biochemistry. Springer, New York, USA. pp. 313–392.

Unno, T., S. -J. Kim, R. A. Kanaly, J. -H. Ahn, S. -I. Kang and H. G. Hur. 2007. Metabolic characterization of newly isolated *Pseudomonas nitroreducens* Jin1 growing on eugenol and isoeugenol. J. Agric. Food Chem. 55: 8556–8561.

Van Beilen, J. B., R. Holtackers, D. Lüscher, U. Bauer, B. Witholt and W. A. Duetz. 2005. Biocatalytic production of perillyl alcohol from limonene by using a novel *Mycobacterium* sp. cytochrome P450 alkane hydroxylase expressed in *Pseudomonas putida*. Appl Environ Microbiol. 71: 1737–44.

Van Dyk, M. S., E. Van Rensburg and N. Moleleki. 1998. Hydroxylation of ()limonene, (–) α-pinene and (–)β-pinene by a *Hormonema* sp. Biotechnol. Lett. 20: 431–436.

Vandamme, E. J. 2003. Bioflavors and fragrances via fungi and their enzymes. Fungal Diversity. 13: 153–166.

Wang, C., C. K. Chin, C. T. Ho, C. F. Hwang, J. J. Polashock and C. E. Martin. 1996. Changes of fatty acids and fatty acid-derived flavor compounds by expressing the yeast Δ-9 desaturase gene in tomato. J. Agric. Food Chem. 44: 3399–3402.

Wang, Z., F. Lie, E. Lim, K. Li and Z. Li. 2009. Regio- and stereoselective allylic hydroxylation of D-limonene to (+)-trans-carveol with *Cellulosimicrobium cellulans* EB-8-4. Adv. Synth. Catal. 351: 1849–1856.

Wijffels, R. H., M. J. Barbosa and M. H. M. Eppink. 2010. Microalgae for the production of bulk chemicals and biofuels. Biofuels Bioprod. Biorefin. 4: 287–295.

Wu, S., U. Zorn, U. Krings and R. G. Berger. 2007. Volatiles from submerged and surface cultured beefsteak fungus *Fistulina hepatica*. Flavour Frag. J. 22: 53–60.

Xiao, Z., N. Xie, P. Liu, D. Hua and P. Xu. 2006. Tetramethylpyrazine production from glucose by a newly isolated *Bacillus* mutant. Appl. Microbiol. Biotechnol. 73: 512–518.

Xu, H., S. Jia and J. Liu. 2011. Development of a mutant strain of *Bacillus subtilis* showing enhanced production of acetoin. African J. Biotechnol. 10: 779–788.

Xu, P., D. Hua, L. Song, C. Ma, Z. Zhang, Y. Du, H. Chen, L. Gan, Z. Wei and Y. Zeng. 2009. *Streptomyces* strain and the method for converting ferulic acid to vanillin by using the same. US Patent Application n° US 2009/0186399 A1.

Xu, P., D. Hua and C. Ma. 2007. Microbial transformation of propenylbenzenes for natural flavour production. Trends in Biotechnol. 25: 571–576.

Yamauchi, H., O. Akita, T. Obata, T. Amachi, S. Hara and K. Yoshizawa. 1989. Production and application of a fruity odor in a solid-state culture of *Neurospora* sp. using pregelatinize polished rice. Agric. Biol. Chem. 53: 2881–2886.

Ye, X., S. Al-Babili, A. Klöti, J. Zhang, P. B. P. Lucca and I. Potrykus. 2000. Engineering the provitamin A (β-carotene) biosynthetic pathway into (carotenoid-free) rice endosperm. Sci. 287: 303–305.

Zhao, L. -Q., Z. -H. Sun, P. Zheng and J. -Y. He. 2006. Biotransformation of isoeugenol to vanillin by *Bacillus fusiformis* CGMCC1347 with the addition of resin HD-8. Process Biochem. 41: 1673–1676.

Zorn, H., S. Langhoff, M. Scheibner and R. G. Berger. 2003. Cleavage of β,β-carotene to flavor compounds by fungi. Appl. Microbiol. Biotechnol. 62: 331–336.

Zorn, H., M. A. Fraatz, S. J. L. Reimer, M. Takenberg, U. Krings, R. G. Berger and S. Marx. 2012. Enzymatic synthesis of nootkatone. US Patent Application N° US 2012/0045806 A1.

6

Engineering Applied to the Wine Fermentation Industry

Francisco López,[1] Alejandra Urtubia Urbina[2] and Jose Ricardo Pérez-Correa[3],*

ABSTRACT

Developing more efficient and better components as well as processes are the main concerns of the engineering disciplines. Therefore, they have multiple applications in the wine industry. In this chapter we focus on wine fermentations, which in our opinion is the heart of a wine cellar. We will first describe some simple kinetic models which simulate the evolution of the fermentation. We will then explain how to use simple models to design and manage the cooling demand in a cellar, illustrating the procedure with a real case study. An EXCEL spreadsheet with the calculation details of this case study can be requested from the authors.

[1] Departament d'Enginyeria Química, Facultat d'Enologia, Universitat Rovira i Virgili, Av. Països Catalans 26, Campus Sescelades, 43007 Tarragona, Spain.
Email: francisco.lopez@urv.cat
[2] Departamento de Ingeniería Química y Ambiental, Universidad Técnica Federico Santa María, Av. España 1680, Valparaíso, Chile.
Email: alejandra.urtubia@usm.cl
[3] ASIS-UC Interdisciplinary Research Program on Tasty, Safe and Healthy Foods, Departamento de Ingeniería Química y Bioproceso, Facultad de Ingeniería, Pontificia Universidad Católica de Chile, Vicuña Mackenna 4860, Macul, Santiago, Chile.
Email: perez@ing.puc.cl
* Corresponding author

In addition to being the most important operation in a winery, wine fermentation is a complex process where often the expected results are not achieved and sometimes the process fails completely. Therefore, this chapter not only mentions the standard methods to monitor and control wine making, but also discusses how spectroscopic methods can help achieve more reproducible and better fermentations. In a case study we illustrate how these methods are applied to monitor fructose and ethanol during a complete wine fermentation.

Introduction

All industrial bioprocesses are permanently challenged to reduce costs, increase productivity, minimize pollution and ensure product quality. Applying engineering methods in the design and operation of the winemaking process is essential to achieve these objectives. For example, strict control of the temperature in the different stages of wine elaboration is a common and widespread practice. The thermal conditioning of the grape on its arrival is already indispensable in some wineries, and even the grape harvest process in warm regions is adapted to this requirement (harvesting very early or at night). Pre-fermentative operations such as cryo-maceration, racking in white wines and skin maceration in red wines also require temperature control. Moreover, fermentation, conditioning, stabilization and storage of the finished wine, require specific temperatures that must be maintained. Therefore, it is increasingly important to carefully consider the energy management in wineries during the initial design phase as well as during the production phase. In both phases, the process engineer establishes and analyses mass and energy balances to provide solutions for specific requirements of the winery.

On the other hand, the quality of the raw material and the handling of the vineyard are not the only factors that must be considered to ensure the quality of the wine. How the alcoholic fermentation is carried out is also a significant factor. For example, sluggish and stuck fermentations are chronic problems in wineries all over the world that significantly degrade wine quality. These are caused by fermentation mishandlings when the yeast is subjected to several isolated or simultaneous stress factors, such as extreme values of pH, temperature, dissolved oxygen and osmotic pressure, or extreme concentrations of nutrients, ethanol and toxic fatty acid (Alexandre and Charpentier 1998, Blateyron and Sablayrolles 2001, Malherbe et al. 2007). Indeed, after yeast faces a given stressor (or a combination of them), there is a period of adjustment when it is possible to minimize its impact, avoiding fermentation problems (Bisson and Butzke 2000). It is important to know the root cause of the stress and to take timely corrective actions in order to recover a fermentation that has been exposed to stress. Therefore, close supervision and appropriate decisions by the winemaker are necessary to ensure a normal fermentation.

Alternatively, a well-designed automatic control system enables a more efficient and reproducible fermentation, reducing the risks of stuck and sluggish fermentations. In turn, all automatic control systems require reliable monitoring of the key process variables (Vojinovic et al. 2006). Monitoring can be classified as off-line, at-line or on-line, depending on the location of the sensor on the bioreactor (Lourenço et al. 2012). Off-line monitoring implies carrying the sample to the laboratory for analysis, which delays the measurements. In at-line monitoring, the samples are not sent to the laboratory; instead, they are processed in the vicinity of the bioreactor, reducing measurement delay. Finally, on-line monitoring is performed directly in the bioreactor and the results are instantly sent to the control room; this is the type of monitoring that an automatic control system requires.

In this chapter we first discuss several mathematical models of varying complexity that describe the kinetics of a wine fermentation. However, we focus on simple models that are sufficient to design and manage the cooling system in a commercial winery. We then propose a rational method to specify the maximum cooling required by a given winery, and we apply the method to a real case study. Next, we describe briefly how wine fermentations are usually controlled in a standard winery. On-line fermentation monitoring using spectroscopic methods is then discussed. The applicability of these methods for on-line monitoring of fructose and ethanol during the entire fermentation from juice to wine is illustrated with a case study.

Methods Details

Modeling wine fermentations

From a process engineering point of view, the following reaction provides a simple representation of the wine fermentation:

$$C_6H_{12}O_6 \rightarrow 2CH_3CH_2OH + 2CO_2 + \Delta H'_{ferm} \tag{1}$$

Where $\Delta H'_{ferm}$ is the energy involved in the fermentation, with a value of 167.1 kJ/mol; 61.0 kJ/mol are attributed to the maintenance needs of the yeasts, while 106.1 kJ/mol are released as heat during fermentation (ΔH_F). In actuality, the value of 100.3 kJ/mol proposed by Boulton et al. 1996 is more accepted for this liberated heat value. Thus, the total heat given off during fermentation is easily estimated knowing the initial concentration of fermentable sugars in the grape must. However, the difficulty is to know the rate at which that heat is released, and therefore, the cooling rate that is required to keep the temperature during fermentation under control.

This raises the need for modeling the kinetics of fermentation to design the cooling system.

Several kinetic models of the wine fermentation have been proposed in the literature. The model proposed by Del Nobile et al. 2003 predicts the evolution of biomass, ethanol, nitrogen and sugar from inoculation to death. Wine fermentations with a *Saccharomyces cerevisiae* strain at 25°C and at two initial nitrogen concentrations were used for validation. Also, Scaglia and associates (Scaglia et al. 2009) described a phenomenological model for an isothermal fermentation of a Syrah must in a batch bioreactor under controlled laboratory conditions. Similarly, a model developed for small scale bioreactors (Malherbe et al. 2004) includes several interactions between explicative variables and many parameters that need to be identified. Alternatively, a simple and yet complete model (Coleman et al. 2007) can predict flask-scale white wine fermentations duration and evolution accurately under several conditions of temperature (11 to 35°C), initial sugar (265 to 300 g/L) and initial nitrogen (70 to 350 mg/L). Pizarro and associates (Pizarro et al. 2007) have developed a hybrid model combining sugar uptake kinetics with a metabolic network, which accurately predicts glucose, fructose and ethanol evolution in an even wider range of operating conditions, both at laboratory and industrial scale (Assar et al. 2012). A combined model based on mathematical and statistical analysis of three wine fermentation models (Coleman et al. 2007, Pizarro et al. 2007 and Scaglia et al. 2009) was developed by Assar et al. 2012. This combined model matches a large set of experimental observations over a wide range of initial conditions.

For example, Palacios and associates (Palacios et al. 2009) used the model of Coleman and associates (Coleman et al. 2007) to predict the energy requirements during the different stages of the elaboration of white wine in a typical Argentinean winery (291 g/L initial sugar, 283 mg/L initial nitrogen and 0.15 g/L of yeast). They analyzed the fermentation process for different fermenters, noting that the cooling requirement varies linearly with their volumes. However, complete kinetic models are sometimes difficult to develop for a commercial winery. In practice, a simple model is sufficient to estimate the cooling requirements and to help the winemaker take appropriate operating decisions to minimize energy consumption. Some simple models that are easy to develop for any winery are described below.

Boulton Model (Boulton 1978). This model considers that the fermentation rate is constant and therefore the heat release too. The fermentation rate is derived from the initial sugars content and the estimated fermentation duration. Boulton estimated that the average sugar consumption rate is 2°Brix/day for white wines and 4 to 6°Brix/day for red wines, resulting in an average heat release rate of 460 kJ/m³h for white wines and 1360 kJ/m³h for red wines.

Flanzy Model (Flanzy 2000). This model also considers that the fermentation rate is constant and distinguishes between white and red wines. However, the fermentation rates are expressed in terms of sugars consumption: 2 kg/m³h for white wines and 7 kg/m³h for red wines.

Hidalgo-Tagores model (Hidalgo-Tagores 2003). Another option for a quick estimate is to calculate the total heat released and distribute it evenly during the fermentation. The calculation is given by:

$$P_{fer} = 1000 \frac{V_F S_0 \Delta H_F}{M 24 t_{fer}} \qquad (2)$$

where:

P_{fer}: heat rate released by the fermentation (kJ/h)

S_0:   initial sugars concentration (kg/m³)

V_F: must volume in the fermenter (m³)

ΔH_F: heat of fermentation (100.5 kJ/mol glucose)

M: molecular weight of glucose (180 g/mol)

t_{fer}:   fermentation length (days)

El Haloui Model (El Haloui et al. 1987). This model considers the relationship between the sugar consumed and the carbon dioxide or ethanol formed during fermentation. With this model, Colombié et al. 2007 and Palacios et al. 2009 estimated the heat released rate using the kinetic models of Malherbe et al. 2004 and Coleman et al. 2007, respectively.

Lopez and Secanell Model (Lopez and Secanell 1992). This model, derived from microvinifications data, correlates the heat released rate with the fermentation temperature, the initial sugar concentration and the total acidity of the grape must. The equation they developed is:

$$P_{fer} = 4187 \frac{k_1}{k_1 - k_2} \left(e^{-k_2 t} - e^{-k_1 t} \right) \qquad (3)$$

where:

P_{fer}: heat rate released by the fermentation (kJ/m³h)

k_1: parameter depending on the temperature and the total acidity

k_2: parameter depending on the temperature and the initial concentration of sugars

t: fermentation time (h)

Parameters k_1 and k_2 are calculated using the following correlations:

$$k_1 = K_{10} e^{-\frac{E_a}{RT} + K_{11} A_t}$$

$$k_2 = \frac{K_{20}}{S_0} e^{-\frac{E'_a}{RT}}$$

(4)

where:

T: fermentation temperature (K)

A_T: total acidity of the must (g/L tartaric acid)

S_0:   initial sugars concentration (kg/m³)

K_{10}:   dimensionless constant (4.1646·10¹⁴)

K_{11}:   constant (0.04848 L/g)

K_{20}:   constant (35.70 g/L)

E_a/R: constant (1.1206·10⁴ K)

E'_a/R: constant (480.875 K)

R: gas constant (8.314 J/mol·K)

Q_{fer}:   total heat released (kJ/m³)

The total heat released during the fermentation (Q_{fer}) can be obtained from integration of equation (3):

$$Q_{fer} = 4187 \frac{k_1}{k_1 - k_2} \left(\frac{1}{k_1} e^{-k_1 t} - \frac{1}{k_2} e^{-k_2 t} + \frac{k_1 - k_2}{k_1 k_2} \right)$$

(5)

This model considers a variable fermentation rate; hence, the heat released rate depends heavily on the must temperature and the progress of the fermentation.

Cooling requirements in a winery

This is a very important aspect to consider in the design and operation of a wine cellar. Using standard process engineering procedures, here we describe how to estimate these needs.

In a standard winery, cooling needs are considerable for all grape juice operations, i.e., in pre-fermentative operations such as racking and maceration, in fermentation and in tartaric stabilization. In this chapter, we will focus on those stages that are relevant during the harvesting period, i.e., pre-fermentative operations and fermentation. The objective is to analyze the impact of cooling on the global process of the elaboration of wine, stressing its strong influence on the design of a winery and its operation.

This analysis will allow winemakers to manage the harvest better and to establish operating conditions rationally.

First, we explain how to estimate the cooling needs prior to fermentation, i.e., racking in the case of white grape musts and maceration in the case of red grape musts. For this, an energy balance is applied,

$$E_{must} = \rho_{must} V_{must} C_{Pmust} \left(T_i - T_f\right) + E_{wall\_losses} \tag{6}$$

where:

E_{must}:   heat energy removed during racking or maceration (kJ)

E_{wall_losses}: heat transferred through the walls of the cooling tank (kJ)

ρ_{must}:   must density (kg/m³)

V_{must}:   volume of must to cool (m³)

C_{Pmust}:   specific heat capacity of the must (kJ/kg·°C)

T_i: must temperature before cooling (°C)

T_f: must temperature after cooling (°C)

The energy balance (eq. 6) provides the total heat that must be removed from the grape juice or grape pulp for pre-fermentative treatments. The winemaker or winery designer must specify the time needed to carry out this cooling. In turn, this time will determine the size of the heat exchanger to carry out this operation as well as to establish the required cooling rate. Therefore, equation (6) should be expressed in terms cooling rates using the flow rate of grape juice or grape pulp,

$$P_{must} = \rho_{must} m_{must} C_{Pmust} \left(T_i - T_f\right) + P_{wall\_losses} \tag{7}$$

where:

P_{must}:   cooling rate before racking or maceration (kJ/h)

P_{wall_losses}: heat transferred through the walls of the cooling tank (kJ/h)

ρ_{must}:   must density (kg/m³)

m_{must}:   flow rate of must to cool (m³/h)

C_{Pmost}:   specific heat capacity of the must (kJ/kg·°C)

T_i: must temperature before cooling (°C)

T_f: must temperature after cooling (°C)

From equation (7) we see that the flow rate of grape juice, the inlet grape temperature and the required temperature for prefermentative operations are key design and operating variables that mainly define the cooling needs at this stage of the winemaking process.

Fermentation is the second stage in a cellar where cooling is important since it is a strongly exothermic process. If the tank is not cooled, there is a high risk that the fermentation may get stuck due to the impossibility of the yeast to survive such stressful conditions.

The energy balance in this case is given by:

$$E_{acum} = E_{gen} + E_{wall} + E_{evap} + Q_{cool} \tag{8}$$

where:

E_{acum}: energy accumulated during fermentation (kJ)

E_{gen}: heat generated during fermentation (kJ)

E_{wall}: heat transferred through the walls of the fermenter (kJ)

E_{evap}: energy lost through evaporation/release of water, ethanol and CO_2 (kJ)

Q_{cool}: total cooling required for controlling the fermentation temperature (kJ)

However, for design and analysis it is more convenient to express equation (8) in terms of rates,

$$P_{acum} = P_{gen} + P_{wall} + P_{evap} + P_{cool} \tag{9}$$

where:

P_{acum}: accumulation rate of energy during fermentation (kJ/h)

P_{gen}: rate of heat generation during fermentation (kJ/h)

P_{wall}: heat transfer rate through the walls of the fermenter (kJ/h)

P_{evap}: rate of energy loss by evaporation/release of water, ethanol and CO_2 (kJ)

P_{cool}: cooling rate required to control the fermentation temperature (kJ/h)

To estimate the rate of heat generated during fermentation, any of the kinetic models described above can be used.

The accumulation term is given by,

$$P_{acum} = \rho_{must} V_F C_{Pmust} \frac{dT}{dt} \tag{10}$$

where:

ρ_{must}: must density (kg/m³)

V_F: volume of must in the fermenter (m³)

C_{Pmust}: specific heat capacity of the must (kJ/kg·°C)

T: must temperature in the fermenter (°C)

t: fermentation time (h)

The standard heat transfer equation can be used to estimate the heat exchange rate through the walls,

$$P_{wall} = UA(T - T_E)$$

(11)

where:

U: overall heat transfer coefficient (kJ/m²hK)

A: heat transfer area (m²)

T_E: external temperature (°C)

T: fermentation temperature (°C)

The heat transfer area depends on the geometry of the fermenter. The overall heat transfer coefficient can be determined by appropriate correlations (Incropera & De Witt 1996) or simple approximated procedures (Colombié et al. 2007); however, a rough estimated U value is good enough in most cases. Boulton et al. 1996 proposed a U value of 16.72 (kJ/m²h·°C) for the standard case of stainless steel tanks with external static air, containing either must or wine. If the tanks are outside in direct contact with sun rays, Flanzy (Flanzy 2000) suggests incorporating the contribution of radiation (400 w/m² in winter and 800 w/m² in the summer); however, the fermenter area should be taken into account to adjust these values.

The energy lost by evaporation can be estimated using different models (Williams and Boulton 1983, Vannobel 1988) depending on the release rate of CO_2; these models estimate that loss at approximately 4% of the energy produced during fermentation. However other authors (Boulton et al. 1996, Ribéreau-Gayon et al. 1998, Flanzy 2000) estimated these losses as 10% of the energy produced during fermentation.

The thermal capacity of a must with 200 g/L of sugar can be estimated as 3.70 (kJ/kg·°C) and for the corresponding wine as 3.62 (kJ/kg·°C). Using these values Colombié and associates (Colombié et al. 2007) derived the following linear expression applicable to any fermenting must:

$$C_{Pmust} = 3.62 + 4.178 \cdot 10^{-4} (S_0 - 2.17c_{EtOH})$$

(12)

where:

S_0:   initial sugar concentration (kg/m³)

C_{EtOH}: ethanol concentration in the must (kg/m³)

The density of the grape must, which depends on the progress of the fermentation and the must temperature, can be calculated using the equation of El Haloui and associates (El Haloui et al. 1987):

$$\rho_{must} = -1.085c_{EtOH} + 0.405S_0 - 0.031T + 996.925$$

(13)

For rapid estimations, the specific heat and density can be approximated to that of water, representing an over-prediction between 5% and 15%. However, Boulton and associates (Boulton et al. 1996) recommend more accurate values: for musts C_p = 3.8 (kJ/kg·°C) and density = 1090 kg/m³; for wines C_p = 4.5 (kJ/kg·°C) and density = 996 kg/m³.

Case Study: Cooperative Vilarodona

The methods mentioned above to estimate cooling needs for a typical co-operative winery in Spain are applied in the following case study. In addition, various design and operation options to reduce these cooling needs without sacrificing product quality are assessed.

Important variables to consider in an energy analysis of a winery are: the *inlet temperature* and the *rate of entry* of the harvest, and the *total amount of grape* that the winery processes. The rate of entry is affected by the ripeness of the grape, the period of maturation of each variety and cultural aspects. We discuss the impact these aspects have in the energy needs of a co-operative winery in Tarragona (Spain).

Typical harvest

In this winery, more than 11 million kg of grapes were processed during the 2009 campaign, according to the pattern detailed in Table 1.

In addition, Fig. 1 presents the daily entrance of grape in the cooperative during the harvest. There are two main white varieties (Macabeo and Parellada) which are collected in different periods. Normally, smaller quantities of other varieties are collected at the same time; for example, Tempranillo is collected simultaneously with Macabeo. In addition, there are days with very small entrance or no entrance at all. These low entrances are mainly due to holidays in the middle of the harvest. In turn, there are days with very high entrance due to the presence of family and friends that

Table 1. 2009 harvest.

Variety	Input (kg)
Macabeo	4,605,985
Parellada	5,136,140
Chardonnay	103,795
Total white	**9,845,920**
Tempranillo	1,438,615
Cabernet Sauvignon	65,313
Total red	**1,503,928**
TOTAL	**11,349,848**

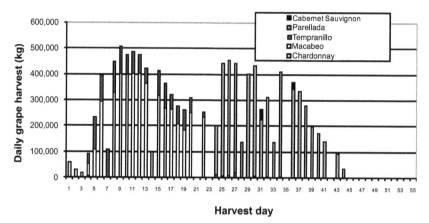

Fig. 1. Daily rate of entrance.

help harvesting. All these factors plus the evolution of grape ripening, cause large fluctuations in the daily arrivals of grapes in the cellar.

Next, the energy consumption for a typical harvest is calculated and the impact that the winery behavior has on energy costs (mostly cooling) is analyzed. In addition, we discuss those measures that could be taken in order to minimize these costs.

Estimation of cooling needs

First, design and operating conditions should be established, as summarized in Table 2 for this case study. In our analysis, we used equation (7) to estimate the cooling needs before fermentation and the model of Lopez and Secanell (Lopez and Secanell 1992) for fermentation, considering variable cooling needs. Losses by evaporation, heat transfer through the fermenters and heat exchanger walls, and cooling of red grape juice/pulp were not taken into account in our analysis. The total generated fermentation heat (calculated by equation 5) has been distributed according to the values listed in Table 3 for a standard fermentation of 9 days.

Figure 2 shows how cooling needs vary throughout the harvest. These variations are mainly due to the white grape juice cooling prior to clarification and the must cooling during fermentation. In addition, the figure shows the cooling needs estimated with the model by Hidalgo-Tagores (Hidalgo-Tagores 2003) (eq. 2) for an average fermentation length of 9 days. When the estimations from both models are compared, no significant difference can be appreciated. Therefore, even the simplest models are useful for the designer and winemaker to estimate the cooling requirements and minimize energy costs in the winery.

Table 2. Design and operating conditions, and physical properties.

Variable/Parameter	units	value
Initial sugars concentration of red grape juice, S_0	(kg/m^3)	240
Initial sugars concentration of white grape juice, S_0	$(kg/m)^3)$	180
Density of white grape juice	$(kg/m)^3)$	1050
Density of red grape juice	(kg/m^3)	1150
Specific heat capacity of the must	$(kJ/kg \cdot ^\circ C)$	3.8
White grape juice yield, L_{juice}/kg_{grape}	(%)	85
Red grape juice yield, L_{juice}/kg_{grape}	(%)	85
Total acidity of white grape juice, $g_{Tartaric\ acid}/L_{juice}$	(g/L)	7.0
Total acidity of red grape juice, $g_{Tartaric\ acid}/L_{juice}$	(g/L)	7.0
Inlet temperature of white grapes	$(^\circ C)$	24
Inlet temperature of red grapes	$(^\circ C)$	24
Red wine fermentation temperature	$(^\circ C)$	23
White wine fermentation temperature	$(^\circ C)$	18
White grape juice cooling time	(h)	12
Fermentation length	(day)	9

Table 3. Distribution of the total heat of fermentation.

Day of fermentation	Fraction of heat generated from the total heat of fermentation
1	0.10
2	0.18
3	0.18
4	0.15
5	0.12
6	0.10
7	0.07
8	0.06
9	0.04

Optimizing harvest entry

The grapes rate of entrance has a remarkable influence on the cooling needs; hence, savings can be derived by properly managing the daily volumes of incoming grape shipments. For example, if the grape entrance is rationalized to average values of 260,000 kg/day for Macabeo, 270,000 kg/day for Parellada and 80,000 kg/day for Tempranillo, the resulting daily cooling needs are much less variable as shown in Fig. 3.

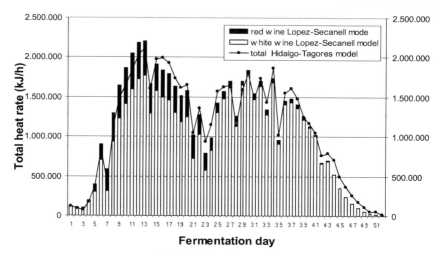

Fig. 2. Daily cooling needs.

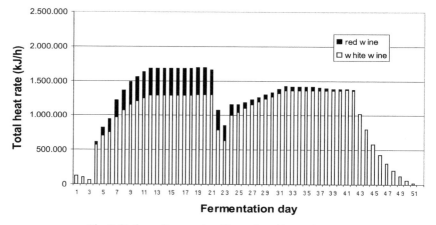

Fig. 3. Daily cooling needs with a rational harvest entry.

Comparing Figs. 2 and 3 shows that by simply rationalizing the harvest entry, the maximum daily cooling can be reduced from 2,200,000 (kJ/h) to 1,690,000 (kJ/h), representing a saving of 23%. Given that this high cooling is only required for 15 days, portable cooling equipment might be used for these days, reducing the cooling capacity of the winery to just 1,425,000 (kJ/h), i.e., an additional 12% savings in installed power.

Moreover, if part of the harvest is carried out at night, the cooling needs can be reduced even further since the grapes, and therefore the grape juice or pulp, will enter the winery at a lower temperature. We tested this idea by reducing the average entry temperature from 24°C to 22°C. In this case, maximum cooling needs were reduced to 1,275,000 (kJ/h), resulting in a total saving of 42%. Furthermore, if white wines are fermented at a slightly higher temperature, say 20°C instead of 18°C, maximum cooling needs would become 1,175,000 (kJ/h), obtaining a total saving of 47%.

In short, the use of a simple model for the estimation of the cooling requirements allowed us to significantly improve the design and operation of our winery. We have seen that good design and good management can reduce energy costs by close to 50%. All calculations above were performed with an EXCEL spreadsheet that can be requested from the authors.

Monitoring and Control of Wine Fermentations

The standard case

Next, we briefly describe how a standard wine fermentation is currently managed. We will then explain how commercially available technology can be used to improve the operation of this process significantly.

Although off-line monitoring, as commonly used in wine fermentations, provides accurate measurements, there is a delay of several hours between the sample and the analytical results, thus greatly hindering the operation, often causing slow or stuck fermentations, both generating lower quality wines. Even more, in a standard wine cellar, most of the analyses of key process variables are performed daily.

To define the initial settings, before fermentation the grape juice is characterized through the following analysis: soluble solids (°Brix), total acidity, assimilable nitrogen, total and free SO_2. During fermentation, traditional monitoring (OIV Standard Methods 2009) includes periodic measurements (between 1 and 3 times per day) of density, soluble solids (°Brix), total and volatile acidity, assimilable nitrogen, reducing sugars and ethanol. It is essential that temperature and pH be maintained in the range defined by the winemaker, given the strong influence these variables have on the normal development of the fermentation.

To ensure a wine with desired characteristics, the fermentation kinetics should be controlled (Sablayrolles 2009). Several factors allow this control: yeast strain, addition of nutrients (oxygen, nitrogen) and temperature. Given its high fermentation capability, *Saccharomyces cerevisiae* is by far the most popular yeast, with more than 200 species commercially available with different fermentation properties. To increase the rate of fermentation, nitrogen can be added (Bely et al.1990, Colombié et al. 2007) as ammonium salts (diamin phosphate, DAP). In turn, adequate additions of oxygen allow higher yeast concentrations, prolonging its fermentative capacity (Ribéreau-Gayon 1999); however, timing and doses of oxygen are still highly empirical (Pszczolkowski et al. 2001). Although sparsely used in commercial wineries, managing fermentation temperature can be extremely effective to control the fermentation rate. High temperatures accelerate the fermentation rate, and temperature also directly influences the assimilation of amino acid by yeasts and the evolution of short-chain fatty acids which are fermentation inhibitors. In actuality, temperature management is subjected to the availability of the operator and the speed in decision-making by the winemaker. It is handled empirically and almost exclusively to avoid high values that would increase the risk of getting a stuck or sluggish fermentation (Pszczolkowski et al. 2001). However, there are few wineries that use automatic temperature control.

In summary, existing methods for measuring must composition during fermentation are sub-optimal for the demands of the modern wine industry, where speed and cost factors must be considered.

Advanced monitoring and control

Although off-line measurements can be useful to develop predictive mathematical models that help improve the performance of the fermentation through a pre-programmed control, these models are not able to adequately reproduce the complexity of industrial fermentations, especially in extreme situations that generate operating problems. Effective control of wine-making requires up-to-date information, diagnosing the state of the process in real time and hence taking timely corrective actions that will prevent complications. Thus, on-line monitoring of the fermentation rate and the main factors that could cause stuck or sluggish fermentation would greatly facilitate the work of the winemaker.

There are several established methods for on-line measurement of key process variables, such as density, ethanol concentration and generated CO_2 (Sablayrolles 2009), so as to monitor the fermentation rate. In addition, advanced techniques are being increasingly used to achieve more complete monitoring of the process. These are called advanced because they provide fast measurements simply, with minimum preconditioning of the sample, even in some cases using non-invasive or non-destructive methods. An example of this is the monitoring of a given metabolite through specific biosensors (Piermarini et al. 2011). There are many techniques such as infrared spectroscopy (IR), ultrasound or magnetic resonance (Resa et al. 2009, Schöck 2010), that allow the simultaneous monitoring of several metabolites. Furthermore, electronic noses and tongues provide sensory and organoleptic assessments (Cozzolino et al. 2006, Buratti et al. 2011).

Monitoring fermentations using one or more of the above techniques and applying data processing techniques (chemometrics, data mining) to identify fermentation patterns would allow timely detection and diagnosis of abnormal behavior (Urtubia et al. 2007, Urtubia et al. 2012).

Infrared spectroscopy

Given the increasing market demands for consistent, high quality and differentiating wines, the industry needs to incorporate monitoring and control techniques that are fast, reliable and robust. Normally used analytical methods require preparation, expensive equipment (HPLC chromatography of gases, etc.) and sometimes multiple steps of purification, which obviously impede results from being obtained quickly. In addition, the traditional methods used in the wine industry for measuring key process variables such as temperature and density are slow and unreliable, since they are subjected to analytical and human errors, and moreover, most wineries continue to handle this information on paper.

Infrared spectroscopy (IR) is an analytical technique used off-line at the laboratory in many wineries, although it can be applied on-line directly in fermentation tanks also. Since infrared radiation interacts with all molecules through vibrational and rotational molecular excitation, IR is particularly useful to identify and quantify chemical functional groups in a sample by analysis of the absorption pattern. The absorption wavelength patterns are useful to identify the molecules present in the sample, while the intensity of this absorption pattern is related to their concentration in the sample (Burns and Ciurczak 2001).

The infrared spectrum is usually divided into three regions: near infrared (NIR) from 13000 cm^{-1} to 4000 (800–2500 nm); medium infrared (MIR) from 4000 cm^{-1} to 400 (2500–25000 nm); and far infrared (FIR) from 400 to 40 cm^{-1} (25000–250000 nm). In the NIR region, all the absorption bands are overtones or combinations of overtones originated by the fundamental frequencies observed in the MIR region (Burns and Ciurczak 2001). However, superimposed vibration states in the MIR region which look narrow and strong, typically translate into extremely broad and weak absorption bands in the NIR region. On the other hand, the MIR region covers a wider spectrum characteristic which, depending on the sample, in some cases is difficult to analyze (Sablayrolles 2009).

The vibrational frequencies of the molecules are observed in the MIR region. This radiation is absorbed by specific atomic bonds between given organic molecules, especially those containing a large number of -OH, -CH and -NH chemical bonds. Splitting the MIR spectrum into two regions simplifies the spectral analysis. The region between 1500 and 4000 cm^{-1} contains many fundamental stretching bands that can be associated with specific atomic pairs (Christy et al. 2001). The fingerprint region, between 1500 and 400 cm^{-1}, contains a mixture of vibrational sources, which sometimes makes it difficult to assign all the absorption bands. However, rather than identifying individual functional groups, many of the bands in the fingerprint region characterize an entire molecular structure. Consequently, a compound can be identified by its fingerprint when compared with reference spectra (Perkin Elmer 1993).

Applications

This technology offers several advantages: it provides direct analysis of organic compounds, requires little or no sample preparation; it is fast (less than 2 minutes), easy to use, noninvasive and nondestructive. It is also highly accurate, easy to automate and capable of measuring on-line many components simultaneously. There are currently several low-cost IR instruments (diode array monochromator based spectrophotometers) on the market, plus fast computers and efficient chemometric software; hence, this technology is increasingly accessible.

Both NIR and MIR spectroscopy have advantages and limitations; therefore the best choice depends on the expected application. For example, NIR spectroscopy is easier to implement with optical fiber and associated instrumentation is robust and low-cost. Several off-line, at-line and on-line NIR instruments are commercially available. On the other hand, available MIR instruments are expensive, heavy (30–70 kg) and only of the off-line and at-line type; however, MIR is widely recognized as a fast and nondestructive technique to analyze various types of samples.

Specific MIR and NIR on-line and at-line instruments for wine analysis such as Foss, Brukker optics and Thermo have been in the market for quite a while. Even though these instruments are specifically designed for wine samples, they cannot be used directly and immediately to analyze all types of wine. Generally, they come with several calibrations obtained with a given set of sample wines that do not necessarily include the kind of wines produced in a given winery (country, grape variety, terroir, etc.). To obtain reliable measurements, new calibrations with samples of the type of wine to be measured should be developed.

Several NIR and MIR applications which are relevant to the wine industry have been reported: anthocyanins in red grapes (Cozzolino et al. 2006), phenolic compounds in fermenting musts and wines (AWRI report 2009), and reducing sugars in grapes, fermenting musts and ageing wine (Fernández-Novales et al. 2009, Herrera et al. 2003). Blanco et al. 2004 used NIR spectroscopy and multivariate analysis to achieve a fast and non-destructive determination of ethanol, glucose, biomass, glycerin and acidity in the fermentation of *Saccharomyces cerevisiae*. They employed an off-line technique, although these authors are testing a new probe for fast and ideally on-line measurements. Other studies describe the use of VIS-NIR spectroscopy and chemometrics to monitor alcoholic fermentations (Cozzolino et al. 2006), identify phenolic compounds (Cozzolino et al.

2004), total anthocyanins (Janik et al. 2007) and even determine elemental composition (Cozzolino et al. 2008).

In turn, MIR spectroscopy has been used to control the designation of origin, to monitor wine ageing (Palma and Barroso 2002), to classify dry extracts of red wine according to its geographical origin (Picque et al. 2001), to discriminate red wines based on the analysis of phenolic extracts (Edelman et al. 2001) and to analyze polysaccharide extracts of white wines (Coimbra et al. 2002). Also, Fayolle and associates (Fayolle et al. 1996) studied the effect of temperature on the monitoring of several metabolites during the alcoholic fermentation of an artificial must. Others have used MIR to measure glucose, fructose, glycerol, ethanol, some organic acids; total SO_2, total phenols and acidity (total and volatile) in wines and musts (Schindler et al. 1998, Dubernet and Dubernet 2000, Patz et al. 1999, 2004, Urtubia et al. 2004).

If reliable measurements are required, specific calibrations for the intended application should be developed. These are mathematical relationships between the spectrum and the metabolite concentrations obtained with reference analyses, which can be difficult, expensive and slow to get. Calibrations must be validated with samples of the type of wines or musts that will be measured.

In summary, IR spectroscopy allows a simple, fast, low-cost analysis throughout the chain of wine production, starting from the grapes and ending with the bottled wine.

Case Study: Monitoring of Fructose and Ethanol During Wine Fermentation

Here we describe how MIR spectroscopy could be used to monitor fructose and ethanol during fermentation of Chilean red wines from grape juice to wine. This covers a wide range of fructose and ethanol concentrations that should be considered for developing the calibration. The instrument used in this case study was an FT-IR detector DTGS KBr, with spectral resolution of 0.5 cm^{-1}, working ranges of 200–740 nm and 1350–28500 nm, i.e., includes NIR and MIR regions. This instrument came with a set of calibrations to measure different compounds in grape juice and wine samples. No calibration for fermenting musts was included; hence measurements would be expected to have biases as the alcoholic fermentation progresses. Thus, a new calibration and validation was developed with samples covering a wide range of Chilean fermenting musts from the 2002 harvest taken in a winery located in the Maipo Valley. The samples were obtained from 4 batches of Cabernet Sauvignon, one of Syrah, one of Carmenère, one Merlot and one of Pinot Noir. Between 30 and 35 samples per batch were collected, giving a total of 273 samples with their respective reference analysis by HPLC. Partial least squares (PLS) were used to develop the calibrations; the optimal number of PLS factors and the standard error of cross-validation (SECV) were obtained by cross-validation. A total of 200 samples covering the whole fermentation were used in calibrations, while the remaining 73 samples were used for validation. Results are shown in Table 4.

Table 4. Calibration performance.

Compound	Calibration			Concentration range
	r²	PLS factors	SECV	
Fructose	0,994	9	4.9 g/L	0–133 g/L
Ethanol	0,99	9	1.1% v/v	0–15.4% v/v

Figure 4 compares the developed calibration (*CR*) with those included in the instrument (*Must* and *Wine*) and the reference analysis (HPLC) values, for monitoring a wine fermentation. For both metabolites the *Must* calibration provides accurate measurements at the beginning of the fermentation, while the *Wine* calibration is accurate at the end of the fermentation. However, the *CR* calibration provides values closer to those of the reference analysis during the whole fermentation.

Conclusions

In this chapter we discussed how engineering methods can be applied to enhance the management of a commercial winery as well as to improve the performance and reliability of a wine fermentation. In particular, with relatively little effort, we applied simple models to find several strategies to improve the design and operation of a given winery that would reduce nearly 50% of the peak energy consumption. Additionally, we explained how IR instrumentation can be used to monitor the evolution of fructose and ethanol during a wine fermentation, although this application requires much more effort. This monitoring is fast, requires minimal pre-treatment and allows better control of the process, reducing the risks of sluggish or stuck fermentations.

Engineering methods involving technology, process knowledge and winemaker's experience are useful to face the challenges of the wine industry nowadays, i.e., reduce costs, obtain better products, being energy efficient and environmentally friendly.

(a)

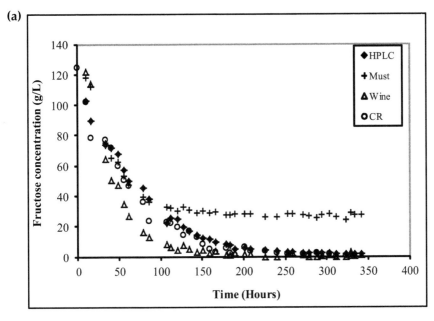

(b)

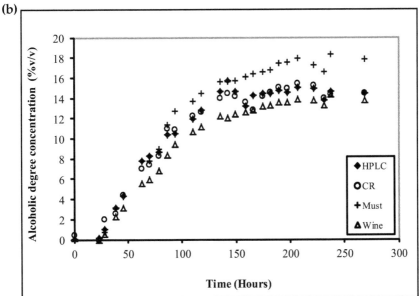

Fig. 4. Comparison between calibrations (*CR, grape juice* and *wine)* and the reference analysis for fructose (a) and ethanol (b) (Urtubia et al. 2004).

References

Alexandre, H. and C. Charpentier. 1998. Biochemical aspects of stuck and sluggish fermentation in grape must. J. Ind. Microbiol. Biotechnol. 20: 20–27.

Assar, R., F. Vargas and D. J. Sherman. 2012. Reconciling Competing Models: A Case Study of Wine Fermentation Kinetics. Algebraic and Numeric Biology. Lecture Notes in Computer Science. *In:* K. Horimoto, M. Nakatsui and N. Popov (eds.). Springer, Berlin, Heidelberg.

Australian Wine Research Institute AWRI. 2009. Near Infrared spectroscopy in the Australian grape and wine sector. http://www.awri.com.au/wp-content/uploads/nir_fact_sheet.pdf.

Bely, M., J. M. Sablayrolles and P. Barré. 1990. Automatic Detection of Assimilable Nitrogen Deficiencies during Alcoholic Fermentation in Oenological Conditions. J. Ferment. Bioeng. 70: 246–252.

Bisson, L. and C. Butzke. 2000. Diagnosis and rectification of stuck and sluggish fermentations. Am. J. Enol. Viticult. 51: 168–177.

Blanco, M., A. Peinado and J. Mas. 2004. Analytical Monitoring of Alcoholic Fermentation Using NIR Spectroscopy. Biotech. Bioeng. 88: 536–542.

Blateyron, L. and J.M. Sablayrolles. 2001. Stuck and Slow Fermentations in Enology: Statistical Study of Causes and Effectiveness of Combined Additions of Oxygen and Diammonium Phosphate. J. Biosci. Bioeng. 91: 184–189.

Boulton, R. 1978. A kinetic model for the control of wine fermentation. Biotechnol. Bioeng. Symp. 9: 167–172.

Boulton, R., V. L. Singleton, L. Bisson and R. Kundee. 1996. Principles and practices of winemaking. Chapman and Hall, New York, USA.

Buratti, S., D. Ballabio, G. Giovanelli, C. Zuluanga, A. Moles, S. Benedetti and N. Sinelly. 2011. Monitoring of alcoholic fermentation using NIR and MIR spectroscopy combined with electronic nose and electronic tongue. Anal. Chim. Acta. 697: 67–74.

Burns, D. and E. Ciurczak. 2001. Handbook of Near-Infrared Analysis. Marcel Dekker, Inc. New York, Basel, Hong Kong.

Christy, A., Y. Ozaki and V. Gregoriou. 2001. Modern Fourier Transform Infrared Spectroscopy. Elsevier Barcelona, Spain.

Coimbra, M., F. Goncalves, A. Barros and I. Delgadillo. 2002. Fourier Transform Infrared Spectroscopy and Chemometric analysis of White Wine Polysaccharide Extracts. J. Agric. Food. Chem. 50: 3405–3411.

Coleman, M. C., R. Fish and D. E. Block. 2007. Temperature-Dependent Kinetic Model for Nitrogen-Limited Wine Fermentations. Appl. Environ. Microbiol. 73: 5875–5884.

Colombié, S., S. Malherbe and J. M. Sablayrolles. 2007. Modeling of heat transfer in tanks during wine-making fermentation. Food Control 18: 953–960.

Cozzolino, D., M. Parker, R. Dambergs, M. Herderich and M. Gishen. 2006. Chemometrics and Visible-Near Infrared spectroscopic monitoring of red wine fermentation in a pilot scale. Biotechnol. Bioeng. 95: 1101–1107.

Cozzolino, D., M. Kwiatkowski, R. Dambergs, W. Cynkar, L. Janik, G. Skouroumounis and M. Gishen. 2008. Analysis of elements in wine using near infrared spectroscopy and partial least squares regression. Talanta. 74: 711–716.

Cozzolino, D., M. Kwiatkowski, M. Parker, W. Cynkar, R. Dambergs, M. Gishen and M. Herderich. 2004. Prediction of phenolic compounds in red wine fermentations by visible and near infrared spectroscopy. Anal. Chim. Acta. 513: 73–80.

Del Nobile, M.A., D. D'Amato, C. Altieri, M. R. Corbo and M. Sinigaglia. 2003. Modeling the Yeast Growth-Cycle in a Model Wine System. J. Food Sci. 68: 2080–2085.

Dubernet, M. and M. Dubernet. 2000. Utilisation de l'analyse infrarouge à transformée de Fourier pour l'analyse oenologique de routine. Revue Française d'Oenologie 181: 10–13.

Edelman, A., J. Diewok, K. Scuster and V. Lendl. 2001. Rapid Method for the Discrimination of Red Wine Cultivates Based on Mid-Infrared Spectroscopy of Phenolic Wine Extracts. J. Agric. Food Chem. 49: 1139–1145.

El Haloui, N., D. Picque and G. Corrieu. 1987. Mesures physiques permettant le suivi biologique de la fermentation alcoolique en oenologie. Sciences des Aliments 7: 241–265.

Fayolle, F., D. Picque, B. Pereet, E. Latrille and G. Corrieu. 1996. Determination of Major Compounds of Alcoholic Fermentation by Middle-Infrared Spectroscopy: Study of Temperature Effects and Calibration Methods. Appl. Spectrosc. 50: 1325–1330.

Fernández-Novales, J., M. López, M. Sánchez, J. Morales and V. González-Caballero. 2009. Shortwave-near infrared spectroscopy for determination of reducing sugar content during grape ripening, winemaking, and aging of whites and red wines. Food Res. Int. 42: 285–291.

Flanzy, C. 2000. Enología: Fundamentos científicos y tecnológicos. AMV Ediciones y Mundi prensa Madrid España.

Herrera, J., A. Guesalaga and E. Agosin. 2003. Shortwave-near infrared spectroscopy for non-destructive determination of maturity of wine grapes. Meas. Sci. Technol. 14: 689–697.

Hidalgo-Tagores, J. 2003. Tratado de Enología, Mundi Prensa Libros S.A. Madrid España.

International Organization of Vine and Wine (OIV). 2009. Compendium of international methods of wine and must analysis.

Incropera, F. P. and D. P. De Witt. 1996. Introduction to Heat Transfer. John Wiley & Sons, New York. U.S.A.

Janik, L., D. Cozzolino, R. Dambergs, W. Cynkar and M. Gishen. 2007. The prediction of total anthocyanin concentration in red-grape homogenates using visible-near-infrared spectroscopy and artificial neural networks. Anal. Chim. Acta. 594: 107–118.

Lopez, A. and P. Secanell. 1992. A simple mathematical empirical model for estimating the rate of heat generation during fermentation in white wine making. Int. J. Refrig. 15: 1–5.

Lourenco, N., J. Lopes, C. Almeida, M. Sarraguca and H. Pinheiro. 2012. Bioreactor Monitoring with spectroscopy and chemometrics: a review. Anal. Bioanal. Chem. (in press).

Malherbe, S., F. Bauer and M. Du Toit. 2007. Understanding Problem Fermentation—A review. S. Afr. J. Enol. Vitic. 28: 169–186.

Malherbe, S., V. Fromion, N. Hilgert and J.M. Sablayrolles. 2004. Modeling the effects of assimilable nitrogen and temperature on fermentation kinetics in enological conditions. Biotechnol. Bioeng. 86: 261–272.

Palacios, C. A., S. M. Udaquiola and R. Rodríguez. 2009. Modelo matemático para la predicción de las necesidades de frío durante la producción de vino. Ciencia, Docencia y Tecnología. 38: 205–226.

Palma, M. and C. Barroso. 2002. Application of FT-IR spectroscopy to the characterization and classification of wines, brandies and other distilled drinks. Talanta. 58: 265–271.

Patz, C., A. Blieke, R. Ristow and H. Dietrich. 2004. Application of FT-MIR spectrometry in wine analysis. Anal. Chim. Acta. 513: 81–89.

Patz, C., A. David, K. Thente, P. Kürbel and H. Dietrich. 1999. Wine Analysis with FTIR Spectrometry. Vitic. Enol. Sci. 54: 80–87.

Perkin Elmer. 1992–1993. Infrared Spectroscopy Tutorial and Reference IRTutor.

Picque, D., T. Cattenoz and G. Corrieu. 2001. Classification of red wines analysed by middle infrared spectroscoscopy of dry extract according to their geographical origin. J. Int. Sci. Vigne Vin. 35: 165–170.

Piermarini, S., G. Volpe, M. Esti, M. Simonetti and G. Palleschi. 2011. Real time monitoring of alcoholic fermentation with low-cost amperometric biosensors. Food Chem. 127: 749–754.

Pizarro, F., C. Varela, C. Martabit, C. Bruno, J. R. Pérez-Correa and E. Agosin. 2007. Coupling Kinetic Expressions and Metabolic Networks for Predicting Wine Fermentations. Biotechnol. Bioeng. 98: 986–998.

Pszczólkowski, P., P. Carriles, M. Cumsille and M. Maklouf. 2001. Reflexiones sobre la madurez de Cosecha y las condiciones de vinificación, con relación a la Problemática de

Fermentaciones Alcohólicas Lentas y/o paralizante, en Chile, Facultad de Agronomía. Pontificia Universidad Católica de Chile.

Resa, P., L. Elvira, F. Montero De Espinosa, R. González and J. Barcenilla. 2009. On-line ultrasonic velocity monitoring of alcoholic fermentation kinetics. Bioprocess Biosys. Eng. 32: 321–331.

Ribereau-Gayón, P., D. Dubourdieu, B. Donèche and A. Lonvaud. 1998. Traité d'oenologie. Dunod, Buenos Aires Argentina.

Ribéreau-Gayon, P. 1999. Reflexions sur les Causes et les Conséquences des Arrêts de la Fermentation Alcoholique en Vinification. J. Int. Sci. Vigne Vin. 33: 39–48.

Sablayrolles, J. M. 2009. Control of alcoholic fermentation in winemaking: Current situation and prospect. Food Res. Int. 42: 418–424.

Scaglia, G. J., P. M. Aballay, C. A. Mengual, M. D. Vallejo and O. A. Ortiz. 2009. Improved phenomenological model for an isothermal winemaking fermentation. Food Control. 20: 887–895.

Schindler, R., R. Vonach, B. Lendl and R. Kellner. 1998. A rapid automated method for wine analysis based upon sequential injection (SI)-FTIR spectrometry. Fresenius J. Anal. Chem. 362: 130–136.

Schöck, T. and T. Becker. 2010. Sensor array for the combined analysis of water-sugar-ethanol mixtures in yeast fermentations by ultrasound. Food Control. 21: 362–369.

Urtubia, A., G. Hernández and J. M. Roger. 2012. Detection of abnormal fermentations in wine process by multivariate statistics and pattern recognition techniques. J. Biotechnol. 159: 336–341.

Urtubia, A., J. R. Pérez-Correa, A. Soto and P. Pszczólkowski. 2007. Using data mining techniques to predict industrial wine problem fermentations. Food Control. 18: 1512–1517.

Urtubia, A., J. R. Pérez-Correa, M. Meurens and E. Agosin. 2004. Monitoring large scale wine fermentations with infrared spectroscopy. Talanta. 64: 778–784.

Vannobel, C. 1988. Refroidissement d'un moût en fermentation par évaporation naturelle d'eau et d'alcool. Connaissance de la Vigne et du Vin. 22: 169–187.

Vojinovíc, V., J. M. Cabral and L. Fonseca. 2006. Real-time bioprocess monitoring. Part I: *in situ* sensors. Sens. and Actuators B. 114: 1083–1091.

Williams, L. A. and R. Boulton. 1983. Modeling and prediction of evaporative ethanol loss during wine fermentations. Am. J. Enol. Viticult. 34: 234–242.

7

Applications of Plant Growth-Promoting Bacteria for Plant and Soil Systems

M.L.E. Reed[1],* and Bernard R. Glick[2]

ABSTRACT

Plant growth-promoting bacteria (PGPB) are typically soil bacteria that stimulate plant growth that can be used in a variety of ways when plant growth enhancements are required, often in association with plant roots, and sometimes on leaves or flowers, or within plant tissues. The most intensively studied use of PGPB has been in agriculture and horticulture where a number of PGPB formulations are currently available and are widely used as commercial products to enhance crop production, both under contained greenhouse as well as field conditions. In recent years, there has been only limited use of PGPB in forest regeneration, however there is increasing interest in using these bacteria in the phytoremediation of contaminated soils. Despite that, the use of genetically engineered PGPB under field conditions is currently extremely rare. As the biochemical and genetic mechanisms of plant growth promotion by these bacteria continue to be better understood, it is likely that this increasing understanding will translate into the development of more efficacious strains of PGPB.

---
[1] Monsanto Canada Inc., 810-180 Kent Street, Ottawa, Ontario, Canada, K1P 0B6.
[2] Department of Biology, University of Waterloo, Waterloo, Ontario, Canada, N2L 3G1.
* Correspondence author: lucy.reed@monsanto.com

Introduction

The world's population currently includes around 7 billion people and is expected to increase to approximately 8 billion some time around the year 2020. As a direct consequence of increases in both environmental damage and worldwide population pressure, global food production will need to become more efficient to feed all of the world's people. Thus, it is essential that agricultural productivity be significantly increased within the next few decades. Motivated by increasing demand, and by the awareness of the environmental and human health damage that can occur as a consequence of overuse of pesticides and fertilizers, agricultural practice is moving to a more sustainable and environmentally favourable approach. This includes both the increasing use of transgenic plants (e.g., http://www.isaaa.org/inbrief/default.asp) and plant growth-promoting bacteria (Reed and Glick 2004) as a part of mainstream agricultural practice.

PGPB typically occupy one of several niches with respect to their plant host including i) living inside of plant tissues (bacterial endophytes), ii) occupying specific structures on the plant (e.g., symbiotic bacteria such as rhizobia with root nodules) or iii) colonizing the area around plant roots (i.e., rhizosphere bacteria). Moreover, PGPB may promote plant growth using any of a wide variety of both direct and indirect mechanisms (Glick 1995), with different bacteria employing various combinations of these mechanisms. Some of the direct mechanisms of plant growth employed by PGPB include the provision of bioavailable phosphorus for plant uptake, nitrogen fixation for plant use, sequestration of iron for plants by siderophores, production of plant hormones like auxins, cytokinins and gibberellins, and lowering of plant ethylene levels (Glick 1995, Glick et al. 1999). The indirect mechanisms used by PGPB involve decreasing or preventing some of the damage to plants that occurs as a consequence of various phytopathogens. These indirect mechanisms include antibiotic production, reduction of the iron available to phytopathogens in the rhizosphere, synthesis of fungus inhibiting metabolites and cell wall-lysing enzymes, synthesis of hydrogen cyanide, induced systemic resistance and competition with detrimental microorganisms for sites on plant roots. Most of these direct and indirect mechanisms are well understood, and are readily selectable or genetically manipulatable so it should be possible to develop and utilize optimized PGPB in a wide range of applications in agriculture, horticulture, forestry and environmental remediation.

Applications of PGPB in Agriculture and Horticulture

By far, the major uses of PGPB have been in agriculture and horticulture (Table 1). Following the early inconsistent results on the use PGPB pioneered

Table 1. Selected examples of responses to PGPB inoculation on crops, fruits, vegetables and trees.

Bacteria	Plant	Experimental conditions	Results	Reference
Achromobacter peichaudii	Tomato	Growth chamber	– selected salt-tolerant PGPB isolate increased plant water use efficiency – significant dry weight increase (53%) in 172 mM NaCl	Mayak et al. 2004a
Achromobacter sp. *Azospirillum* sp. *Burkholdiera* sp.	Sunflower	Growth chamber	– increased dry shoot weights of 58 to 77% – enhanced N uptake of 62 to 140% – no significant increase of K uptake	Ambrosini et al. 2012
Acinetobacter baumannii CD-1 *Bacillus megaeorium* GC subgroup A *Bacillus subtilus* BA-142 *Pantoea agglomerans* FF	Tomato, cucumber	Greenhouse	– increased mineral content of tomato and cucumber fruit – FF produced highest fruit number per plant, fruit weight per plant, plant length, and dry weight	Dursun et al. 2010
Acinetobacter rhizospharae	Pea, chickpea, maize, barley	Greenhouse and field	– greenhouse testing showed significant increases in growth of all plant species – pea plant in field test showed significant increases in yield of 33%	Gulati et al. 2009
Agrobacterium sp. *Phyllobacterium* sp. *Pseudomonas* sp. *Vario-vovax* sp.	Canola	Growth chamber	– significantly increased root dry weight from 11 to 52% – largest promotion effect by *Phyllobacterium* sp.	Bertrand et al. 2001
Agrobacterium radiobacter	Beech, scotch pine	Greenhouse	– biomass of beech increased up to 235% – biomass of pine up to 15%	Leyval and Berthelin 1989
Arthrobacter citreus *Pseudomonas fluorescens* *Pseudomonas putida*	Black spruce, jack pine, white spruce	Greenhouse	– increased height and biomass	Beall and Tipping 1989

Table 1. contd....

Table 1. contd.

Bacteria	Plant	Experimental conditions	Results	Reference
Arthrobacter citreus *Pseudomonas putida biovar B* *Pseudomonas fluorescens* *Serratia liquefaciens*	Canola (*Brassica campestris* L. and *Brassica napus* L.)	Field and greenhouse	– in greenhouse, selected strains produce 57% increase in yield – in field, select strains increase seedling emergence and vigor – yield increase from 6 to 13% over two year test period	Kloepper et al. 1988
Arthrobacter oxydans *Pseudomonas aurefaciens*	Douglas fir	Greenhouse and field	– increased height and biomass up to 68% – increased branch and root weight – variability in response depending on ecotype	Chanway and Holl 1994
Azospirillum sp.	Wheat, Maize	Field	– for wheat, increases of yield from 15 to 30%, and increases in yield of 50–60% when fertilized, over seven years – increases of maize yield from 15 to 25% observed, and with fertilization, yield increased up to 40%, over six years	Okon and Labandera-Gonzalez 1994
Azospirillum sp.	Maize	Field	– increased yield from 6.7 to 75.1%	Kapulnik et al. 1981
Azospirillum sp.	Wheat	Field	– changes in yield from –9.6 to 14.8%	Reynders and Vlassak 1982
Azospirillum sp.	Wheat	Field	– changes in yield from –15.8 to 31%	Baldani et al. 1987
Azospirillum sp.	Millet	Field	– changes in yield from –12.1 to 31.7%	Kloepper et al. 1989
Azospirillum sp.	Mustard	Field	– increased yield from 16 to 128%	Kloepper et al. 1989
Azospirillum sp.	Rice	Field	– increased yield from 4.9 to 15.5%	Kloepper et al. 1989

Azospirillum sp.	Maize	– significant increases in yield in light soils and with moderate nitrogen fertilization – in fields with no nitrogen fertilization, yield increased over uninoculated plants, but not statistically significant – in fields with high nitrogen fertilization, no growth enhancement effect	Okon and Labandera-Gonzalez 1994
Azospirillum sp.	Maize	– no effect of inoculation on plant yields when soils are heavy and high in nitrogen content – in light soils low in nitrogen fertilization, yield increase from 11 to 14%	Fallik and Okon 1996
Azospirillum sp.	Maize	– significant increase in dry matter yield – increased magnesium content	Hernandez et al. 1997
Azospirillum sp.	Sorghum	– increased yield from 20.5 to 30.5%	Kapulnik et al. 1981
Azospirillum sp.	Sorghum	– increased yield from 12 to 18.5%	Sarig et al. 1998
Azospirillum sp.	Greenhouse	– increased activity of glutamate dehydrogenase and glutamine synthetase – increased N content in leaves and roots	Ribaudo et al. 2001
Azospirillum sp.	Wheat	Greenhouse	– increased biomass, grain yield, protein content and plant nitrogen content – speculated that nitrogen uptake likely mechanism of plant growth promotion	Saubidet et al. 2002
Azospirillum sp.	Maize and soybean	– dry shoot yield was not enhanced with inoculation for both maize and soybean – significant differences seen among differing soil types	Laditi et al. 2012
Azospirillum sp. *Azotobacter* sp.	Sunflower	– increased oil and protein content in grain – increased vegetative growth	Akbari et al. 2011
Azospirillum sp. *Azotobacter* sp.	Maize	– increase in dry matter yield with seed inoculation	Sharifi et al. 2011

Table 1. contd....

Table 1. contd.

Bacteria	Plant	Experimental conditions	Results	Reference
Azospirillum sp. B510	Rice	Greenhouse and field	– greenhouse testing showed significant increases in leaf length and biomass – in field, significant increase in panicles and tiller numbers, but not seed weight	Isawa et al. 2010
Azospirillum brasilense Cd, Az-39	Wheat	Field	– increased plant yield, especially with Az-29	Caceres et al. 1996
Azospirillum brasilense NO40	Rice	Field	– increased yield by 15–20% in two test locations	Omar et al. 1989
Azospirillum brasilense Sp245	Wheat	Growth chamber hydroponics	– growth promotion not affected by fungicide Tebuconazole – growth enhancement effect determined after 72 hr growth	Pereyra et al. 2009
Azospirillum brasilense Sp-111	Wheat	Field	– yield increases of 1.3 to 2 fold over five years – variable results due to climactic conditions	Okon and Labandera-Gonzalez 1994
Azospirillum brasilense Sp-245, Sp-107st	Wheat	Field	– significant increases of grain yield and plant nitrogen content – strain Sp-245 most effective on wheat	Boddey and Dobereiner 1988
Azospirillum brasilense Cd Az-39 *Azospirillum lipoferum* Az-30	Millet	Field	– increased yield up to 30% and 21% over two years	Di Ciocco and Rodriguez-Caceres 1994
Azospirillum brasilense Cd *Azospirillum lipoferum* Br-17	Maize	Field	– consistently increased yield at intermediate soil fertility – replaced 35–40% of nitrogen fertilizer requirements	Okon and Labandera-Gonzalez 1994
Azospirillum brasilense Sp 245 *Azospirillum irakense* KBCl	Winter wheat, Maize	Field	– plant growth promotion effect inhibited by over fertilization with nitrogen – in plots with low nitrogen, higher yields not obtained	Dobbelaere et al. 2001

Azospirillum brasilense *Azospirillum lipoferum* *Azotobacter chroococcum* *Pseudomonas fluorescens*	Maize	Field	– combined inoculation with all PGPB species had highest promoting effect on corn phenology	Hamidi et al. 2009
Azospirillum brasilense **DSM 1690** *Azospirillum lipoferum* **DSM 1691** *Pseudomonas fluorescens* **R-93, DSM 50090** *Pseudomonas putida* **DSM 291, R-168**	Maize	Pot experiments and field	– greater stimulatory effect in unsterilized soil – all bacterial strains significantly increased seed yield (up to 44%), plant height (up to 22%) and leaf area (up to 73%).	Gholami et al. 2009
Azospirillum brasilense **Sp 245** *Bacillus megatorum* **M-3** *Bacillus subtilis* **OSU-142** *Burkholderia gladii* **BA-7**	Grapevine	Greenhouse	– increased growth promotion, especially with Sp 245, which significantly increased (10%) leaf chlorophyll concentrations – OSU-142 significantly increased vegetative development (21% root dry weight) and mineral nutrient uptake	Sabir et al. 2011
Azospirillum brasilense *Psuedomonas pumilus*	*Prosopis articulate* *Parkinsonia microphylla* *Parkinsonia florida*	Screenhouse	– no significant response by *P. florida* – for other two species, survival increased, as well as leaf gas exchange	Bashan et al. 2009
Azospirillum lipoferum	Sunflower	Greenhouse	– strains originally selected from plant rhizosphere – increased germination response	Fages and Arsac 1991
Azospirillum lipoferum **B1(AZ1), B2(AZ9), B3(AZ45)**	Wheat	Pot experiment	– test performed in drought-stress conditions – significant yield increases of 43% and 100% over control, respectively in 50% and 25% of field capacity moisture with B3 inoculation – strain B2 had best survival in low-moisture conditions	Arzanesh et al. 2011

Table 1. contd....

Table 1. contd.

Bacteria	Plant	Experimental conditions	Results	Reference
Azospirillum lipoferum **CRT1**	Maize	Field	– growth promotion effect observed, despite rapid decrease of bacterial density of introduced bacteria – plant height, primary root length and root fresh weight all enhanced by the addition of the bacteria	Jacoud et al. 1998
Azospirillum lipoferum **CRT1**	Maize	Field	– average grain yields and N content higher for the inoculated plants, but not statistically significant – larger root systems, and lower grain moisture	Dobbelaere et al. 2001
Azospirillum lipoferum **CRT-1**	Maize	Field	– positive response of yield to inoculation regardless of cultivar or soil type	Fages 1994
Azospirillum lipoferum **N7** *Gluconacetobacter azotocaptans* **DS1** *Pseudomonas putida* **CQ179**	Maize	Greenhouse	– significantly increased root and shoot weights	Mehnaz and Lazarovits 2006
Azospirillum lipoferum *Pseudomonas fluorescens* *Pseudomonas putida*	Maize	Field	– increased root biomass of roots significantly (59–23%)	Adjanohoun et al. 2011
Azospirillum lipoferum *Pseudomonas putida*	Maize	Field	– increased grain yield in both optimal and drought stress conditions	Moslemi et al. 2011
Azospirillum zeae **N7** *Burkholderia phytofirmans* **E24** *Enterobacter cloacae* **CR1** *Pseudomonas pudita* **CR7** *Sphingobacterium canadense* **E24, CR 11** *Streptophomonas matlophilia* **CR3**	Maize	Greenhouse and field	– significant increases in root and shoot dry weight with sterilized sand of 10–32% – in two year field study, no significant yield increase over control plots	Mehnaz et al. 2010

Organism	Plant	Experiment	Effects	Reference
Azotobacter sp. Bacillus sp. Enterobacter sp. Xanthobacter sp.	Rice	Field	– increased total dry matter yield, grain yield, and nitrogen accumulation by 6 to 24% over two years – yield increases likely due to increase in root length, leaf area and chlorophyll content	Alam et al. 2001
Azotobacter chroococcum	Barley	Growth chamber assays	– increased seed germination and seedling development – no change germination rate, but increased root lengths – addition of nitrate decreases plant stimulation effect – inconsistent results, authors conclude Azotobacter is not a reliable inoculant	Harper and Lynch 1979
Azotobacter chroococcum	Quercus serrata	Outdoor pot assay	– biomass increase up to 38%	Pandey et al. 1986
Azotobacter chroococcum Bacillus megaterium	Eucalyptus	Potted plant experiment	– increased biomass up to 44%	Mohammed and Prasad 1998
Azotobacter chroococcum Pseudomonas corrugata	Amaranthus paniculatus, Eleusine coracana		– increased plant growth and nitrogen content – hypothesized that growth promotion effect is due to stimulation of native bacterial communities	Pandey et al. 1999
Bacillus spp. (selected strains)	Cocoa	Growth chamber	– high abiotic stress tolerance (10% NaCl, 50 C temperature) – increased seedling vigour	Thomas et al. 2011
Bacillus sp.	Sorghum	Field	– increased yield from 15.3 to 33%	Broadbent et al. 1977
Bacillus sp. M-3 Bacillus sp. OSU-142	Raspberry	Field	– significantly increased yield (g fruit/plant, 34 to 75%), cane length (up to 15%), cluster per cane (up to 36%) – increased N, P and Ca content in leaves	Orhan et al. 2006
Bacillus sp. M-3 Bacillus sp. OSU-142 Microbacterium sp. FS01	Apple	Field	– significantly increased yield (kg/tree, 26 to 88%), fruit weight (14 to 25%), shoot length (16 to 30%) and shoot diameter (16 to 18%) – increased N, P K and Ca in tree leaves	Karlidag et al. 2007

Table 1. contd....

Table 1. contd.

Bacteria	Plant	Experimental conditions	Results	Reference
Bacillus sp. M-3 **Bacillus sp. OSU-142** **Pseudomonas sp. BA-8**	Strawberry	Field	– root and/or shoot inoculation significantly increased yield per plant (up to 26%) – root applications significantly increased total sugar (10%), soluble and solid contents (3%) – no effect on fruit weight and pH	Pirlak and Kose 2009
Bacillus sp. M-3 **Bacillus sp. OSU-142** **Pseudomonas sp. BA-8**	Strawberry	Field	– increased fruit yield of 10 to 30%	Esitken et al. 2010
Bacillus sp. M-3 **Bacillus sp. OSU-142** **Burkholderia sp. OSU-7** **Pseudomonas sp. BA-8**	Apple	Field	– one year old apple tree roots inoculated – fruit yield significantly increased (137%) OSU-142 – selected isolates increased shoot yield (up to 30%), shoot length (up to 59%) and shoot diameter (16%) – no significant increases in trunk diameter	Aslantasa et al. 2007
Bacillus sp. OSU-142	Apricot	Field	– fruit yield increase per tree of 30% in first year following treatment and 90% in second year	Karlidag et al. 2010
Bacillus sp. **Pseudomonas sp.** **Serratia liqufaciens**	Maize	Greenhouse	– increased yield from 8 to 14 % – *S. liquefaciens* and *Pseudomonas* sp. gives highest stimulation effect in different soils	Lalande et al. 1989
Bacillus amyliquefaciens IN937a **Bacillus pumilis T4**	Tomato	Greenhouse	– equivalent increased plant growth in 75% recommended fertilization rate, as 100% fertilization rate with no PGPB	Adesemoye et al. 2009
Bacillus amyliquefaciens IN937a **Bacillus pumilis T4**	Tomato	Greenhouse	– significantly increased ^{15}N uptake – significant dry weight increase	Adesemoye et al. 2010

Bacillus amyliquefaciens IN937 *Bacillus cereus* C4 *Bacillus pumilis* INR7, SE34 *Bacillus subtilis* GB03	Tomato, Pepper	Field	– significant increases of stem diameter, stem area, leaf surface area, weights of roots and shoots and number of leaves, over two years – transplant vigor and fruit yield improved – pathogen numbers and disease not reduced in tomatoes or pepper with the exception of reduction of galling in pepper by root-knot nematode	Kokalis-Burelle et al. 2002
Bacillus amyloliquefaciens GB99 *Paenobacillus macerans* GB122	Pepper	Field	– no impact on field populations of green peach aphid (*Myzus persicae*) – increase in yield (1.7–2.3 times greater) in first harvest – no yield increase in subsequent harvests	Boutard-Hunt et al. 2009
Bacillus cereus C1L	Maize	Field and greenhouse	– significant increases in plant height and dry weight – protection against Southern corn leaf blight	Huang et al. 2010
Bacillus cereus *Bacillus subtilis* *Enterobacter aerogenes* *Enterobacter agglomerans* *Pseudomonas putida*	Cucumber	*In vitro* and greenhouse	– most strains increased root length in *Pythium*-infected plants *in vitro* – in greenhouse, increased weight of cucumber plants by 29%, fruit yield by 14% and fruit number by 50% by *B. subtilis*	Uthede et al. 1999
Bacillus cereus RC18 *Bacillus licheniformis* RC08 *Bacillus megaterum* RC07 *Bacillus subtilus* RC11 *Bacillus* OSU-142 *Bacillus* M-13 *Pseudomonas putida* RC06 *Paenibacillus polymyxa* RC05, RC14	Wheat Spinach	Greenhouse	– for wheat, all bacterial strains increased fresh shoot weight by 16–54%, and leaf area size by 6–47% – in spinach all bacterial strains increased fresh show weight by 2–53% and leaf area size by 5–49%	Cakmakci et al. 2007b
Bacillus cereus S18 *Pseudomonas* sp. W34	Lettuce, Tomato	Pot experiment	– significantly reduced galling and enhanced seedling biomass in soils infested with *Meloidogyne incognita* – *B. cereus* S18 increases yield up to 9% compared to the control	Hoffmann-Hergarten et al. 1998

Table 1. contd....

Table 1. contd.

Bacteria	Plant	Experimental conditions	Results	Reference
Bacillus licheniformis CECT 5105 *Bacillus pumilis* CECT 5106	Silver spruce	Greenhouse	– significant increase in above-ground plant growth – no increase in root system development – increased N content in plant	Probanza et al. 2002
Bacillus licheniformis *Phylobacterium* sp.	Mangrove	Greenhouse	– doubling of N content in plant – increased leaf development	Bashan and Holguin 2002, Rojas et al. 2001
Bacillus licheniformis RC02 *Rhodobacter capsulatus* RC04 *Paenibacillus polymyxa* RC05 *Pseudomonas putida* RC06 *Bacillus* OSU-142 *Bacillus megaterum* RC01 *Bacillus* M-13	Barley	Greenhouse	– increased root weight 18%–32% – increased shoot weight 29%–54%	Cakmakci et al. 2007a
Bacillus macauensis 1PC-11	Squash	Greenhouse	– significant reduction in Phytophthora blight (*Phytophthora capsi*) disease in three separate trials	Zhang et al. 2010
Bacillus polymyxa	Douglas fir, lodgepole pine, white spruce	Growth chamber and greenhouse	– lodgepole pine had significant increases in root dry weight and number and length of secondary roots – pine had increased in root growth, emergence, height, weight – white spruce showed increased seedling emergence	Chanway et al. 1991
Bacillus polymyxa	Western hemlock	Greenhouse	– increased seedling height and biomass up to 30% – growth promotion effects differ with biovar type	Chanway 1995
Bacillus polymyxa	Wheat	Field	– increased plant yield	Caceres et al. 1996

Bacillus polymyxa L6	Lodgepole pine	Greenhouse	– co-inoculation with mycorrhizae increased root and shoot biomass – no effect with only bacterial inoculation	Chanway and Holl 1991
Bacillus polymyxa L6	Lodgepole pine	Growth chamber	– significant increases in shoot and root weights up to 35% at 6 weeks of growth	Holl and Chanway 1992
Bacillus polymyxa *Burkholderia* sp. *Pseudomonas* sp.	Sugar beet	Field	– significantly increased root yield (6.1 to 13.0%) and sugar yield (2.3 to 7.8%) – yields further enhanced by N, P and NP applications	Cakmakci et al. 2001
Bacillus polymyxa BcP26 *Mycobacterium phlei* MbP18 *Pseudomonas alcaligenes* PsA15	Maize	Pot experiments	– greater increase of plant growth stimulation in nutrient-deficient soil	Egamberdiveya 2007
Bacillus polymyxa *Pseudomonas fluorscens*	Loblolly pine, slash pine	Greenhouse	– significantly increased seedling emergence rate and biomass – reduced damping-off in seedlings – some strains increase loblolly pine root length	Enebak et al. 1998
Bacillus polymyxa *Pseudomonas fluorscens*	Hybrid spruce	Field	– increased seedling dry weight up to 57% at 5 of 9 test sites – increased dry weight of mature plants at 4 of 9 test sites – some growth inhibition at some sites	Chanway et al. 2000
Bacillus polymyxa *Staphylococcus hominis*	Hybrid spruce	Greenhouse	– significant growth increased up to 59%	O'Neill et al. 1992
Bacillus pumilus SE34 *Bacillus subtilus* GB03	Rice	Greenhouse	– protection from bacterial leaf blight (*Xanthomonas oryzae* pv. *oryzae*), 50–71% protection – most effective protection from liquid inoculation or talc powder delivery system	Chithrashree et al. 2011
Bacillus pumilus *Pseudomonas fluorescens* **Pf5**	Highbush blueberry		– increased leaf area and stem diameter	de Silva et al. 2000

Table 1. contd....

Table 1. contd.

Bacteria	Plant	Experimental conditions	Results	Reference
Bacillus subtilis A-13	Peanut	Field	– yield increases up to 37%, 22 of 24 test sites produced positive results – plant responses most positive when subjected to stress like limited water, poor nutrition, cold temperatures – plant disease reduced	Turner and Backman 1991
Bacillus subtilis AF 1	Pigeon pea	Field	– chitin supplemented peat formulation with AF 1 seed treatment increased emergence and dry weight by 29 % and 33% – cell suspension of AF 1 as seed treatment, increased emergence and dry weight by 21% and 30%	Manjula and Podile 2005
Bacillus subtilis B2	Onion	Growth chamber	– significantly increased shoot dry weight (12–94%), dry root weight (13–100%) and shoot height (12–40%) over controls – rhizosphere populations of inoculated bacteria decreased throughout study, but growth promotion effect still observed	Reddy and Rahe 1989
Bacillus subtilis RC11 **Paenibacillus polymxa RC105** **Paenibacillus polymxa RP23/2** **Pseudomonas putida RF29/2**	Italian ryegrass	Field	– RC21 and RP24/3 produced greater dry weight yield – all strains increased protein content	Yolcu et al. 2011
Bacillus subtilis **Pseudomonas spp.**	Hybrid spruce	Greenhouse and field	– significant increases in biomass at all test sites of 10 to 234%, and decreased transplant injury with *Pseudomonas* inoculation – *B. subtilis* strain ineffective	Shishido and Chanway 2000

Beijerinckia mobilis *Clostridium* sp.	Beet, Barley, Wheat, Red radish, Cucumber	Lab experiments and greenhouse	– increased biomass plant production by 1.5 to 2.5 times with mineral fertilization – positive increases seen with seed germination rate, mean plant length and plant weight – cucumber cultivars demonstrated variability in seed germination response	Polyanskaya et al. 2000
Burkholderia vietnamiensis TVV75	Rice	Outdoor pot and field trials	– plants inoculated and transplanted at day 24 – increased shoot weight (up to 33%), root weight (up to 55%) and leaf surface (up to 30%) observed – end grain yield increase of 13–22% – grain weight significantly increased due to inoculation	Tran Van et al. 2000
Enterobacter cloacae CAL3	Tomato, Pepper, Mung bean	Greenhouse	– positive growth response by seedling of all three plant species, especially tomato, where no exogenous mineral nutrients added – early stimulation effect on seedlings observed	Mayak et al. 2001
Hydrogenophaga pseudoflava *Pseudomonas putida*	Hybrid spruce	Field	– *H. psuedoflava* increased biomass of seedlings – *P. putida* increased seedling biomass in two trials, but inhibitory effects in other trials	Chanway and Holl 1993
Methylobacterium fujisawaense	Canola	Pouch assay	– promotion of root elongation	Madhaiyan et al. 2006
Paenibacillus polymyxa B5, B6	Peanut	Greenhouse and field	– both strains suppressed crown rot disease (*Aspergillus niger*) in field and greenhouse conditions – strain B5 had greater biofilm-producing capacity and was superior in disease suppression	Haggag and Timmusk 2007
Pseudomonas spp.	Potato	Field	– changes in yield from –10 to 37%	Howie and Echandi 1983
Pseudomonas spp.	Potato	Field	– changes in yield from –9 to 20%	Geels et al. 1986

Table 1. contd....

Table 1. contd.

Bacteria	Plant	Experimental conditions	Results	Reference
Pseudomonas spp.	Apple	Greenhouse and field	– increased growth in seedlings up to 65% and root biomass up to 179% – biocontrol effects observed against some pathogenic fungi	Caesar and Burr 1987
Pseudomonas spp.	Potato	Field	– changes in yield from –14 to 33%	Kloepper et al. 1989
Pseudomonas spp.	Rice	Field	– increased yield from 3 to 160%	Kloepper et al. 1989
Pseudomonas spp. (fluorescent Strains)	Winter wheat	Field	– biocontrol effects observed against 'take all' (*Gaeumannomyces graminis*) – 27% yield increase	De Freitas and Germida 1990
Pseudomonas sp.	Potato	Growth chamber	– significant increases in root dry weight (44 to 201%), stem length (26 to 28%), lignin up to 43%, and enhanced stem hair formation (55 to 110%)	Frommel et al. 1991
Pseudomonas spp.	Lettuce, Cucumber, Tomato, Canola	Hydroponic growth chamber	– increased root and shoot weights for all plants tested – most significant increased growth responses in lettuce, tomato and cucumber	Van Peer and Schippers 1998
Pseudomonas spp. (fluorescent strains) A1, B10, TL3, BK1, E6	Potato	Greenhouse and field	– treated seed pieces produce larger root systems in greenhouse – significantly increased yield in all test fields, however strains promote plant growth differently depending on soil type – early plant responses correlated to increased yields	Kloepper et al. 1980

Pseudomonas spp. (fluorescent strains) A1, B2, B4, E6, RV3, SH5	Sugar beet	Greenhouse and field trials	– increased seedling mass by all strains in greenhouse – effects of inoculation positive on yield in field trials, but results variable from site to site – hypothesized that promotion effect due to antagonism of plant disease	Suslow and Schroth 1982
Pseudomonas sp. DW1	Eggplant	Greenhouse	– significant increase of growth and Ca2+ content, and leaf superoxide dismutase activity in high salt conditions	Fu et al. 2010
Pseudomonas sp. 7NSK2	Maize, Barley, Wheat	Field	– increased yield from 15 to 25%	Iswandi et al. 1987
Pseudomonas sp. PsJN	Potato	Greenhouse and field trials	– in greenhouse, increase of whole plant dry weight, results not influenced by soil sterility – in field, early emergence stimulated and significant tuber yield increases in 3 of 4 trials	Frommel et al. 1993
Pseudomonas aerugenosa N39 *Pseudomonas putida* N21 *Serratia proteamaculans* M35	Wheat	Feild	– N21 was most effective, significant increases in plant height (52%), root length (60%), grain yield (76%), 100–grain weight (19%) and straw yield (67%)	Zahir et al. 2009
Pseudomonas aurantiaca SR1	Wheat, Maize	Field	– wheat yield increases in grain (36%) and root volume (23–27%) – maize yield increases in grain (11%) and root volume (36–42%) – higher yields with inoculation at fertilization rates (40 kg/ha) lower than conventionally applied (80–100 kg/ha)	Rosas et al. 2009
Pseudomonas cepacia *Pseudomonas fluorescens* *Pseudomonas putida*	Winter wheat	Potted plants in growth chamber	– biocontrol against *Rhizoctonia solani* and *Leptosphaera maculans* – strains differentially stimulate growth o fplant parts – increased plant growth in less fertile soil	De Freitas and Germida 1990

Table 1. contd....

Table 1. contd.

Bacteria	Plant	Experimental conditions	Results	Reference
Pseudomonas cepacia **MR85, R85** *Pseudomonas putida* **MR111, R105**	Winter wheat	Field	– bacteria inoculated on plants able to overwinter on roots, with levels reaching 10^4 to 10^8 CFU/g – significantly increased wheat grain yields at several locations, however overall results were not significant due to variability of results between trials	De Freitas and Germida 1992b
Pseudomonas cepacia **R55, R85** *Pseudomonas putida* **R104**	Winter wheat	Growth chamber	– antagonism demonstrated against *Rhizoctonia solani* – increased dry weight of inoculated plants (62–78%) grown in *R. solani* infected soil – dry root weight increased by 92–128% and shoot dry weight increased by 28–48%	De Freitas and Germida 1991
Pseudomonas cepacia **R85** *Pseudomonas fluorescens* **R104, R105** *Pseudomonas putida* **R111**	Winter wheat	Potted plants in growth chamber	– two soil types tested at low temperatures (5°C) – response of wheat nutrient uptake to inoculation dependant on soil composition – grain yield enhanced 46–75% in fertile soil	De Freitas and Germida 1992a
Pseudomonas chlororaphis **2E3, O6**	Spring wheat	Field and laboratory	– increased emergence at two different sites by 8 to 6% – strong inhibition of *Fusarium culmorum* – promotion effect not seen in soils free of *Fusarium* infection	Kropp et al. 1996
Pseudomonas corrugate **13** *Pseudomonas fluorescens* **63–49, 63–28, 15** *Serratia plymthica* **R1GC4**	Cucumber	Field	– strain 63–49 significantly increased fruit numbers by 12% and fruit weight by 18% – strains 13, 15, R1GC4 slightly increased yields – in *Pythium* infected soils, yield increased up to 18% with addition of strains 63–49 and 63–28	McCullagh et al. 1996
Pseudomonas fluorescens	Winter wheat	Field and growth chamber	– in growth chamber, seedling height promotion seen – in *Pythium*-contaminated sites, significant increases in stand, plant height, number of heads, and grain yield	Weller and Cook 1986

Organism	Plant	Condition	Results	Reference
Pseudomonas fluorescens Aur6	*Pinus halepensis* Mill. *Quercus coccifera* L.	Greenhouse	– inadequate moisture levels, *P. halepensis* and *Q. coccifera* seedlings had significantly increased photochemical efficiency and electron transport rate – in drought conditions, inoculation of *P. halepensis* showed negative results for stomatal conductance and pre-dawn shoot water potential – in drought conditions, inoculation of *P. halepensis* showed increased stomatal conductance and pre-dawn shoot water potential	Rincon et al. 2008
Pseudomonas fluorescens *Pseudomonas fluorescens* (biotype F)	Wheat	Pot experiments and field	– strains less effective with increasing N, P and K fertilization rates (significant linear correlations in fresh biomass) – wheat biomass increases of 30%, 26%, 26%, 21% and 22%, at respective fertilizer rates of 0%, 25%, 50%, 75% and 100% – biotype F less effective	Shaharoona et al. 2008
Pseudomonas fluorescens 63-28, R17-FP2, QP5, R15-A4	Tomato	Greenhouse	– in favorable light conditions, fruit yields increased by 5.6 to 9.4% – in unfavorable light conditions, yields increased up to 18.2%	Gagné et al. 1993
Pseudomonas fluorescens *Pseudomonas putida* TL3, BK1	Potato	Field	– in dry soils, no growth promotion effect observed and low bacterial survival – in normal conditions, statistically significant increases of yield of 14-33% in 5 of 9 plots	Burr et al. 1978
Pseudomonas psuedoalcaligens MSC4 *Psuedomonas putida* MSC1	Chickpea	Pot experiments	– in salt stress conditions (300 mM), growth promotion observed in leaf size, lateral roots, number of leaves, and number of fruit	Patel et al. 2012
Pseudomonas putida UW4	Canola	Growth chamber	– significant improvement in plant growth in inhibitory levels of salt (1 mol/L) and low temperature (10°C) compared to ACC deaminase-minus mutant strain inoculation or no inoculation	Cheng et al. 2007

Table 1. contd....

Table 1. contd.

Bacteria	Plant	Experimental conditions	Results	Reference
Pseudomonas putida GR12-2	Canola	Greenhouse	– non-nitrogen fixing mutants provide greater root elongation effects and greater phosphate uptake	Lifshitz et al. 1987
Pseudomonas putida GR12-2	Canola, Lettuce, Tomato, Barley, Wheat, Oat	Growth chamber	– in dicot plants, root elongation stimulated – in monocot plants, little to no growth promotion observed – difference due to sensitivity differences to ethylene, as strain GR12-2 contains gene to reduce ethylene synthesis (i.e., ACC deaminase)	Hall et al. 1996
Pseudomonas putida Rs-198	Cotton	Greenhouse and field trials	– plant height increased by 13%, fresh weight by 31% and dry weight by 10% in greenhouse testing – germination increased by 16% and stand health by 22% in field testing	Yao et al. 2010
Pseudomonas putida W4P63	Potato	Field	– increased yield from 10.2 to 11.7% – potato soft rot (*Erwinia carotovora*) suppressed	Xu and Gross 1986
Pseudomonas syringae pv. *Phaseolicola*	Bean	Greenhouse	– poor establishment of test pathogen in plants inoculated with *P. syringae* – increased amounts of protein in inoculated plants	Alstrom 1995
Serratia liquefaciens 2-68 *Serratia proteamaculans* 102	Soybean	Field	– treatment effects of both bacterial strains not significant over two years	Pan et al. 2002
Variovorax paradoxus 5C-2	Lettuce	Greenhouse and field	– biomass accumulation greater (37%) in reduced moisture conditions – effects not seen in long term field trials – effects not dependent on inoculation concentration	Teijeiro et al. 2011
Xanhomonas maltophila	Sunflower	Lab and greenhouse	– increased germination rate	Fages and Arsac 1991
Unclassified PGPB isolates (10 spp.)	Rice	Pot experiment	– significant increases in plant height, root length, dry weight and germination	Ashrafuzzaman et al. 2009

by researchers in the former Soviet Union and India in the early to the middle part of the 20th century, many scientists questioned whether the use of these bacteria might ever be efficacious. However, beginning with the work of Kloepper et al. (1980), scientists began detailed mechanistic studies of rhizospheric PGPB. This work was undertaken at more or less the same time that other scientists were focusing their efforts on understanding the details of how rhizobia nodulate their host plants (Long et al. 1982) and then provide them with fixed nitrogen (Sprent 1986). From then until the present time, modern biotechnology including recombinant DNA technology, the polymerase chain reaction and DNA sequencing technology (Glick et al. 2010) provided researchers with many powerful tools with which to dissect and understand the intricate workings of PGPB.

The results of some of the numerous studies (in laboratories, growth chambers, greenhouses and fields), of the impact of free-living rhizobacteria on various crop plants, conducted over approximately the last thirty years or so, are summarized in Table 1. This table summarizes the results of a large number of studies and indicates that PGPB confer a wide range of benefits onto treated plants including increases in germination rates, root growth, yield (including grain), leaf area, chlorophyll content, magnesium content, nitrogen content, protein content, hydraulic activity, tolerance to drought, shoot and root weights, and delayed leaf senescence. Moreover, many biocontrol PGPB confer increased phytopathogen/disease resistance onto treated plants. In addition to the many studies with free-living PGPB, there is a very large body of literature that addresses the numerous and extensive studies that have been performed with symbiotic PGPB with an emphasis on *Rhizobia* spp. providing fixed nitrogen to host legumes (e.g., Deaker et al. 2004, Jones et al. 2007, Baset Mia and Shamsuddin 2010, Oldroyd et al. 2011).

The more widespread commercial use of PGPB to increase crop yield has been limited due to the variability and inconsistency of results between laboratory, greenhouse and field studies (Mishustin and Naumova 1962). Soil is an unpredictable environment and the hypothesized outcomes expected are not always obtainable (Bashan 1998). Climatic variability can have a large impact on the effectiveness of PGPB (Okon and Labandera-Gonzalez 1994), however, unfavorable growth conditions in the field can be common in agriculture. Even though there is the possibility of a significant amount of variability in terms of promoting plant growth in the field, if a positive effect of a PGPB is seen on a specific crop in greenhouse studies and the mechanistic basis for that result is understood at a fundamental level, there is a high probability that the selected PGPB will behave as expected under field conditions. Some studies have suggested that plant growth promotion effects that are seen early in plant development often translate into higher yields (Kloepper et al. 1988, Glick et al. 1997, Hoffmann-

Hergarten et al. 1998, Polyanskaya et al. 2000). However, there is also some evidence of late season grain weight increases following treatment of rice plants with PGPB (Tran Van et al. 2000).

A limited number of free living PGPB are available commercially (Table 2). Some of these products are biocontrol agents which contribute indirectly to the growth promotion of crops (Chet and Chernin 2002) while other commercial free living PGPB stimulate plant growth directly. In addition, a small but growing fraction of worldwide legume growth is currently promoted by various commercially available strains of *Rhizobia* spp.

There is contradictory information on the effectiveness of PGPB on plants in soils simultaneously treated with fertilizers. This is particularly the case for bacteria like *Azospirillum* spp. that actively fix nitrogen since the presence of high levels of nitrogen in the soil generally inhibit bacterial nitrogen fixation. However, nitrogen fixation probably plays only a minor role in the mechanism used by *Azospirillum* spp. and similar organisms to promote plant growth so that, given our current understanding, this concern seems exaggerated.

Several studies have suggested that different soil types can dramatically influence the effectiveness of PGPR (e.g., Kloepper et al. 1980, Timmusk et al. 2011). For example, in one study, results suggested that the less fertile the soil, the greater the plant growth stimulation by the added PGPB (De Freitas and Germida 1990). In another study, it was observed when comparing bacteria isolated from two separate sites, that a greater percentage of the rhizosphere bacterial population contained traits that facilitated plant growth when the bacteria were isolated from the more environmentally stressed soil (Timmusk et al. 2011). These observations are consistent with the notion that PGPB do little for plants when the plants are grown under ideal conditions, and these bacteria are most effective in poor soil or otherwise stressful conditions.

In some instances, specific strains of bacteria may promote bacterial growth only in certain crops. This may reflect a variety of different factors. For example, certain bacteria such as *Azospirillum* spp. preferentially utilize organic acids as carbon sources, and these acids are commonly found in root exudates from C4 plants such as maize and sorghum (Bashan and Lavanony 1990). There are rhizosphere bacteria and bacterial endophytes that can bind to the roots of a range of plant species, however, some rhizosphere bacteria and bacterial endophytes exhibit preferences for binding to the roots of some plants rather than others.

Table 2. Examples of commercial products for plant growth promotion using free-living bacteria.

Bacterial ingredient	Product	Company	Intended Crop
Azotobacter chroococcum, Bacillus megaterium, Bacillus mucilaginous	Biogreen	AgroPro, LLC	Field crops
Bacillus firmus	BioNem L	Lidochem, Inc.	Turf
Bacillus spp., *Streptomyces* spp., *Pseudomonas* spp.	Compete	Plant Health Care, Inc.	Field crops
Bacillus pumilus GB34	Yield Shield	Bayer CropScience	Soybean
Bacillus pumilus QST 2808	Sonata AS	AgraQuest, Inc.	Food and field crops
Bacillus subtilis var. *amyloliquifaciens* FZB24	Taegro	Earth BioSciences	Field crops
Bacillus subtilis GB03	Companion	Growth Products	Turf, greenhouse, nursery crops, ornamental, food and forage crops
Bacillus subtilis GB03	Kodiak	Bayer Crop Science	Cotton, vegetables, cereals
Bacillus subtilis (with *Bradyrhizobium japonicum*)	HiStick N/T, Turbo-N	Becker Underwood, Inc.	Soybean
Bacillus subtilis, (with *Bradyrhizobium japonicum*)	Patrol N/T	United Agri Products Canada, Inc.	Soybean
Bacillus subtilis MBI 600	Subtilex	Microbio, Ltd.	Field crops
Bacillus subtilis QST 713	Serenade	AgraQuest, Inc.	Vegetables, fruit, nut and vine crops
Burkholderia cepacia type Wisconsin	Deny	Market VI LLC	Field crops
Delftia acidovorans	BioBoost	Brett-Young Seeds, Ltd.	Canola
Pantoea agglomerans E325	BlightBan C9-1	Nufarm Agricultural, Inc.	Apple and pear
Pseudomonas fluorescens A506	BlightBan A506	Nufarm Agricultural, Inc.	Apple and pear
Pseudomonas syringae	Bio-Save	Jet Harvest Solutions	Citrus, pome fruit and potato
Streptomyces griseoviridis K61	Mycostop	AgBio Development, Inc.	Field, ornamental and vegetable crops
Streptomyces lydicus WYEC 108	Actinovate	Natural Industries, Inc.	Fruit, nut, vegetable and ornamental crops

The number of PGPB cells that are applied to a plant in field applications is important for effective plant growth promotion (Boddey and Dobereiner 1988). Many researchers have reported using up to 10^8 bacteria per seed (Weller and Cook 1986, Okon et al. 1988, De Freitas and Germida 1991, Di Ciocco and Rodriguez-Caceres 1994, Fages 1994, Tran Van et al. 2000) or up to 10^9 bacteria/g of inoculant (Lalande et al. 1989, De Freitas and Germida 1992a, 1992b, Fallik and Okon 1996). While there may be a minimal number of bacteria that need to be inoculated onto any particular plant, large numbers of bacteria are sometimes inhibitory to the germination and growth of seeds or plants (Chanway 1997). The negative effects of large bacterial inocula are often attributed to excessive levels of the plant hormone indoleacetic acid (IAA) being secreted by the bacterial inocula (Bashan and Levanony 1990, Holguin and Glick 2001).

There are a few other points of interest related to the agricultural uses of PGPB with respect to plant stressors. Regarding issues of irrigation, it has been shown that some strains of PGPB can obviate irrigation problems by reducing some of the negative effects of irrigation on crops with water that contains high salt and other dissolved minerals (Hamaoui et al. 2001). This protective effect reflects the lowering of plant ethylene levels, that are elevated as a consequence of salt stress, by 1-aminocyclopropane-1-carboxylate (ACC) deaminase-containing PGPB (Mayak et al. 2004a, Saravankumar and Samiyappan 2006, Cheng et al. 2007, Nadeem et al. 2007, Yue et al. 2007, Siddikee et al. 2011). In addition to salt stress, ACC deaminase-containing bacteria can lower plant ethylene levels and hence growth inhibition that occurs as a consequence of a variety of stresses (Glick 2004) including flooding (Grichko and Glick 2001, Li et al. 2012), drought (Mayak et al. 2004b, Belimov et al. 2009), metals (Burd et al. 1998, Burd et al. 2000, Belimov et al. 2001, Belimov et al. 2005, Reed and Glick 2005, Reed et al. 2005, Farwell et al. 2007), organic contaminants (Huang et al. 2004 a and b, Reed and Glick 2005) and phytopathogens (Wang et al. 2000, Hao et al. 2007, Toklikishvili et al. 2010, Hao et al. 2011). The mechanism of lowering the amount of stress ethylene in plants is most effective with plants that are more susceptible to the inhibitory effects of ethylene, such as dicotyledonous plants (Hall et al. 1996).

A variety of methods exist for the delivery of bacteria to crops in the field. Bacteria may be delivered as wet inoculants, peat-based inoculants, adsorbed onto inert materials or encapsulated within other materials. Wet inoculants generally have good stability and reasonable lifetimes, however, unless they are produced locally, cell suspensions need to be shipped, often long distances. Peat-based inoculants are quite common, are relatively inexpensive and offer good stability, however, peat may be quite variable in terms of quality, availability, nutrients and the presence of other organisms.

In addition, heat sterilization of peat can sometimes release substances toxic to the bacterial inocula (Bashan 1998). Adsorption of bacteria onto inert materials like talc, kaolinite, lignite, bentonite or vermiculite has also been used with some success. In Japan, it is common for bacterially coated seeds to be encapsulated in dissolvable calcium carbonate. Finally, although it is still considered to be an experimental approach, some bacterial inoculants have been encapsulated in alginate polymer microbeads (Bashan et al. 2002, Bashan et al. 2006).

A number of free living bacteria that promote the rhizobial-legume symbiosis have been identified and characterized (e.g., Xu et al. 1994, Andrade et al. 1998, Marek- Kozaczuk et al. 2000). These free living bacteria are thought to act by decreasing the interference in the nodulation process by other soil microorganisms or by lowering the increased (and inhibitory) ethylene levels that occur as a consequence of infection of the plant by a rhizobial strain (Ma et al. 2002, Ma et al. 2003, Ma et al. 2004).

Applications of PGPB in forestry

Research on the use of PGPB in forestry has been and continues to be much less widespread than research for agricultural applications (examples are included in Table 1). This, in part, reflects the need of academic scientists to publish regularly, a reality that is at odds with the fact that most trees grow much more slowly than crop plants. Thus, forestry in general and the effect of PGPB on trees in particular is an area of scientific investigation that is somewhat neglected by academic scientists. Both governments and the forestry industry need to facilitate this type of much needed research thereby ultimately benefiting the commercial forestry sector, as well as reforestation efforts worldwide.

When evaluating the performance of the inoculation of PGPB on tree species, the major objective is the increase in biomass due to inoculation. Seedling emergence and reduction in seedling transplant injury during the transfer from the nursery to the field are also important (Shishido and Chanway 2000). While some tree types are very effective in rapid seed germination, translating those germination results into the successful establishment of adult trees is difficult (Zaady and Perevoltsky 1995). Also, some PGPB are sensitive to low pH conditions, which should be taken into account considering some forest soils are acidic (Brown 1974).

The overwintering survival of PGPB is quite important for trees intended for the colder regions of the world and particularly since trees are perennial plants in contrast to many agricultural crops. Data indicates that many PGPR can over-winter on the roots of field-planted trees. For example, it was observed that from one year to the next, there was a decrease of approximately two orders of magnitude in the inoculated bacterial populations, however, the benefits of inoculation were seen the next year (Chanway et al. 2000).

Similar to the specificity that is sometimes observed in agricultural crops, a particular bacterial strain may promote growth only in certain tree species, or even subspecies (Enebak et al. 1998, Shishido and Chanway 2000). However, there are also some broad-host-range bacterial strains that consistently promote the growth of many pine varieties as well as other tree species (Holl and Chanway 1992, Chanway 1995).

Applications of PGPB for environmental remediation

Phytoremediation may be defined as the use of plants to extract, degrade or stabilize hazardous substances that are present in the environment (Cunningham and Berti 1993, Cunningham et al. 1995, Cunningham and Ow 1996). Ideally, plants that are used as a part of a phytoremediation protocol should have the ability to accumulate high amounts of a target contaminant compared to its concentration in the environment, and should also be able to produce a significant amount of biomass. However, very often the growth of plants in the presence of high levels of contaminants is inhibited as a consequence of the inherent toxicity of those contaminants, leading to a relatively low level of plant biomass even when the plants are able to accumulate a high concentration of the target contaminant. Clearly, if plants that are used for phytoremediation are able to grow well, the rate of site remediation/detoxification will be significantly enhanced. As suggested in the previous section, one way to facilitate plant growth in the presence of either metal or organic contaminants is through the use PGPB that contain specific biological activities (Gamalero et al. 2009, Gerhardt et al. 2009, Glick 2010, Glick and Stearns 2011). For both metals and organic compounds, there isn't a single set of ideal conditions that works well for all phytoremediation experiments. Rather, a number of variables must be considered including differences in: plant type, soil composition, endogenous bacteria, the nature, concentration and range of the contaminants, the temperature range, and the type and physiological state of the added bacteria. Nevertheless, examination of the data from a large number of experiments suggests that certain conditions may facilitate the phytoremediation of a wide range of environmental contaminants (Glick 2010). These conditions include the use of bacteria (resistant to the toxic effects of the target contaminants) that: (i) promote plant growth through the provision of IAA, (ii) in the case of organic contaminants, possess biodegrative activity that can degrade soil contaminants, (iii) possess the enzyme ACC deaminase that can lower plant ethylene levels that typically are a direct consequence of the stressful conditions, and (iv) are endophytic (i.e., colonize the inner tissues of the plant) rather than binding exclusively to the plant rhizosphere. As seen in Table 3, there are now a limited number of phytoremediation field studies in addition to the numerous controlled studies in greenhouses and/or growth chambers. In addition, since the earliest experiments that employed PGPB as part of a phytoremediation protocol, an increasing number of both different PGPB and different plants have been tested and found to be effective. All of the existing data is consistent with the conclusion that PGPB inoculation technology is a useful and important adjunct in phytoremediation protocols. In the case of rhizosphere bacteria that degrade contaminants but are not PGPB, plant roots serve only as a site for contaminant breakdown by these bacteria (Anderson et al. 1993) so that these examples are not included in Table 3.

At the present time, phytoremediation protocols, including those that utilize PGPB, that target organic contaminants are reasonably advanced to the point where commercialization is imminent. On the other hand, as a consequence of the limited bioavailability of metals in the environment, considerable research remains to be done before this becomes a commercial reality. Improvement of phytoremediation protocols utilizing PGPB might include genetically engineering some of these bacterial strains. However, there are a significant number of political and regulatory hurdles that must be overcome, despite the lack of obvious scientific problems, before the deliberate release of genetically engineered bacteria is deemed to be acceptable in many countries of the world.

Some plants such as barley, tomato, canola (*Brassica campestris*) and Indian mustard (*Brassica juncea*) have been reported to not accumulate more contaminants (metal) per gram of plant biomass upon the addition of PGPB, even though the total biomass increases (Belimov et al. 1998a, Burd et al. 1998, Nie et al. 2002). However, in other studies, with maize and *Thlaspi caerulescens* (Hoflich and Metz 1997, Whiting et al. 2001), it was shown that bacterial inoculation increased the uptake of metals by these plants. In addition, de Souza et al. (1999) found increased selenium accumulation by *Brassica juncea* after inoculation. To explain these apparently contradictory results, it appears that the level of contaminants used in some studies affected the amount of metal uptake so that increased metal accumulation following the addition of PGPB occurs only at low metal concentrations and not at the higher levels that inhibit plant growth.

In recent years, scientists have genetically engineered a variety of different plants with the aim of improving their performance in phytoremediation protocols (Cherian and Oliveira 2005, Eapen et al. 2007, Doty 2008, Van Aken 2008, James and Strand 2009). These transgenic plants have typically been engineered to be more efficient at either taking up or breaking down various environmental contaminants. While this work has been successful in the laboratory and potentially extends the usefulness of

Table 3. Results of PGPB tested for phytoremediation.

Bacteria	Plant	Contaminant	Conditions	Results of inoculation	Reference
Acinetobacter sp. CC30 *Acinetobacter* sp. CC33 *Enterobacter sakazakii* CC24 *Pseudomonas putida* CC22	Sunflower	Copper	Pot experiment	– CC30 significantly increased Cu uptake to plant roots, improved photosynthetic pigment content in leaves	Rojas-Tapias et al. 2011
Acinetobacter calcoaceticus A6 *Pseudomonas putida* A1 *Stenotrophomonas maltophilia* A2	Oat	Copper	Greenhouse and field	– increased uptake of Cu from contaminated soils – 327 to 404 g/ha Cu removal with inoculation	Andreazza et al. 2010
Agrobacterium radiobacter D14	Poplar	Arsenic	Greenhouse pot experiment	– in soil amended with 300 mg/kg As, inoculated had more As removal (11%) than uninoculated – As concentrations were higher in roots (229%), stems (113%), and leaves (291%) – greater translocation of As from roots to above-ground tissues	Wang et al. 2011
Agrobacterium radiobacter 10 *Arthrobacter mysorens* 7 *Azospirillum lipoferum* 137 *Flavobacterium* sp. L30	Barley	Cadmium, Lead	Pot experiments in greenhouse	– *Flavobacterium* sp. L30 negatively sensitive to cadmium – increased grain yield with *Flavobacterium* sp. L30 and *A.mysorens* 7 inoculation – enhanced lead accumulation by plants inoculated with *A. radiobacter* 10 and *A. mysorens* 7 – significantly increased growth in plants inoculated by all strains at higher cadmium concentrations	Belimov et al. 1998a

Table 3. contd....

Table 3. contd.

Bacteria	Plant	Contaminant	Conditions	Results of inoculation	Reference
Agrobacterium radiobacter 10 *Arthrobacter mysorens* 7 *Azospirillum lipoferum* 137 *Flavobacterium* sp. L30	Barley	134Cesium	Pot experiments in greenhouse	– *Flavobacterium* sp. L30 increases 134Cs uptake by barley, but not significantly, due to increased plant biomass – *A. lipoferum* 137 significantly decreases the total accumulation of 134Cs	Belimov et al. 1998b
Agrobacterium radiobacter 10 *Arthrobacter mysorens* 7 *Azospirillum lipoferum* 137 *Flavobacterium* sp. L30	Barley	Cadmium	Pot experiments in growth chamber	– increased absorption of essential nutrients from contaminated growth medium – small stimulation of root length and biomass in contaminated growth medium – *A. lipoferum* 137 increased concentration of cadmium in roots, but no change in cadmium uptake by plants inoculated with other strains	Belimov and Dietz 2000
Agrobacterium sp. *Pseudomonas* sp. *Stenotrophomas* sp.	Maize, Rye, Pea, Lupin	Cadmium, Copper, Lead, Nickel, Zinc, Chromium	Pot experiments in growth chamber	– bacteria stimulates growth of maize and increases metal uptake by maize, effect more pronounced on more weakly contaminated soils compared to heavily-contaminated soils – no effect in growth and metal uptake in lupin, pea and rye	Hoflich and Metz 1997
Azospirillum brasilense Cd *Bacillus pumilis* ES4 *Bacillus pumilis* RIZO1	*Atriplex lentiformis*	Acidic high metal content mine tailings, neutral low-metal content mine tailings	Greenhouse	– significant enhancement of plant growth, germination, root length, dry weight in both tailing types – ES4 most effective	de-Bashan et al. 2010

Microorganism	Plant	Pollutant	Experiment	Results	Reference
Azospirillum brasilense Cd *Enterobacter cloacae* CAL 2 *Pseudomonas putida* UW3	Tall fescue	Polycyclic aromatic hydrocarbons (PAHs)	Pot experiments in growth chamber	– accelerated and more complete PAH removal from the soil – effectiveness of PAH removal enhanced in combination with landfarmed soil (mechanically cultivated) and inoculation with PAH-degrading bacteria	Huang et al. 2004a
Azospirillum brasilense Cd *Enterobacter cloacae* CAL2 *Pseudomonas putida* UW3	Kentucky bluegrass, Tall fescue, Wild rye	PAHs	Pot experiments in growth chamber	– increased PAH removal from soil – germination of all three plant types increased dramatically in PAH-spiked soil with inoculation – root biomass significantly increased in all plant types	Huang et al. 2004b
Bacillus sp. XIII	*Axonopus affinis*	Cadmium, nickel, zinc	Growth chamber	– growth maintained in increasing heavy metal concentrations	Cardon et al. 2010
Bacillus megaterum JL35 *Burkholderia* sp. GL12 *Sphingomonas* sp. YM22	Maize	Copper	Pot experiments	– increased root weight (48–83%) and increased shoot weight (33–56%) – increased Cu content in roots (69 to 107%) and shoot (16 to 86%) – increased Cu removal from growth medium (63 to 94%)	Sheng et al. 2012
Bacillus megaterum Bm4C *Pseudomonas* sp. PS29C	Mustard	Nickel	Pot experiments in greenhouse	– inoculation had no difference on Ni accumulation in roots and shoots – significant increases in shoot length (23–29%), fresh weight (22–42%) and dry weight (17–28%)	Rajikumar and Freitas 2008
Burkholderia sp. **CBMB40** *Methylobacterium oryzae* **CBMB20**	Tomato	Nickel and Cadmium	Pot experiments	– reduction in accumulation of Ni(II) and Cd(II) in roots and shoots – significant plant growth	Madhaiyan et al. 2007

Table 3. contd....

Table 3. contd.

Bacteria	Plant	Contaminant	Conditions	Results of inoculation	Reference
Burkholderia **sp. D54**	*Amaranthus cruentus, Phytolacca americana*	Cesium	Growth chamber	– significant increases in biomass for A. cruentus of 22 to 139% – significant increases in biomass for P. Americana of 14 to 254% – significant increases in Cs content in plant tissues	Tang et al. 2011
Burkholderia **sp. D54**	*Sedum alfredi*	Multiple metal contaminated soil	Pot experiments in growth chamber	– significant increases in root and shoot (120%) weight – increased accumulation of Cd in roots (20%) and shoots (20%) – no increases in Pb or Zn accumulation	Guo et al. 2011
Burkholderia **sp. J62**	Maize and tomato	Lead, cadmium	Pot experiments (indoor and outdoor)	– isolate found to have multiple resistance to heavy metals – biomass of tomato and maize significantly increased – greater Zn and Cd uptake in tomato (192 and 191%, respectively) – greater Zn and Cd uptake in maize (38 and 5%, respectively)	Jiang et al. 2008
Burkholderia **sp. J62**	Tomato	Lead, cadmium	Pot experiments (indoor and outdoor)	– increase of Cd in above-ground tissues (92 to 113%) – increase of Pb in above-ground tissues (73 to 79%)	He et al. 2009
Burkholderia cepacia	*Sedum alfredii*	Cadmium, zinc	Hydroponic growth chamber	– with Zn treatment significant increases in plant growth (110%), metal uptake (96%) and metal translocation (135%). – with Cd treatment significant increases in plant growth (56%), metal uptake (243%) and metal translocation (296%).	Li et al. 2007

Organism	Plant	Contaminant	Setup	Results	Reference
Enterobacter asburiae PSI3	Mung bean	Cadmium	Pot experiments in growth chamber	– increased root growth over uninoculated control (72 to 85%) in 500 mg/kg Cd^{2+}	Kavita et al. 2008
Enterobacter cancerogenes *Microbacterium saperdae* *Pseudomonas monteilii*	*Thlaspi caerulescens* *Thlaspi arvense*	Zinc	Pot experiments in growth chamber	– in *T. caerulescens*, two-fold increase of zinc concentration in roots and four-fold increase of zinc accumulation in shoots – increased shoot biomass of *T. caerulescens* – no increased growth or metal accumulation for *T. arvense*	Whiting et al. 2001
Enterobacter cloacae CAL2	Canola	Arsenate	Pot experiments in growth chamber	– slight inhibitory effect of CAL2 on germination of canola in presence of arsenate – partnered with transgenic plants, bacteria induced significantly higher root and shoot weights in plants – no increase of arsenate concentration by roots of plants	Nie et al. 2002
Gordonia sp. S2RP-17	Maize	Diesel contamination (Total petroleum hydrocarbon, TPH)	Greenhouse	– root length, shoot and root biomass significantly increased – TPH removal from soil 96% (compared with 85% uninoculated maize or 13% no treatment)	Hong et al. 2011
Kluyvera ascorbata SUD165	Canola, Tomato	Nickel	Pot experiments in growth chamber	– for tomato and canola, roots and shoots protected from toxicity – significantly deceased ethylene production by plants – no increase or change in nickel uptake in plant material	Burd et al. 1998

Table 3. contd....

Table 3. contd.

Bacteria	Plant	Contaminant	Conditions	Results of inoculation	Reference
Kluyvera ascorbata SUD165/26, SUD165	Indian mustard, Canola, Tomato	Nickel, Lead, Zinc	Pot experiments in growth chamber	– both strains decrease some plant growth inhibition by the metals – no increase of metal uptake with either strain over noninoculated plants	Burd et al. 2000
Microbacterium oxydans AY509223	*Alyssum murale*	Nickel	Pot experiments in greenhouse	– significant increases in Ni uptake in low, medium and high Ni-containing soils, 36, 39 and 28%, respectively – increased foliar Ni by 56%, 65% and 38%, respectively	Abou-Shanab et al. 2006
Mycobacterium sp. ACC14 *Pseudomonas fluorescens* ACC9 *Pseudomonas tolaasii* ACC23	Canola	Cadmium	Pot experiments	– growth promotion of plants grown in 15µgCd^{2}g^{-1} soil – no increased accumulation of metals in soil, but total uptake increased due to increased biomass of plants	Dell'Amico et al. 2008
Pseudomonas sp. TLC 6-6.5-4	Maize, sunflower	Multiple metal contaminated soil	Growth chamber	– bacterial cells demonstrated high bioaccumulation of Zn (up to 16 mg/g dry weight) and Pb (81 mg/g dry weight) – inoculation significantly increased Cu accumulation in maize (> 100%) and sunflower and total maize biomass	Li and Ramakrishna 2011
Pseudomonas aeruginosa KUCd1	Mustard, pumpkin	Cadmium	Pot experiment in growth chamber	– stimulated plant growth – for pumpkin, accumulation of Cd was reduced by 59% in roots and 47% in shoots – for mustard, accumulation of Cd was reduced by 52% in roots and 37% in shoots	Sinha and Mukherjee 2008

Pseudomonas aeruginosa **MKRh3**	Chickpea	Cadmium	Pot experiment in greenhouse	– reduced cadmium uptake in leaves – enhanced plant growth (e.g., root dry weight increased 204%) and rooting	Ganesan 2008
Pseudomonas brassicacearum **Am3** *Pseudomonas marginalis* **Dp1** *Rhodococcus* sp. **Fp2**	Pea	Cadmium	Pot experiment in greenhouse	– only Fp2 (not demonstrating ACC deaminase activity) did not stimulate growth, other strains did – AM3 and Dp1 increased root and shoot growth in some plant varieties in Cd-spiked soil up to 40%	Safronova et al. 2006
Pseudomonas-montelli	Sorghum	Cadmium	Growth chamber	– significant increases in shoots and total plant biomass (up to >200%) in 9 out of 10 isolates – significant increase in Cd uptake by plants	Duponnois et al. 2006
Pseudomonas putida **UW4**	Canola	Metal-contaminated soil (primarily Ni)	Field	– low and high flood stress also a parameter – canola biomass increased by 31–38% – no effect on Ni concentration in shoot	Farwell et al. 2007
Pseudomonas sp. **UW3** *Pseudomonas putida* **UW4**	Annual ryegrass, tall fescue, fall rye, barley	Total petroleum hydrocarbons	Three-year field experiment	– reduction (130 g/kg to 50 g/kg over three years) of high molecular-weight TPHs in soil consistently enhanced by PGPB	Gurska et al. 2009
Pseudomonas putida **HS-2**	Canola	Nickel	Pot experiment in growth chamber	– increase in plant biomass – increase in Ni uptake in roots (51 to 107%) and shoots (89 to 89%)	Rodriguez et al. 2008

Table 3. contd....

Table 3. contd.

Bacteria	Plant	Contaminant	Conditions	Results of inoculation	Reference
Serratia sp. SY5	Maize	Cadmium and copper	Hydroponic culture	– significant increase in primary and radicular root lengths in high Cu concentrations (15 mg/L) – significant increase in radicular root length in high Cd concentrations (15 mg/L)	Koo and Cho 2009
Staphylococcusarlettae Cr11	Wheat	Chromium(VI)	Pot experiments	– Cr 11 highly tolerant to Cr(VI), 2000 to 5000 mg/L, and cross-tolerant to other toxic heavy metals – significant increase in germination, root and shoot length and dry weight	Sagar et al. 2012

different plants in phytoremediation protocols, the effectiveness of these engineered plants might be further increased by pairing them with PGPB that contain some of the traits indicated above. However, this approach remains to be tested.

Conclusion

The use of PGPB shows strong promise to replace the use of chemicals to facilitate plant growth enhancement for a variety of different applications. In the last thirty years or so, a great deal of research on PGPB has been conducted by thousands of laboratories worldwide so that our knowledge and fundamental understanding of these organisms have reached the point where their more widespread commercial use is feasible. The basic studies were predicated on the notion that it is necessary to first understand the fundamental genetic and biochemical mechanisms that govern the relationship between PGPB and plants before using them on a massive scale in the environment. While there is still a lot to be done, the fundamental knowledge that has been developed until now is a cause for optimism about the potential efficacy of using PGPB to move to more sustainable practices in improving crop productivity. With increasing populations and decreasing resource availability, PGPB are poised to play ever increasing roles in agriculture, horticulture, forestry and environmental restoration practices.

References

Abou-Shanab, R. A. I., J. S. Angle and R. L. Chaney. 2006. Bacterial inoculants affecting nickel uptake by *Alyssum murale* from low, moderate and high Ni soils. Siol Biol. Biochem. 38: 2882–2889.

Adesemoye, A. O., H. A. Torbert and J. W. Kloepper. 2009. Plant growth-promoting rhizobacteria allow reduced application rates of chemical fertilizers. Microb. Ecol. 58: 921–929.

Adesemoye, A. O., H. A. Torbert and J. W. Kloepper. 2010. Increased plant uptake of nitrogen from ^{15}N-depleted fertilizer using plant growth-promoting rhizobacteria. Appl. Soil Ecol. 46: 54–58.

Adjanohoun, A., M. Allagbe, P. A. Noumavo, H. Gotoechan-Hodonou, R. Sikirou, K. K. Dossa, R. GleleKakai, S. O. Kotchoni and L. Baba-Moussa. 2011. Effects of plant growth promoting rhizobacteria on field grown maize. J. Animal Plant Sci. 11: 1457–1465.

Akbari, P., A. Ghalavand, A. M. M. Sanavy, M. A. Alikhani and S. S. Kalkhoran. 2011. Comparison of different nutritional levels and the effect of plant growth promoting rhizobacteria (PGPR) on the grain yield and quality of sunflower. Austral. J. Crop Sci. 5: 1570–1576.

Alam, M. S., Z. J. Cui, T. Yamagishi and R. Ishii. 2001. Grain yield and related physiological characteristics of rice plants *Oryza sativa* L. inoculated with free-living rhizobacteria. Plant Prod. Sci. 4: 125–130.

Alstrom, S. 1995. Evidence of disease resistance induced by rhizosphere pseudomonad against *Pseudomonas syringae* pv. *phaseolicola*. J. Gen. Appl. Microbiol. 41: 315–325.

Ambrosini, A., A. Beneduzi, T. Stefanski, F. G. Pinheiro, L. K. Vargas and L. M. P. Passaglia. 2012. Screening of plant growth promoting Rhizobacteria isolated from sunflower (*Helianthus annuus* L.). Plant and Soil. DOI 10. 1007/s11104-011-1079-1.

Anderson, T. A., E. A. Guthrie and B. T. Walton. 1993. Bioremediation in the rhizosphere: plant roots and associated microbes clean contaminated soil. Environ. Sci. Technol. 27: 2630–2636.

Andrade, G., F. A. A. M. De Leij and J. M. Lynch. 1998. Plant mediated interactions between *Pseudomonas fluorescens*, *Rhizobium leguminosarum* and arbuscular mycorrhizae on pea. Lett. Appl. Microbiol. 26: 311–316.

Andreazza, R., B. C. Okeke, M. R. Lambais, L. Bortolon, G. W. Bastos de Melo and F. A. de Oliveira Camargoa. 2010. Bacterial stimulation of copper phytoaccumulation by bioaugmentation with rhizosphere bacteria. Chemosphere. 81: 1149–1154.

Arzanesh, M. H., H. A. Alikhani, K. Khavazi, H. A. Rahimian and M. Miransari. 2011. Wheat (*Triticum aestivum* L.) growth enhancement by *Azospirillum* sp. under drought stress. World J. Microbiol. Biotechnol. 27: 197–205.

Ashrafuzzaman, M., F. A. Hossen, M. R. Ismail, A. Hoque, M. Z. Islam, S. M. Shahidullah and S. Meon. 2009. Efficiency of plant growth-promoting rhizobacteria (PGPR) for the enhancement of rice growth. African J. Biotechnol. 8: 1247–1252.

Aslantasa, R., R. Cakmakci and F. Sahin. 2007. Effect of plant growth promoting rhizobacteria on young apple tree growth and fruit yield under orchard conditions. Scientia Horticulturae. 111: 371–377.

Baldani, V. L. D., J. I. Baldani and J. Döbereiner. 1987. Inoculation on field grown wheat *Triticum aestivum* with *Azospirillum* spp. in Brazil. Biol. Fert. Soils. 4: 37–40.

Bashan, Y. 1998. Inoculants of plant growth-promoting bacteria for use in agriculture. Biotechnol. Adv. 16: 729–770.

Bashan, Y., J. P. Hernandez., L. A. Leyva and M. Bacilio. 2002. Alginate microbeads as inoculant carrier for plant growth-promoting bacteria. Biol. Fertil. Soils. 35: 359–368.

Bashan, Y., J. J. Bustillos, L. A. Leyva, J. P. Hernandez and M. Bacilio. 2006. Increase in auxiliary photoprotective photosynthetic pigments in wheat seedlings induced by *Azospirillum brasilense*. Biol. Fertil Soils. 42: 279–285.

Bashan, Y. and G. Holguin. 2002. Plant growth-promoting bacteria: A potential tool for arid mangrove reforestation. Trees. 16: 159–166.

Bashan, Y. and H. Levanony. 1990. Current status of *Azospirillum* inoculation technology: *Azospirillum* as a challenge for agriculture. Can. J. Microbiol. 36: 591–608.

Bashan, Y., B. Salazar and M. E. Puente. 2009. Responses of native legume desert trees used for reforestation in the Sonoran Desert to plant growth-promoting microorganisms in screen house. Biol. Fert. Soils. 45: 655–622.

Baset Mia, M. A. and Z. H. Shamsuddin. 2010. Rhizobium as a crop enhancer and biofertilizer for increased cereal production. Afr. J. Biotechnol. 9: 6001–6009.

Beall, F. and B. Tipping. 1989. Plant growth-promoting rhizobacteria in forestry. Abstr. 177. For. Res. Market. Proc., Ont. For. Res. Com., Toronto, USA.

Belimov, A. A. and K. Dietz. 2000. Effect of associative bacteria on element composition of barley seedlings grown in solution culture at toxic cadmium concentrations. Microbiol. Res. 155: 113–121.

Belimov, A. A., V. I. Safranova, T. A. Sergeyeva, T. N. Egorova, V. A. Matveyeva, V. E. Tsyganov, A. Y. Borisov, I. A. Tikhonovich, C. Kluge, A. Preisfeld, K. J. Dietz and V. V. Stepanok. 2001. Characterization of plant growth promoting rhizobacteria isolated from polluted soils and containing 1-aminocyclopropane-1-carboxylate deaminase. Can. J. Microbiol. 47: 642–52.

Belimov, A. A., N. Hontzeas, V. I. Safronova, S. V. Demchinskaya, G. Piluzza, S. Bullitta and B. R. Glick. 2005. Cadmium-tolerant plant growth-promoting rhizobacteria associated with the roots of Indian mustard (*Brassica juncea* L. Czern.). Soil Biol. Biochem. 37: 241–250.

Belimov, A. A., A. M. Kunakova, A. P. Kozhemiakov, V. V. Stepanok and L. Y. Yudkin. 1998a. Effect of associative bacteria on barley grown in heavy metal contaminated soil. *In:* Proceedings of the International Symposium on Agro-environmental issues and future strategies: Towards the 21st Century. Faisalakad, Pakistan, May 25–30.

Belimov, A. A., A. M. Kunakova, N. D. Vasilyeva, T. S. Kovatcheva, V. F. Dritchko, S. N. Kuzovatov, I. R. Trushkina and Y. V. Alekseyev. 1998b. Accumulation of radionuclides by associative bacteria and the uptake of ^{134}Cs by the inoculated barley plants. *In:* Malik et al. [eds.]. Nitrogen Fixation with Non-Legumes. Kluwer Academic Publishers, Great Britain. pp. 275–280.

Belimov, A. A., I. C. Dodd, N. Hontzeas, J. C. Theobold, V. I. Safronova and W. J. Davies. 2009. Rhizosphere bacteria containing 1-aminocyclopropane-1-carboxylate deaminase increase yields of plants grown in drying soil via both local and systemic hormone signaling. New Phytol. 181: 413–423.

Bertrand, H., R. Nalin, R. Bally and J. C. Cleyet-Marel. 2001. Isolation and identification of the most efficient plant growth-promoting bacteria associated with canola (*Brassica napus*). Biol. Fert. Soils. 33: 152–156.

Boddey, R. M. and J. Dobereiner. 1988. Nitrogen fixation associated with grasses and cereals: recent results and perspectives for future research. Plant. Soil. 108: 53–65.

Boutard-Hunt, C., C. D. Smart, J. Thaler and B. A. Nault. 2009. Impact of plant growth-promoting rhizobacteria and natural enemies on *Myzus periscae* (Hemiptera: Aphididae) infestations in pepper. Horticult. Entomol. 102: 2183–2191.

Broadbent, P., K. F. Baker, N. Franks and J. Holland. 1977. Effect of *Bacillus* spp. on increased growth of seedlings in steamed and in non-treated soil. Phytopathol. 67: 1027–1034.

Brown, M. E. 1974. Seed and root bacterization. Ann. Rev. Phytopathol. 12: 181–197.

Burd, G. I., D. G. Dixon and B. R. Glick. 2000. Plant growth-promoting bacteria that decrease heavy metal toxicity in plants. Can. J. Microbiol. 46: 237–245.

Burd, G. I., D. G. Dixon and B. R. Glick. 1998. A plant-growth promoting bacterium that decreases nickel toxicity in seedlings. Appl. Environ. Microbiol. 64: 3663–3668.

Burr, T. J., M. N. Schroth and T. Suslow. 1978. Increased potato yields by treatment of seed pieces with specific strains of *Pseudomonas fluorescens* and *P. putida*. Phytopathol. 68: 1377–1383.

Caesar, A. J. and T. J. Burr. 1987. Growth promotion of apple seedlings and rootstocks by specific strains of bacteria. Phytopathol. 77: 1583–1588.

Caceres, E. A. R., G. G. Anta, J. R. Lopex, C. A. Di Ciocco, J. C. P. Basurco and J. L. Parada. 1996. Response of field-grown wheat to inoculation with *Azospirillum brasilense* and *Bacillus polymyxa* in the semiarid region of Argentina. Arid. Soil. Res. Rehab. 10: 13–20.

Cakmakci, R., F. Kantar and F. Sahin. 2001. Effect of N2-fixing bacterial inoculations on yield of sugar beet and barley. J. Plant. Nutr. Soil. Sci. 164: 527–531.

Cakmakci, R., M. F. Donmez and U. Erdogan. 2007a. The effect of plant growth promoting rhizobacteria on barley seedling growth, nutrient uptake, some soil properties, and bacterial counts. Turk. J. Ag. Forest. 31: 189–199.

Cakmakci, R., M. Erat, U. Erdogan. and M. F. Donmez. 2007b. The influence of plant growth-promoting rhizobacteria on growth and enzyme activities in wheat and spinach plants. J. Plant Nut. Soil Sci. 170: 288–295.

Cardon, D. L., S. M. Villafan, A. Rodriguez Tovar, S. Perez Jimenez, L. A. Guerrero Zuniga, M. A. Amezcua Allieri, N. O. Perez and A. Rodríguez Dorantes. 2010. Growth response and heavy metals tolerance of *Axonopus affinis*, inoculated with plant growth promoting rhizobacteria. African J. Biotechnol. 9: 8772–8782.

Chanway, C. P. 1995. Differential response of western hemlock from low and high elevations to inoculation with plant growth promoting *Bacillus polymyxa*. Soil. Biol. Biochem. 27: 767–775.

Chanway, C. P. 1997. Inoculation of tree roots with plant growth promoting rhizobacteria: An emerging technology for reforestation. For. Sci. 43: 99–112.

Chanway, C. P. and F. B. Holl. 1991. Biomass increase and associative nitrogen fixation of mycorrhizal *Pinus contorta* Dougl. seedlings inoculated with a plant growth promoting *Bacillus* strain. Can. J. Microbiol. 69: 507–511.

Chanway, C. P. and F. B. Holl. 1993. First year performance of spruce seedlings after inoculation with plant growth promoting rhizobacteria. Can. J. Microbiol. 39: 520–527.

Chanway, C. P. and F. B. Holl. 1994. Ecological growth response specificity of two Douglas-fir ecotypes inoculated with coexistent beneficial rhizosphere bacteria. Can. J. Bot. 72: 582–586.

Chanway, C. P., R. A. Radley and F. B. Holl. 1991. Inoculation of conifer seed with plant growth promoting *Bacillus* strains causes increased seedling emergence and biomass. Soil Biol. Biochem. 23: 575–580.

Chanway, C. P., M. Shishido, J. Nairn, S. Jungwirth, J. Markham, G. Xiao and F. B. Holl. 2000. Endophytic colonization and field responses of hybrid spruce seedlings after inoculation with plant growth-promoting rhizobacteria. For. Ecol. Manage. 133: 81–88.

Cherian, S. and M. M. Oliveira. 2005. Transgenic plants in phytoremediation: recent advances and new possibilities. Environ. Sci. Technol. 39: 9377–9390.

Chet, I. and L. Chernin. 2002. Biocontrol, microbial agents in soil. *In:* G. Bitton [ed.]. Encyclopedia of Environmental Microbiology. John Wiley and Sons Inc., New York, USA. pp. 450–465.

Cheng, Z., E. Park and B. R. Glick. 2007. 1-Aminocyclopropane-1-carboxylate (ACC) deaminase from *Pseudomonas putida* UW4 facilitates the growth of canola in the presence of salt. Can. J. Microbiol. 53: 912–918.

Chithrashree, C., A. C. Udayashankar, S. Chandra Nayaka, M. S. Reddy and C. Srinivas. 2011. Plant growth-promoting rhizobacteria mediate induced systemic resistance in rice against bacterial leaf blight caused by *Xanthomonas oryzae* pv. *oryzae*. Biol. Control. 59: 114–122.

Cunningham, S. D. and W. R Berti. 1993. Remediation of contaminated soils with green plants: an overview. *In vitro* Cell. Dev. Biol. 29P: 207–212.

Cunningham, S. D. and D. W. Ow. 1996. Promises and Prospects of phytoremediation. Plant Physiol. 110: 715–719.

Cunningham, S. D., W. R. Berti and J. W. Huang. 1995. Phytoremediation of contaminated soils. Trends Biotechnol. 13: 393–397.

Deaker, R., R. J. Roughley and I. R. Kennedy. 2004. Legume seed inoculation technology: a review. Soil Biol. Biochem. 36: 1275–1288.

de-Bashan, L. E., J. P. Hernandez, Y. Bashan and R. M. Maier. 2010. *Bacillus pumilus* ES4: Candidate plant growth-promoting bacterium to enhance establishment of plants in mine tailings. Environmental Exp. Bot. 69: 343–352.

De Freitas, J. R. and J. J. Germida. 1990. Plant growth promoting rhizobacteria for winter wheat. Can. J. Microbiol. 36: 265–272.

De Freitas, J. R. and J. J. Germida. 1991. *Pseudomonas cepacia* and *Pseudomonas putida* as winter wheat inoculants for biocontrol of *Rhizoctonia solani*. Can. J. Microbiol. 37: 780–784.

De Freitas, J. R. and J. J. Germida. 1992a. Growth promotion of winter wheat by fluorescent pseudomonads under growth chamber conditions. Soil. Biol. Biochem. 24: 1127–1135.

De Freitas, J. R. and J. J. Germida. 1992b. Growth promotion of winter wheat by fluorescent pseudomonads under field conditions. Soil. Biol. Biochem. 24: 1137–1146.

Dell'Amico, E., L. Cavalca and V. Andreoni. 2008. Improvement of *Brassica napus* growth under cadmium stress by cadmium-resistant rhizobacteria. Soil Biol. Biochem. 40: 74–84.

de Silva, A., K. Patterson, C. Rothrock and J. Moore. 2000. Growth promotion of highbush blueberry by fungal and bacterial inoculants. Hort. Sci. 35: 1228–1230.

de Souza, M. P., D. Chu, M. Zhao, A. M. Zayed, S. E. Ruzin, D. Schichnes and N. Terry. 1999. Rhizosphere bacteria enhance selenium accumulation and volatiliation by Indian mustard. Plant. Physiol. 119: 565–573.

Di Ciocco, C. A. and E. Rodriguez-Caceres. 1994. Field inoculation of *Setaria italica* with *Azospirillum* spp. in Argentine humid pampas. Field Crop Res. 37: 253–257.

Dobbelaere, S., A. Croonenborghs, A. Thys, D. Ptacek, J. Vanderleyden, P. Dutto, C. Labandera-Gonzalez, M. Caballero, J. F. Aguirre, Y. Kapulnik, S. Brener, S. Burdman, D. Kadouri, S. Sarig and Y. Okon. 2001. Responses of agronomically important crops to inoculation with *Azospirillum*. Aust J. Plant Physiol. 28: 871–879.

Doty, S. L. 2008. Enhancing phytoremediation through the use of transgenics and endophytes. New Phytol. 179: 318–333.

Duponnois, R., M. Kisa, K. Assigbetse, Y. Prin, J. Thioulouse, M. Issartel, P. Moulin and M. Lepage. 2006. Fluorescent pseudomonads occuring in *Macrotermes subhyalinus* mound structures decrease Cd toxicity and improve its accumulation in sorghum plants. Sci. Total Env. 370: 391–400.

Dursun, A., M. Ekinci and M. F. Donmez. 2010. Effects of foliar application of plant growth promoting bacterium on chemical contents, yield and growth of tomato (*Lycopersicon esculentum* L.) and cucumber (*Cucumis sativus* L.). Pak. J Bot. 42: 3349–3356.

Eapen, S., S. Singh and S. F. D'Souza. 2007. Advances in development of transgenic plants for remediation of xenobiotic pollutants. Biotechnol. Adv. 25: 442–451.

Egamberdiyeva, D. 2007. The effect of plant growth promoting bacteria on growth and nutrient uptake of maize in two different soils. Appl. Soil Ecol. 36: 184–189.

Enebak, S. A., G. Wei and J. W. Kloepper. 1998. Effects of plant growth-promoting rhizobacteria on loblolly and slash pine seedlings. For. Sci. 44: 139–144.

Esitken, A., H. E. Yildiz, S. Ercisli, M. F. Donmez, M. Turan and A. Gunes. 2010. Effects of plant growth promoting bacteria (PGPB) on yield, growth and nutrient contents of organically grown strawberry. Scientia Horticulturae. 124: 62–66.

Fages, J. 1994. *Azospirillum* inoculants and field experiments. *In:* Okon Y. (ed.), Azospirillum-Plant associations. CRC Press, Boca Raton, Florida. pp. 87–110.

Fages, J. and J. F Arsac. 1991. Sunflower inoculation with *Azospirillum* and other plant growth promoting rhizobacteria. Plant. Soil. 137: 87–90.

Fallik, E. and Y. Okon. 1996. The response of maize (*Zea mays*) to *Azospirillum* inoculation in various types of soils in the field. World J. Microbiol. Biotechnol. 12: 511–515.

Farwell, A. J., S. Vessely, V. Nero, H. Rodriguez, K. McCormack, S. Shah, D. G. Dixon and B. R. Glick. 2007. Tolerance of transgenic canola plants (*Brassica napus*) amended with plant growth-promoting bacteria to flooding stress at a metal-contaminated field site. Environ. Pollut. 147: 540–545.

Frommel, M. I., J. Nowak and G. Lazarovitis. 1991. Growth enhancement and developmental modifications of *in vitro* grown potato (*Solanum tuberosum* ssp. *tuberosum*). Plant Physiol. 96: 928–936.

Frommel, M. I., J. Nowak and G. Lazarovitis. 1993. Treatment of potato tubers with a growth promoting *Pseudomonas* sp.: Plant growth responses and bacterium distribution in the rhizosphere. Plant Soil. 150: 51–60.

Fu, Q., C. Liu, N. Ding, Y. Lin and B. Guo. 2010. Ameliorative effects of inoculation with the plant growth-promoting rhizobacterium *Pseudomonas* sp. DW1 on growth of eggplant (*Solanum melongena* L.) seedlings under salt stress. Ag. Water Management. 97: 1994–2000.

Gagné, S., L. Dehbi, D. Le Quéré, F. Cayer, J. Morin, R. Lemay and N. Fournier. 1993. Increase of greenhouse tomato fruit yields by plant growth-promoting rhizobacteria (PGPR) inoculated into the peat-based growing media. Soil Biol. Biochem. 25: 269–272.

Gamalero, E., G. Berta and B. R. Glick. 2009. Effects of plant growth promoting bacteria and AM fungi on the response of plants to heavy metal stress. Can. J. Microbiol. 55: 501–514.

Ganesan, V. 2008. Rhizoremediation of cadmium soil using a cadmium-resistant plant growth-promoting rhizopseudomonad. Curr. Microbiol. 56: 403–407.

Geels, F. P., J. G. Lamers, O. Hoekstra and B. Schippers. 1986. Potato plant response to seed tuber bacterization in the field in various rotations. Neth. J. Plant Pathol. 92: 257–272.

Gerhardt, K. E., X. D. Huang, B. R. Glick and B. M. Greenberg. 2009. Phytoremediation and rhizoremediation of organic soil contaminants: potential and challenges. Plant Sci. 176: 20–30.

Gholami, A., S. Shahsavani and S. Nezarat. 2009. The effect of growth promoting rhizobacteria (PGPR) on germination, seedling growth and yield of maize. Int. J. Biological Life Sci. 5: 35–40.

Glick, B. R. 2010. Using soil bacteria to facilitate phytoremediation. Biotechnol. Adv. 28: 367–374.

Glick, B. R., J. J. Pasternak and C. L. Patten. 2010. "Molecular Biotechnology" fourth edition, Amer. Soc. for Microbiol., Washington, D. C.

Glick, B. R. and J. C. Stearns. 2011. Making phytoremediation work better: Maximizing a plant's growth potential in the midst of adversity. Internat. J. Phytorem. 13 (S1): 4–16 (2011).

Glick, B. R., C. L. Patten, G. Holguin and D. M. Penrose. 1999. Biochemical and genetic mechanisms used by plant growth-promoting bacteria. Imperial College Press, London, UK.

Glick, B. R., C. Liu, S. Ghosh and E. B. Dumbroff. 1997. Early development of canola seedling in the presence of the plant growth-promoting rhizobacterium *Pseudomonas putida* GR12-2. Soil Biol. Biochem. 29: 1233–1239.

Glick, B. R. 1995. The enhancement of plant growth by free-living bacteria. Can. J. Microbiol. 41: 109–117.

Glick, B. R. 2004. Bacterial ACC deaminase and the alleviation of plant stress. Adv. Appl. Microbiol. 56: 291–312.

Grichko, V. P. and B. R. Glick. 2001. Amelioration of flooding stress by ACC deaminase-containing plant growth-promoting bacteria. Plant Physiol. Biochem. 39: 11–17.

Gulati, A., P. Vyas, P. Rahi and R. C. Kasana. 2009. Plant growth-promoting and rhizosphere-competent *Acinetobacter rhizosphaerae* Strain BIHB 723 from the cold deserts of the Himalayas. Curr. Microbiol. 58: 371–377.

Guo, J., S. Tang, X. Ju, Y. Ding, S. Liao and N. Song. 2011. Effects of inoculation of a plant growth promoting rhizobacterium *Burkholderia* sp. D54 on plant growth and metal uptake by a hyperaccumulator *Sedum alfredii* Hance grown on multiple metal contaminated soil. World J. Microbiol. Biotechnol. 27: 2835–2844.

Gurska, J., W. Wang, K. E. Gerhardt, A. M. Khalid, D. M. Isherwood, X. D. Huang, B. R. Glick and B. M. Greenberg. 2009. Three year field test of a plant growth promoting rhizobacteria enhanced phytoremediation system at a land farm for treatment of hydrocarbon waste. Environ. Sci. Technol. 43: 4472–4479.

Haggag, W. M. and S. Timmusk. 2007. Colonization of peanut roots by biofilm-forming *Paenibacillus polymyxa* initiates biocontrol against crown rot disease. J. Appl. Microbiol. 104: 961–969.

Hall, J. A., D. Peirson, S. Ghosh and B. R. Glick. 1996. Root elongation in various agronomic crops by the plant growth promoting rhizobacterium *Pseudomonas putida* GR12-2. Isr. J. Plant Sci. 44: 37–42.

Hamaoui, B., J. M. Abbadi, S. Burdman, A. Rashid, S. Sarig and Y. Okon. 2001. Effects of inoculation with *Azospirillum brasilense* on chickpeas (*Cicer arietinum*) and faba beans (*Vicia faba*) under different growth conditions. Agronomie 21: 553–560.

Hamidi, A., R. Chaokan, A. Asgharzadeh, M. Dehghaoshoar, A. Ghalavand and M. J. Malakouti. 2009. Effect of plant growth promoting rhizobacteria (PGPR) on phenology of late maturity maize (*Zea mays* L.) hybrids. Iranian J. Crop. Sci. 11: 249–270.

Hao, Y., T. C. Charles and B. R. Glick. 2007. ACC deaminase from plant growth promoting bacteria affects crown gall development. Can. J. Microbiol. 53: 1291–1299.

Hao, Y., T. C. Charles and B. R. Glick. 2011. An ACC deaminase containing *A. tumefaciens* strain D3 shows biocontrol activity to crown gall disease. Can. J. Microbiol. 57: 278–286.

Harper, S. H. T. and J. M. Lynch. 1979. Effects of *Azotobacter chroococcum* on barley seed germination and seedling development. J. Gen. Microbiol. 112: 45–51.

He, L. Y., Z. J. Chen, G. D. Ren, Y. F. Zhang, M. Qian and X. F. Sheng. 2009. Increased cadmium and lead uptake of a cadmium hyperaccumulator tomato by cadmium-resistant bacteria. Ecotox. Env. Safety. 72: 1343–1348.

Hernandez, Y., J. Sogo and M. Sarmiento. 1997. *Azospirillum* inoculation on *Zea mays*. Cuban J. Agr. Sci. 31: 203–209.

Hoffmann-Hergarten, S., M. K. Gulati and R. A. Sikora. 1998. Yield response and biological control of *Meloidogyne incognita* on lettuce and tomato with rhizobacteria. J. Plant Dis. Protect. 105: 349–358.

Hoflich, G. and R. Metz. 1997. Interactions of plant-microorganism associations in heavy metal containing soils from sewage farms. Bodenkultur 48: 239–247.

Holguin, G. and B. R. Glick. 2001. Expression of the ACC deaminase gene from *Enterobacter cloacae* UW4 in *Azospirillum brasilense*. Microb. Ecol. 41: 281–288.

Holl, F. B. and C. P. Chanway. 1992. Rhizosphere colonization and seedling growth promotion of lodgepole pine by *Bacillus polymyxa*. Can. J. Microbiol. 38: 303–308.

Hong, S. H., H. W. Ryu, J. Kim and K. S. Cho. 2011. Rhizoremediation of diesel-contaminated soil using the plant growth-promoting rhizobacterium *Gordonia* sp. S2RP-17. Biodegradation. 22: 593–601.

Howie, W. J. and E. Echandi. 1983. Rhizobacteria: Influence of cultivar and soil type on plant growth and yield of potato. Soil Biol. Biochem. 15: 127–132.

Huang, C. J., K. H. Yang, Y. H. Liu, Y. J. Lin and C. Y. Chen. 2010. Suppression of southern corn leaf blight by a plant growth-promoting rhizobacterium *Bacillus cereus* C1L. Annals of Appl. Biol. 157: 4553.

Huang, X. D., Y. El-Alawi, D. M. Penrose, B. R. Glick and B. M. Greenberg. 2004a. Multi-process phytoremediation system for removal of polycyclic aromatic hydrocarbons from contaminated soils. Environ. Pollut. 130: 465–476.

Huang, X. D., Y. El-Alawi, D. M., Penrose, B. R. Glick and B. M. Greenberg. 2004b. Responses of plants to creosote during phytoremediation and their significance for remediation processes. Environ. Pollut. 130: 453–463.

Iswandi, A., P. Bossier, J. Vandenbeele and W. Verstraete. 1987. Effect of seed inoculation with the rhizopseudomonad strain7NSK2 on the root microbiota of maize (*Zea mays*) and barley (*Hordeum vulgare*). Biol. Fert. Soils 3: 153–158.

Jacoud, C., D. Faure, P. Wadoux and R. Bally. 1998. Development of a strain-specific probe to follow inoculated *Azospirillum lipoferum* CRT1 maize root development by inoculation. FEMS Microbiol. Ecol. 27: 43–51.

James, C. A. and S. E. Strand. 2009. Phytoremediation of small organic compounds using transgenic plants. Curr. Opin. Biotechnol. 20: 237–241.

Isawa, T., M. Yasuda, H. Awazaki, K. Minamisawa, S. Shinozaki and H. Nakashita. 2010. *Azospirillum* sp. Strain B510 enhances rice growth and yield. Microbes Environ. 25: 58–61.

Jiang, C. Y., X. F. Sheng, M. Qian and Q. Y. Wang. 2008. Isolation and characterization of a heavy metal-resistant *Burkholderia* sp. from heavy metal-contaminated paddy field soil and its potential in promoting plant growth and heavy metal accumulation in metal-polluted soil. Chemosphere. 72: 157–164.

Jones, K. M., H. Kobayashi, B. W. Davies, M. E. Taga and G. C. Walker. 2007. How rhizobial symbionts invade plants: the Sinorhizobium-Medicago model. Nat. Rev. Microbiol. 5: 619–633.

Kapulnik, Y., S. Sarig, I. Nur, J. Okon, J. Kigel and V. Henis. 1981. Yield increases in summer cereal crops of Israel in fields inoculated with *Azospirillum*. Experientia Agricola. 17: 179–187.

Karlidag, H., A. Esitken, S. Ercisli and M. F. Donmez. 2010. The Use of PGPR (Plant Growth Promoting Rhizobacteria) in Organic Apricot Production. Proc. XIVth IS on Apricot Breeding and Culture. Ed. C. Xiloyannis. Acta Hort. 862, ISHS 2010. pp. 309–312.

Karlidag, H., A. Esitken, M. Turan and F. Sahin. 2007. Effects of root inoculation of plant growth promoting rhizobacteria (PGPR) on yield, growth and nutrient element contents of leaves of apple. Scientia Horticulturae. 114: 16–20.

Kavita, B., S. Shukla, G. N. Kumar and G. Archana. 2008. Amelioration of phytotoxic effects of Cd on mung bean seedlings by gluconic acid secreting rhizobacterium *Enterobacter asburiae* PSI3 and implication of role of organic acid. World J. Microbiol. Biotechnol. 24: 2965–2972.

Kloepper, J. W., R. Lifshitz and R. M. Zablotowicz. 1989. Free-living bacterial inocula for enhancing crop productivity. Trends Biotechnol. 7: 39–43.

Kloepper, J. W., D. J. Hume, F. M. Scher, C. Singleton, B. Tipping, M. Laliberté, K. Frauley, T. Kutchaw, C. Simonson, R. Lifshitz, I. Zaleska and L. Lee. 1988. Plant growth-promoting rhizobacteria on canola (rapeseed). Plant Dis. 72: 42–46.

Kloepper, J. W., J. Leong, M. Teintze and M. N. Schroth. 1980. Enhanced plant growth by siderophores produced by plant growth-promoting rhizobacteria. Nature. 286: 885–886.

Kloepper, J. W., M. N. Schoth and T. D. Miller. 1980. Effects of rhizosphere colonization by plant growth-promoting rhizobacteria on potato plant development and yield. Phytopathol. 70: 1078–1082.

Kokalis-Burelle, N., E. N. Vavrina, E. N. Rosskopf and R. A. Shelby. 2002. Field evaluation of plant growth-promoting rhizobacteria amended transplant mixes and soil solarization for tomato and pepper production in Florida. Plant Soil. 238: 257–266.

Koo, S. Y. and K. S. Cho. 2009. Isolation and characterization of a plant growth-promoting rhizobacterium, *Serratia* sp. SY5. J. Microbiol. Biotechnol. 19: 1431–143.

Kropp, B. R., E. Thomas, J. I. Pounder and A. J. Anderson. 1996. Increased emergence of spring wheat after inoculation with *Pseudomonas chlororaphis* isolate 2E3 under field and laboratory conditions. Biol. Fert. Soil. 23: 200–206.

Laditi, M. A., O. C. Nwoke, M. Jemo, R. C. Abaidoo and A. A. Ogunjobi. 2012. Evaluation of microbial inoculants as biofertilizers for the improvement of growth and yield of soybean and maize crops in savanna soils. African J. Ag. Res. 7: 405–413.

Lalande, R., N. Bissonette, D. Coutlée and H. Antoun. 1989. Identification of rhizobacteria from maize and determination of the plant-growth promoting potential. Plant Soil. 115: 7–11.

Leyval, C. and J. Berthelin. 1989. Influence of acid-producing *Agrobacterium* and *Laccaria laccata* on pine and beech growth, nutrient uptake and exudation. Agr. Ecosyst. Environ. 28: 313–319.

Li, J., J. Sun, Y. Yang, S. Guo and B. R. Glick. 2012. Identification of hypoxic-responsive proteins in cucumber using a proteomic approach. Plant Physiol. Biochem. 51: 74–80.

Li, K. and W. Ramakrishna. 2011. Effect of multiple metal-resistant bacteria from contaminated lake sediments on metal accumulation and plant growth. J. Hazardous Mat. 189: 531–539.

Li, W. C., Z. H. Ye and M. H. Wong. 2007. Effects of bacteria on enhanced metal uptake of the Cd/Zn-hyperaccumulating plant, *Sedum alfredii*. J. Experiment. Bot. 58: 4173–4182.

Lifshitz, R., J. W. Kloepper, M. Kozlowski, C. Simonson, E. M. Tipping and I. Zaleska. 1987. Growth promotion of canola (rapeseed) seedlings by a strain of *Pseudomonas putida* under gnototropic conditions. Can. J. Microbiol. 23: 390–395.

Long, S. R., W. J. Buikema and F. M. Ausubel. 1982. Cloning of *Rhizobium meliloti* nodulation genes by direct complementation of Nod⁻ mutants. Nature. 298: 485–488.

Ma, W., D. M. Penrose and B. R. Glick. 2002. Strategies used by rhizobia to lower plant ethylene levels and increase nodulation. Can. J. Microbiol. 48: 947–954.

Ma, W., F. C. Guinel and B. R. Glick. 2003. The *Rhizobium leguminosarum* bv. *viciae* ACC deaminase protein promotes the nodulation of pea plants. Appl. Environ. Microbiol. 69: 4396–4402.

Ma, W., T. C. Charles and B. R. Glick. 2004. Expression of an exogenous 1-aminocyclopropane-1-carboxylate deaminase gene in *Sinorhizobium meliloti* increases its ability to nodulate alfalfa. Appl. Environ. Microbiol. 70: 5891–5897.

Madhaiyan, M., S. Poonguzhali and T. Sa. 2007. Metal tolerating methylotrophic bacteria reduces nickel and cadmium toxicity and promotes plant growth of tomato (*Lycopersicon esculentum* L.). Chemosphere. 69: 220–228.

Madhaiyan, M., P. Selvaraj, J. Ryu and T. Sa. 2006. Regulation of ethylene levels in canola (*Brassica campestris*) by 1-aminocyclopropane-1-carboxylate deaminase-containing *Methylobacterium fujisawaense*. Planta. 224: 268–278.

Manjula, K. and A. R. Podile. 2005. Increase in seedling emergence and dry weight of pigeon pea in the field with chitin-supplemented formulations of *Bacillus subtilus* AF 1. World J. Microbiol. Biotech. 21: 1057–1062.

Marek-Kozaczuk, M., J. Kopcinska, B. Lotocka, W. Golinowski and A. Skorupska. 2000. Infection of clover by plant growth promoting *Pseudomonas fluorescens* strain 267 and *Rhizobium leguminosarum* bv. *trifolii* studied by mTn-*gus*A. Antonie vanLeeuwenhoek. 78: 1–11.

Mayak, S., T. Tirosh and B. R. Glick. 2001. Stimulation of the growth of tomato, pepper and mung bean plants by the plant growth-promoting bacterium *Enterobacter cloacae* CAL3. Biol. Agr. Hort. 19: 261–274.

Mayak, S., T. Tirosh and B. R. Glick. 2004a. Plant growth-promoting bacteria that confer resistance in tomato to salt stress. Plant Physiol. Biochem. 42: 565–572.

Mayak, S., T. Tirosh and B. R. Glick. 2004b. Plant growth-promoting bacteria that confer resistance to water stress in tomato and pepper. Plant Sci. 166: 525–530.

McCullagh, M., R. Utkhede, J. G. Menzies, Z. K. Punja and T. C. Paulitz. 1996. Evaluation of plant growth promoting rhizobacteria for biological control of *Pythium* root rot of cucumbers grown in rockwool and effects on yield. Europ. J. Plant Pathol. 102: 747–755.

Mehnaz, S., T. Kowalik, B. Reynolds and G. Lazarovits. 2010. Growth promoting effects of corn (*Zea mays*) bacterial isolates under greenhouse and field conditions. Soil Biol. and Biochem. 42: 1848–1856.

Mehnaz, S. and G. Lazarovits. 2006. Inoculation effects of *Pseudomonas putida, Gluconacetobacter azotocaptans,* and *Azospirillum lipoferum* on corn plant growth under greenhouse conditions. Microbial Ecol. 51: 526–335.

Mishustin, E. N. and A. N. Naumova. 1962. Bacterial fertilizers: Their effectiveness and mode of action. Mikrobiologiya. 31: 543–555.

Mohammad, G. and R. Prasad. 1988. Influence of microbial fertilizers on biomass accumulation in polypotted *Eucalyptus camaldulensis* Dehn. seedlings. J. Trop. For. 4: 47–77.

Moslemi, Z., D. Habibi, A. Asgharzadeh, M. R. Ardakani, A. Mohammadi and A. Sakari. 2011. Effects of super absorbent polymer and plant growth promoting rhizobacteria on yield and yield components of maize under drought stress and normal conditions. African J. Ag. Res. 6: 4471–4476.

Nadeem, S. M., Z. A. Zahair, M. Naveed and M. Arshad. 2007. Preliminary investigation on inducing salt tolerance in maize through inoculation with rhizobacteria containing ACC deaminase activity. Can. J. Microbiol. 53: 1141–1149.

Nie, L., S. Shah, A. Rashid, G. I. Burd, D. G. Dixon and B. R. Glick. 2002. Phytoremediation of arsenate contaminated soil by transgenic canola and the plant growth-promoting bacterium *Enterobacter cloacae* CAL2. Plant Physiol. Biochem. 40: 355–361.

Okon, Y., Y. Kapulnik and S. Sarig. 1988. Field inoculation studies with *Azospirillium* in Israel. *In:* N. S. Subba Rao [eds.]. BiologicalNitrogen Fixation Recent Developments. Oxford and IBH Publishing Co. New Delhi, India. pp. 175–195.

Okon, Y. and C. A. Labandera-Gonzalez. 1994. Agronomic applications of *Azospirillum*: an evaluation of 20 years worldwide field inoculation. Soil Biol. Biochem. 26: 1591–1601.

Oldroyd, G. E. D., J. D. Murray, P. S. Poole and J. A. Downie. 2011. The rules of engagement in the legume-rhizobial symbiosis. Annu. Rev. Genet. 45: 119–144.

Omar, N., T. Heulin, P. Weinhard and M. N. Alaa El-Din. 1989. Field inoculation of rice with *in vitro* selected plant growth promoting-rhizobacteria. Agronomie 9: 803–808.

O'Neill, G. A., C. P. Chanway, P. E. Axelrood, R. A. Radley and F. B. Holl. 1992. Growth response specificity of spruce inoculated with coexistent rhizosphere bacteria. Can. J. Bot. 70: 2347–2353.

Orhan, E., A. Esitken, S. Ercisli, M. Turan and F. Sahin. 2006. Effects of plant growth promoting rhizobacteria (PGPR) on yield, growth and nutrient contents in organically growing raspberry. Scientia Horticulturae. 111: 38–43.

Patel, D., C. K. Jha, N. Tank and M. Saraf. 2012. Growth enhancement of chickpea in saline soils using plant growth-promoting rhizobacteria. J. Plant Growth Regul. 31: 53–62.

Pan, B., J. K. Vessey and D. L. Smith. 2002. Response of field-grown soybean to co-inoculation with the plant growth promoting rhizobacteria *Serratia proteamaculans* or *Serratia liquefaciens* and *Bradyrhizobium japonicum* pre-incubated with genistien. Europ. J. Agronomy 17: 143–153.

Pandey, A., A. Durgapal, M. Joshi and L. M. S. Palni. 1999. Influence of *Pseudomonas corrugate* inoculation on root colonization and growth promotion of two important hill crops. Microbiol. Res. 154: 259–266.

Pandey, R. K., R. K. Bahl and P. R. T. Rao. 1986. Growth stimulation effects of nitrogen fixing bacteria (biofertilizer) on oak seedlings. Ind. For. 112: 75–79. Parke J. L. 1991. Root colonization by indigenous and introduced microorganisms. *In:* Keister D. L. and P. B. Cregan [eds.]. The Rhizosphere and Plant Growth. Kluwer Academic Publishers, Dordrecht, the Netherlands. pp. 33–42.

Pereyra, M. A., F. M. Ballesteros, C. M. Creus, R. J. Sueldo and C. A. Barassi. 2009. Seedlings growth promotion by *Azospirillum brasilense* under normal and drought conditions remains unaltered in Tebuconazole-treated wheat seeds. Eruo. J. Soil Biol. 45: 20–27.

Pirlak, L. and M. Kose. 2009. Effects of plant growth promoting rhizobacteria on yield and some fruit properties of strawberry. J. Plant Nutrition. 32: 1173–1184.

Polyanskaya, L. M., O. T. Vedina, L. V. Lysak and D. G. Zvyagintev. 2000. The growth-promoting effect of *Beijerinckia mobilis* and *Clostridium* sp. cultures on some agricultural crops. Microbiol. 71: 109–115.

Probanza, A., J. A. Lucas Garcia, M. Ruiz Palomino, B. Ramos and F. J. Gutiérrez Mañero. 2002. *Pinus pinea* L. seedling growth and bacterial rhizosphere structure after inoculation with PGPR *Bacillus* (*B. licheniformis* CECT 5106 and *B. pumilis* CECT 5105). Appl. Soil Ecol. 20: 75–84.

Rajkumar, M. and H. Freitas. 2008. Effects of inoculation of plant-growth promoting bacteria on Ni uptake by Indian mustard. 2008. Bioresource Tech. 99: 3491–3498.

Reddy, M. S. and J. E. Rahe. 1989. Growth effects associated with seed bacterization not correlated with populations of *Bacillus subtilis* inoculant in onion seedling rhizospheres. Soil Biol. Biochem. 21: 373–378.

Reed, M. L. E. and B. R. Glick. 2004. Applications of free living plant growth-promoting rhizobacteria. Anton. Van Leeuwenhoek. 86: 1–25.

Reed, M. L. E and B. R. Glick. 2005. Growth of canola (*Brassica napus*) in the presence of plant growth-promoting bacteria and either copper or polycyclic aromatic hydrocarbons. Can. J. Microbiol. 51: 1061–1069.

Reed, M. L. E., B. Warner and B. R. Glick. 2005. Plant growth-promoting bacteria facilitate the growth of the common reed *Phragmites australis* in the presence of copper or polycyclic aromatic hydrocarbons. Curr. Microbiol. 51: 425–429.

Reynders, L. and K. Vlassak. 1982. Use of *Azospirillum brasilens*eas biofertilizer in intensive wheat cropping. Plant Soil. 66: 217–223.

Ribaudo, C. M., D. P. Rondanini, J. A. Cura and A. A. Fraschina. 2001. Response of *Zea mays* to the inoculation with *Azospirillum* on nitrogen metabolism under greenhouse conditions. Biol. Plantarum. 44: 631–634.

Rincón, A., F. Valladares, T. E. Gimeno and J. J. Pueyo. 2008. Water stress responses of two Mediterranean tree species influenced by native soil microorganisms and inoculation with a plant growth promoting rhizobacterium. Tree Physiol. 28: 1693–1701.

Rodriguez, H., S. Vessely, S. Shah and B. R. Glick. 2008. Effect of a nickel-tolerant ACC deaminase-producing *Pseudomonas* strain on growth of nontransformed and transgenic canola plants. Curr. Microbiol. 57: 170–174.

Rojas, A., G. Holguin, B. R. Glick and Y. Bashan. 2001. Synergism between *Phyllobacterium* sp. (N$_2$-fixer) and *Bacillus licheniformis* (P-solubilizer), both from a semi-arid mangrove rhizosphere. FEMS Microbiol. Ecol. 35: 181–187.

Rojas-Tapias, D. F., R. R. Bonilla and J. Dussán. 2011. Effect of inoculation with plant growth-promoting bacteria on growth and copper uptake by sunflowers. Water Air Soil Pollution. 223: 643–653.

Rosas, S. B., G. Avanzini, E. Carlier, C. Pasluosta, N. Pastor and M. Rovera. 2009. Root colonization and growth promotion of wheat and maize by *Pseudomonas aurantiaca* SR1. Soil Biol. Biochem. 41: 1802–1806.

Sabir, A., M. A. Yazici, Z. Kara and F. Sahin. 2011. Growth and mineral acquisition response of grapevine rootstocks (*Vitis* spp.) to inoculation with different strains of plant growth-promoting rhizobacteria (PGPR). J. Sci. Food Agric. DOI 10. 1002/jsfa. 5600.

Safronova, V. I., V. V. Stepanok, G. L. Engqvist, Y. V. Alekseyev and A. A. Belimov. 2006. Root-associated bacteria containing 1-aminocyclopropane-1-carboxylate deaminase improve growth and nutrient uptake by pea genotypes cultivated in cadmium supplemented soil. Biol. Fertil. Soils. 42: 267–272.

Sagar, S., A. Dwivedi, S. Yadav, M. Tripathi and S. D. Kaistha. 2012. Hexavalent chromium reduction and plant growth promotion by *Staphylococcusarlettae* strain Cr11. Chemosphere. 86: 847–852.

Saravanakumar, D. and R. Samiyappan. 2006. ACC deaminase from *Pseudomonas fluorescens* mediated saline resistance in ground nut (*Arachis hypogea*) plants. J. Appl. Microbiol. 102: 1283–1292.

Sarig, S., A. Blum and Y. Okon. 1998. Improvement of the water status and yield of field grown grain sorghum (*Sorghum bicolour)* by inoculation with *Azospirillum brasilense*. J. Agr. Sci. 110: 271–277.

Saubidet, M. I., N. Fatta and A. J. Barneix. 2002. The effect of inoculation with *Azospirillum brasilense* on growth and nitrogen utilization by wheat plants. Plant Soil. 245: 215–222.

Shaharoona, B., M. Naveed, M. Arshad and Z. A. Zahir. 2008. Fertilizer-dependent efficiency of Pseudomonads for improving growth, yield, and nutrient use efficiency of wheat (*Triticum aestivum* L.). Appl. Microbiol. Biotechnol. 79: 147–155.

Sharifi, R. S., K. Khavazi and A. Gholipouri. 2011. Effect of seed priming with plant growth promoting rhizobacteria (PGPR) on dry matter accumulation and yield of maize (*Zea mays* L.) hybrids. Int. Res. J. Biochem. Bioinformat. 3: 76–83.

Sheng, X., L. Sun, Z. Huang, L. He, W. Zhang and Z. Chen. 2012. Promotion of growth and Cu accumulation of bio-energy crop (*Zea mays*) by bacteria: Implications for energy plant biomass production and phytoremediation. J. Environ. Management. 103: 58–64.

Shishido, M. and C. P. Chanway. 2000. Colonization and growth of outplanted spruce seedlings pre-inoculated with plant growth promoting rhizobacteria in the greenhouse. Can. J. For. Res. 30: 848–854.

Siddikee, M. A., B. R. Glick, P. S. Chauhan, W. J. Yim and T. Sa. 2011. Enhancement of growth and salt tolerance of red pepper seedlings (*Capsicum annuum* L.) by regulating stress ethylene synthesis with halotolerant bacteria containing ACC deaminase activity, Plant Physiol. Biochem. 49: 427–434.

Sinha, S. and S. K. Mukherjee. 2008. Cadmium-induced siderophore production by a high Cd-resistant bacterial strain relieved Cd toxicity in plants through root colonization. Curr. Microbiol. 56: 55–60.

Sprent, J. I. 1986. Benefits of Rhizobium to agriculture. Trends Biotechnol. 4: 124–129.

Suslow, T. V. and M. N. Schroth. 1982. Rhizobacteria of sugar beets: Effects of seed application and root colonization on yield. Phytopathol. 72: 199–206.

Tang, S., S. Liao, J. Guo, Z. Song, R. Wang and X. Zhou. 2011. Growth and cesium uptake responses of *Phytolacca americana* Linn. and *Amaranthus cruentus* L. grown on cesium contaminated soil to elevated CO_2 or inoculation with a plant growth promoting rhizobacterium *Burkholderia* sp. D54, or in combination. J. Hazard Mater. 30: 188–197.

Teijeiro, R. G., I. C. Dodd, E. D. Elphinstone, V. I. Safronova and A. A. Belimov. 2011. From seed to salad: impacts of ACC deaminase-containing plant growth promoting rhizobacteria on lettuce growth and development. Proc. Vth IS on Seed, Transplant and Stand

Establishment of Hort. Crops. Eds. J. A. Pascual and F. Perez-Alfocca. Acta Hort. 898, ISHS 2011. pp. 245–252.

Thomas, L., A. Gupta, M. Gopal, P. George and G. V. Thomas. 2011. Efficacy of rhizospheric *Bacillus* spp. for growth promotion in *Theobroma cacao* L. seedlings. J. Plantation Crops. 39: 19–25.

Timmusk, S., V. Paalme, T. Pavlicek, J. Bergquist, A. Vangala, T. Danilas and E. Nevo. 2011. Bacterial distribution in the rhizosphere of wild barley under contrasting microclimates. PLoS One 6(3) e17968.

Toklikishvili, N., N. Dandurishvili, M. Tediashvili, N. Giorgobiani, E. Szegedi, B. R. Glick, A. Vainstein and L. Chernin. 2010. Inhibitory effect of ACC deaminase-producing bacteria on crown gall formation in tomato plants infected by *Agrobacterium tumefaciens* or *A. vitis*. Plant Pathol. 59: 1023–1030.

Tran Van, V., O. Berge, S. Ngo Ke, J. Balandreau and T. Heulin. 2000. Repeated beneficial effects of rice inoculation with a strain of *Burkholeria vietnamiensis* on early and late yield components in low fertility sulphate acid soils of Vietnam. Plant Soil. 281: 273–284.

Turner, J. T. and P. A. Backman. 1991. Factors relating to peanut yield increases after seed treatment with *Bacillus subtilis*. Plant Disease. 75: 347–353.

Uthede, R. S., C. A. Koch and J. G. Menzies. 1999. Rhizobacterial growth and yield promotion of cucumber plants inoculated with *Pythium aphanidermatum*. Can. J. Plant Pathol. 21: 265–271.

Van Aken, B. 2008. Transgenic plants for phytoremediation: helping nature to clean up environmental pollution. Trends Biotechnol. 26: 225–227.

Van Peer, R. and B. Schippers. 1989. Plant growth responses to bacterization with selected *Pseudomonas* spp. strains and rhizosphere microbial development in hydroponic cultures. Can. J. Microbiol. 35: 456–463.

Wang, C., E. Knill, B. R. Glick and G. Défago. 2000. Effect of transferring 1-aminocyclopropane-1-carboxylic acid (ACC) deaminase genes into *Pseudomonas fluorescens* strain CHA0 and its derivative CHA96 on their plant growth promoting and disease suppressive capacities. Can. J. Microbiol. 46: 898–907.

Wang, Q., D. Xiong, P. Zhao, X. Yu, B. Tu and G. Wang. 2011. Effect of applying an arsenic-resistant and plant growth-promoting rhizobacterium to enhance soil arsenic phytoremediation by *Populus deltoides* LH05-17. Appl. Microbiol. 111: 1065–1074.

Weller, D. M. and R. J. Cook. 1986. Increased growth of wheat by seed treatment with fluorescent pseudomonads, and implications of *Pythium* control. Can. J. Microbiol. 8: 328–334.

Whiting, S. N., M. P. De Souza and N. Terry. 2001. Rhizosphere bacteria mobilize Zn for hyperaccumulation by *Thlaspi caerulescens*. Environ. Sci. Technol. 35: 3144–3150.

Xu, G. W. and D. C. Gross. 1986. Field evaluations of the interactions among fluorescent Pseudomonads, *Erwinia carotovora*, and potato yields. Phytopathol. 76: 423–430.

Xu, Y., A. Yokota, H. Sanada, M. Hisamatsu, M. Araki, H. Cho, Morinaga and Y. Murooka. 1994. *Enterobacter cloacae* A105, isolated from the surface of root nodules of *Astragalus sinicus* cv. Japan, stimulates nodulation by *Rhizobium huakuii* bv. *renge*. J. Ferment. Bioeng. 77: 630–635.

Yao, L., Z. Wu, Y. Zheng, I. Kaleem and C. Li. 2010. Growth promotion and protection against salt stress by *Pseudomonas putida* Rs-198 on cotton. Eur. J. Soil Biol. 46: 49–54.

Yolcu, H., M. Turan, A. Lithourgidis, R. Cakmakci and A. Koc. 2011. Effects of plant growth-promoting rhizobacteria and manure on yield and quality characteristics of Italian ryegrass under semi arid conditions. Austral. J. Crop Sci. 5: 1730–1736.

Yue, H. T., W. P. Mo, C. Li, Y. Y. Zheng and H. Li. 2007. The salt stress relief and growth promotion effect of Rs-5 on cotton. Plant Soil. 297: 139–145.

Zaady, E. and A. Perevoltsky. 1995. Enhancement of growth and establishment of oak seedlings (*Quercus ithaburensis* Decaisne) by inoculation with *Azospirillum brasilense*. For. Ecol. Manage. 72: 81–83.

Zahir, Z. A., U. Ghani, M. Naveed, S. M. Nadeem and H. N. Asghar. 2009. Comparative effectiveness of *Pseudomonas* and *Serratia* sp. containing ACC-deaminase for improving growth and yield of wheat (*Triticum aestivum* L.) under salt-stressed conditions. Arch. Microbiol. 191: 415–424.

Zhang, S., T. L. White, M. C. Martinez, J. A. McInroy, J. W. Kloepper and W. Klassen. 2010. Evaluation of plant growth-promoting rhizobacteria for control of Phytophthora blight on squash under greenhouse conditions. Biological Cont. 53: 128–135.

8

Engineering of Microbes for Plant and Soil Systems

Kieran J. Germaine, Martina McGuinness and
*David N. Dowling**

ABSTRACT

Over the last twenty-five years, a significant amount of bio-environmental research has focused on the potential application of microorganisms as biofertilisers and/or bioconrol agents for use in agriculture and in the bioremediation of contaminated sites. These microorganisms, when used to inoculate plant or soil systems, can accelerate certain microbial processes in the soil which augment the extent of availability of nutrients in a form easily assimilated by plants. Use of biofertilisers is one of the important components of integrated nutrient management as they facilitate a cost effective renewable source of plant nutrients which supplement chemical fertilizers contributing to sustainable agriculture. Several microorganisms are currently being marketed commercially as biofertilisers for crop plants. Unfortunately, field experiments show that these microorganisms are not always as efficient in the field as they are in laboratory or greenhouse experiments.

Envirocore Research Centre, Department of Science and Health, Institute of Technology Carlow, Kilkenny Road, Carlow, Ireland.
  Emails: germaink@itcarlow.ie; Martina.McGuinness@itcarlow.ie; David.Dowling@itcarlow.ie
* Corresponding author

This chapter reviews the attempts to improve reliability of these microorganisms in the field through genetic modification of their desired traits and of their survival ability.

Introduction

A microbe that is modified using genetic engineering techniques is referred to as a genetically modified microorganism (GMM) or, alternatively, a genetically engineered microbe (GEM). Microbes can be engineered so as to (i) improve expression of beneficial traits, and (ii) be traceable in plant and soil systems. Studies carried out under controlled laboratory, growth chamber or greenhouse conditions facilitate the understanding of some of the underlying mechanisms of GMMs. However, field studies are necessary for evaluation of the ultimate potential of GMMs in plant and soil systems since local environmental factors appear to play a critical role in the survival of microbes (both unmodified parent strains and engineered strains) and in the expression of beneficial traits. It is important, therefore, that field release experiments be carried out to further investigate the factors influencing the introduction of engineered microbes into plant and soil systems.

Not surprisingly, there are concerns about the effects that release of large numbers of genetically modified bacteria might have on indigenous microbiota in the environment. Several field releases of modified bacteria have been carried out to investigate if the inoculant bacteria affected numbers, composition and activities of indigenous microbiota or if they exchanged genetic material with indigenous species. Studies were also carried out to determine the effect of modified versus unmodified bacteria on the environment. These studies will be discussed in detail later in this chapter. Monitoring inoculant bacteria after their release in field studies requires differentiation of inoculant bacteria from the many indigenous bacteria that can be genetically and/or phenotypically very similar. The introduction of gene markers normally uncommon in the soil population, such as *lacZY, gusA, luxAB, luc, gfp* or *xylE*, into inoculant bacteria has facilitated monitoring of the released bacteria, particularly when they are introduced into antibiotic resistant derivatives, as has usually been the case. Amarger (2002) reported that there were no differences between the survival, spread, persistence in the field, and ecological impacts of engineered microbes and unmodified parent strains and suggested that introduction of modified bacteria does not appear to be problematic in terms of hazard or risk to the environment. However, since only approximately 10% of microbes in environmental samples are culturable in a laboratory, it is possible that culture independent methods might be more sensitive to changes in indigenous microbes. More recently, Yamamoto-Tamura et al. (2011) reported that when they used culture-independent methods [denaturing gradient gel electrophoresis (DGGE) and the polymerase chain reaction (PCR)] as opposed to traditional culture-dependent methods they could show changes in the bacterial, but not fungal, communities in the soil.

There are also concerns about the possibility of genetic interactions between engineered microbes and indigenous microbes in the environment. A number of studies listed by Amarger (2002) involving field releases of strains specifically designed for assessing the transfer of plasmids under natural conditions have provided evidence that, under certain circumstances and with certain bacteria, genes could be transferred from the indigenous bacterial population to the introduced bacteria. However, the changes induced in the population were found to be transient in the absence of selective pressure on the transferred trait.

Release of genetically engineered organisms remains controversial at international level. While some governments support genetic engineering, others prohibit the use of the technology and require labeling of products containing genetically modified ingredients. However, a number of studies carried out to date have shown that when appropriate regulations are adhered to, GMMs can be applied safely in agriculture (Armarger 2002, Morrissey et al. 2002). As a precaution, chromosomal insertion of genes in inoculant bacteria can be used to avoid horizontal transfer of genes introduced via plasmids into plant and soil systems.

Da and Deng (2003) suggested that genetic engineering, but not mutation of specific genes, weakened the bacteria's ability to survive in a soil environment. This impact, however, was not as prominent as the impact resulting from the presence of host plants or variations in soil properties. They studied (i) the effects of mutation of two acid phosphatase genes on survival and persistence of the rhizospheric bacteria, *Sinorhizobium meliloti* 104A14 in plant growth units using experimental soil from Oklahoma, USA, (ii) the effect of alfalfa growth on the establishment and persistence of genetically engineered *S. meliloti* in the soil environment, and (iii) the impact of genetically engineered *S. meliloti* on the soil microbial populations and community structure in the alfalfa rhizosphere (Da and Deng 2003). Da and Deng (2003) showed that the survival and persistence of *S. meliloti* 104A14 was enhanced in soils with adequate nutrient supply, high organic matter, optimal pH, and host plants. In this study, the two genes, *NapD* and *NapE*, were inactivated by inserting an 8Kb transposon Tn5-B20, which contains a promoterless *lacZ* and a kanamycin resistant gene as described by Simon et al. (1989).

Da and Deng (2003) also showed that the total culturable microbial population in soil doubled after six weeks of alfalfa growth and inoculation of genetically engineered *S. meliloti* when compared with the unmodified parent strain. However, microbial community structure was not altered.

Reports in the literature on this subject are inconsistent, De Leij et al. (1995) reported that the total bacterial population was not affected by inoculation with the wild-type or genetically engineered *P. fluorescens*, Bej et al. (1991) reported an increase in total bacterial population after inoculation with a genetically engineered *Burkholderia cepacia* and suggested that the introduced genetically engineered microbes not only displaced a proportion of the indigenous population, but that they also occupied a wider niche or colonized more intensely in the rhizosphere. However, *Sinorhizobia* are not known to be competitive in the rhizosphere. Da and Deng (2003) suggest that genetically engineered microbes stimulated the interaction between alfalfa roots and *Sinorhizobia* which resulted in changes in the pattern of root exudates. As a result, stimulated growth of the indigenous microbes was observed as evidenced by higher nodule numbers after *NapD* and *NapE* mutant inoculation when compared with wildtype.

Liu et al. (2010a) introduced a *Pseudomonas* strain genetically engineered to degrade polychlorinated biphenyls (PCBs) into the rhizosphere of pea (*Pisum sativum*) plants grown in PCB contaminated soil. Using community level metabolic profiling they showed that there was a significant shift in the activity of the microbial community in the soil compared to that of the unmodified parent strain. However, they also showed that this same shift in activity could be achieved by the introduction of a wildtype strain with natural PCB degrading ability. Therefore, the shift in community activity was due to the enhanced degradation of PCBs, which occurred whether the introduced PCB degrading strain was genetically modified or not.

Genetically Engineered Plant Growth Promoting Bacteria

Successful crop growth depends on the genetic make-up of the plant, adequate availability of nutrients, the presence of beneficial microbes and the absence of phytopathogens. The use of chemical fertilisers and pesticides has facilitated substantial increases in crop yields. However, the cost of chemical fertilisers is closely linked to the cost of fossil fuels. High fossil fuel prices and depleting phosphate resources are resulting in significant increases in the cost of fertiliser production. Since fossil fuel prices are only set to increase it will be necessary to find alternatives to chemical fertilisers and pesticides in order to meet the growing demands for food and energy production in an economically sustainable manner. The use of beneficial microbes as 'biofertilisers' has enormous potential for reducing the cost of crop production and alleviate the negative environmental effects of fertiliser and pesticide use (Germaine et al. 2010).

Plant beneficial bacteria, also known as Plant Growth Promoting Rhizobacteria (PGPR) positively influence plant growth promotion in two ways: (i) directly by producing phytohormones such as auxins, lowering plant stress ethylene levels by 1-aminocyclopropane-1-carboxylate (ACC) deaminase activity or by helping plants acquire nutrients, e.g., via nitrogen (N_2) fixation, phosphate solubilisation or iron chelation (Spaepen et al. 2009), or (ii) indirectly by preventing pathogen infections via release of antimicrobial agents, by outcompeting pathogens, or by establishing the plant's systemic resistance (Hardoim et al. 2008). However, with the exception of rhizobia, there has been only limited commercial use of plant growth-promoting bacteria in agriculture and forestry. The main reason for this is the reported variability and inconsistency of results between laboratory, greenhouse, and field trials. Long et al. (2008) concluded that natural plant associated bacteria with plant growth promoting traits do not have consistent and predictable effects on the growth and fitness of all host plants. These inconsistent and irreproducible results may reflect variations in inoculant-crop compatibility, soil composition and moisture content, and, perhaps most importantly, an incomplete understanding of the mechanisms employed by plant growth-promoting bacteria to facilitate plant growth. Recently, biotechnology and genetic engineering have been used as tools to further our understanding of the mechanisms behind plant growth promotion and to enhance their effects on plants.

Nitrogen (N) fixation is the ability to fix atmospheric nitrogen and supply it in a usable form to the host plant and is the most economically important plant growth promotion trait. N fixing bacteria can supply over 155Kg nitrogen/hectare in some agricultural regions that grow leguminous crops (Crews and Peoples 2004) and can result in nitrogen residues remaining in the soil in sufficient quantities to supply the needs of follow on non-leguminous crops in the same field. This rotation of legume and non-legume crops is particularly important for farmers in developing countries, where chemical fertilisers may be prohibitively expensive. The N fixing bacteria that form symbiotic associations with these leguminous crops are collectively called rhizobia and currently the term is used to describe 44 species of plant nodulating bacteria dispersed among 11 genera of the α- and β-proteobacteria (Cummings et al. 2006). These bacteria colonise and infect the roots of legumes, induce nodule formation, form bacteroids (cell wall free bacteria), and fix N in the oxygen limited environment of the nodule. The most commercially important N fixing bacteria are *Rhizobium*, *Sinorhizobium* and *Bradyrhizobium* which are sold as crop inoculants. The economic value due to increased crop yields and the reduction in fertiliser costs has led to significant research into the development of enhanced rhizobia inoculation systems, survival in soil, root colonisation and nodule formation and N fixing activity.

Genetic modifications carried on rhizobia have attempted to improve strains used as inocula through enhancing (i) their ability to successfully compete with indigenous soil rhizobia for increased nodule formation and

(ii) their capacity to fix N (Amarger 2002). Robleto et al. (1998) introduced the genes for trifolitoxin production into *Rhizobium etli*. Trifolitoxin is an antibiotic to which many non-trifolitoxin producing strains are sensitive to. Robleto et al. (1998) found that strains carrying the trifolitoxin genes were more competitive and persisted longer in soil than did non-trifolitoxin producing strains. They showed that over 2 years, 20% more nodules contained the trifolitoxin producing strains. Van Dillewijn et al. (2001) introduced an over-expressed *put*A gene into *S. meliloti*. This is a metabolic gene, involved in root colonization. Once again, larger populations of the modified strain were initially identified in nodules. Orikasa et al. (2010) showed that insertion of a highly active catalase (*vkt*A) gene in *Rhizobium leguminosarum* resulted in increased nitrogen-fixing activity of nodules, 1.7 to 2.3 times that of the wildtype bacteria. The levels of H_2O_2 in these nodules were decreased by around 73%, while those of leghemoglobins were increased by 1.2–2.1 times compared with the parent. However, this study showed that there was no increase in crop yield. Genetically modified nitrogen fixing strains have been commercialized both in the US and in Australia. *Sinorhizobium meliloti* strain RMBPC-2 has been genetically enhanced through the insertion of a *nif*A gene to increase nitrogen fixing activity, the *dct*ABD genes which include the regulation and structural genes necessary for bacteroid uptake of C_4-dicarboxylic acids from the plant and a spectinomycin resistance gene as a genetic marker. This strain was first commercialized in 1997 and has resulted in significant increases in the yield of alfalfa, although these increases were very much dependent on environmental conditions.

After N, phosphorus (P) is the most essential macronutrient required by plants for growth. Inorganic phosphates occur in soil, mostly in insoluble mineral complexes, some of them appearing after the application of chemical fertilizers. These precipitated forms of P cannot be absorbed by plants. Therefore, the ability to convert insoluble phosphates to a form accessible to the plants, e.g., orthophosphate, is an important trait for a PGPR strain. There are numerous mechanisms through which bacteria can solubilise phosphates. These include the production of simple organic acids such as gluconic or α-keto acids which simply re-dissolve insoluble phosphates and the production of phosphatases and phytases which liberate phosphates from organic matter (Rodriguez et al. 2006). Introduction or over-expression of genes involved in soil phosphate solubilisation (both organic and inorganic) in natural rhizosphere bacteria is a very attractive approach for improving the capacity of microorganisms to work as inoculants. Insertion of phosphate-solubilising genes into microorganisms that do not have this capability may avoid the current need to mix two populations of bacteria for inoculation (N fixers and phosphate-solubilisers) (Bashan et al. 2000). Glucose dehydrogenase is a membrane bound enzyme encoded by the *gcd* gene and converts glucose into gluconic acid (Rodriquez et al. 2006). Recently, the *gcd* gene from *E. coli* was cloned under the control of phosphate transport gene promoters and then inserted into *Azotobacter* (Sashidhar and Podile 2009). This modified strain showed enhanced phosphate solubilisation and resulted in enhanced plant growth when inoculated into sorghum. Rodriguez et al. (2000) cloned the pyrroloquinoline quinone (PQQ) synthesis genes from *Erwinia herbicola*, and transferred them into two rhizobacterial strains, *B. cepacia* IS-16 and *Pseudomonas* sp. PSS recipient cells. This trait is also involved in mineralisation of phosphate and after this gene was transferred to both rhizobacterial strains, they produced larger zones of clearing in plates with insoluble phosphate as the mineral phosphate source, in comparison with those of parental strains without PQQ. However, whether insertion of PQQ improved the plant growth promotion abilities of the engineered bacteria was not addressed by the authors (Rodriguez et al. 2000).

There are numerous other plant growth promoting traits that can be targeted in genetic engineering experiments for improving the performance of microbial biofertilisers, e.g., many rhizosphere associated bacteria can produce the plant growth hormone indole-3-acetic acid (IAA) which stimulates growth of the meristematic tissues of the roots and shoots. The most common metabolic pathway for producing IAA uses tryptophan as a precursor (Spaepen et al. 2009). In the indole-3-acetamide (IAM) pathway, tryptophan is transformed to IAM by the enzyme tryptophan-2-monooxygenase, encoded by the *iaa*M gene. IAM is then converted to IAA by IAM hydrolase, encoded by the *iaa*H gene. The genes *iaa*M and *iaa*H have been cloned, sequenced and characterized from *Pseudomonas*, *Pantoea*, *Rhizobium* and *Bradyrhizobium* (Gafni et al. 1997, Yamada et al. 1985). Recently, the *iaa*M gene has been inserted into both *Rhizobium* and *Sinorhizobium*, which when inoculated into vetch and clover plants led to enhanced N fixing activity and increased plant dry weight (Imperlini et al. 2009, Camerini et al. 2008). Kochar et al. (2011) showed that heterologous expression of the indole-3-pyruvate decarboxylase/phenylpyruvate decarboxylase gene (*ipdC/ppdC*) of the IPyA pathway from the PGPR strain *Azospirillum brasilense* SM led to enhancement of the IAA level. This recombinant strain was subsequently found to enhance root growth and development in sorghum.

Similarly, many rhizobacteria have been found to reduce plant growth inhibiting signals within plants. Ethylene is a plant stress hormone and its production in the plant is induced under a variety of stress conditions, such as drought, exposure of toxic compounds, pathogen infection and mechanical wounding (Shaharoona et al. 2006). These stress conditions stimulate the production of 1-aminocyclopropane-1-carboxylate (ACC) in the plant, which is subsequently converted to ethylene. High ethylene levels inhibit plant growth. Many plant bacteria have been found to produce ACC deaminase which degrades ACC into ammonium (NH_4) and α-ketoglutarate (Davoud et al. 2010). A number of studies have cloned ACC deaminase genes and inserted them into plant associated bacteria, which when inoculated into plants resulted in increased plant growth (Davoud et al. 2010, Govindasamy et al. 2008, Naveed et al. 2008).

The potential to exploit soil and plant microorganisms as plant growth promoters is exciting. Whether or not this will be achieved through better management of soil microbial communities, by development of more effective microbial inoculants, through the genetic manipulation of specific organisms or with a combination of these approaches remains to be seen. However, genetic manipulation appears to offer the best hope of creating reliable and reproducible results in the field and should contribute to increased acceptance of microbial biofertiliser technology by farmers worldwide.

Genetically Engineered Bacteria for Plant Protection and Biocontrol

Phytopathogens can decrease crop yields by 25–100%. Many of the chemical pesticides used to control fungal and bacterial diseases of plants are hazardous to animals and humans and persist and accumulate in the environment. Agriculture relies heavily on the use of these chemicals in order to maximise crop yield. However, the widespread use of chemical pesticides has resulted in severe environmental pollution, and many pathogens are developing resistance to existing chemicals. In addition, many pesticides are now banned from use in the EU under the Sustainable Use Directive (91/414). Given the negative environmental impact and cost of these chemicals, biological approaches offer many potential advantages in developing sustainable agriculture. Biological approaches currently being developed to control phytopathogens include (i) the use of biocontrol microbes that can suppress or prevent phytopathogen damage, and (ii) the development of transgenic plants that are resistant to one or more pathogenic agents.

Plant associated bacteria can control the damage to plants from phytopathogens by a number of different mechanisms including outcompeting the phytopathogen, physical displacement of the phytopathogen, secretion of siderophores to prevent pathogens in the intermediate vicinity from proliferating, synthesis of antibiotics, synthesis of a variety of small molecules that can inhibit phytopathogen growth, production of enzymes that lyse or inhibit the phytopathogen and stimulation of the systemic resistance of the plant (Hofte and Altier 2010).

Plant associated *Pseudomonas* spp. are especially suited to synthesizing a wide range of anti-microbial compounds including 2,4 diacetylphloroglucinol (2,4 DAPG), phenazine, hydrogen cyanide, pyoluteorin and pyrrolnitrin (Dwivedi and Johri 2003). 2,4-DAPG in particular has shown much potential for controlling plant pathogens, including take-all disease caused by *Gaeumannomyces graminis*, dampening off disease caused by *Pythium ultimum* and plant pests such as some plant pathogenic nematodes (De Souza et al. 2003, Meyer et al. 2009, Tilak et al. 2010).

However, biocontrol effectiveness just like plant growth promotion is strongly influenced by environmental factors, which can result in inconsistent suppression of phytopathogens and pests in the field. Considerable research efforts have been made in order to develop and enhance biocontrol activity of microbes in order to make the technology more reliable. Recently, Ohno et al. (2011) showed that a genetically engineered rhizospheric bacterium, *Pseudomonas putida* strain 101-9, harbouring the chitinase-expression vector pKAC9-p07, showed over production of enzyme chitinase and suppressed the damping-off of cucumber seedlings caused by the plant pathogenic fungus, *Rhizoctonia solani*. Chitinases are useful in the biological control of soil-borne fungal pathogens because they degrade chitin, a major component of fungal cell walls (Bartnicki-Garcia 1968). As previously mentioned, in a follow-up study this group reported that using culture-independent methods they detected a difference in bacterial communities in the soil after introduction of the modified strain (Yamamoto-Tamura et al. 2011). Zhou et al. (2005) cloned the *phl*ACBDE operon, encoding 2,4 DAPG production, into a number of *Pseudomonas* strains. All of these transgenic strains showed enhanced biocontrol activity against plant microbial pathogens and two strains provided significantly better protection against wheat take-all disease caused by *Gaeumannomyces graminis* and tomato bacterial wilt caused by *Ralstonia solanacearum* in greenhouse studies. Table 1 shows some other examples of genetic modifications that introduce new biocontrol traits or enhance existing traits in plant associated bacteria.

The first commercially available genetically modified biocontrol agent was a modified strain of *Agrobacterium radiobacter* strain K84 which has been marketed in Australia since 1989. This strain is used as a means of controlling *Agrobacterium tumefaciens*, the causative agent of crown gall disease which affects stone fruit trees and almonds. The antibiotic agrocin 84, produced by *A. radiobacter*, is normally toxic to agrobacteria carrying a nopaline/agrocinopine A type Ti plasmid (Kerr 1989, McClure et al. 1994). However, agrocin 84 resistant strains of the pathogen *A. tumefaciens* can develop if the plasmid carrying the genes for the biosynthesis of agrocin 84 is accidentally transferred from *A. radiobacter*. To avoid this possibility, the region of DNA responsible for plasmid transfer was removed from the agrocin 84 plasmid. This mutant designated *A. radiobacter* strain K1026 was

Table 1. Other genetic modifications in plant associated bacteria that have led to improved biocontrol activity.

Biocontrol organism	Genetic Modification	Mode of action	Target control organism	Reference
Lysobacter enzymogenes strain OH11	Insertion of *aiiA* gene under the P_{hrp} promoter	Inactivation the AHL quorum-sensing signal	*Agrobacterium tumefaciens, Pectobacterium carotovorum, Pseudomonas syringae*	Qian et al. 2010
Xanthomonas campestris	Deletion of *hrp* genes involved in virulence and pathogenicity	Competitive displacement of phytopathogenic *Xanthomonas*	*Xanthomonas campestris*	Moss et al. 2007
Pseudomonas fluorescens strain SBW25 EeZY6KX	Insertion of *phl* operon	Synthesis of 2,4 diacetylphoroglucinol	*Pythium ultimum*	Bainton et al. 2004
Pseudomonas fluorescens strain Q8r1-96	Insertion of *phz*ABCDEFG genes	Synthesis of phenazine-1-carboxylic acid	*Pythium, Rhizoctonia*	Huang et al. 2004
Enterobacter cloacae strain SJ-10	Insertion of *pta* gene	Synthesis of an insecticidal plant lectin	*Sogatella furcifera* Horvath (white backed planthopper)	Zhang et al. 2011
Bacillus thuringiensis strain SFC24	Insertion of the *hrpZPss* gene	Synthesis and secretion of harpin$_{Pss}$ which induces the plant immune response	*Phaeoisariopsis personata*	Anil and Podile 2012

constructed so that it can no longer transfer the modified agrocin plasmid to pathogenic agrobacteria, thereby retaining its effectiveness as a biocontrol agent (Jones et al. 1988). This strain is commercially available under the trade name 'NoGall' and is also a registered biocontrol agent in the US.

Another interesting example of a biocontrol microbe that has been genetically modified and is commercially available is a phylosphere colonising *Pseudomonas fluorescens* strain. The *Cry*1Ac and *Cry*1C genes encoding the δ-endotoxin proteins from *Bacillus thuringiensis* were transferred into this *Pseudomonas* strain (Panetta 1993). This endotoxin was found to persist longer in the environment when expressed inside *Pseudomonas* cells than in their original host *B. thuringiensis*, due to the fact that when the *Pseudomonas* cells died their cell walls remained intact whereas the *Bacillus* cell walls disintegrate (Soares and Quick 1990). In this case, the *Pseudomonas* cells are killed so that effectively the dead cell acts as a capsule to deliver the toxin. This product is sold under the trade name 'MVP Bioinsecticide' and used to reduce crop damage from the Diamondback moth (*Plutella xylostella*).

Genetic enhancement of biocontrol strains has mainly focused on the introduction of new biocontrol traits into plant associated strains. However, to be a successful biocontrol agent, the microbe must have the ability to survive, compete and maintain a high population in the target crop. To this end the organism must be adaptable to the environmental conditions in the region that the crop is being grown. Future research should also include improving not only those traits involving colonization and effectiveness, but also traits involved in stress responses such as desiccation and nutrient starvation. The addition of one or more traits associated with improved stress responses, competitiveness and biocontrol activity may lead to new biocontrol products with improved effectiveness and reliability. Also the genetic manipulation of the host crops for root-associated traits to enhance establishment and proliferation of beneficial microorganisms will help biocontrol agents in the plant biosphere (Singh et al. 2011).

Genetically Engineered Rhizobacteria for Bioremediation/ Phytoremediation

The cost of cleaning up environmental pollutants is estimated to be in the range of $US25–50 billion annually (Tsao 2003, Glass 1999). Unfortunately, this high cost contributes to the abandonment worldwide of a large number of polluted commercial sites or brownfields. Traditional technologies routinely used for the remediation of contaminated environmental soil include excavation, transport to specialized landfills, incineration, stabilization and vitrification. Recently, however, there has been much interest in bioremediation technologies which use plants and microorganisms (including bacteria) to degrade toxic contaminants in environmental soil into less-toxic and/or non-toxic substances. Bioremediation technologies offer many advantages over traditional remediation technologies as they can be applied *in situ* without the need for removal and transport of contaminated soil, are usually less expensive and less labour-intensive relying on solar energy, have a lower carbon footprint, and have a high level of public acceptance. The rhizosphere in particular is a hotspot for the degradation of pollutants due to the enhanced microbial activity associated with plant roots.

However, the degradation rate of a pollutant is very much dependent on its physical and chemical properties. Compounds with high water solubility tend to be rapidly removed from the rhizosphere by the plant, whereas those with very low water solubility tend to be immobile and absorb onto soil particles reducing their bioavailability to the degrading microbes (Trapp and Karlson 2001). Indeed, in some cases degradation pathways may be lacking altogether. Using biotechnology, plant associated bacteria (rhizospheric and/or endophytic) can be engineered, via natural gene transfer or recombinant DNA technology, to produce specific enzymes, capable of degrading toxic organic pollutants found in the environment. Genetic engineering of endophytic and rhizospheric bacteria for use in plant-associated degradation of toxic compounds in soil is considered one of the most promising new technologies for remediation of contaminated environmental sites (Dzantor 2007).

Studies using two genetically modified strains of the rhizospheric bacteria *Pseudomonas fluorescens* F113, i.e., *Pseudomonas fluorescens* F113*rif bph* (with a single chromosomal insertion of the *bph* operon) (Brazil et al. 1995) and *Pseudomonas fluorescens* F113:1180 (with a single chromosomal insertion of the *bph* operon under the control of the *Sinorhizobium meliloti* nod regulatory system) (Villacieros et al. 2005) reported that (i) the modified rhizospheric bacteria colonized roots as effectively as the wildtype rhizospheric bacteria, (ii) *bph* genes were expressed *in situ* in soil, and (iii) the modified rhizospheric bacteria could degrade PCBs more efficiently than the wildtype rhizospheric bacteria, indicating considerable potential for the manipulation of the rhizosphere as a useful strategy for bioremediation. *Pseudomonas fluorescens* F113:1180 does not contain antibiotic resistance genes from the vector making this strain more suitable for *in situ* applications. Since the *bph* element in *Pseudomonas fluorescens* F113:1180 is stable, lateral transfer of the *bph* element to a homologous recipient would not be expected to occur at detectable frequencies in the rhizosphere (Ramos et al. 1994).

However, because toxic organic compounds or their metabolites can enter the root xylem from the soil before they are degraded (McCrady et al. 1987), plant-associated endophytes genetically enhanced to degrade toxic organic compounds appear to offer more potential than rhizospheric bacteria. Endophytic bacteria can be isolated from host plants of interest (e.g., plants native to a geographical region) and genetically enhanced to contain degradation pathways or genes to degrade target contaminants before being re-inoculated back into the host plant for bioremediation purposes. Germaine et al. (2009) reported that a genetically enhanced endophytic strain of the poplar endophyte *Pseudomonas putida* VM1441, i.e., *Pseudomonas putida* VM1441 (pNAH7), could protect inoculated pea plants from the toxic effects of naphthalene. They also showed that inoculation of plants with this strain facilitated higher (40%) naphthalene degradation rates compared with uninoculated plants in artificially contaminated soil. Barac et al. (2004) reported that a genetically enhanced endophytic strain of the soil bacterium *Burkholderia cepacia* G4 could increase inoculated yellow lupine plant tolerance to toluene, and decrease phytovolatilization of toluene from the plant into the atmosphere by 50–70% in laboratory scale experiments. In this study, the plasmid, pTOM, which encodes a pathway for the degradation of toluene, was transferred via conjugation to the natural endophyte, providing the genes for toluene degradation. Later, Taghavi et al. (2009) extended this work to poplar trees and showed that this degradative plasmid, pTOM, could transfer naturally, via horizontal gene transfer, to a number of different endophytes *in planta*, promoting more efficient degradation of toluene in poplar plants. Horizontal gene transfer results in the natural microbial population having the capacity to degrade environmental pollutants without the need to establish the inoculants strain long-term. Endophytes that have been engineered by horizontal gene transfer, have the distinct advantage that they are not considered to be genetically modified microorganisms and are exempt from current international and national GM legislation thus facilitating the testing of these microorganisms in the field at an accelerated pace (assuming of course that the transferrable genetic element is itself not genetically modified).

Brennan (2010) constructed bacteria expressing a specific bacterial glutathione-S-transferase (GST) isolated from *Burkholderia xenovorans* LB400, BphK[LB400] [wildtype and mutant (Ala180Pro)], capable of dehalogenating toxic chlorinated organic pesticides, was shown to protect inoculated pea plants from the effects of a chlorinated organic pesticide, chloromequat chloride. Previously, it had been shown that mutating the conserved amino acid at position 180 in BphK[LB400] from Ala to Pro resulted in an approximate two-fold increase in GST activity towards a number of chlorinated organic substrates tested including commonly used pesticides, and 2,4-D (McGuinness et al. 2007, McGuinness et al. 2006, Gilmartin et al. 2003). These data suggest that BphK[LB400] [wildtype and mutant (Ala180Pro)], when inserted into endophytic or rhizospheric bacteria, could have potential for bioremediation of chlorinated organic pollutants in environmental soil.

Although much field work remains to be done, GMMs, as well as their unmodified parent strains, show much potential for sustainable bioremediation of the environment in the future.

Genetically Engineered Microbial Biosensors used in the Plant Biosphere

Non-specific reporter genes such as genes encoding green fluorescent proteins (*gfp*), luciferases (*lux*), and β-glucuronidase (*gus*), have been extensively used in plant-microbe studies to monitor the colonisation of plants by inoculated microbial strains (Larrainzar et al. 2005). However, such studies yield little or no information on the environmental conditions (nutrient availability, pH, etc.) that the microbes experience in the plant biosphere. Conventional analytical techniques can provide information on the overall conditions of a particular parameter. However, conditions on plant surfaces are extremely heterogenous and studies using non-specific reporter genes have shown that microbes tend to colonise specific locations on and within their host plant. Therefore, miniature sensors are necessary to examine the environmental conditions either on a mm or μm scale.

A biosensor is an analytical device composed of a biological sensing element in intimate contact with a physical transducer (optical, mass or electrochemical) which together translate the concentration of an analyte to a measurable electrical signal. Microbes are ideal tools to use as biosensing elements due to the ease at which they can be engineered to consume and degrade new substrates (Lei et al. 2006). Whole cell microbial biosensors couple microorganisms with a transducer to enable accurate, sensitive and rapid detection of target analytes in microscale environments. Today, most whole cell microbial biosensors rely on inducible promoter genes fused to reporter systems such as *gfp*, *lux* or a chromogen catalysing enzyme, e.g., *lacZ* or *gusA* (Liu et al. 2010b). These biosensors have been used for a number of useful applications in plant-microbe studies to detect microscale distributions of rhizosphere compounds such as nutrients, metals and organic exudates or contaminants.

Microbial biosensors have been particularly useful for examining specific nutrient availability and distribution in the rhizosphere and in the phylosphere (Miller et al. 2001). Carbohydrate and amino acid detecting biosensors have been used to identify major areas of rhizodeposition of exudates such as sucrose, galactosides, tryphophan and glutamine in the root zone (Bringhurst et al. 2001, Tessaro et al. 1999, Jaeger et al. 1999, Yeomans et al. 1999). In the study by Jarger et al. (1999) they created an *Erwinia herbicola* tryptophan-reporter strain with a fusion between the *aat*l gene (encoding a tryptophan aminotransferase) and an *ina*Z ice nucleation reporter. When this biosensor was inoculated into an annual grass (*Avena barbata*), the reporter showed significant induction in older root segments, but not at the root tip. DeAngelis et al. (2005) fused the nitrate-regulated promoter of *narG* gene to a *gfp* reporter and inserted the contruct into *Enterobacter cloacae*. Significantly lower nitrate abundance was detected by the *E. cloacae* reporter in the rhizosphere compared to in bulk soil, indicating plant competition for nitrate. DeAngelis et al. (2007) also constructed a highly sensitive whole cell biosensor, *Agrobacterium tumefaciens* (pAHL-Ice) and used it in greenhouse experiments to demonstrate the increased N-acyl homoserine lactone (AHL) availability in intact rhizospheric microbial communities compared to that in bulk soil. Nutrient availability in the phylosphere has also been examined using microbial biosensors. Leveau and Lindow (2001) constructed a microbial biosensor to determine fructose availability on the leaf surface. They used short-half-life variants of the *gfp* reporter gene with a fructose-responsive *fruB* promoter. Their study revealed that those cells that consumed fructose were generally in localised sites on the leaf surface, suggesting that nutrients may be relatively abundant in only a few locations. However, a high variability in available Fe^{+3} on leaves was observed when a *gfp*-based iron biosensor strain of *P. syringae* was used, with many cells experiencing no iron limitation Joyner and Lindow (2000).

Microbial biosensors are particularly useful tools when carrying out environmental risk assessments on polluted sites or for monitoring the efficiency of a remediation strategy, due to their ability to detect only the biologically relevant (bio-available) fraction of the contaminant (Tecan and van der Meer 2008). Numerous xenobiotic detecting biosensors have been constructed and they include sensors for detecting nitrate, phosphate, naphthalene, toluene, polyaromatic hydrocarbons (PAHs), copper, arsenite and mercury (Fu et al. 2008, Li et al. 2008, Kohlmeier et al. 2008, Brandt et al. 2006, Trang et al. 2005, Taylor et al. 2004, Keane et al. 2002, Schreiter et al. 2001). Liu et al. (2010b) constructed PCB biosensor strains using the rhizobacterial strain *Pseudomonas fluorescens* F113. These strains were constructed by the chromosomal insertion of a promoter reporter construct, xylSPm::gfpmut3 into *P. fluorescens* F113rif (a non PCB degrader) and *P. fluorescens* F113rifPCB (genetically engineered to degrade PCBs). The pm promoter in this construct is derived from the TOL plasmid and regulates the meta-pathway of aromatic hydrocarbon degradation. The meta-pathway is induced by chloro-benzoic acid derivatives (which are also end products of the PCB degradation pathway). Induction of the pm promoter in the xylSPm::gfpmut3 construct drives the expression of the *gfp* which is detectable using fluorescent spectroscopy or epi-fuorescent microscopy. When these PCB degrading biosensors were inoculated on pea seeds and subsequently grown in PCB contaminated soils, *gfp* expressing cells could be visualised colonising the roots of the pea plants thereby demonstrating active PCB degradation in the plant rhizosphere (Fig. 1).

Biosensors offer the possibility of determining not only specific chemicals but also their biological effects, such as toxicity, cytotoxicity, genotoxicity or endocrine disrupting effects, i.e., relevant information which in some instances is more meaningful than the chemical composition itself. Biosensors can provide both total and bioavailable/bioaccessible pollutant concentrations. Despite these advantages, the application of biosensors in the environmental field is still limited in comparison to medical or pharmaceutical applications and the majority of the systems developed are prototypes that still need to be validated in the field (Rodrigeuz-Mozaz et al. 2006).

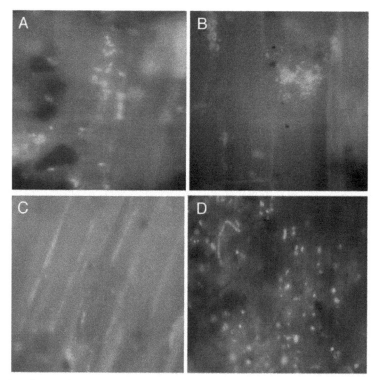

Fig. 1. *Pseudomonas fluorescens* strain F113PCB *gfp* biosensor responding to the presence of PCBs in the rhizosphere of *Pisum sativum* (reproduced from Liu et al. 2010b with permission).

Color image of this figure appears in the color plate section at the end of the book.

Conclusion

Although a considerable amount of attention has been diverted towards the possibility of genetically manipulating plants so that they are able to withstand a number of different environmental challenges including insects, viruses, fungi, bacteria, xenobiotics and weather (Greenberg and Glick 1993) each variety of each crop species has to be protected against a wide range of possible environmental challenges. In contrast, it should be possible using a combination of traditional mutagenesis, selection, and genetic engineering, to develop microbial inoculants that are effective plant growth promoting and biocontrol agents against a range of different phytopathogens and xenobiotics and which can be used with a number of different crop plants. There are several advantages of developing genetically-modified microbes over transgenic plants for improving plant performance: (i) with current technologies, it is far easier to modify a bacterium than complex higher organisms, (ii) several plant growth-promoting traits can be combined in a single organism, and (iii) instead of engineering crop by crop, a single, engineered inoculant can be used for several crops, especially when using a wide-host range genus like *Azospirillum*. While more field release studies are needed to fully understand the impact genetically engineered microbes may have in contributing to a sustainable environment, initial studies suggest that GMMs can be introduced into the environment without significantly affecting the ecosystem.

Wider adoption of microbial inoculants, for plant growth promotion, biocontrol or bioremediation, has been hindered by the inconsistent and low levels of success that are sometimes seen in the field. However, genetic modification of promising microbes, so that they meet economically acceptable targets and are reliable and consistent, remains a realistic goal. Microbes do have significant potential to contribute to sustainable agriculture and bioremediation of the environment but this will require further discovery of novel bacterial strains, genetic improvement of both microbe and their plant hosts, and improved inocula delivery systems. Recent plant microbiome studies (Lundberg et al. 2012) have shown the immense microbial diversity associated plants and further supports the claim that less than 10% of the total diversity of microbes have been described to date. This suggests considerable potential for construction and application in the field of, as yet, unidentified engineered microbes in the future.

References

Amarger, N. 2002. Genetically modified bacteria in agriculture. Biochimie. 84: 1061–1072.

Anil, K. and A. R. Podile. 2012. Harpin$_{Pss}$-mediated enhancement in growth and biological control of late leaf spot in groundnut by a chlorothalonil-tolerant *Bacillus thuringiensis* SFC24. Microbiological Research. 167: 194–198.

Bainton, N. J., J. M. Lynch, D. Naseby and J. A. Way. 2004. Survival and ecological fitness of *Pseudomonas fluorescens* genetically engineered with dual biocontrol mechanisms. Microb. Ecol. 48: 349–357.

Barac, T., S. Taghavi, B. Borresman, A. Provoost, L. Oeyen, J. Colpaert, J. Vangronsveld and D. van der Lelie. 2004. Engineered endophytic bacteria improve phytoremediation of water-soluble, volatile, organic pollutants. Nat. Biotechnol. 22, 583.

Bartnicki-Garcia, S. 1968. Cell wall chemistry, morphogenesis and taxonomy of fungi. Annu. Rev. Microbiol. 22: 87–108.

Bashan, Y., M. Moreno and E. Troyo. 2000. Growth promotion of the seawater-irrigated oil seed halophyte *Salicornia bigelovii* inoculated with mangrove rhizosphere bacteria and halotolerant *Azospirillum* spp. Biol. Fertil. Soils. 32: 265–272.

Bej, A. K., M. Perlin and R. M. Atlas. 1991. Effect of introducing genetically engineered microorganisms on soil microbial community diversity. FEMS Microbiol. Ecol. 86: 169–176.

Brandt, K. K., P. E. Holm and O. Nybroe. 2006. Bioavailability and toxicity of soil particle-associated copper as determined by two bioluminescent *Pseudomonas fluorescens* biosensor strains. Environ. Toxicol. Chem. 25: 1738–1741.

Brazil, G. M., L. Kenefick, M. Callanan, A. Haro, V. De Lorenzo, D. N. Dowling and F. O'Gara. 1995. Construction of a rhizosphere pseudomonad with potential to degrade polychlorinated biphenyls and detection of *bph* gene expression in the rhizosphere. Appl. Environ. Microbiol. 61: 1946–1952.

Brennan, E. 2010. Purification and kinetic analysis of a specific glutathione transferase (BphK$_{LB400}$) involved in the detoxification of PCBs. PhD Thesis. Insitute of Technology, Carlow. Carlow, Ireland.

Bringhurst, R. M., Z. G. Cardon and D. J. Gage. 2001. Galactosides in the rhizosphere: Utilization by *Sinorhizobium meliloti* and development of a biosensor. PNAS. 98: 4540–4545.

Camerini, S., B. Senatore, E. Lonardo, E. Imperlini, C. Bianco, G. Moschetti, G. Rotino, B. Campion and R. Defez. 2008. Introduction of a novel pathway for IAA biosynthesis to rhizobia alters vetch root nodule development. Archives of Microbiology. 90: 67–77.

Crews, T. E. and M. B. Peoples. 2004. Legume versus fertilizer sources of nitrogen: ecological trade offs and human needs. Agriculture Ecosystems & Environment. 102: 279–297.

Cummings, S. P., D. R. Humphry, S. R. Santos, M. Andrews and E. K. James. 2006. The potential and pitfalls of exploiting nitrogen fixing bacteria in agricultural soils as a substitute for inorganic fertiliser. Environmental Biotechnology. 2: 1–10.

Da, H. N. and S. P. Deng. 2003. Survival and persistence of genetically modified *Sinorhizobium meliloti* in soil. Applied Soil Ecology. 22: 1–14.

Davoud, F., A. Naser, S. B. Nemat and Y. Bagher. 2010. Cloning and characterization of a plasmid encoded ACC deaminase from an indigenous *Pseudomonas fluorescens* FY32. Current Microbiology. 61: 37–43.

De Leij, F., E. J. Sutton, J. M. Whipps, J. S. Fenton and J. M. Lynch. 1995. Impact of field release of genetically modified *Pseudomonas fluorescens* on indigenous microbial populations of wheat. Appl. Environ. Microbiol. 61: 3443–3453.

De Souza, J. T., D. M. Weller and J. M. Raaijmakers. 2003. Frequency, diversity, and activity of 2,4 diacetylphloroglucinol-producing fluorescent *Pseudomonas* spp. in Dutch Take-all decline Soils. Phytopathology. 93: 54–63.

DeAngelis, K. M., P. Ji, M. K. Firestone and S. E. Lindow. 2005. Two novel bacterial biosensors for detection of nitrate availability in the rhizosphere. Appl. Environ Microbiol. 71: 8537–8547.

DeAngelis, K. M., M. K. Firestone and S. E. Lindow. 2007. Sensitive whole-cell biosensor suitable for detecting a variety of N-acyl homoserine lactones in intact rhizosphere microbial communities. Appl. Environ. Microbiol. 73: 3724–3727.

Dwivedi, D. and B. N. Johri. 2003. Antifungals from fluorescent pseudomonads: biosynthesis and regulation. Curr. Sci. 85: 1693–1703.

Dzantor, E. K. 2007. Phytoremediation: the state of rhizosphere 'engineering' for accelerated rhizodegradation of xenobiotic contaminants. J. Chem. Technol. Biotechnol. 82: 228–232.

Fu, Y., W. Chen and Q. Huang. 2008. Construction of two lux-tagged Hg^{2+} specific biosensors and their luminescence performance. Appl. Microbiol. Biotechnol. 79: 363–370.

Gafni, Y., S. Manulis, T. Kunik, A. Lichter, I. Barash and Y. Ophir. 1997. Characterization of the auxin synthesis genes of *Erwinia herbicola* pv. *Gypsophilae*. Israel Journal of Plant Sciences 45: 279–284.

Germaine, K. J., E. Keogh, D. Ryan and D. N. Dowling. 2009. Bacterial endophyte-mediated naphthalene phytoprotection and phytoremediation. FEMS Microbiol. Lett. 296: 226–234.

Germaine, K. J., S. Chhabra, B. Song, D. Brazil and D. N. Dowling. 2010. Microbes and sustainable production of biofuel crops: a nitrogen perspective. Biofuels. 1: 877–888.

Gilmartin, N., D. Ryan, O. Sherlock and D. N. Dowling. 2003. BphK shows dechlorination activity against 4-chlorobenzoate, an end product of bph-promoted degradation of PCBs. FEMS Microbiol. Letts. 222: 251–255.

Glass, D. U. S. and International Markets for Phytoremediation 1999–2000, D. Glass Associates, Inc. : Needham, MA, USA, 1999.

Govindasamy, V., M. Senthilkumar, K. Kishore Gaikwad and K. Annapurna. 2008. Isolation and characterization of ACC deaminase gene from two plant growth promoting rhizobacteria. Current Microbiology. 57: 312–317.

Greenberg, B. M. and B. R. Glick. 1993. The use of recombinant DNA technology to produce genetically modified plants. *In:* Methods in Plant Molecular Biology and Biotechnology, B. R. Glick and J. E. Thompson [eds.]. CRC Press, Boca Raton. pp. 1–10

Hardoim P. R., L. S. van Overbeek and J. D. van Elsas. 2008. Properties of bacterial endophytes and their proposed role in plant growth. Trends in Microbiology 16: 463–471.

Höfte, M. and N. Altier. 2010. Fluorescent pseudomonads as biocontrol agents for sustainable agricultural systems. Research in Microbiology. 161: 464–471.

Huang, Z., R. F. Bonsall, D. V. Mavrodi, D. M. Weller and L. S. Thomashow. 2004. Transformation of *Pseudomonas fluorescens* with genes for biosynthesis of phenazine-1-carboxylic acid improves biocontrol of *rhizoctonia* root rot and *in situ* antibiotic production. FEMS Microbiol Ecol. 49: 243–51.

Imperlini, E., C. Bianco, E. Lonardo, S. Camerini, M. Cermola, G. Moschetti and R. Defez. 2009. Effects of indole-3-acetic acid on *Sinorhizobium meliloti* survival and on symbiotic nitrogen fixation and stem dry weight production. Applied Microbiology and Biotechnology. 83: 727–738.

Jaeger, C. H., S. E. Lindow, W. Miller, E. Clark and M. K. Firestone. 1999. Mapping of sugar and amino acid availability in soil around roots with bacterial sensors of sucrose and tryptophan Appl. Environ. Microbiol. 65: 2685–2690.

Jones, D. A., M. H. Ryder, B. G. Clare, S. K. Farrand and A. Kerr. 1988. Construction of a Tra- deletion mutant of pAgK84 to safeguard the biological control of crown gall. Mol. Gen. Genet. 212: 207–214.

Joyner, D. C. and S. E. Lindow. 2000. Heterogeneity of iron bioavailability on plants assessed with a whole-cell GFP-based bacterial biosensor. Microbiology. 146: 2435–2445.

Keane, A., P. Phoenix, S. Ghoshal and P. C. K. Lau. 2002. Exposing culprit organic pollutants: A review. J. Microbiol. Methods. 49: 103–119.

Kerr, A. 1989. Commercial release of a genetically engineered bacterium for the control of crown gall. Agric. Sci. 2: 41–44.

Kochar, M., A. Upadhyay and S. Srivastava. 2011. Indole-3-acetic acid biosynthesis in the biocontrol strain *Pseudomonas fluorescens* Psd and plant growth regulation by hormone over expression. Research in Microbiology. 162: 426–435.

Kohlmeier, S., M. Mancuso, U. Deepthike, R. Tecon, J. van der Meer, H. Harms and M. Wells. 2008. Comparison of naphthalene bioavailability determined by whole cell biosensing and availability determined by extraction with Tenax. Environ. Pollut. 156: 803–808.

Larrainzar, E., F. O'Gara and J. P. Morrissey. 2005. Application of autofluorecsent proteins for *in situ* studies in microbial ecology. Annual Review of Microbiology. 59: 257–277.

Lei, Y., W. Chen and A. Mulchandani. 2006. Microbial biosensors. Analytica Chimica Acta. 568: 200–210.

Leveau, J. H. J. and S. E. Lindow. 2001. Appetite of an epiphyte: quantitative monitoring of bacterial sugar consumption in the phyllosphere. Proc. Natl. Acad. Sci. USA. 98: 3446–3453.

Li, Y., F. Li, C. Ho and V. H. Liao. 2008. Construction and comparison of fluorescence and bioluminescence bacterial biosensors for the detection of bioavailable toluene and related compounds. Environ. Pollut. 152: 123–129.

Liu, X., K. J. Germaine, D. Ryan and D. N. Dowling. 2010a. Genetically modified *Pseudomonas* biosensing biodegraders to detect PCB and chlorobenzoate bioavailability and biodegradation in contaminated soils. Bioengineered Bugs. 1: 1–9.

Liu, X., K. J. Germaine, D. Ryan and D. N. Dowling. 2010b. Whole-Cell fluorescent biosensors for bioavailability and biodegradation of polychlorinated biphenyls. Sensors. 10: 1377–1398.

Long, H. H., D. D. Schmidt and I. T. Baldwin. 2008. Native bacterial endophytes promote host growth in a species-specific manner, phytohormone manipulations do not result in common growth responses. PLoS ONE 3(7):e2702doi:10. 1371/journal. pone. 0002702

Lundberg, D. S., S. L. Lebeis, S. H. Paredes, S. Yourstone, J. Gehring, S. Malfatti, J. Tremblay, A. Engelbrektson, V. Kunin, T. Glavina del Rio, R. C. Edgar,T. Eickhorst, R. E. Ley, P. Hugenholtz, S. Green Tringe and J. L. Dangl. 2012. Defining the core *Arabidopsis thaliana* root microbiome. Nature. 488: 86–90.

McClure, N. C., A. R. Ahmadi and B. G. Clare. 1994. The role of agrocin 434 produced by *Agrobacterium* strain K84 and derivatives in the biological control of *Agrobacterium* biovar 2 pathogens. *In:* Improving plant productivity with rhizosphere bacteria, M. H. Ryder, P. M. Stephens and G. D. Bowen (eds.). pp. 125–127.

McCrady, J., C. McFarlane and F. Lindstrom. 1987. The transport and affinity of substituted benzenes in soybean stems. J. Exp. Bot. 38: 1875–1890.

McGuinness, M., C. Ivory, N. Gilmartin and D. N. Dowling. 2006. Investigation of substrate specificity of wildtype and mutant BphK^{LB400} (a glutathione S-transferase) from *Burkholderia* LB400. Int. Biodet. Biodegrad. 58: 203–208.

McGuinness, M. C., V. Mazurkiewicz, E. Brennan and D. N. Dowling. 2007. Dechlorination of pesticides by a specific bacterial glutathione S-transferase, BphK^{LB400}: Potential for bioremediation. Eng. Life Sci. 7: 611–615.

McGuinness, M. and D. N. Dowling. 2009. Plant-associated bacterial degradation of toxic organic compounds in soil. Int. J. Env. Res. Pub. Health. 6: 2226–2247.

Meyer, S., J. M. Halbrendt, L. K. Carta, A. M. Skantar, T. Liu, H. M. E. Abdelnabby and B. T. Vinyard. 2009. Toxicity of 2,4-diacetylphloroglucinol (DAPG) to Plant-parasitic and Bacterial-feeding Nematodes J. Nematol. 41: 274–280.

Miller, W. G., M. T. Brandl, B. Quiñone and S. E. Lindow. 2001. Biological sensor for sucrose availability: relative sensitivities of various reporter genes. Appl. Environ. Microbiol. 67: 1308–1317.

Morrissey, J. P., U. F. Walsh, A. O'Donnell, Y. Moenne-Loccoz and F. O'Gara. 2002. Exploitation of genetically modified inoculants for industrial ecology applications. Antonie van Leeuwenhoek. 81: 599–606.

Moss, W. P., J. M. Byrne, H. L. Campbell, P. Ji, U. Bonas, J. B. Jones and M. Wilson. 2007. Biological control of bacterial spot of tomato using *hrp* mutants of *Xanthomonas campestris* pv. Vesicatoria Biological Control. 41: 199–206.

Naveed, M., M. Khalid, D. L. Jones, R. Ahmad and Z. A. Zahir. 2008. Relative efficacy of *Pseudomonas* spp. Containing ACC-deaminase for improving growth and yield of maize (*Zea mays*) in the presence of organic fertiliser. Pakistan Journal of Botany. 40: 1243–1251.

Ohno, M., S. Kataoka, K. Yamamoto-Tamura, T. K. Akutsu and A. Hasebe. 2011. Biological control of *Rhizoctonia* damping-off of cucumber by a transformed *Pseudomonas putida* strain expressing a chitinase from a marine bacterium. JARC. 45: 91–98.

Orikasa, Y., Y. Nodasaka, T. Ohyama, H. Okuyama, N. Ichise, I. Yumoto, N. Morita, M. Wei, and T. Ohwada. 2010. Enhancement of the nitrogen fixation efficiency of genetically-engineered *Rhizobium* with high catalase activity. J. Biosci. Bioeng. 110: 397–402.

Schreiter, P. P., O. Gillor, A. Post, S. Belkin, R. D. Schmid and T. T Bachmann. 2001. Monitoring of phosphorus bioavailability in water by an immobilized luminescent cyanobacterial reporter strain. Biosens. Bioelectron. 16: 811–818.

Panetta, J. D. 1993. Engineered Microbes, the Cellcap system. *In:* L. Kim (ed.). Advanced Engineered Pesticides. Marcel Dekker, New York. pp 379–3.

Qian, G. L., J. Q. Fan, D. F. Chen, Y. J. Kang, B. Han, B. S. Hu and F. Q. Liu. 2010. Reducing *Pectobacterium* virulence by expression of an N-acyl homoserine lactonase gene Plpp-aiiA in *Lysobacter enzymogenes* strain OH11. Biological Control. 52: 17–23.

Ramos, J. L., E. Diaz, D. N. Dowling, V. deLorenzo, S. Molin, F. O'Gara, C. Ramos and K. N. Timmis. 1994. The behavior of bacteria designed for biodegradation. Biotechnol. 12: 1349–1356.

Robleto, E. A., K. Kmiecik, E. S. Oplinger, J. Nienhuis and E. W. Triplett. 1998. Trifolitoxin production increases nodulation competitiveness of *Rhizobium etli* CE3 under agricultural conditions. App. Environ. Microbial. 64: 2630–2633.

Rodriguez, H., R. Fraga, T. Gonzalez and Y. Bashan. 2006. Genetics of phosphate solubilization and its potential applications for improving plant growth-promoting bacteria. Plant and Soil. 287: 15–21.

Rodriguez, H., T. Gonzalez and G. Selman. 2000. Expression of a mineral phosphate solubilizing gene from *Erwinia herbicola* in two rhizobacterial strains. Journal of Biotechnology. 84: 155–161.

Rodriguez-Mozaz, S., M. J. Lopez de Alda and D. Barcelo. 2006. Biosensors as useful tools for environmental analysis and monitoring. Anal. Bioanal. Chem. 386: 1025–1041.

Sashidhar, B. and A. R. Podile. 2009. Transgenic expression of glucose dehydrogenase in *Azotobacter vinelandii* enhances mineral phosphate solubilisation and growth of sorghum seedlings. Microbial. Biotechnology. 2: 521–529.

Shaharoona, B., M. Arshad, Z. A. Zahir and A. Khalid. 2006. Performance of *Pseudomonas* spp. containing ACC-deaminase for improving growth and yield of maize (*Zea mays* L.) in the presence of nitrogenous fertilizer. Soil Biology and Biochemistry. 38: 2971–2975.

Simon, R., J. Quandt and W. Klipp. 1989. New derivatives of transposon Tn5 suitable for mobilization of replicons, generation of operon fusions and induction of genes in gram-negative bacteria. Gene. 80: 161–169.

Singh, J. S., V. C. Pandey and D. P. Singh. 2011. Efficient soil microorganisms: A new dimension for sustainable agriculture and environmental development. Agriculture, Ecosystems & Environment. 140: 339–353.

Soares, G. G. and T. C. Quick. 1990. MVP, A novel bioinsecticide for control of the Diamondback Moth. Chapter 15: In Proceedings of the Second International Workshop, 10–14 Dec, Taiwan.

Spaepen, S., J. Vanderleyden and Y. Okon. 2009. Plant growth-promoting actions of rhizobacteria. Advances in Botanical Research. 51: 283–320.

Taghavi, S., C. Garafola, S. Monchy, L. Newman, A. Hoffman, N. Weyens, T. Barac, J. Vangronsveld and D. van der Lelie. 2009. Genome survey and characterization of endophytic bacteria exhibiting a beneficial effect on growth and development of poplar. Appl. Environ. Microbiol. 75: 748–757.

Taylor, C. J., L. A. Bain, D. J. Richardson, S. Spiro and D. A. Russell. 2004. Construction of whole-cell gene reporter for the fluorescent bioassay of nitrate. Anal. Biochem. 328: 60–66.

Tecan, R. and J. R. van der Meer. 2008. Bacterial biosensors for measuring availability of environmental pollutants. Sensors. 8: 4062–4090.

Tessaro, M. J., S. S. M. Soliman and M. N. Raizada. 1999. Bacterial whole-cell biosensor for glutamine with applications for quantifying and visualizing glutamine in plants. Environ. Microbiol. 65: 2685–2690.

Tilak, K. V. B. R., K. K. Pal and R. Dey. 2010. Microbes for sustainable agriculture. *In:* Biological Control. K. International Pvt. Ltd., New Delhi. Chapter 14: pp. 117–119.

Trang, P. T. K, M. Berg, P. H. Viet, N. V. Mui and J. R. van der Meer. 2005. Bacterial bioassay for rapid and accurate analysis of arsenic in highly variable groundwater samples. Environ. Sci. Technol. 19: 7625–7630.

Tsao, D. T. 2003. Overview of Phytotechnologies. *In:* T. Scheper and D. T. Tsao (eds.). Phytoremediation. Springer-Verlag, Berlin, Germany. pp. 1–7.

Trapp, S. and U. Karlson. 2001. Aspects of phytoremediation of organic compounds. Journal of Soils and Sediments. 1: 37–43.

Van Dillewijun, P., M. J. Soto, P. J. Villadas and N. Toro. 2001. Construction and environmental release of a *Sinorhizobium meliloti* strain genetically modified to be more competitive for alfalfa nodulation. Appl. Environ. Microbiol. 67: 3860–3865.

Villacieros, M., C. Whelan, M. Mackova, J. Molgaard, M. Sánchez-Contreras, J. Lloret, D. Aguirre de Cárcer, R. I. Oruezábal, L. Bolanos, T. Macek, U. Karlson, D. N. Dowling, M. Martín and R. Rivilla. 2005. Polychlorinated biphenyl rhizoremediation by *Pseudomonas fluorescens* F113 derivatives, using a *Sinorhizobium meliloti* nod system to drive bph gene expression. Appl. Environ. Microbiol. 71: 2687–2694.

Yamada, T., C. J. Palm, B. Brooks and T. Kosuge. 1985. Nucleotide sequences of the *Pseudomonas savastanoi* indoleacetic acid genes show homology with *Agrobacterium tumefaciens* T-DNA. Proceedings of the National Academy of Sciences. 82: 6522–6526.

Yamamoto-Tamura, K., M. Ohno, T. Fujii, S. Kataoka, S. Numata, M. Nakajima, A. Hasebe and K. Akutsu. 2011. Assessment of the effects of genetically modified *Pseudomonas* spp. expressing chitinase on the soil microbial community in the cucumber rhizosphere. JARC. 45: 377–383.

Yeomans, C., F. Porteous, E. Paterson, A. A. Meharg and K. Killham. 1999. Assessment of *lux* marked *Pseudomonas fluorescens* for reporting on organic carbon compounds. FEMS Microbiol. Lett. 176: 79–83.

Zhang, X. F., J. Li, G. Qi, K. Wen, J. Lu and X. Zhao. 2011. Insecticidal effect of recombinant endophytic bacterium containing *Pinellia ternata* agglutinin against white backed planthopper, *Sogatella furcifera*. Crop Protection. 30: 1478–1484.

Zhou, H., H. Wei, X. Liu, Y. Wang, L. Zhang and W. Tang. 2005. Improving biocontrol activity of *Pseudomonas fluorescens* through chromosomal integration of 2,4-diacetylphloroglucinol biosynthesis genes. Chinese Science Bulletin. 50: 777–781.

9

Engineering the Rhizosphere for Agricultural and Environmental Sustainability

Mélanie M. Paulin and *Martin Filion**

ABSTRACT

Reliance on chemical-based components to meet food supply demands has allowed unprecedented agricultural productivity in the past century. Concerns that these agricultural practices are unsustainable are however increasing, and in addition to the negative impacts these chemicals may have on the environment, they will most probably be unable to provide long-term solutions for food production. This chapter addresses various aspects of rhizosphere engineering that maximize beneficial plant-microbe interactions regarded as a valuable strategy in achieving sustainable crop production. Special emphasis is given to the use of plant growth-promoting rhizobacteria as microbial inoculants to promote plant health and development. The potential of plant growth-promoting rhizobacteria to be used as biofertilizers to improve plant nutrition and as biopesticides to control phytopathogenic microorganisms causing plant diseases is discussed.

Department of Biology, Université de Moncton, 18 Antonine-Maillet, Moncton , NB, E1A 3E9, Canada.
Emails: melanie.m.paulin@umoncton.ca; martin.filion@umoncton.ca
* Corresponding author

Introduction

According to United Nations projections, the global population will reach 9 billion by 2050 and will likely exceed 10 billion by 2100 (United Nations Department of Economic and Social Affairs 2011). Historically, increases in global food demand have been met by intensification of agriculture or by bringing new land into cultivation (Fitter 2012). The agricultural revolution of the twentieth century termed the green revolution saved hundreds of millions from starvation by modernizing agricultural management practices, developing better irrigation systems and creating new high-yielding varieties (HYV) of seed. While the productivity of most plants, especially cereal crops, significantly increased, the development of HYV meant that traditional varieties of seed were replaced by only a few selected varieties leading to increased crop homogeneity and vulnerability to diseases and pests. These new varieties generally required higher rates of chemical fertilizer intake and relied heavily on the use of chemical pesticides and herbicides. While our reliance on chemical-based components to meet food supply demands has allowed unprecedented agricultural productivity, there are increasing concerns that conventional agricultural practices are unsustainable and will be unable to provide long-term solutions for food production. Amongst other problems, intensive agriculture has led to increased water use, soil erosion, loss of soil fertility, presence of residual pesticides in soil and the environment and increased pesticide and herbicide resistance. Furthermore, availability of new farmland suitable for cultivation is increasingly poor due to urbanization, salinization, desertification, soil degradation and other forms of land loss. A new approach is clearly needed.

Soil is a resource that must be developed and managed in a sustainable manner not only to ensure food safety, but also because it is a critical component of our ecosystems. In addition to food production, soil is essential in other processes, for example biogeochemical cycling of organic carbon, nitrogen and phosphorus and in water storage and purification. The challenge we face today is to find novel ways of maintaining and even increasing current levels of food production in order to meet future demand while also protecting the environment and ensuring sustainability of agriculture practices for the long-term. One possible avenue is through technological innovations in plant and soil science which, if approached from the right perspective, may have the potential to reduce our reliance on agrochemicals.

As a habitat, soil is a highly diverse environment and its microbial inhabiting community is one of high biological diversity. The potential for crop productivity is largely determined by the environment that soil provides for root growth. Several important biological processes that are crucial for plant growth and health are mediated by microorganisms that are under the influence of plant roots. The rhizosphere, a narrow soil zone of only few millimeters surrounding plant roots, is a unique microenvironment high in nutrient content and microbial activity which the plant itself helps to create through releases of nutrients such as root exudates, mucilage, secretions, lysates and dead cell material (Lynch and Whipps 1990, Somers et al. 2004). The structural and functional diversity of rhizospheric microbial communities is also influenced by several biotic and abiotic factors such as soil quality, plant species, climate, human activities and agricultural treatments among others (Berg and Smalla 2009). The rhizosphere microbiota plays an important role in plant growth and health and its relationship with the host plant can have either a beneficial, harmful, or neutral effect.

Plant growth-promoting rhizobacteria (PGPR) are free-living beneficial bacteria that colonize the rhizosphere and contribute to increased growth and improved plant yields (Lugtenberg and Kamilova 2009, Kloepper and Schroth 1978). While they can be classified according to different criteria, PGPR generally function in three ways, by synthesizing compounds for the plants, facilitating nutrient uptake from soil, and reducing or preventing plant diseases (Hayat et al. 2010). PGPR can stimulate plant growth directly through promotion of nutrition or synthesis of phytohormones, or indirectly by protecting the plant against soil-borne pathogens through antagonism or induction of plant defense responses (Somers et al. 2004). PGPR have also been categorized according to their location within plant cells (intracellular PGPR or iPGPR) or outside plant cells in the rhizosphere or on the rhizoplane (extracellular PGPR or ePGPR) (Gray and Smith 2005). iPGPR generally refer to endophytes such as *Rhizobia* including bacteria from the genera *Allorhizobium, Azorhizobium, Bradyrhizobium, Mesorhizobium, Rhizobium*, and *Sinorhizobium* (Vessey 2003) that can symbiotically fix atmospheric N_2 in leguminous plants within specialized nodular structures of the root cells. ePGPR include bacteria from the genera *Agrobacterium, Arthrobacter, Azotobacter, Azospirillum, Bacillus, Burkholderia, Caulobacter, Chromobacterium, Erwinia, Flavobacterium, Micrococcus, Pseudomonas* and *Serratia* (Bhattacharyya and Jha 2012, Gray and Smith 2005). Some PGPR mechanisms of action include (i) the ability to produce ACC deaminase to reduce ethylene levels in the roots of developing plants which increase root length and growth (Dey et al. 2004, Penrose and Glick 2003), (ii) modulation of phytohormone synthesis such as indole-3-acetic acid (IAA), ethylene, cytokinins and gibberellins (Dey et al. 2004, Patten and Glick 2002), (iii) symbiotic and asymbiotic nitrogen fixation (Kennedy et al. 2004, Vessey 2003, Vanrhijn and Vanderleyden 1995), (iv) solubilization and mineralization of nutrients such as phosphates (Richardson et al. 2009, Richardson and Simpson 2011), and (v) antagonism against plant pathogens by production of antibiotics, siderophores, chitinases, cyanide, and β-1,3-

glucanase (Raaijmakers and Mazzola 2012, Hayat et al. 2010). Ultimately, PGPR often act through a combination of different mechanisms.

Engineering the Rhizosphere for Sustainable Agriculture

Understanding the complexity of interactions in the rhizosphere holds the key to creating ideal conditions for promoting plant growth in order to maximize microbiologically mediated processes that are beneficial to the plant while decreasing potentially harmful interactions. Manipulating the rhizosphere to exploit or enhance this innate genetic potential through manipulation of root/soil interactions will most probably play a key role in the future development of sustainable agricultural practices despite decreased resource inputs. Ultimately, rhizosphere engineering aims to reduce our reliance on agrochemicals by replacing or complementing their actions with those of beneficial microbes (or through the use of biostimulants or transgenic plants) (Ryan et al. 2009). Plants have naturally evolved several strategies to modify their rhizosphere in order to lessen the impact of environmental stresses, such as in the case of soils naturally suppressive to soil-borne pathogens where certain crop species select populations of antagonistic antibiotic-producing beneficial bacterial strains (Ryan et al. 2009).

One approach to rhizosphere engineering is to select beneficial microbial populations that promote plant growth and plant health. Generally, the concept of rhizosphere engineering through microbial population management can be approached in three different ways, either by (i) exploiting natural processes, (ii) genetic modification of microbial biological activities, or (iii) exogenous amendments through soil or plant inoculations.

The first approach is exemplified in soils that are naturally suppressive to soil-borne plant pathogens such as *Rhizoctonia, Fusarium, Pythium, Phytophthora* and *Gaeumannomyces*. In most suppressive soils, the resident microbiota is the primary factor contributing to disease suppression. Pasteurization of these soils most often leads to loss of suppressiveness and suppressiveness can be transferred to disease conducive soils by mixing them with 0.1–10% of the suppressive soil (Haas and Defago 2005, Mendes et al. 2011, Weller et al. 2002). Disease suppressive soils have been reported in various plant-pathogen systems such as take-all disease of wheat caused by *Gaeumannomyces graminis* var. *tritici*, Rhizoctonia damping-off disease of sugar beet, Fusarium wilt diseases in several plant species and common scab of potato caused by *Streptomyces* spp. (Haas and Defago 2005, Weller et al. 2002). While the mechanisms responsible for naturally disease suppressive soils are not clear, the presence of several PGPR producing antimicrobial compounds has been linked to the ability to suppress diseases (Raaijmakers et al. 1997, Weller et al. 2002). Recent work has demonstrated that the soil microbiome as a whole is what drives disease-suppressive ability rather than individual taxon or groups of soil microbes and that plants can exploit microbial consortia from soil for protection against pathogen infection (Mendes et al. 2011).

Rhizosphere engineering can also be approached by genetic modification of microbiological activities through direct manipulation of genes influencing rhizosphere functions. Genetically modified plants and genetically modified microbial inoculants have been developed and tested under laboratory and field conditions (Morrissey et al. 2002). For example, cloning and expression of genes implicated in mineral and organic phosphate solubilization led to the engineering of rhizobacterial strains with improved capacity for phosphate solubilization (Rodríguez and Fraga 1999, Rodriguez et al. 2006). Some have advocated the advantages of developing genetically modified PGPR over transgenic plants to achieve improved plant performance, citing that genetic modification of bacteria is much easier to achieve than manipulation of more complex organisms such as plants, and that several plant growth-promoting traits can be combined into a single organism, leading to engineered inoculants that have the possibility to be used on several crops instead of engineering plants crop by crop (Rodriguez et al. 2006). Even though studies have shown that genetically modified microorganisms can be applied safely in agriculture if appropriate regulations are followed (Amarger 2002, Morrissey et al. 2002), the release of genetically modified organisms remains controversial and is prohibited in some countries. Commercial exploitation requires compliance with regulatory requirements and the successful utilization of microbial inoculants genetically engineered for improved plant growth promotion relies on the demonstration that they pose no risk to the environment following their introduction in the rhizosphere.

Lastly, rhizosphere engineering can be regarded as application of microbial inoculants to promote plant growth with the aim of achieving the best combinations of beneficial bacteria as a valuable part of sustainable agriculture production systems. The potential of PGPR inoculants to contribute to the development of sustainable agriculture has been recognized due to their multifaceted ability to fulfill diverse plant growth promoting functions. Over the past decades, several reports of significant increases in growth and yield of agronomically important crops have been recorded due to inoculation with PGPR (Biswas et al. 2000, Biari et al. 2008, Vessey 2003, Dobbelaere et al. 2001). Based on their various mechanisms of action, potential PGPR inoculants can be categorized as biofertilizers involved in improving nutrient acquisition (Vessey 2003), biopesticides that suppress plant diseases (Berg 2009), phytostimulators that produce phytohormones (Lugtenberg and Kamilova 2009, Somers et al. 2004), and more recently as rhizoremediators involved in bioremediation of contaminated soils in association with plants (Zhuang et al. 2007). Furthermore, developing microbial inoculants that respond to different stress conditions such as salinity, drought, waterlogging, heavy metals and pathogenicity will be of emerging importance in the future (Berg 2009). Although all important for their potential role in the development of sustainable agricultural practices, PGPR inoculants falling under the categories of biofertilizers and biopesticides have been selected for further discussion.

PGPR as Biofertilizers for Integrated Nutrient Management

Harvesting crops removes nutrients from soils and sustainable agriculture requires that these nutrients be somehow replaced to maintain soil fertility (Tilman 1998, Vitousek et al. 1997). Prior to the green revolution, farmers traditionally replenished soil fertility by application of manure (organic fertilizer), crop rotations and by non-cultivating fields for a few years to allow gradual growth of natural vegetation. Intensification of agriculture has required important increases in fertilizer inputs such as nitrogen (N), phosphorus (P), and potassium (K) which has been largely met by application of inorganic fertilizers. However, this has come at a cost. The continuous use of fertilizers has been associated with negative impacts on biodiversity (McLaughlin and Mineau 1995) and biogeochemical cycles, degradation of soil and downstream water quality, eutrophication of fresh water and coastal marine ecosystems, environmental pollution, etc. (Vitousek et al. 1997). Modern agriculture relies heavily on chemical fertilizers to maintain productivity, and while their use is stabilizing in Europe and Northern America, projections indicate that global demand for inorganic fertilizers will increase due to developing and emerging economies such as China, India and Africa (Cordell et al. 2009). However, current practices are probably not sustainable (Fitter 2012). For instance, modern agriculture requires regular inputs of P fertilizer in order to replenish P that is removed from soil during growth and harvest of crops. P fertilizer comes from mined phosphate rock which is a non-renewable resource with only approximately 50–100 years of known reserves left (Cordell et al. 2009). Nitrogen, the major limiting nutrient for plant growth, is produced through an energy-intensive method requiring fossil fuels (Jensen and Hauggaard-Nielsen 2003). Furthermore, fertilizers often have low use efficiency in relation to application rates; meaning that higher amounts of fertilizer do not directly translate into higher nutrient uptake and plant growth. Large fractions of applied fertilizers are lost, causing leaching and running off of essential nutrients. Keeping in mind the high energy cost of fertilizer production, stock depletion, low efficiency in fertilizer uptake and associated environmental costs, it is not surprising that research activities aimed at better using fertilizers have been increasing over the past decades. Although eliminating the use of inorganic fertilizers is probably not feasible, one approach is to reduce the loss of nutrients in cropping systems without necessarily increasing inorganic fertilizer use. Integrated nutrient management calls for a more efficient and sustainable approach that combines the use of traditional fertilizers with biofertilizers (Adesemoye and Kloepper 2009).

Biofertilizers have been defined as products which contain living microorganisms that promote plant growth by increasing availability/supply of primary nutrients to the host (Vessey 2003). Three major groups of microbial inoculants include: (i) PGPR, (ii) arbuscular mycorrhiza fungi (AMF), and (iii) the well-known nitrogen-fixing rhizobia associated with legume hosts, which are typically not considered PGPR or are designated as iPGPR (Adesemoye and Kloepper 2009). The following sections will generally focus on recent examples of successful PGPR used as inoculants to enhance nutrient availability and uptake.

Biological N_2 fixation

One aspect of integrated nutrient management is to exploit biological N_2 fixation (BNF) to complement and hopefully eventually reduce the dependence on chemical nitrogen fertilizers. While N_2 gas is abundant in our atmosphere, it must first be reduced to ammonia before it can be utilized by plants. Nitrogen-fixing bacteria (or diazotrophs) such as *Rhizobium* and *Bradyrhizobium* have long been studied for their ability to fix N_2 for their leguminous host plants by forming nodules on the roots of plants such as soybean, pea and alfalfa in which they convert N_2 to ammonia. Commercial inoculants have been available since the late 19th century (Vessey 2003) and legume-rhizobia symbiosis has been studied extensively with many excellent reviews on the subject (Oldroyd et al. 2011, Graham and Vance 2000, Cooper 2007) and will therefore not be further discussed. While symbiotic N_2 fixation is mostly limited to legume crops, BNF has also been noted in non-legume crops where bacteria form associative relationships with plants (associative N_2-fixers) or by N_2 fixation by free living heterotrophs that typically do not interact with other organisms (Vessey 2003).

Sugarcane has been a model system for studying biological N_2 fixation. In Brazil, long-term continuous cultivation of sugarcane using only low inputs of N fertilizers led to the discovery that associated N_2-fixing bacteria were significantly contributing to plant host nutrition. Although organisms responsible for BNF in this system have not been fully characterized and the mechanisms of plant growth promotion are not well elucidated, several N_2-fixing bacteria have been isolated from the root surface or rhizosphere including *Beijerinckia* spp., *Azospirillum* spp., *Azotobacter* spp., *Bacillus* spp., *Derxia* spp., *Enterobacter* spp. and *Erwinia* spp., but also endophytic

N_2-fixers such as *Gluconacetobacter diazotrophicus* (formerly *Acetobacter diazotrophicus*), *Herbaspirillum* spp. and *Burkholderia* spp. (Boddey et al. 2003). Endophytic diazotrophs have received a lot of attention in relation to non-leguminous plants. Brazilian sugarcane varieties can obtain up to 72% of their nitrogen requirement through BNF (Boddey et al. 2003, Urquiaga et al. 1992, Yoneyama et al. 1997). While the amount of biologically fixed N_2 provided to the plant host varies considerably, it is believed to depend on several factors such as plant variety and growth stage, inoculation method, and environmental conditions. Nevertheless, endophyte inoculants have the potential to increase BNF and contribute to plant growth. Sugarcane plantlets inoculated with wild-type *G. diazotrophicus* had a higher total N content and grew generally better 60 days after planting than uninoculated plants or plants inoculated with a *nifD*- mutant strain. In this case, $^{15}N_2$ incorporation experiments demonstrated that *G. diazotrophicus* actively fixed N_2 in sugarcane plants (Sevilla et al. 2001). However, the diazotroph/ sugarcane association may be far more complex than BNF by only a single strain. In addition to previously characterized *Gluconacetobacter* spp. and *Burkholderia* spp., *nifH* expression from a wide diversity of previously uncharacterized *Ideonella/Herbaspirillum* related phylotypes as well as *Bradyrhizobium* sp. and *Rhizobium* sp. was detected in Brazilian sugarcane and may also contribute to nitrogen fixation (Fischer et al. 2012). Over two seasons of sugarcane growth, Thaweenut and colleagues (Thaweenut et al. 2011) detected expression of *nifH* sequences in sugarcane plant tissues similar to those of *Bradyrhizobium* sp., *Azorhizobium caulinodans*, *Sinorhizobium fredii*, *Derxia gummosa*, *Nostoc commune* and *Beijerinckia derxii*. In these examples, BNF may be the result of a bacterial consortium and successful strategies to increase the nitrogen plant content may need to rely on mixed inocula. Indeed, successful field trials of mixed bacterial inocula consisting of *Gluconacetobacter diazotrophicus*, *Herbaspirillum seropedicae*, *H. rubrisubalbicans*, *Azospirillum amazonense*, and *Burkholderia tropica* have been conducted using *in vitro* sugarcane inoculation methods (Fischer et al. 2012, Oliveira et al. 2006, Oliveira et al. 2009). In addition to sugarcane, substantial benefits of BNF related to PGPR plant associations have also been reported in other non-legume crops such as the association between rice and *Herbaspirillum seropedicae* (Gyaneshwar et al. 2002, James et al. 2002) and wheat and *Klebsiella pneumoniae* (Iniguez et al. 2004). The encouraging results obtained in the Brazilian sugarcane plant system exemplifies the potential improvements in plant nitrogen status that can be achieved using PGPR as bioinoculants and should help to fuel research efforts in various other cropping systems.

Apart from direct inoculation of N_2-fixing strains to enhance plant nutrition, another way to use PGPR in the context of integrated nutrient management and biofertilization is to use them as inoculants that can enhance uptake of applied N fertilizers, therefore counteracting application rate increases by improving use efficiency. Few studies have concretely demonstrated the enhanced N fertilizer uptake by inoculated PGPR. A recent example of treatment with PGPR leading to increased N fertilizer uptake by the host plant was demonstrated by Adesemoye and colleagues in greenhouse tomato experiments (Adesemoye et al. 2010). By using a mixture of *Bacillus amyloliquefaciens* strain IN937a and *Bacillus pumilus* strain T4, a significant increase in uptake of N fertilizer in tomato was observed by tracking ^{15}N-depleted ammonium sulfate. In this case, improved N nutrition was not related to BNF but to greater uptake of applied N fertilizer by the host plant.

Phosphorus solubilization

Phosphorus (P) is an essential macronutrient for plant growth and microorganisms play a central role in soil P cycling. While present in soil at levels of 400–1200 mg kg^{-1}, only small amounts of soluble P are available to the plant, typically at levels of 1 mg kg^{-1} or less (Rodríguez and Fraga 1999). Plants uptake P from the soil as orthophosphate anions (HPO$_4^{2-}$ and H$_2$PO$_4^{1-}$) but due to its limited bioavailability in soil, chemical P fertilizers are usually added to fulfill plant nutrient requirements, although as in the case of N fertilizers, use efficiency is rather low. Soon after applying P fertilizers to soil, large portions of soluble inorganic phosphate are immobilized by fixation and precipitation and P becomes unavailable to plants (Rodríguez and Fraga 1999). This low use efficiency leads to the use of large amounts of P fertilizers and subsequent leaching in the environment. Research focused towards developing a more effective utilization of phosphates is thus of considerable interest.

The dynamics of soil P is characterized by physicochemical (sorption-desorption) and biological (immobilization-mineralization) processes (Khan et al. 2009). The availability of P to plants can be increased by various microbial processes such as mineralization of organic phosphate or by solubilization of inorganic phosphate by production of acids (Richardson and Simpson 2011). Among the most powerful phosphate solubilizers are bacteria belonging to the genera *Pseudomonas*, *Bacillus*, *Rhizobium* and *Enterobacter* and fungi belonging to the genera *Penicillium* and *Aspergillus* (Whitelaw 1999). Bacteria capable of mediating these processes are referred to as phosphate solubilizing bacteria (PSB). Several PSB are found in soil, but the amount of P that they release is generally not sufficient for substantial plant growth increases (Rodríguez and Fraga 1999). A higher proportion of PSB are found in the rhizosphere where some may also be acting as PGPR through production of other metabolites beneficial to the plants that improve plant growth and nutrition. For this reason, the real contribution of phosphate solubilizing activity by PSB in the rhizosphere for improving plant growth and nutrition is not always clear.

In sunflower, the most important effect of PSB inoculation was observed in conjunction with 100 kg P_2O_5 ha^{-1} of fertilizer when highest oil yields were obtained (Ekin 2010). However, the highest seed yield obtained with 100 kg P_2O_5 ha^{-1} of fertilizer alone could also be achieved by combining inoculation of PSB *Bacillus* M-13 with only half the level of fertilizer used (50 kg P_2O_5 ha^{-1}). The authors noted that while the combined effect of PSB inoculation with P fertilizers did increase P content by 28% over the control, increased plant growth and seed quality may also be attributed to other plant growth-promoting traits, thus complementing the effect of enhanced phosphate availability. Indeed, several PSB may also carry other plant growth-promoting traits. While the potential benefits of PSB on improving crop productivity is well recognized (Rodríguez and Fraga 1999), their use as commercial biofertilizer inoculants has remained limited. Part of the road to success lies in the combined use of multiple plant growth promoting traits to improve plant nutrition. One study aimed at comparing the effectiveness of *Bacillus* spp. simultaneously carrying traits for P-solubilization and ACC deaminase activity with *Bacillus* spp. carrying only one of these traits for improving growth, yield and P uptake in wheat. Bacterial strains with dual traits were more effective than single trait strains under soil conditions to improve growth, wheat yield and increase P use efficiency of P fertilizer (Baig et al. 2012).

In other studies, co-inoculations of multiple PGPR have enhanced performance in comparison to single strain inoculations. Co-inoculations of bacteria with phosphate solubilizing activity and nitrogen fixing activity have been reported to be more effective than single strain inoculations on promoting plant growth and providing adequate nutrition to various crops (Sahin et al. 2004, Valverde et al. 2006, Madhaiyan et al. 2010). For instance, co-inoculation with *Pseudomonas chlororaphis* (PSB strain) and *Arthrobacter pascens* (NFB strain) resulted in the highest plant height, shoot and root dry weight, P and N uptake in walnut seedlings under shade house conditions (Yu et al. 2012). However, other tested PSB and NFB strain combinations failed to increase these parameters, hence effects of inoculation with PSB and NFB strains should always be corroborated under greenhouse and field conditions in order to obtain the most accurate conclusions.

A few studies, such as the one performed by Elkoca et al. 2008 (Elkoca et al. 2008), have even proposed that combinations of selected inocula may completely alleviate the use of NP fertilizers. Chickpea inoculated with strains of *Rhizobium*, N_2-fixing *Bacillus subtilis* OSU-142, and P-solubilizing *B. megaterium* M-3, especially in dual or triple combinations, increased all parameters measured compared to treatments with N, P or NP fertilizers without inoculation. However, most studies are pointing towards using biofertilizers to complement the use of chemical fertilizers in an integrated nutrient management strategy. This approach aims at reducing the impact of fertilizer overuse by promoting low chemical input and improving nutrient-use efficiency without sacrificing high crop productivity or long-term sustainability. A mixture of PGPR strains of *Bacillus amyloliquefaciens* IN937a and *Bacillus pumilus* T4 with the arbuscular mycorrhiza fungus (AMF) *Glomus intraradices* was evaluated for its ability to improve fertilizer use efficiency in a greenhouse study on tomato. When supplemented with the mixture of PGPR and AMF, fertilizer could be reduced to 75% of the recommended rates and still produce plant height, shoot dry weight, root dry weight, yield, and nutrient uptake (including N and P uptake) that were statistically equivalent to application of the full fertilizer rate without inoculants (Adesemoye et al. 2009).

Facilitating uptake of other elements

Inoculation of PGPR can also influence the uptake of many other nutrient elements in addition to N and P. Plant nutrient uptakes of N, P, K, Fe, Zn, Cu and Mn in maize were all significantly influenced by application of PGPR belonging to the genera *Azotobacter* and *Azospirullum* (Biari et al. 2008). Combinations of the PGPR *Bacillus* M3, *Bacillus* OSU-142 and *Microbacterium* FS01improved uptake of the mineral nutrients Ca, K, Fe, Mn, Cu and Zn in apple leaves (Karlidag et al. 2007). Authors indicated that this increase might be explained by a decrease in soil pH as a result of organic acids production by plants and bacteria in the rhizosphere, thus stimulating bioavailability of nutrients. One of the essential criteria in facilitating the transport of most of these nutrients is solubilization of unavailable forms of nutrients (Glick 1995). Iron for instance is an essential nutrient for both plants and microorganisms but remains relatively insoluble in soil, especially in calcareous soil, and is mostly found in the form of Fe(III) oxides and hydroxides. Availability of Fe is therefore often severely limited in soil and efficient Fe users (plants and rhizospheric bacteria) have evolved strategies to improve Fe uptake (Romheld and Marschner 1986, Hindt and Guerinot 2012), although some of these strategies may still be insufficient under Fe-limiting conditions. Under conditions of low iron availability, most aerobic and facultative anaerobic microorganisms produce siderophores, low molecular weight Fe^{3+} chelators that sequester ferric ions from the environment and are taken up by the microbial cells (Bakker et al. 2007). Fluorescent pseudomonads found at a high density in the rhizosphere are known to synthesize a siderophore called pyoverdine. Vansuyt and colleagues (Vansuyt et al. 2007) evaluated the impact of Fe-pyoverdine on iron content of *Arabidopsis thaliana* compared to that of Fe-EDTA. They reported that iron chelated to pyoverdine was incorporated in a more efficient way than when chelated to EDTA, leading to plants with significantly higher iron content and increased plant growth. Others also demonstrated an increase in iron content and uptake in peanut, cotton, cucumber, and *Arabidospsis* plants supplemented with Fe-pyoverdine (Bar-Ness et al. 1991). However, improved plant iron nutrition by bacterial Fe^{3+} siderophore uptake is not always the case. In certain studies, bacterial siderophores have proven inefficient at improving Fe uptake as in the case of pseudobactin and ferrioxamine B in oat and maize where authors concluded that these bacterial siderophores were in competition with the plants for available Fe (Bar-Ness et al. 1992). Other studies proposed that increased iron uptake by bacterial activity may not necessarily be a direct consequence of bacterial siderophore uptake. The PGPR *Bacillus subtilis* GB03 activated *Arabidospsis* plant's own iron acquisition machinery to increase assimilation of metal ions in an experimental set-up that excluded the possibility that bacterial siderophores contributed to increased iron uptake (Zhang et al. 2009). In fact, the majority of research on siderophore-producing PGPR have instead focused on their biocontrol traits since siderophore-producing PGPR are mostly associated with the ability to deprive surrounding microflora of iron, thus exerting antagonistic effects against phytopathogens. While the mechanisms by which soil microbes may contribute to plant Fe acquisition remain unclear, many still suggest an important role played by these organisms as highlighted by studies where plants grown in sterile soil showed poor growth and lower Fe content compared to plants grown in non-sterile conditions (Masalha et al. 2000, Jin et al. 2006, Rroço et al. 2003).

PGPR as Biocontrol Agents

Control of plant pathogens and disease development is a prerequisite to crop health and productivity. Since the 1960's, farmers have increasingly relied on the use of synthetic agrichemicals to combat pests and pathogens (Gullino et al. 2000). Environmental pollution and increasing levels of resistance among pathogens are creating a need to focus research efforts on developing alternative forms of disease control and to reduce reliance on the use of synthetic pesticides, which are increasingly being subjected to strict regulations. Using PGPR to control plant diseases is a form of biological control than can be part of sustainable soil-pathogen management. Some PGPR have the ability to stimulate growth indirectly by protecting the plant against soil-borne pathogens. This can occur through direct antagonistic interactions with phytopathogens or through induction of host resistance's mechanisms (Singh et al. 2011). Interactions that lead to biocontrol can be complex and antagonistic microorganisms that suppress pathogens can do so by one or more mechanisms such as competition for nutrients, competition for iron (through the ability to produce siderophores), niche exclusion, production of lytic enzymes, induction of systemic resistance and direct pathogen inhibition by the production of antimicrobial metabolites (Weller et al. 2007). The mechanisms used for biocontrol of phytopathogens have been extensively reviewed elsewhere (Raaijmakers and Mazzola 2012, Haas and Defago 2005, Whipps 2001). Biocontrol bacteria have been identified belonging to different genera, including *Pseudomonas, Bacillus, Burkholderia, Agrobacterium, Arthrobacter, Azotobacter, Collimonas, Pantoea, Serratia, Stenotrophomonas* and *Streptomyces* (Raaijmakers and Mazzola 2012). Of these, *Pseudomonas* spp. are probably the most dominant and studied organisms for their broad spectrum antagonistic activity against a large number of phytopathogens and for their effectiveness in colonizing the rhizosphere. The ability to synthesize a variety of bioactive molecules for biological control of plant pathogens has been recognized in various *Pseudomonas* spp., which can in part explain their high research interest and potential for use in agricultural practices. Of particular interest is the production of siderophores, 2,4-diacetylphloroglucinol (2,4-DAPG), pyrrolnitrin, pyoluteorin, phenazines, hydrogen cyanide (HCN), 2,5-dialkylresorcinol, quinolones, gluconic acid, rhamnolipids and various lipopeptides (Raaijmakers and Mazzola 2012).

One of the key biocontrol mechanisms used by pseudomonads is antibiosis and many of the antibiotics produced by biocontrol bacteria have broad spectrum activity. Several potential biocontrol agents have first been selected based on their *in vitro* capacity of antibiosis against pathogens. For example, *Pseudomonas* sp. LBUM300 was originally isolated from the rhizosphere of strawberry and selected for further testing of its biocontrol ability based on *in vitro* inhibition assays of the pathogens *Phytophthora cactorum* and *Verticillium dahlia* (Paulin et al. 2009). This strain was found to produce both 2,4-DAPG and HCN, antifungal metabolites that can play a key role in the suppression of diseases caused by various soil-borne fungal pathogens. It was recently shown that under *in planta* conditions, *Pseudomonas* sp. LBUM300 applied as an inoculant also contributes to the biological control of bacterial canker of tomato caused by the bacterium *Clavibacter michiganensis* subsp. *michiganensis* (*Cmm*). Four weeks following inoculation of tomato plants, disease symptoms were significantly reduced in treatments inoculated with *Pseudomonas* sp. LBUM300, while those inoculated with isogenic mutants deficient in production of 2,4-DAPG or HCN could not significantly reduce symptoms (Lanteigne et al. 2012). In this plant-pathogen system, the development of biocontrol agents is of particular importance since other approaches such as chemical-based compounds or disease resistant commercial tomato cultivars have so far been unsuccessful for reliably and consistently controlling bacterial canker of tomato. Similarly, *Pseudomonas* sp. LBUM223 was also first isolated from the rhizosphere of strawberry and selected based on its ability to produce phenazine-1-carboxylic acid, another commonly produced antibiotic that inhibits the growth of certain plant pathogens and contributes to biological control. *Pseudomonas* sp. LBUM223 was later found to limit pathogen growth and control common scab of potato caused by *Streptomyces scabies* (St-Onge et al. 2011).

It has been noted that pseudomonads such as *P. fluorescens* typically have the interesting ability to protect plants in more than one pathosystem, meaning that each biocontrol strain may protect more than one plant species from distinct pathogens (Couillerot et al. 2009). In addition to the production of various antimicrobial metabolites, *P. fluorescens* are also well adapted to survive in soil and colonize plant roots (Kiely et al. 2006), which are essential traits for successful biocontrol agents. Some have proposed to use biocontrol agents along with chemical pesticides as integrated partners in plant disease control. Various fungicides in conjunction with the PGPR biocontrol agent *P. fluorescens* CW2 were tested in greenhouse tomato experiments to evaluate the potential of an integrated biological and chemical strategy for the control of *Pythium ultimum* (Trow) causing damping-off disease on tomato. Combined treatments of *P. fluorescens* CW2 with fungicides led to significant improvements in disease control and improved plant growth as indicated by shoot and root dry weights (Salman and Abuamsha 2012). However, successful examples of biocontrol *P. fluorescens* used as plant/soil inoculants without associated chemical control have also been demonstrated in field trials. *P. fluorescens* was able to significantly improve plant establishment and harvest yield in winter wheat infested with *Microdochium nivale* causing blight disease. Plant survival numbers increased by up to 48% and yield by 26.5% in different field trials and showed consistent plant protection over long periods of the growing seasons (Amein et al. 2008).

Commercialization of Microbial Inoculants

Successful application of biofertilizers for improving plant nutrition and biocontrol agents for biological control of plant diseases requires knowledge of when and where microbial inoculants can be profitable and better awareness of the costs and benefits these inocula can provide (McSpadden Gardener and Fravel 2002). Integrated management systems should aim to minimize economic and environmental costs in order to achieve sustainable agricultural production. Several challenges accompany development of new microbial inoculants if they are to be effective at a commercial level. For instance, formulations of PGPR inoculants that can ensure survival and activity in the field along with compatibility with chemical treatments allowing integrated management approaches are necessary for product effectiveness (Barreto Figueiredo et al. 2010). In addition to technical and scientific considerations, a number of other obstacles related to regulatory/ registration procedures, business management and marketing must also be overcome when developing new products destined for application to agricultural fields. Furthermore, risk assessment must confirm that microbial-based inoculants have no negative impact on the environment and/or human health (Mark et al. 2006).

Considerations for producing PGPR biofertilizers

PGPR biofertilizers have not yet undergone extensive commercialization in comparison to rhizobia inoculants in legumes that have been used in agriculture for over a century. Lack of consistent results in different host cultivars and field sites undoubtedly contributed to the factors preventing expansion of commercialized biofertilizer products. For these and other reasons, the commercial availability of PGPR biofertilizers is low and *Azospirillum* spp. mostly dominates the PGPR biofertilizer market (Fuentes-Ramirez and Caballero-Mellado 2006) with inoculants available for a variety of crops in Europe and Africa (Dobbelaere et al. 2001, Vessey 2003). Commercial inoculants composed of *Azotobacter* spp. and *Azospirillum* spp. have also been developed and are marketed for use as biofertilizers in rice and corn (Banayo et al. 2012). While few commercial microbial inoculants advertised as biofertilizers are available, many have been developed for their plant growth promoting properties which may in turn improve plant nutrition. Special emphasis has been given to *Bacillus* and *Pseudomonas*-based formulations with several being registered on the US market (Barreto Figueiredo et al. 2010).

Considerations for producing PGPR biocontrol inoculants

A number of *Pseudomonas*-based biocontrol inoculants are now commercially available targeting pathogens such as *Erwinia amylovora*, *Fusarium oxysporum*, *Pythium* spp., *Rhizoctonia solani*, etc. (Mark et al. 2006). In the United States, microbial inoculants and the antimicrobial metabolites they produce fall under the category of biopesticides. In comparison to biofertilizers, microbial inoculants registered and developed as biopesticides are several and successfully commercialized inoculants include bacteria belonging to the genera *Agrobacterium*, *Bacillus*, *Pseudomonas* and *Streptomyces*, as well as fungi belonging to the genera *Ampelomyces*, *Candida*, *Coniothyrium*, and *Trichoderma* (McSpadden Gardener and Fravel 2002). However, major factors still limit the development of biocontrol inoculants for widespread use in agriculture. With regards to formulation of biocontrol agents for commercial use, one weakness of *Pseudomonas*-based biocontrol agents for example, is their inability to produce resting spores which can seriously limit their shelf-life in comparison to other biocontrol bacteria such as spore-producing *Bacillus* spp. Formulation of biocontrol agents has been defined not only as a technique to preserve and deliver antagonists to their targets but also as a method to improve their biocontrol activity (Burges and Jones 1998). Although several hurdles are to be surmounted, progress is still being made as demonstrated by formulation of *P. fluorescens* B5 that underwent a pelleting process on sugar beet seeds with manganese sulphate and zinc sulphate as formulation additives. This formulation significantly increased the biocontrol activity of *P. fluorescens* B5 against the fungal pathogen *Pythium ultimum* var. *ultimum* without negatively affecting sugar beet seedlings, causing an increase in plant height and fresh weight of treated seedlings. Even after one year of storage at 5°C, no significant decline in *P. fluorescens* B5 survival rate was observed (Wiyono et al. 2008).

Conclusion

A majority of producers are now interested in biologically-based management of soil fertility and health, particularly in the context of an integrated management approach or for organic production. PGPR-based microbial inoculants that can fulfil diverse functions for plants can lead to sustainable practices that will lessen negative impacts on the environment. Selection of appropriate PGPR for use as inoculants as part of an effective rhizosphere engineering strategy still requires more knowledge on the underlying mechanisms related to plant growth promotion. The development and production of microbial inoculants will benefit from current and future research aimed at achieving a better understanding of chemical and biological interactions occurring in the rhizosphere, which will in turn help to maximize the use of plant growth promoting activities through better management of soil microbial populations. While achieving environmental and agricultural sustainability is an enormous task, it is not insurmountable and will undoubtedly require multifaceted and interdisciplinary strategies that will combine innovations in plant and soil science with approaches aimed at preserving our ecosystems.

References

Adesemoye, A. and J. Kloepper. 2009. Plant–microbes interactions in enhanced fertilizer-use efficiency. Appl. Microbiol. Biot. 85: 1–12.

Adesemoye, A., H. Torbert and J. Kloepper. 2009. Plant growth-promoting rhizobacteria allow reduced application rates of chemical fertilizers. Microb. Ecol. 58: 921–929.

Adesemoye, A. O., H. A. Torbert and J. W. Kloepper. 2010. Increased plant uptake of nitrogen from ¹⁵N-depleted fertilizer using plant growth-promoting rhizobacteria. Appl. Soil Ecol. 46: 54–58.

Amarger, N. 2002. Genetically modified bacteria in agriculture. Biochimie. 84: 1061–1072.

Amein, T., Z. Omer and C. Welch. 2008. Application and evaluation of *Pseudomonas* strains for biocontrol of wheat seedling blight. Crop Prot. 27: 532–536.

Baig, K., M. Arshad, B. Shaharoona, A. Khalid and I. Ahmed. 2012. Comparative effectiveness of *Bacillus* spp. possessing either dual or single growth-promoting traits for improving phosphorus uptake, growth and yield of wheat (*Triticum aestivum* L.). Ann. Microbiol. 62: 1109–1119.

Bakker, P. A., C. M. Pieterse and L. C. van Loon. 2007. Induced systemic resistance by *Fluorescent pseudomonas* spp. Phytopathology. 97: 239–243.

Banayo, N. P. M., P. C. S. Cruz, E. A. Aguilar, R. B. Badayos and S. M. Haefele. 2012. Evaluation of biofertilizers in irrigated rice: effects on grain yield at different fertilizer rates. Agriculture. 2: 73–86.

Bar-Ness, E., Y. Chen, Y. Hadar, H. Marschner and V. Römheld. 1991. Siderophores of *Pseudomonas putida* as an iron source for dicot and monocot plants. Plant Soil. 130: 231–241.

Bar-Ness, E., Y. Hadar, Y. Chen, V. Romheld and H. Marschner. 1992. Short-term effects of rhizosphere microorganisms on Fe uptake from microbial siderophores by maize and oat. Plant Physiol. 100: 451–456.

Barreto Figueiredo, M. d. V., L. Seldin, F. F. de Araujo and R. d. L. Ramos Mariano. 2010. Plant growth promoting rhizobacteria: fundamentals and applications. *In: Plant Growth and Health Promoting Bacteria*, edited by D. K. Maheshwari: Springer, 233 Spring Street, New York, Ny 10013, United States. 19–43.

Berg, G. 2009. Plant–microbe interactions promoting plant growth and health: perspectives for controlled use of microorganisms in agriculture. Appl. Microbiol. Biot. 84: 11–18.

Berg, G. and K. Smalla. 2009. Plant species and soil type cooperatively shape the structure and function of microbial communities in the rhizosphere. FEMS Microbiol. Ecol. 68: 1–13.

Bhattacharyya, P. and D. Jha. 2012. Plant growth-promoting rhizobacteria (PGPR): emergence in agriculture. World J. Microb. Biot. 28: 1327–1350.

Biari, A., A. Gholami and H. A. Rahmani. 2008. Growth promotion and enhanced nutrient uptake of maize (*Zea mays* L.) by application of plant growth promoting rhizobacteria in arid region of Iran. J. Biol. Sci. 8: 1015–1020.

Biswas, J. C., J. K. Ladha, F. B. Dazzo, Y. G. Yanni and B. G. Rolfe. 2000. Rhizobial inoculation influences seedling vigor and yield of rice. Agron. J. 92: 880–886.

Boddey, R. M., S. Urquiaga, B. J. R. Alves and V. Reis. 2003. Endophytic nitrogen fixation in sugarcane: present knowledge and future applications. Plant Soil. 252: 139–149.

Burges, H. D. and K. A. Jones. 1998. Trends in formulation of microorganisms and future research requirements. *In Formulation of Microbial Biopesticides, Beneficial Microorganisms, Nematodes and Seed Treatment*, edited by H. D. Burges. Dordrecht: Kluwer Academic Publication. 311–332.

Cooper, J. E. 2007. Early interactions between legumes and rhizobia: disclosing complexity in a molecular dialogue. J. Appl. Microbiol. 103: 1355–1365.

Cordell, D., J.-O. Drangert and S. White. 2009. The story of phosphorus: global food security and food for thought. Global Env. Change. 19: 292–305.

Couillerot, O., C. Prigent-Combaret, J. Caballero-Mellado and Y. Moenne-Loccoz. 2009. *Pseudomonas fluorescens* and closely-related fluorescent pseudomonads as biocontrol agents of soil-borne phytopathogens. Lett. Appl. Microbiol. 48: 505–512.

Dey, R., K. K. Pal, D. M. Bhatt and S. M. Chauhan. 2004. Growth promotion and yield enhancement of peanut (*Arachis hypogaea* L.) by application of plant growth-promoting rhizobacteria. Microbiol. Res. 159: 371–394.

Dobbelaere, S., A. Croonenborghs, A. Thys, D. Ptacek, J. Vanderleyden, P. Dutto, C. Labandera-Gonzalez, J. Caballero-Mellado, J. F. Aguirre, Y. Kapulnik, S. Brener, S. Burdman, D. Kadouri, S. Sarig and Y. Okon. 2001. Responses of agronomically important crops to inoculation with *Azospirillum*. Aust. J. Plant Physiol. 28: 871–879.

Ekin, Z. 2010. Performance of phosphate solubilizing bacteria for improving growth and yield of sunflower (*Helianthus annuus* L.) in the presence of phosphorus fertilizer. Afr. J. Biotechnol. 9: 3794–3800.

Elkoca, E., F. Kantar and F. Sahin. 2008. Influence of nitrogen fixing and phosphorus solubilizing bacteria on the nodulation, plant growth, and yield of chickpea. J. Plant Nutr. 31: 157–171.

Fischer, D., B. Pfitzner, M. Schmid, J. Simões-Araújo, V. Reis, W. Pereira, E. Ormeño-Orrillo, B. Hai, A. Hofmann, M. Schloter, E. Martinez-Romero, J. Baldani and A. Hartmann. 2012. Molecular characterisation of the diazotrophic bacterial community in uninoculated and inoculated field-grown sugarcane (*Saccharum* sp.). Plant Soil. 356: 83–99.

Fitter, A. 2012. Why plant science matters. New Phytol. 193: 1–2.

Fuentes-Ramirez, L. E. and J. Caballero-Mellado. 2006. Bacterial biofertilizers. *In: PGPR: Biocontrol and Biofertilization*, edited by Z. Siddiqui. Dordrecht: Springer. 143–172.

Glick, B. R. 1995. The enhancement of plant-growth by free-living bacteria. Can. J. Microbiol. 41: 109–117.

Graham, P. H. and C. P. Vance. 2000. Nitrogen fixation in perspective: an overview of research and extension needs. Field Crop Res. 65: 93–106.

Gray, E. J. and D. L. Smith. 2005. Intracellular and extracellular PGPR: commonalities and distinctions in the plant–bacterium signaling processes. Soil Biol. Biochem. 37: 395–412.

Gullino, M. L., P. Leroux and C. M. Smith. 2000. Uses and challenges of novel compounds for plant disease control. Crop Prot. 19: 1–11.

Gyaneshwar, P., E. K. James, P. M. Reddy and J. K. Ladha. 2002. *Herbaspirillum* colonization increases growth and nitrogen accumulation in aluminium-tolerant rice varieties. New Phytol. 154: 131–145.

Haas, D. and G. Defago. 2005. Biological control of soil-borne pathogens by fluorescent pseudomonads. Nat. Rev. Microbiol. 3: 307–319.

Hayat, R., S. Ali, U. Amara, R. Khalid and I. Ahmed. 2010. Soil beneficial bacteria and their role in plant growth promotion: a review. Ann. Microbiol. 60: 579–598.

Hindt, M. N. and M. L. Guerinot. 2012. Getting a sense for signals: Regulation of the plant iron deficiency response. Mol. Cell Res. 1823: 1521–1530.

Iniguez, A. L., Y. Dong and E. W. Triplett. 2004. Nitrogen fixation in wheat provided by *Klebsiella pneumoniae* 342. Mol. Plant Microbe In. 17: 1078–1085.

James, E. K., P. Gyaneshwar, N. Mathan, W. L. Barraquio, P. M. Reddy, P. P. M. Iannetta, F. L. Olivares and J. K. Ladha. 2002. Infection and colonization of rice seedlings by the plant growth-promoting Bacterium *Herbaspirillum seropedicae* Z67. Mol. Plant Microbe In. 15: 894–906.

Jensen, E. S. and H. Hauggaard-Nielsen. 2003. How can increased use of biological N₂ fixation in agriculture benefit the environment? Plant Soil. 252: 177–186.

Jin, C. W., Y. F. He, C. X. Tang, P. Wu and S. J. Zheng. 2006. Mechanisms of microbially enhanced Fe acquisition in red clover (*Trifolium pratense* L.). Plant Cell Environ. 29: 888–897.

Karlidag, H., A. Esitken, M. Turan and F. Sahin. 2007. Effects of root inoculation of plant growth promoting rhizobacteria (PGPR) on yield, growth and nutrient element contents of leaves of apple. Sci. Horticulturae. 114: 16–20.

Kennedy, I. R., A. Choudhury and M. L. Kecskes. 2004. Non-symbiotic bacterial diazotrophs in crop-farming systems: can their potential for plant growth promotion be better exploited? Soil Biol. Biochem. 36: 1229–1244.

Khan, A. A., G. Jilani, M. S. Akhtar, S. M. S. Naqvi and M. Rasheed. 2009. Phosphorus solubilizing bacteria: occurrence, mechanisms and their role in crop production. J. Agric. Biol. Sci. 1: 48–58.

Kiely, P. D., J. M. Haynes, C. H. Higgins, A. Franks, G. L. Mark, J. P. Morrissey and F. O'Gara. 2006. Exploiting new systems-based strategies to elucidate plant-bacterial interactions in the rhizosphere. Microbial. Ecol. 51: 257–266.

Kloepper, J. and M. Schroth. 1978. Plant growth-promoting rhizobacteria on radishes. *In Proceedings of the 4th international conference on plant pathogenic bacteria*, edited by Gilbert-Clarey. Tours. 879–882.

Lanteigne, C., V. J. Gadkar, T. Wallon, A. Novinscak and M. Filion. 2012. Production of DAPG and HCN by *Pseudomonas* sp. LBUM300 contributes to the biological control of bacterial canker of tomato. Phytopathology. 102: 967–973.

Lugtenberg, B. and F. Kamilova. 2009. Plant-Growth-Promoting Rhizobacteria. Annu. Rev. Microbiol. 63: 541–556.

Lynch, J. M. and J. M. Whipps. 1990. Substrate flow in the rhizosphere. Plant Soil. 129: 1–10.

Madhaiyan, M., S. Poonguzhali, B. G. Kang, Y. J. Lee, J. B. Chung and T. M. Sa. 2010. Effect of co-inoculation of methylotrophic *Methylobacterium oryzae* with *Azospirillum brasilense* and *Burkholderia pyrrocinia* on the growth and nutrient uptake of tomato, red pepper and rice. Plant Soil. 328: 71–82.

Mark, G. L., J. P. Morrissey, P. Higgins and F. O'Gara. 2006. Molecular-based strategies to exploit Pseudomonas biocontrol strains for environmental biotechnology applications. FEMS Microbiol. Ecol. 56: 167–177.

Masalha, J., H. Kosegarten, Ö. Elmaci and K. Mengel. 2000. The central role of microbial activity for iron acquisition in maize and sunflower. Biol. Fertil. Soils. 30: 433–439.

McLaughlin, A. and P. Mineau. 1995. The impact of agricultural practices on biodiversity. Agric. Eco. Environ. 55: 201–212.

McSpadden Gardener, B. and D. Fravel. 2002. Biological control of plant pathogens: research, commercialization, and application in the USA. In Plant Health Progress.

Mendes, R., M. Kruijt, I. de Bruijn, E. Dekkers, M. van der Voort, J. H. M. Schneider, Y. M. Piceno, T. Z. DeSantis, G. L. Andersen, P. A. H. M. Bakker and J. M. Raaijmakers. 2011. Deciphering the rhizosphere microbiome for disease-suppressive bacteria. Science. 332: 1097–1100.

Morrissey, J. P., U. F. Walsh, A. O'Donnell, Y. Moenne-Loccoz and F. O'Gara. 2002. Exploitation of genetically modified inoculants for industrial ecology applications. Antonie Van Leeuwenhoek Int. J. Gen. Mol. Microbiol. 81: 599–606.

Oldroyd, G. E. D., J. D. Murray, P. S. Poole and J. A. Downie. 2011. The rules of engagement in the legume-rhizobial symbiosis. Ann. Rev. Genet. 45: 119–144.

Oliveira, A. L., E. L. Canuto, S. Urquiaga, V. M. Reis and J. I. Baldani. 2006. Yield of micropropagated sugarcane varieties in different soil types following inoculation with diazotrophic bacteria. Plant Soil. 284: 23–32.

Oliveira, A. L. M., M. Stoffels, M. Schmid, V. M. Reis, J. I. Baldani and A. Hartmann. 2009. Colonization of sugarcane plantlets by mixed inoculations with diazotrophic bacteria. European J. Soil Biol. 45: 106–113.

Patten, C. L. and B. R. Glick. 2002. Role of *Pseudomonas putida* indole-acetic acid in development of the host plant root system. Appl. Environ. Microbiol. 68: 3795–3801.

Paulin, M. M., A. Novinscak, M. St-Arnaud, C. Goyer, N. J. DeCoste, J. P. Prive, J. Owen and M. Filion. 2009. Transcriptional activity of antifungal metabolite-encoding genes *phlD* and *hcnBC* in *Pseudomonas* spp. using qRT-PCR. FEMS Microbiol. Ecol. 68: 212–222.

Penrose, D. M. and B. R. Glick. 2003. Methods for isolating and characterizing ACC deaminase-containing plant growth-promoting rhizobacteria. Physiol. Plantarum. 118: 10–15.

Raaijmakers, J. M. and M. Mazzola. 2012. Diversity and natural functions of antibiotics produced by beneficial and plant pathogenic bacteria. Annu. Rev. Phytopathol. 50: 403–424.

Raaijmakers, J. M., D. M. Weller and L. S. Thomashow. 1997. Frequency of antibiotic-producing *Pseudomonas* spp. in natural environments. Appl. Environ. Microbiol. 63: 881–887.

Richardson, A., J.-M. Barea, A. McNeill and C. Prigent-Combaret. 2009. Acquisition of phosphorus and nitrogen in the rhizosphere and plant growth promotion by microorganisms. Plant Soil. 321: 305–339.

Richardson, A. E. and R. J. Simpson. 2011. Soil microorganisms mediating phosphorus availability update on microbial phosphorus. Plant Physiol. 156: 989–996.

Rodríguez, H. and R. Fraga. 1999. Phosphate solubilizing bacteria and their role in plant growth promotion. Biotechnol. Adv. 17: 319–339.

Rodriguez, H., R. Fraga, T. Gonzalez and Y. Bashan. 2006. Genetics of phosphate solubilization and its potential applications for improving plant growth-promoting bacteria. Plant and Soil. 287: 15–21.

Romheld, V. and H. Marschner. 1986. Evidence for a specific uptake system for iron phytosiderophores in roots of grasses. Plant Physiol. 80: 175–180.

Rroço, E., H. Kosegarten, F. Harizaj, J. Imani and K. Mengel. 2003. The importance of soil microbial activity for the supply of iron to sorghum and rape. Eur. J. Agron. 19: 487–493.

Ryan, P. R., Y. Dessaux, L. S. Thomashow and D. M. Weller. 2009. Rhizosphere engineering and management for sustainable agriculture. Plant Soil. 321: 363–383.

Sahin, F., R. Cakmakci and F. Kantar. 2004. Sugar beet and barley yields in relation to inoculation with N(2)-fixing and phosphate solubilizing bacteria. Plant Soil. 265: 123–129.

Salman, M. and R. Abuamsha. 2012. Potential for integrated biological and chemical control of damping-off disease caused by *Pythium ultimum* in tomato. Biocontrol. 57: 711–718.

Sevilla, M., R. H. Burris, N. Gunapala and C. Kennedy. 2001. Comparison of benefit to sugarcane plant growth and $^{15}N_2$ incorporation following inoculation of sterile plants with *Acetobacter diazotrophicus* wild-type and *Nif⁻* mutant strains. Mol. Plant Microbe. In. 14: 358–366.

Singh, J. S., V. C. Pandey and D. P. Singh. 2011. Efficient soil microorganisms: A new dimension for sustainable agriculture and environmental development. Agr. Ecosys. Environ. 140: 339–353.

Somers, E., J. Vanderleyden and M. Srinivasan. 2004. Rhizosphere bacterial signalling: A love parade beneath our feet. Crit. Rev. Microbiol. 30: 205–240.

St-Onge, R., V. J. Gadkar, T. Arseneault, C. Goyer and M. Filion. 2011. The ability of *Pseudomonas* sp. LBUM 223 to produce phenazine-1-carboxylic acid affects the growth of *Streptomyces scabies*, the expression of thaxtomin biosynthesis genes and the biological control potential against common scab of potato. FEMS Microbiol. Ecol. 75: 173–183.

Thaweenut, N., Y. Hachisuka, S. Ando, S. Yanagisawa and T. Yoneyama. 2011. Two seasons' study on *nifH* gene expression and nitrogen fixation by diazotrophic endophytes in sugarcane (*Saccharum* spp. hybrids): expression of *nifH* genes similar to those of rhizobia. Plant Soil. 338: 435–449.

Tilman, D. 1998. The greening of the green revolution. Nature. 396: 211–212.

United Nations Department of Economic and Social Affairs, P. D. 2011. World population prospects: the 2010 revision, highlights and advance tables. working paper no. ESA/P/WP.220.

Urquiaga, S., K. H. S. Cruz and R. M. Boddey. 1992. Contribution of nitrogen-fixation to sugarcane-N-15 and nitrogen-balance estimates. Soil Sci. Soc. Amer. J. 56: 105–114.

Valverde, A., A. Burgos, T. Fiscella, R. Rivas, E. Velazquez, C. Rodriguez-Barrueco, E. Cervantes, M. Chamber and J. M. Igual. 2006. Differential effects of coinoculations with *Pseudomonas jessenii* PS06 (a phosphate-solubilizing bacterium) and *Mesorhizobium ciceri* C-2/2 strains on the growth and seed yield of chickpea under greenhouse and field conditions. Plant Soil. 287: 43–50.

Vanrhijn, P. and J. Vanderleyden. 1995. The Rhizobium-Plant Symbiosis. Microbiol. Rev. 59: 124–142.

Vansuyt, G., A. Robin, J.-F. Briat, C. Curie and P. Lemanceau. 2007. Iron acquisition from Fe-pyoverdine by Arabidopsis thaliana. Mol. Plant Microb. In. 20: 441–447.

Vessey, J. K. 2003. Plant growth promoting rhizobacteria as biofertilizers. Plant Soil. 255: 571–586.

Vitousek, P. M., H. A. Mooney, J. Lubchenco and J. M. Melillo. 1997. Human domination of Earth's ecosystems. Science. 277: 494–499.

Weller, D. M., B. B. Landa, O. V. Mavrodi, K. L. Schroeder, L. De La Fuente, S. B. Bankhead, R. A. Molar, R. F. Bonsall, D. V. Mavrodi and L. S. Thomashow. 2007. Role of 2,4-

diacetylphloroglucinol-producing fluorescent *Pseudomonas* spp. in the defense of plant roots. Plant Biol. 9: 4–20.

Weller, D. M., J. M. Raaijmakers, B. B. M. Gardener and L. S. Thomashow. 2002. Microbial populations responsible for specific soil suppressiveness to plant pathogens. Annu. Rev. Phytopathol. 40: 309–348.

Whipps, J. M. 2001. Microbial interactions and biocontrol in the rhizosphere. J. Exp. Bot. 52: 487–511.

Whitelaw, M. A. 1999. Growth promotion of plants inoculated with phosphate-solubilizing fungi. In *Advances in Agronomy*, edited by L. S. Donald: Academic Press. 99–151.

Wiyono, S., D. F. Schulz and G. A. Wolf. 2008. Improvement of the formulation and antagonistic activity of *Pseudomonas fluorescens* B5 through selective additives in the pelleting process. Biol. Control. 46: 348–357.

Yoneyama, T., T. Muraoka, T. H. Kim, E. V. Dacanay and Y. Nakanishi. 1997. The natural ^{15}N abundance of sugarcane and neighbouring plants in Brazil, the Philippines and Miyako (Japan). Plant Soil. 189: 239–244.

Yu, X., X. Liu, T. H. Zhu, G. H. Liu and C. Mao. 2012. Co-inoculation with phosphate-solubilzing and nitrogen-fixing bacteria on solubilization of rock phosphate and their effect on growth promotion and nutrient uptake by walnut. Eur. J. Soil Biol. 50: 112–117.

Zhang, H., Y. Sun, X. Xie, M. S. Kim, S. E. Dowd and P. W. Pare. 2009. A soil bacterium regulates plant acquisition of iron via deficiency-inducible mechanisms. Plant J. 58: 568–577.

Zhuang, X. L., J. Chen, H. Shim and Z. H. Bai. 2007. New advances in plant growth-promoting rhizobacteria for bioremediation. Environ. Inter. 33: 406–413.

10

Microbial Engineering for Developing Stress Tolerant Cultivars
An Innovative Approach

Vijai K. Gupta,[1,*] *Lallan P. Yadava,*[2,*] *Anthonia O' Donovan,*[1] *Gauri D. Sharma,*[3] *Maria G. Tuohy*[1] and *Sanjiv K. Maheshwari*[4]

ABSTRACT

The effect of biotic and abiotic stresses in crop production is considerable. Abiotic stress encompasses many environmental factors. Amongst the major abiotic stresses are oxidative stress and water deficit.

Excess salinity is a major problem for agriculture in dry parts of the world. Scientists have used biotechnology to develop plants with enhanced tolerance to salty conditions. Revelation to high concentrations

---

[1] Molecular Glycobiotechnology Group, Department of Biochemistry, School of Natural Sciences, National University of Ireland Galway, Galway, Ireland.
[2] Genetics & Plant Breeding Department, Central Institute for Subtropical Horticulture, Rahmankhera, P.O. Kakori, Lucknow, UP, India.
[3] Bilaspur University, Chattisgarh, India.
[4] Department of Biotechnology, College of Engineering & Technology IFTM University, Moradabad, UP, India.
* Corresponding authors: vijai.gupta@nuigalway.ie, drlpy.spa@gmail.com

of environmental NaCl exerts two stress effects on living cells; increasing the osmotic pressure and the concentration of inorganic ions. Salt stress dramatically suppresses the photosynthetic activity in cells of phototrophic organisms, such as cyanobacteria. During salt adaptation, cyanobacterial cells accumulate osmoprotectors, export excessive Na$^+$ with the help of Na$^+$/H$^+$ antiporters, and actively absorb K$^+$ with the help of K$^+$-transporting systems. These physiological processes are accompanied by induction or suppression of several genes involved in salt adaptation. Researchers have noticed that plants with high tolerance to salt possess naturally high levels of a substance called glycine betaine. Genetically modified plants with enhanced glycine betaine production have increased tolerance to salty conditions.

The exploitation of microbial genes to alter the function of gene products in transgenic plants provides novel opportunities to assess their biological role in a stress response. This review considers the main mechanisms responsible for the resistance of microbial cells to salt and hyperosmotic stresses. Special emphasis is placed on recent achievements in studying the genetic control of salt resistance and regulation of gene expression during adaptation of cyanobacteria to salt and hyperosmotic stresses. The genetic transformation using microbes provides an additional tool for crop breeders who wish to introduce value-added traits into the plant species cultivars that serve as vital food sources and a means of generating export income for producing nations.

Introduction

The immense challenge of food security faced worldwide has directed researchers and scientists towards genetic engineering. Microbial gene engineering, involves modification of qualitative and quantitative traits in an organism by transferring desired genes from one species to another. This stratagem is referred to as the transgenic advance. In contrast to conventional breeding, the transgenic approach allows the incorporation of only the specific cloned genes into an organism and restricts the transfer of undesirable genes from the donor organism. Through microbial engineering, pyramiding of genes with similar effects can also be achieved. Rapid advances in recombinant-DNA technology and development of precise and proficient gene-transfer protocols have resulted in efficient transformation and generation of transgenic lines in a number of crop species (Gosal et al. 2009). Recent studies reported several miRNAs associated with abiotic stress responses. Recognizing abiotic stress-associated microRNAs (miRNAs) and understanding their function will help develop new approaches for improvement of plant stress tolerance (Barrera-Figueroa et al. 2012, Saini et al. 2012). It is well known that miRNAs are also important in regulating plant-microbe interaction during nitrogen fixation by *Rhizobium* and tumor formation by *Agrobacterium*.

Stress, notably extremes in temperature, photon irradiance, and supplies of water and inorganic solutes, frequently limit growth and productivity of major crops growing under either natural or cultivated conditions. Stress may be even more severe for crops, including forest trees, planted in marginal areas. Placid stresses are likely to occur frequently and result in losses in growth.

Oxidative stress occurs when plants are exposed to various forms of environmental stress (Krause 1994). World crop production is limited largely by environmental stress (Blum 1988). For example, Dudal (cited in Blum 1988) estimated that only 10% of the world's arable land is free from some form of stress. The main factor responsible for the difference between potential and actual yield is environmental stress (Cardwell 1982). Thus, while a specific genetic solution may not be appropriate or available for all situations, the incorporation of stress resistance must be part of the primary goal of all breeding programmes while potential and actual yields are significantly separated.

Growth and development of plants are adversely affected by a range of external factors. These include factors that can be grouped together as temperature stresses (chilling, heat and freezing), which in turn belong to a larger subgroup that can be categorised as stresses that result from water deficit. In this chapter we also emphasise the point that most abiotic and biotic stresses directly or indirectly lead to the production of free radicals and reactive oxygen species (ROS), creating oxidative stress. These stress factors include, most notably, desiccation, nutrient deprivation, exposure to ultraviolet radiation (UVR), and oxidative stress encountered both from exposure to UV-A radiation and from the plant defence response. Severe stresses, however, can lead to catastrophic losses. Breeders have long addressed such problems, normally selecting for stability of performance over a range of environments, using extensive testing and an intricate biometrical approach. This traditional approach to breeding is becoming limited and new methods are required. Increased understanding of how the interaction of chemical and physical environments reduces plant development and yield opens the door to a combination of breeding, physiological and biotechnological approaches to plant modification in a comprehensive strategy for improving resistance to environmental stress. This chapter will consider three types of stresses: water stress, which will be considered mainly in terms of drought stress; temperature stress; and salt stress. Each of these three stresses represents a broad topic with extensive literature and a wealth of experimental detail. All have common components in the sense of having mechanisms for stress tolerance and avoidance. However, there is no single mechanism by which multiple stresses are alleviated. All indications are that the ability to withstand stressful environments is controlled by a number of genes, and this multigenic character imposes limits to subsequent manipulation.

Genetic engineering involves direct changes in DNA (see Box 1) by addition of genes from micro-organisms, which may be known as microbial engineering. Microbial engineering has many applications, such as medical application, industrial application, environmental application and agricultural application. Recently a new application of genetic engineering has been developed, a boon for agricultural industry in the form of transgenic plants (see Box 1). Transgenic plants carry desirable traits like disease resistance, insect resistance, herbicides resistance and stress tolerance. This, eventually, may also be used for increasing photosynthetic efficiency, nitrogen fixing ability, improved storage protection, hybrid crops, crops for food processing and molecular farming. In this chapter, we will look at the effects of environmental conditions such as temperature, water availability and salinity on crop plants.

Sundin (2003) reported that growth and survival of *Pseudomonas syringae* in association with both the external and internal leaf environment is correlated with an ability to tolerate and/or escape environmental and plant-associated stresses. Information on particular physiological and genetic responses of *P. syringae* to the environmental stress associated with leaf colonization has increased in recent years. The ability to survive on dry leaf surfaces is a critical ecological adaptation of phyllosphere microbes. Several *P. syringae* traits including exopolysaccharide production, motility, and even methionine biosynthesis, have been shown to increase survival under dry or alternating wet/dry conditions. The UV-B component of solar UVR causes direct DNA damage and cell death; thus, enzymatic mechanisms of DNA repair are essential for the survival of leaf surface-associated *P. syringae*. The rulAB determinant encodes a mutagenic DNA repair system that confers UVR tolerance to *P. syringae* and is required for enhanced survival on UVR-exposed leaf surfaces. rulAB is plasmid-encoded and widely distributed among *P. syringae* pathovars. The bacterial RpoS alternate sigma factor regulates genes in response to oxidative stress including genes encoding superoxide dismutase. Surprisingly, a *P. syringae* pv. syringae B728a sod A sod B mutant exhibited increased UV-A sensitivity but no change in virulence in bean. Increasing knowledge of stress factors and the responses of *P. syringae* to stress will continue to foster a better understanding of *P. syringae* biology and the potential for disease management.

Induced systemic resistance (ISR) by plant-associated bacteria was initially demonstrated using *Pseudomonas* spp. and other Gram-negative bacteria. Several reviews have summarized various aspects of the large volume of literature on *Pseudomonas* spp. as elicitors of ISR. Fewer published accounts of ISR by *Bacillus* spp. are available, and we review this literature for the first time. Published results are summarized showing that specific strains of the species *B. amyloliquefaciens, B. subtilis, B. pasteurii, B. cereus, B. pumilus, B. mycoides,* and *B. sphaericus* elicit significant reductions in the incidence or severity of various diseases on a diversity of hosts. Elicitation of ISR by these strains has been demonstrated in greenhouse or field trials on tomato, bell pepper, muskmelon, watermelon, sugar beet, tobacco, *Arabidopsis thaliana,* cucumber, loblolly pine, and 2 tropical crops (long cayenne pepper and green kuang futsoi). Protection resulting from ISR elicited by *Bacillus* spp. has been reported against leaf-spotting fungal and bacterial pathogens, systemic viruses, a crown-rotting fungal pathogen, root-knot nematodes, and a stem-blight fungal pathogen as well as damping-off, blue mold, and late blight diseases. Reductions in populations of three insect vectors have also been noted in the field: striped and spotted cucumber beetles that transmit cucurbit wilt disease and the silver leaf whitefly that transmits Tomato mottle virus. In most cases, *Bacillus* spp. that elicit ISR also elicit plant growth promotion. Studies on mechanisms indicate that elicitation of ISR by *Bacillus* spp. is associated with ultrastructural changes in plants during pathogen attack and with cytochemical alterations. Investigations into the signal transduction pathways of elicited plants suggest that *Bacillus* spp. activate some of the same pathways as *Pseudomonas* spp. and some additional pathways. For example, ISR elicited by several strains of *Bacillus* spp. is independent of salicylic acid but dependent on jasmonic acid, ethylene, and the regulatory gene NPR1—results that are in agreement with the model for ISR elicited by *Pseudomonas* spp. However, in other cases, ISR elicited by *Bacillus* spp. is dependent on salicylic acid and independent of jasmonic acid and NPR1. In addition, while ISR by *Pseudomonas* spp. does not lead to accumulation of the defence gene PR1 in plants, in some cases, ISR by *Bacillus* spp. does (Kloepper et al. 2004).

In several cases, scientists have used biotechnology to develop plants with enhanced tolerance to salty conditions. Revelation to high concentrations of environmental NaCl exerts two stress effects on living cells; increasing the osmotic pressure and the concentration of inorganic ions. Salt stress dramatically suppresses the photosynthetic activity in cells of phototrophic organisms, such as cyanobacteria. During salt adaptation, cyanobacterial cells accumulate osmoprotectors, export excessive Na^+ with the help of Na^+/H^+ antiporters, and actively absorb K^+ with the help of K^+-transporting systems. These physiological processes are accompanied by induction or suppression of several genes involved in salt adaptation. This review considers the main mechanisms responsible for the resistance of cyanobacterial cells to salt and hyperosmotic stresses. Special emphasis is placed on recent achievements in studying the genetic control of salt resistance and regulation of gene expression during adaptation of cyanobacteria to salt and hyperosmotic stresses. Another approach to engineering salt tolerance uses a protein that takes excess sodium and diverts it into a cellular compartment where it does not harm the cell. Abiotic stress may affect plant response to pathogen attack through induced alterations in growth regulator and gene expression. Abscisic acid (ABA) mediates several plant responses to abiotic stress.

Contemporary breeding strategies have clearly been successful in incorporating degrees of tolerance in a variety of species. In this chapter these successes will be considered along with the current advances in microbial biotechnologies. The new methods of gene identification and isolation, the ability to transfer genes within and between species by various transformation schemes, and selection in microbial engineering will be evaluated for their potential to contribute to the overall management of environmental stress in both the short and long term.

Box 1 DNA and transgenic plants

DNA-Deoxyribonucleic acid is the information carrying hereditary material that comprises genes. A gene is a hereditary determinant of a specific biological function, i.e., a unit of inheritance (DNA) located in a fixed place on the chromosome.

Transgenic plants are those plants in which a foreign gene (s) has/have been introduced and stably integrated into the host DNA. It results in the synthesis of appropriate gene product by the transformed plants.

Abiotic Stress

Basic fundamentals of plant physiology show changes when plants are subjected to many types of oscillation in their physical environment. Therefore you must first try to define stress on the basis of plant physiology before discussing the subject of tolerance to stress. Plants are able to cope with normal variation by virtue of their plasticity. Thereby, they are adapted to function in a fluctuating environment, and normal outdoor changes are countered by internal changes without detriment to growth and development. High temperature and high photon irradiance often accompany low water supply, which can in turn be exacerbated by subsoil mineral toxicities that constrain root growth. Furthermore, one abiotic stress can decrease a plant's ability to resist a second stress. For example, low water supply can make a plant more susceptible to damage from high irradiance due to the plant's reduced ability to reoxidize NADPH and thus maintain an ability to dissipate energy delivered to the photosynthetic light-harvesting reaction centers. If a single abiotic stress is to be identified as the most common in limiting the growth of crops worldwide, it most probably is low water supply (Araus et al. 2002). However, other abiotic stresses, notably salinity and acidity, are becoming increasingly significant in limiting growth of crop plants. The other type of damage occurs from the production of reactive oxygen species (ROS), causing chemical damage to the cellular constituents of plants. Both abiotic and biotic stresses cause the production of ROS either directly, or indirectly.

Plants can produce antioxidants for protection against the cytotoxic species of activated oxygen such as superoxide (O_2^-), hydrogen peroxide (H_2O_2), and the hydroxyl radical (OH^-). Studies of the effects of environmental stresses (i.e., high NaCl levels (Gossett et al. 1994a, 1994b), drought (Dhindsa and Matowe 1981, Burke et al. 1985, McCue and Hanson 1990), temperature and light extremes (Rabinowitch and Fridovich 1983, Wise and Naylor 1987b, Spychalla and Desborough 1990, Baker 1994, Rainwater et al. 1996), mineral deficiencies (Monk and Davies 1989, Cakmak and Marschner 1992, Polle et al. 1992), and herbicide treatment (Harper and Harvey 1978, Dodge 1994)) have shown that oxidative stress often disrupts the homeostasis between production of ROS and the quenching activity of antioxidant enzymes. Expressing candidate genes in specific root tissues such as the outer cell layers also seems to be a promising technique to enhance abiotic stress tolerance (Ghanem et al. 2011).

Temperature Stress

Any temperature beyond the optimum for growth and development may be measured as a stress temperature. The definition of a stress temperature will clearly depend on the stage of growth, and the effects of such stresses will not be equal at all stages of growth. Nearly all stem perturbations occur at the extremes of the growing seasons, namely at germination and at fruit formation. It is at these two extremes that much of the concentration in developing resistance has occurred. Three different types of temperature stresses will be considered, heat stress, chilling stress and freezing stress. Perception at the physiological and molecular levels is probably greatest for heat stress, although for all three stresses the basic information is still meager.

Tolerance to Heat Stress

Limiting plant growth caused by high temperatures is a major environmental factor in subtropical and tropical areas (Radin et al. 1994). At extremes, such stress inhibits photosynthesis, limits carbohydrate accumulation, damages cell membranes and leads to death of cells (Liu and Haung 2000), organs or whole plants. The reproductive process is most sensitive to heat, with the resulting flower abscission, pollen sterility and/or poor fruit set being responsible for large yield reductions. However, even sublethal heat stress, which increases the rate of plant development, can also result in a loss of yield. Thus thermal tolerance must be viewed in terms of the degree of stress and the duration of exposure, as well as in relation to the stage of development at which the stress occurs.

Upon exposure to the stresses, many genes are induced and their products are thought to function as cellular protectants of stress-induced damage (Bray 1997, Thomashow 1999, Shinozaki and Yamaguchi-Shinozaki 2000). Genetic variation in heat tolerance for various attributes exists in a number of crop plants. For example, genetic variation has been demonstrated in tolerance at germination (Zeng and Khan 1984, Soman and Peacock 1985), for growth under heat stress (Mendoza and Estrada 1979, Stamp et al. 1983, Shpiler and Blum 1986), for recovery from heat stress (Coffman 1957), in flowering and fruit or seed set, and in the functions of photosynthesis, translocation and the stability of membranes. The genetic resources for heat tolerance are apparently abundant in a number of crop species such as rice, potatoes, soybean and tomato (Rose et al. 2003), but for a number of other crop plants the extent of accessible genetic variation has scarcely been explored.

To a great extent molecular characterization has been directed towards the heat-shock response in terms of the heat-shock proteins (HSPs). Based on their molecular weight, the major HSPs synthesized by plants belong to five classes: HSP100, HSP90, HSP70, HSP60 (kDa) and small HSPs ($\approx$17 to 30 kDa) (Vierling 1991). This heat-shock response is one of the most highly conserved biological responses known, being present in organisms ranging from bacteria to man. In general, the heat-shock response is characterized by a number of steps which include:

1. a reduction in normal protein synthesis;
2. induction of a new set of mRNAs and the translation of these RNAs into the HSPs;
3. amassing of large amounts of HSP;
4. acquirement of thermo tolerance to otherwise lethal temperatures;
5. gradual decline in HSPs synthesis and a return to normal protein synthesis during prolonged heat treatments.

As mentioned above, plant breeders have been able to manipulate thermal tolerance as an important and heritable agronomic trait. However, the relationship between the thermal tolerance observed in cultivars and the transient heat-shock response in the laboratory is still unclear. It has been established that certain crop species growing in the field can exhibit the synthesis of heat-shock mRNAs (Kimpel and Key 1985) and proteins (Vierling et al. 1988 and 1991, Nguyen et al. 1994, Al-Neimi and Stout 2002, Al-Whaibi 2010) under conditions of heat stress. However, Edelman and Key (cited in Nagao et al. 1990) did not detect qualitative differences in the profiles of heat-shock response in a comparison of soybean seedlings, which show heritable differences in their field thermal tolerance (Nagao et al. 1990), whereas a positive correlation between the degree of thermal tolerance and the contents of low-molecular-mass HSPs was found in wheat (Nguyen et al. 1994). Thereby, it may be concluded that the precise role of the HSPs in the generation of thermal tolerance is still vague, but a subject of intense study.

Tolerance to Cold Stress

Globally, low temperature also is a major limitation of plant growth, and this has a major impact on grasses via, for example, vernalization and low temperature damage at anthesis. Tolerance to low temperatures can be divided arbitrarily into tolerance to chilling and tolerance to freezing. As an operational definition, chilling tolerance is that which comes into play at temperatures above 0°C, whereas freezing tolerance is that exhibited below 0°C. In both cases one of the primary sites of damage by low temperatures is thought to be the membrane.

Chilling stress

Chilling stress is probably the most common environmental stress during germination. The low temperature at germination not only reduces the germination rate, but also affects the subsequent growth. Chilling can also be important during flowering, with consequences such as deformed fruit or the failure to set fruit or seed. Genetic variation in chilling tolerance has been demonstrated for various physiological processes such as photo-synthesis and plasma membrane function, as well as for whole plant processes such as germination, growth and seed set. In practical terms, most information is available for germination and seedling growth. Genetic resources for chilling tolerance are to be found at the margins of a sensitive species range. Ample genetic variation has been found in maize (Mock and Eberhart 1972), while in tomato the introgression of chilling tolerance from exotic germplasm has been successful (Patterson and Payne 1983).

Freezing resistance

Plants are known to differ in their ability to withstand freezing temperatures, but the molecular and genetic bases of this differential are unknown. However, it is known that a prior exposure to low non-freezing temperatures (cold acclimation) increases the tolerance to subsequent freezing. Tolerance to freezing stress has been studied in a wide range of plants, with the most detailed knowledge of the inheritance of the trait coming from wheat.

Studies, made by Nilsson-Ehle, 1912 and Thomashow, 1990, have concluded that frost hardiness is a quantitative trait controlled by several genes. In addition to the number of genes involved, the interactions between the genes can be complex. Outside wheat, frost hardiness has also been studied in oats and barley, as well as in a variety of woody plants including fruit trees, tea plants, pines and roses (Sakai and Larcher 1987). Generally the results indicate a complex interaction of many genes. Frost tolerance of tea stems and winterkill of roses may be exceptions, where only a few genes may be involved.

A large number of biochemical changes take place in a cold-acclimated plant, but exactly which are responsible for the increased tolerance is not known. Much attention has been focused on the plasma membrane since disruption of cellular membranes has been regarded as the primary cause of freezing injury in plants. A characterization of the differences in biophysical properties of plasma membranes from cold-acclimated and non-acclimated plants suggests that the biochemical composition of the membrane is altered during cold hardening. Efforts are being directed to determining a causal relationship between specific membrane changes and freezing tolerance. The results of Steponkus et al. (1988) strongly support the view that changes in plasma lipid composition are causally related to increased tolerance to expansion-induced lysis. However, the overall increase in freezing tolerance of membranes probably involves multiple alterations, each having different effects.

The changes in membranes are only one part of a series of biochemical alterations associated with cold acclimation. Increases in superoxide dismutase activity have been reported to play a role in tolerance to cold in potatoes (*Solanum tuberosum* L.) (Spychalla and Desborough 1990) and maize (*Zea mays* L.), to paraquat in ryegrass (*Lolium perenne* L.) (Harper and Harvey 1978), to salt in chick pea (*Cicer arietinum* L.) (Hernandez et al. 1994) and cotton (Gossett et al. 1994a), and resistance to paraquat in tobacco (*Nicotiana tabacum* L.) (Shaaltiel et al. 1988) and to chilling in spinach (*Spinacea oleracea* L.) (Schoner and Krause 1990).

The appearance of new isozymes and an increase in proline and organic acids, as well as sugar and soluble proteins, also occur. The direct causal relationship between these changes and low-temperature survival is still unclear. Specific changes in gene expression have been demonstrated by the identification of cold-regulated genes. The expression of these genes with the synthesis of their polypeptide products closely parallels freezing tolerance in plants (Thomashow 1990). The current state of knowledge of the molecular genetics of cold acclimatization is unsatisfactory. However, it is clear that the response is distinct from the heat-shock response in terms of the identities of new proteins synthesized and the length of time over which these proteins are synthesized. Much more will have to be learned before a biotechnological manipulation can be considered for this property, and conventional breeding programmes will have to suffice.

Tolerance to Water Deficit Stress

Drought stress

Drought is one of the major abiotic stresses limiting crop productivity around the world. By definition drought is a water deficit stress because the environmental conditions either reduce the soil water potential and/or increase the level of water potential due to hot, dry or windy conditions. During physiological stress, the production of reactive oxygen species is known to increase to such an extent that the normal levels of antioxidant enzymes are not sufficient to prevent the damaging effects of the highly reactive radicals. Cell damage and a reduction in growth rate result. Numerous studies have shown a positive correlation between the activity of antioxidant enzymes and stress caused by temperature extremes, drought, ultraviolet light, mineral deficiencies, herbicide treatment, and excess salt.

Both drought and heat cause water deficiency in plants (Ingram and Bartels 1996). The loss of plant productivity caused by drought stress is often more serious when combined with heat stress (Vanjildorj et al. 2006). Resistance to drought is a complex phenomenon in both its physiology and its genetics. The essential requirement for tolerance to drought is the ability of the plant to continue to function under conditions of water limitation. This can be done by drought avoidance where a short growing season can function as an attribute of drought escape. Such a scenario clearly depends on the environment over the growing season as to whether or not such a mechanism is useful. However, the relative yield advantage of early-maturing genotypes under conditions of water stress has been exploited (Reitz 1974). An alternative to drought avoidance is dehydration avoidance, which is the ability of a plant to maintain sufficient tissue hydration for appropriate functioning of the metabolic processes involved in growth and development under conditions of environmental water stress. Upon exposure to the stresses, many genes are induced and their products are thought to function as cellular protectants of stress-induced damage (Thomashow 1999, Shinozaki and Yamaguchi-Shinozaki 2000).

The role of abscisic acid (ABA) in drought resistance is likely to be crucial. ABA concentration can affect many processes in addition to its possible effect on yield under water stress conditions (Milborrow 1981). It should be noted that some of the cold-regulated genes described in the previous section are also induced by both ABA and drought stress. A small family of ABA-responsive elements (ABRE)-binding bZIP proteins, named as ABRE binding factors (ABFs), has been isolated (Uno et al. 2000). The ABFs interact with the ABRE in the promoter regions of ABA-regulated genes and mediate ABA dependent stress signaling during vegetative growth in *Arabidopsis*. Numerous studies have demonstrated that the plant hormone ABA is essential for the adaptive response to water stress imposed by drought (Xiong et al. 2002) and also that ABA (Box 2) is necessary for protection against high temperature (Gong et al. 1998, Larkindale and Knight 2002).

Genetic variations in the components of drought tolerance have been observed and manipulated. For example, a weak association between pro-line accumulation and yield under stress was observed for *Brassica* spp. (Richards and Thurling 1979), but a natural mutant of barley which accumulated proline did not show an alteration in the plant's water status. Thus, as with many of the components of stress tolerance, it is not just the alteration of a characteristic, but how that alteration occurs in response to a stress environment that is important in evaluating the role of that particular response. Overexpression of gene ABF3, one of the ABF family members, enhances tolerance to drought and low/high temperature in *Arabidopsis* (Kang et al. 2002, Kim et al. 2004), lettuce (Enkhchimeg et al. 2005), increases tolerance to drought stress alone in rice (Oh et al. 2005) and also increases tolerance to drought and heat stresses in bentgrass (*Agrostis mongolica* Roshev) (Vanjildorj et al. 2006).

Salt Stress

Over the past two decades increasing attempts have been made to develop salt-resistant varieties able to grow under saline conditions. However, in spite of this effort, there are few instances of crop cultivars that have been bred for salinity resistance and used as an economic solution in saline eco-systems. Saline fields are inherently variable in their salt distribution, so that most of the yield from such areas is from the non-saline patches, with little from the saline patches. Thus the question arises as to the appropriate strategy for dealing with such a distribution; whether it is better to breed for high yields for the unstressed patches, or for lower yields (Blum 1988).

The effects of salinity on plant processes are threefold, in terms of water stress from the osmotic effects, mineral toxicity of the salt and interruptions to the mineral nutrition of the plant. The multipartite nature of the stress results in difficulty in predicting the extent of stress in a saline environment.

The mechanisms by which plants adapt to saline conditions are typically unique to halophytes and have evolved over a considerable period of time. The lessons learned from a study of the integrated control of halophytes may not be immediately applicable to crop breeding programmes in a piecemeal fashion. Thus, genetic material can be found in saline habitats from which useful salt tolerance may be exploited, but a great deal of additional physiological and genetic characterization needs to be done.

Soil salinity affects plant growth and development by way of osmotic stress, injurious effects of toxic Na^+ and Cl^- ions and to some extent Cl^- and SO_4^{2-} of $Mg2^+$ and nutrient imbalance caused by excess of Na^+ and Cl^- ions. Salinity stress response is multigenic, as a number of processes involved in the tolerance mechanism are affected, such as various compatible solutes/osmolytes, polyamines, reactive oxygen species and antioxidant defence mechanism, ion transport and compartmentalization of injurious ions. Various genes/cDNAs encoding proteins involved in the above-mentioned processes have been identified and isolated. The role of genes/cDNAs encoding proteins involved in regulating other genes/ proteins, signal transduction process involving hormones like ABA, JA and polyamines, and strategies to improve salinity stress tolerance have also been discussed. The problem of osmoregulation has received a great deal of attention. Although the role of proline accumulation in response to a variety of environmental stresses is not clear, proline is an important solute in osmoregulation in halophytes (Stewart and Lee 1974). The roles of individual solutes in osmoregulation of glycophytes are much less clear, although proline (Stewart and Lee 1974), myoinositol (Sacher and Staples 1985) and glycine betaine (Wynn Jones 1981) have all been implicated. The root system is also likely to play an important part in salt tolerance. An increased development of the root system would provide a mechanism to meet the high transpirational demands at low water potential, as well as allow the exploitation of the variability in the soil salt content.

The molecular aspects of salt tolerance are still in the early stages of investigation. It is clear that the profile of proteins synthesized in response to saline stress differs from that in the absence of stress (Ramagopal 1987). However, the role of any of these proteins in a causal relationship to any tolerance that develops is still not confirmed. A major gene which will control salt tolerance is unlikely in view of the continuous nature of tolerance, although the development of tolerant tissue culture lines suggests that such a search may not be hopeless.

Strategies for Improving Stress Tolerance

Recent advances in plant genome mapping and molecular biology techniques offer a new opportunity for understanding the genetics of stress-resistance genes and their contribution to plant performance under stress. These biotechnological advances will provide new tools for breeding in stress environment. Molecular genetic maps have been developed for major crop plants, including rice, wheat, maize, barley, sorghum and potato, which make it possible for scientists to tag desirable traits using known DNA landmarks. Molecular genetic markers allow breeders to track genetic loci controlling stress resistance without having to measure the phenotype, thus reducing the need for extensive field-testing over time and space. Moreover, gene pyramiding or introgression can be done more precisely using molecular tags. Together, molecular genetic markers offer a new strategy known as marker-assisted selection. Another molecular strategy which depends on gene cloning and plant transformation technology, is genetic engineering of selected genes into elite breeding lines. What makes a particular goal attainable or unattainable in genetic engineering experiments is the availability of the following three inputs: (i) the gene of interest, (ii) an effective technique for transferring the desired gene from one species to another and (iii) promoter sequences for regulated expression of that gene. Amongst these, the first is considered a rate-limiting factor. Arrays of stress-induced genes have been isolated. Stress-responsive genes can be analysed following targetted or non-targetted strategy. The targetted approach relies upon the availability of relevant biochemical information (i.e., in terms of defined enzyme, protein, a biochemical reaction or a physiological phenomenon). The non-targetted strategy to obtain a desired gene is indirect. This strategy, for instance, includes differential hybridization and shotgun cloning. The list of genes whose transcription is upregulated in response to stress, is rapidly increasing. Understanding of the mechanisms which regulate gene expression and the ability to transfer genes from other organisms into plants, will expand the ways in which plants can be utilized. To exploit the full potential of these approaches, it is essential that the knowledge is applied to agriculturally and ecologically important microorganism as well as plant species. Different plant species tend to produce different compatible solutes. For example, mannitol is produced in large amounts by celery in response to salt stress, whereas spinach accumulates glycine betaine.

CASE STUDY 1 Glycine betaine production

A quaternary ammonium compound found in at least 10 flowering plants and marine algae, is extemely soluble in water and is electrically neutral over a wide range of physiological pH values, is known as glycine betaine. Levels of glycine-betaine in *Poaceae* species are correlated with salt-tolerance. Highly tolerant *Spartina* and *Distichlis* accumulate the highest levels, moderately tolerant species accumulate intermediate levels and sensitive species accumulate low levels or no glycine-betaine (Rhodes et al. 1989). Glycine-betaine is synthesized from choline in two steps, the first being catalysed by choline monooxygenase leading to synthesis of betaine-aldehyde, which is further oxidized by betaine-aldehyde dehydrogenase. Salinity stress induces both the enzyme activities (Arakawa et al. 1990, Weretilnyk and Hanson

1990). Genetic evidence that glycine-betaine improves salinity tolerance has been obtained for barley and maize (Rhodes et al. 1989, Grumet and Hanson 1986). Isogenic barley lines containing different levels of glycine-betaine have different abilities to adjust osmotically. Transgenic rice plants expressing betaine-aldehyde dehydrogenase converted high levels of exogenously applied betainealdehyde to glycine-betaine than did wild-type plants. The elevated level of glycine-betaine in transgenic plants conferred significant tolerance to salt, cold and heat stress. Huang et al. (2000) reported metabolic limitation in betaine production in transgenic plants. *Arabidopsis thaliana*, *Brassica napus* and *Nicotiana tobbacum* were transformed with bacterial *Choline oxidase* cDNA. The levels of glycine betaine were 18.6, 12.8 and 13.0 μmol g^{-1} dry weight in *A. thaliana*, *B. napus* and *N. tobbacum*, respectively, 10–20 fold lower than the levels found in natural betaine producers. A moderate stress tolerance was noted in some transgenic lines based on relative shoot growth in response to salinity, drought and freezing. However, choline-fed transgenic plants synthesized substantially more glycine-betaine, suggesting that there is need to enhance the endogenous supply of choline to support accumulation of physiologically relevant amount of betaine.

Transformation of planta with the betaine aldehyde dehydrogenase genes from *E. coli* and *Arthrobacter globiformis* permitted the accumulation of glycine betaine when plants were supplied with betaine aldehyde, and conferred resistance to this toxic compound. In a number of cases where glycine betaine accumulation has been achieved, the tolerance to various water-deficit stresses has been determined, including salt, chilling, freezing, heat and drought (Table 3).

Gene Transfer

The development of transformation techniques for many of the major crop species over the past decade has opened many possibilities for developing new varieties incorporating novel genetic material which would not have been available to the traditional breeder. New techniques, such as restriction fragment length polymorphism (RFLP) mapping, may be used in conjunction with conventional breeding programmes to accelerate the process and to reduce the size of the populations required.

There are three parts to a successful strategy to genetically engineer new stress resistances. First is the isolation of appropriate genes of interest and utility. Second, the newly tailored gene or gene complex must be transferred back into the plant of interest. Finally, the novel plants have to be characterized as to their phenotype. The second and third processes can be done relatively efficiently, especially with the advent of the biolistic transformation schemes which have proved to be useful in many important crop species which had previously been intractable. However, the first prerequisite, the identification and isolation of the genes for important stress resistance traits, is likely to prove most difficult. As noted earlier, many of the resistances appear to have complex inheritance patterns, involving genes with unknown functions. Thus the isolation of the genes involved will be difficult. The present transformation schemes are also limited in the amount of information which can be transferred. Although the technology is very powerful, its use is likely to be limited by the quantity and quality of the biological knowledge that is available. Much more fundamental work is needed to understand the biochemical basis of stress resistance and to fully utilize the available technology.

One possible success of the technological approach is seen in the work on superoxide dismutase and catalase. Both of these enzyme systems are needed to protect cells against the damaging effects of active oxygen species, which may be produced during photorespiration (Halliwell 1984), chilling (Wise and Naylor 1987a) and dehydration (Dhindsa and Matowe 1981). Representative cDNA clones have been isolated for superoxide dismutase (Bowler et al. 1989) and catalase (Redinbaugh et al. 1988). The coupling of these genes to the appropriate regulatory sequences may well provide some measure of protection against a limited array of stressful environments, but is unlikely to be a generic solution. The functions of the many other stress-related polypeptides need to be known before their usefulness in the generation of stress-resistant transgenic material can be considered.

Enhancing Transformation

The frequencies at which transgenic plants have been selected and regenerated following *Agrobacterium* transformation vary considerably between plant species. These range from 100% of explants producing transgenic shoots in potato (De Block 1988), to less than 1% in sunflower (Schrammeijer et al. 1990) and asparagus (Conner et al. 1988). Marked variations in transformation frequencies are also common among genotypes of the same crop (Conner et al. 1991).

Identification of suitable *Agrobacterium* strain and plant genotype combinations is important for optimizing transformation rates (see above). The choice of selectable marker system, components of the plant culture media and environment, and explant tissue to be co-cultivated are all important. Since plant tissues often decrease in their tissue culture aptitude following co-cultivation, the cells which interact with *Agrobacterium* must have a high potential for growth and regeneration. Moloney et al. (1989) were able to obtain higher transformation rates than previously reported in *Brassica napus (55% vs. < 10%)* solely by the use of intact petioles from cotyledons instead of other explants. In some instances the co-cultivation of explants with *Agrobacterium* results in a severe hypersensitive response, which dramatically reduces shoot regeneration. By using agarose instead of agar, Charest et al. (1988) reduced tissue necrosis in *Brassica napus*, thereby allowing shoot regeneration and production of transgenic plants from thin layers of epidermal cells.

An essential component of the *Agrobacterium* system is the induction of the *vir* operons. A number of methods have been used to induce these operons, thereby enhancing transformation. For example, preincubation of *A. tumefaciens* with 20 µM acetosyringone increased the trans-formation frequency of *Arabidopsis thaliana* from 2–3 to 64% of explants producing transgenic plants (Sheikholeslam and Weeks 1987). In the monocotyledonous *Dioscorea bulbifera* (yam), *in vitro* tumour production was achieved only after induction of *A. tumefaciens* with wound exudates from potato tuber tissue (Schafer et al. 1987).

Using a cell suspension feeder layer during the co-cultivation period has been found to have beneficial effects on the transformation frequency in some species. For example, in tomato the proportion of explants producing transformation events increased 15% following co-cultivation on media with tobacco cells (Fillatti et al. 1987a). In lettuce, Michelmore et al. (1987) found the presence of a feeder layer, although not absolutely necessary, ensured repeatable high rates of transformation. The use of feeder cells, such as plated tobacco cell suspensions, may enhance transformation via two mechanisms: (i) inducing *Agrobacterium vir* genes (Veluthambi et al. 1988), and (ii) promoting cell division and growth of individual plant cells (Horsch and Jones 1980).

Agricultural Approaches

Plants transformed using *Agrobacterium* have been obtained in a wide range of crops (Table 1). In virtually all instances kanamycin resistance has been used as a selectable marker gene. The information presented in Table 1 reveals that *Agrobacterium*-mediated transformation has been highly successful. However, in many crops only one or very few transgenic plants have been produced (e.g., asparagus, broccoli and walnut). In other crops, considerable research is still required for the efficient production of transgenic plants in agriculturally useful genotypes. Sometimes *Agrobacterium*-mediated transformation of elite genotypes has only operated well in single or very few laboratories (e.g., lettuce, kale, cotton, clover). The methods are routine in solanaceous crops (e.g., potato, tomato and tobacco), with systems also being well developed in oilseed rape, soybean, alfalfa, and lotus.

Table 1. Examples of transgenic crop plants, containing foreign genes, produced via *Agrobacterium*-mediated transformation.

Vegetables	Arable crops	Ornamental and medicinal	Fruit and trees	Pasture crops
Potato (De Block 1988)	Sunflower (Everett et al. 1987)	Petunia (Horsch et al. 1985)	Pepino (Atkinson and Gardner 1991)	Alfalfa/Lucerne (Shahin et al. 1986)
Tomato (McCormick et al. 1986)	Sugarbeet (Gasser and Fraley 1989)	Tobacco (Horsch et al. 1985)	Apple (James et al. 1989; Bondt et al. 1994)	Lotus (Jensen et al. 1986)
Lettuce (Michelmore et al. 1987)	Oilseed rape (Fry et al. 1987)	Kalanchoe (Jia et al. 1989)	Tamarillo (Atkinson et al. 1990)	Stylosanthes (Manners and Way 1989)
Celery (Catlin et al. 1988)	Blackgram (Saini and Jaiwal 2007)	Chrysanthemum (Ledger et al. 1991)	Kiwifruit (Janssen and Gardner 1993)	White clover (White and Greenwood 1987)
Cucumber (Trulson et al. 1986)	Soybean (Hinchee et al. 1988)	Geranium (Butcher et al. 1990)	Walnut (McGranahan et al. 1988)	
Carrot (Scott and Draper 1987)	Cotton (Firoozabadi et al. 1987)	Lisianthus (Deroles et al. 1990)	Populus (Fillatti et al. 1987b)	
Cauliflower (Chakrabarty et al. 2002)	Flax/linseed (Basiran et al. 1987)	Sesbania drummondii (Padmanabhan and Sahi 2008)	Strawberry (Nehra et al. 1990)	
Broccoli (Christey and Earle 1989) Eggplant (11)	Kale (Christey and Sinclair 1990)		Citrus (Hidaka at al. 1990)	
Cabbage (Lee et al. 2004, Rafat 2008)	Mustard (Mathews et al. 1990)		Muskmelon (Tian et al. 2010)	
Asparagus (Conner et al. 1988)	Marula (Margaret and Goyvaerts 2005)		Azadirachta (Naina et al. 1989)	
Eggplant (Guri and Sink 1988)				
Pea (Puonti-Kaerlas et al. 1990)				

The main obstacle to efficient *Agrobacterium*-mediated transformation in many crops is combining selection of transformed cells and their subsequent regeneration. Although each of these components can be achieved independently in most crops, it is often difficult to obtain both *Agrobacterium* transformation and plant regeneration in the same cell. Genetic variation for sensitivity to *Agrobacterium* transformation (see above) and response in cell culture (Conner and Meredith 1989) are well known within crop species. Selection and breeding for enhanced regeneration from tissue culture have proved successful, even after only one or two generations of selection (Conner and Meredith 1989). In many crops it may be necessary to use a genotype specifically developed for transformation. It would then be necessary to incorporate newly introduced genes into elite genotypes or existing cultivars by conventional plant breeding techniques, although this would increase the time to cultivar release.

Despite these limitations considerable progress has already been made towards gene transfer of agriculturally important traits. This principally involves the development of plants with improved resistance to herbicides, insect pests and viral diseases. Many transgenic plants expressing agriculturally useful traits have already reached field testing stage (Table 2).

For the successful integration of *Agrobacterium*-mediated transformation into plant breeding programmes, procedures must be developed for efficient, large-scale selection of transformed cells and rapid regeneration of transgenic plants. Many different independently transformed lines can be screened for those that retain all or most of their previous influential traits, but in-corporate the desired genetic change. This is important for two main reasons.

1. The point of T-DNA integration can interrupt the functioning of existing genes in plant cells via insertional mutagenesis (e.g., Feldmann et al. 1989) and result in position effects on gene expression (see above). Since these events differ among independently transformed lines, plants exhibiting such effects can be discarded.
2. Genetic changes (somaclonal variation) are known to occur during the cell culture and plant regeneration phase associated with plant trans-formation. This variation is unrelated to the specific modification sought, and is accentuated by plant cell culture over prolonged periods. Rapid regeneration of many independently transformed lines will help reduce the frequency of such changes and allow selection of transgenic lines that are phenotypically normal when grown in the field.

The number of independently selected transformed lines that constitute a large collection is unknown, and will no doubt differ for the specific crop and the nature of its breeding system(s). For seed-propagated crops it may

Table 2. Examples of transgenic crop plants resulting from *Agrobacterium*-mediated gene transfer to have reached field testing stage of development.

Transgenic phenotype	Crop plants
Herbicide resistance	
Phosphinothricin	Tobacco, tomato, potato, oilseed rape, sugarbeet, alfalfa, poplar
Glyphosate	Tobacco, tomato, oilseed rape, flax, sugarbeet, alfalfa, soybean, cotton
Bromoxynil	Tobacco, tomato, cotton
Sulphonylurea	Tobacco, tomato, potato, flax
Atrazine	Tobacco
Insect resistance	
Insecticidal BT proteins	Tobacco, tomato, potato, cotton
Proteinaseinhibitors	Tobacco
Virus resistance	
Alfalfa mosaic virus	Tobacco, alfalfa
Tobacco mosaic virus	Tomato
Tomato mosaic virus	Tomato
Potato virus X	Potato
Potato virus Y	Potato
Potato leaf roll virus	Potato
Rhizomaniavirus	Sugarbeet
Cucumber mosaic virus	Cucumber
Other characters	
Heavy metal resistance	Tobacco
Firm fruit	Tomato
Thaumatin production	Potato
Modified seed storage	Oilseed rape proteins

be more efficient to produce fewer transformed lines and then eliminate un-desirable somaclonal changes via conventional plant breeding. However, in clonal crops further breeding would be undesirable, since the genetic integrity of elite clones would be lost. For this reason larger numbers of primary transgenic plants may be important for clonal crops.

Conclusion

It is now technically possible to transfer genes across all taxonomic boundaries into plants-from other plants, animals and microbes—or even to introduce totally artificial genes. This offers considerable potential for genetic improvement of crop plants, especially for disease and pest resistance, and improved quality characteristics. In addition to applications in crop improvement, Agro-bacterium-mediated transformation offers a powerful research tool for studying plant biology, especially the control mechanisms in gene expression and development.

These genes may be plant-derived or could come from other sources. However, the complex inheritance of stress resistance suggests a tailoring

Table 3. Glycine betaine production in transgenic plants.

Osmoprotectant	Transgenes	Crop plants	Accumulation	Stress tolerance
Mannitol	*E. coli mt1D* (Mannitol-1-phosphate dehydrogenase)	*Arabidopsis*	$10 \ \mu g \ g^{-1} FW$	Salt
		Tobacco	$6 \ \mu g \ g^{-1} FW$	Salt
Trehalose	Yeast *tps 1* (trehalose-6-phosphate synthase, T-6-PS)	Tobacco	$3.2 \ \mu g \ g^{-1} DW$	Drought
		Potato		Drought
	E. coli otsA + otsB (T-6-PS and T-6-P phosphatase)	Tobacco	$90 \ \mu g \ g^{-1} FW$	Drought
Fructance	*B. subtalis sacB*	Tobacco	$0.35 \ mg \ g^{-1} FW$	Drought
		Sugar beet	$5 \ \mu g \ g^{-1} DW$	
Glycine betaine	*E. coli betB* (betaine aldehyde dehydrogenase)	Tobacco	Not noticed	Not noticed
	E. coli betA	Tobacco	Not tested	Salt
	betA/betB	Tobacco	$0.035 \ \mu g \ g^{-1} FW$	Chilling, Salt
	betA	Rice	$5.0 \ \mu g \ g^{-1} FW$	Drought, Salt
	A. globiformis codA	*Arabidopsis*	$1.2 \ \mu g \ g^{-1} FW$	Salt, Chilling, Freezing, Heat and Strong light
	codA	Rice	$5.3 \ \mu g \ g^{-1} FW$	Salt, Chilling
	A. pascens cox (choline oxidase)	*Arabidopsis*	$19 \ \mu g \ g^{-1} DW$	Freezing, Salt
	cox	*Brassica napus*	$13 \ \mu g \ g^{-1} DW$	Drought, Salt
	cox	Tobacco	$13 \ \mu g \ g^{-1} DW$	Salt

of any novel combinations to very specific sets of conditions-rather than for a generic type of stress resistance. The molecular manipulations are most likely to be added to the present assortment of breeding methods rather than replacing them in the future. The natural gene-transferring ability of *Agrobacterium* can be exploited to genetically modify plants. The transfer of DNA from *Agrobacterium* to plant cells is a consequence of specific DNA-protein interactions. However, understanding of the precise mechanisms involved is incomplete. Although further research on the processes involved may enhance the efficiency of gene transfer by *Agrobacterium* and extend the use of such a system to a broader range of plant species, it is not crucial for the successful transfer of genes to many crop plants. *Agrobacterium*-mediated transformation offers an exciting, proven approach for the genetic manipulation of crop plants.

References

Al-Niemi, T. S. and R. G. Stout. 2002. Heat shock protein expression in a perennial grass commonly associated with active geothermal areas in western North America. J. Thermal Biol. 27: 547–553.

Al-Whaibi, M. H. 2010. Plant heat-shock proteins: A mini review. Journal of King Saud University (Science). doi:10.1016/j.jksus.2010.06.022.

Arakawa, K., M. Katayama and T. Takabe. 1990. Levels of betaine and betaine aldehyde dehydrogenase activity in the green leaves and etiolated leaves and roots of barley. Plant Cell Physiol. 31: 797–803.

Araus, J. L., G. A. Slafer, M. P. Reynolds and C. Royo. 2002. Plant breeding and drought in C3 cereals: What should we breed for? Ann. Bot. (Lond.). 89: 925–940.

Atkinson, R. G. and R. C. Gardner. 1991. Regeneration of transgenic pepino plants. Plant Cell Rep. 10: 208–212.

Atkinson, R. G., D. Hutching and R. C. Gardner. 1990. Transformation and regeneration of the pepino and the tamarillo. *New Zealand Genetical Society*, 36th Annual Meeting, Abstract 2.

Baker, N. 1994. Chilling stress and photosynthesis. *In:* C. H. Foyer and P. M. Mullineaux [eds.]. Causes of photooxidative stress and amelioration of defense systems in plants, CRC, Boca Raton, FL. pp. 127–154.

Barrera-Figueroa, B. E., Z. Wu and R. Liu. 2012. Abiotic stress-associated microRNAs in plants: discovery, expression, analysis, and evolution. Frontiers in Biology. DOI: 10.1007/s11515-012-1210-6.

Basiran, N., P. Armitage, R. J. Scott and J. Draper. 1987. Genetic transformation of flax (*Linum usitatissimum*): regeneration of transformed shoots via a callus phase. Plant Cell Rep. 6: 396–399.

Blum, A. 1988. Plant Breeding for Stress Environments. CRC Press, Boca Raton, Florida.

Bondt, A., K. Eggermont, P. Druart, M. Vil, I. Goderis, J. Vanderleyden and W. F. Broekaert. 1994. *Agrobacterium*-mediated transformation of apple (*Malus x domestica* Borkh.): an assessment of factors affecting gene transfer efficiency during early transformation steps. Plant Cell Rpt. 13: 587–593.

Bowler, C., T. Allione, M. De Loose, M. van Montagu and D. Inze. 1989. The induction of manganese superoxide dismutase in response to stress in Nicotiana plumbaginifolia. EMBO Journal. 8: 31–38.

Bray, E. A. 1997. Plant responses to water deficit. Trends Plant Sci. 2: 48–54.

Burke, J. J., P. E. Gamble, J. L. Hatfield and J. E. Quisenberry. 1985. Plant morphological and biochemical responses to field water deficits. I. Responses of glutathione reductase activity and paraquat sensitivity. Plant Physiol. 79: 415–419.

Butcher, S. M., S. C. Deroles and S. E. Ledger. 1990. Transformation of Geranium. New Zealand Genetical Society, 36th Annual Meeting.

Cakmak, I. and H. Marschner. 1992. Magnesium deficiency and high light intensity enhance activities of superoxide dismutase, ascorbate peroxidase, and glutathione reductase in bean leaves. Plant Physiol. 98: 1222–1227.

Cardwell, V. B. 1982. Fifty years of Minnesota corn production: sources of yield increases. Agronomy Journal. 74: 984–995.

Catlin, D., O. Ochoa, S. McCormick and C. F. Quirors. 1988. Celery transformation by *Agrobacterium tumefaciens:* cytological and genetic analysis of transgenic plants. Plant Cell Rep. 7: 100–103.

Charest, P. J., L. A. Holbrook, J. Gabard, V. N. Iyer and B. L. Miki. 1988. Agro-bacterium-mediated transformation of thin cell layer explants from *Brassica napus* L. Theoretical and Applied Genetics. 75: 438–445.

Chakrabarty, R., N. Viswakarma, S. R. Bhat, P. B. Kirti, B. D. Singh and V. L. Chopra. 2002. Agrobacterium-mediated transformation of cauliflower: optimization of protocol and development of Bt-transgenic cauliflower. Journal of Biosciences. 27: 495–502.

Christey, M. C. and B. K. Sinclair. 1990. Selection of transformed hairy root lines in *Brassica oleracea, B. napus* and *B. campestris*. Proceedings of the 6th Crucifer Genetics Workshop, Cornell University, Ithaca. p. 20.

Christey, M. C. and E. D. Earle. 1989. Genetic manipulation of *Brassica oleracea* var. Italica (broccoli) via protoplast fusion and transformation. *Australian Society of Plant Physiologists,* 29th Annual Meeting, Abstract 40.

Coffman, F. A. 1957. Factors influencing heat resistance in oats. Agronomy Journal. 49: 368–376.

Conner, A. J. and C. P. Meredith. 1989. Genetic manipulation of plant cells. *In:* A. Marcus [ed.]. The Biochemistry of Plants: A Comprehensive Treatise, vol. 15, Molecular Biology. Academic Press, Orlando. pp. 653–688.

Conner, A. J., M. K. Williams, R. C. Gardner, S. C. Deroles, M. L. Shaw and J. E. Lancaster. 1991. *Agrobacterium*-mediated transformation of New Zealand potato cultivars. New Zealand Journal of Crop and Hortic. Sci. 19: 1–3.

Conner, A. J., M. K. Williams, S. C. Deroles and R. C. Gardner. 1988. Agro-bacterium-mediated transformation of asparagus. *In:* K. S. McWhirter, R. W. Downes and B. J. Reid [eds.]. Ninth Australian Plant Breeding Conference, *Proceedings*. Agricultural Research Institute, Wagga Wagga. pp. 131–132.

De Block, M. 1988. Genotype-independent leaf disc transformation of potato (*Solanum tuberosum*) using *Agrobacterium tumefaciens*. Theoretical and Applied Genetics. 76: 767–774.

Deroles, S., S. Ledger, K. Markham, N. Given and K. Davis. 1990. Changing the colour of *Lisianthus*. New Zealand Genetical Society, 36th Annual Meeting, Abstract 97.

Dhindsa, R. S. and W. Matowe. 1981. Drought tolerance in two mosses: correlated with enzymatic defense against lipid peroxidation. Journal of Exper. Bot. 32: 79–92.

Dodge, A. 1994. Herbicide action and effects on detoxification processes. *In:* C. H. Foyer and P. M. Mullineaux [eds.]. Causes of photooxidative stress and amelioration of defense systems in plants, CRC, Boca Raton, FL. p. 219.

Enkhchimeg, V., T. W. Bae, K. Z. Riu, S. Y. Kim and H. Y. Lee. 2005. Overexpression of *Arabidopsis ABF3* gene enhances tolerance to drought and cold in transgenic lettuce (*Lactuca sativa* L.). Plant Cell Tissue Organ Cult. 83: 41–50.

Everett, N. P., K. E. P. Robinson and D. Mascarenhas. 1987. Genetic engineering of sunflower (*Helianthus annuus*). Bio-Technology. 5: 1201–1204.

Feldmann, K. A., M. D. Marks, M. L. Christanson and R. S. Quatrano. 1989. A dwarf mutant of *Arabidopsis* generated by T-DNA insertion mutagenesis. Science. 243: 1351–1354.

Fillati, J. J., J. Kiser, R. Rose and L. Comai. 1987a. Efficient transfer of a glyphosate tolerance gene into tomato using a binary *Agrobacterium tumefaciens* vector. Bio-Technology. 5: 726–730.

Fillati, J. J., J. Sellmer, J. B. McCown, B. Haissig and L. Comai. 1987b. *Agrobacterium*-mediated transformation and regeneration of Populus. Molecular and General Genetics. 206: 192–201.

Firoozabadi, E., D. L. DeBoer, D. J. Merlo, E. L. Halk, L. N. Amerson, K. E. Rashka and E. E. Murray. 1987. Transformation of cotton (*Gossypium hirsutum* L.) by *Agrobacterium tumefaciens* and regeneration of transgenic plants. Plant Molecular Biology. 10: 105–116.

Fry, J., A. Barnason and R. B. Horsch. 1987. Transformation of *Brassica napus* with *Agrobacterium tumefaciens* based vectors. Plant Cell Rep. 6: 321–325.

Gasser, C. S. and R. T. Fraley. 1989. Genetically engineering plants for crop improvement. Science. 244: 1293–1299.

Ghanem, M. E., I. Hichri, A. C. Smigocki, A. Albacete, M. L. Fauconnier, E. Diatloff, C. M. Andujar, S. Lutts, I. C. Dodd and F. P. Alfocea. 2011. Root-targeted biotechnology to mediate hormonal signaling and improve crop stress tolerance. Plant Cell Rep. 30: 807–823. DOI 10.1007/s00299-011-1005-2.

Gong, M., Y. J. Li and S. Z. Chen. 1998. Abscisic-acid induced thermotolerance in maize seedlings is mediated by calcium and associated with antioxidant systems. J. Plant Physiol. Sci. 153: 488–496.

Gosal, S. S., S. H. Wani and M. S. Kang. 2009. Biotechnology and drought tolerance. J. Crop Improv. 23: 19–54.

Gossett, D. R., E. P. Millhollon and M. C. Lucas. 1994a. Antioxidant response to NaCl stress in salt-tolerant and salt-sensitive cultivars of cotton. Crop Sci. 34: 706–714.

Gossett, D. R., E. P. Millhollon, M. C. Lucas, S. W. Banks and M. M. Marney. 1994b. The effects of NaCl on antioxidant enzyme activities in callus tissue of salttolerant and salt-sensitive cultivars of cotton. Plant Cell Rep. 13: 498–503.

Grumet, R. and A. D. Hanson. 1986. Glycine-betaine accumulation in barley. Aust. J. Plant Physiol. 13: 353–364.

Guri, A. and K. C. Sink. 1988. *Agrobacterium* transformation of eggplant. Journal of Plant Physio. 133: 52–55.

Halliwell, B. 1984. Oxygen derived species and herbicide action. Physiologica Plantarum. 15: 21–24.

Harper, D. B. and B. M. R. Harvey. 1978. Mechanism of paraquat tolerance in perennial ryegrass II. Role of superoxide dismutase, catalase and peroxidase. Plant Cell Environ. 1: 211–215.

Hernandez, J. A., F. J. Corpas, M. Gomez, L. A. Del Rio and F. Sevilla. 1994. Salt stress induced changes in superoxide dismutase isozymes in leaves and mesophyll protoplasts from *Vigna unguiculata* L. New Phytol. 126: 37–44.

Hidaka, T., M. Omura, M. Ugaki, M. Tomiyama, A. Kato, M. Ohshima and F. Motoyoshi. 1990. *Agrobacterium*-mediated transformation and regeneration of Citrus spp. From suspension cells. Japan. J. Breed. 40: 199–207.

Hinchee, M. A. W., D. V. Connor-Wand, C. A. Newell, R. E. McDonnel, S. J. Sato, C. S. Gasser, D. A. Fischhoff, D. Re, R. T. Fraley and R. B. Horsh. 1988. Production of transgenic soybean plants using *Agrobacterium*-mediated DNA transfer. Bio-Technology. 6: 915–922.

Horsch, R. B., J. E. Fry, N. L. Niedermeyer, D. Eichholtz, S. G. Rogers et al. 1985. A simple and general method for transferring genes into plants. Science. 227: 1229–1231.

Horsch, R. B. and G. E. Jones. 1980. A double filter paper technique for plating cultured plant cells. *In vitro*. 16: 103–108.

Huang, J., R. Hirji, L. Adam, K. L. Rozwadowski, J. K. Hammerlindl, W. A. Keller and G. Selvaraj. 2000. Genetic engineering of glycine-betaine production toward enhancing stress tolerance in plants: Metabolic limitations. Plant Physiol. 122: 747–756.

Ingram, J. and D. Bartels. 1996. The molecular basis of dehydration tolerance in plants. Annu Rev. Plant Physiol. Plant Mol. Bio. 47: 377–403.

James, D. J., A. J. Passey, D. J. Barbara and M. Bevan. 1989. Genetic transformation of apple (*Malus pumila* Mill.) using a disarmed Ti-binary vector. Plant Cell Rpt. 7: 658–661.

Janssen, B. and R. C. Gardner. 1993. The use of transient GUS expression to develop an *Agrobacterium*-mediated gene transfer system for kiwifruit. Plant Cell Rpt. 13: 28–31.

Jensen, J. S., K. A. Marcker, L. Otten and J. Schell. 1986. Nodule specific expression of a chimeric soybean leghaemoglobin gene in transgenic Lotus corniculatus. Nature. 321: 669–674.

Jia, S. R., M. Z. Yang, R. Ott and N. H. Chua. 1989. High frequency transformation of *Kalanchoe laciniata*, Plant Cell Reports. 8(6): 336–340.

Kang, J. Y., H. I. Choi, M. Y. Im and S. Y. Kim. 2002. Arabidopsis basic leucine zipper proteins that mediate stress responsive abscisic acid signaling. Plant Cell. 14: 343–357.

Kim, J. B., J. Y. Kang and S. Y. Kim. 2004. Over-expression of a transcription factor regulating ABA-responsive gene expression confers multiple stress tolerance. Plant Biotech. J. 2: 459–466.

Kimpel, J. A. and J. L. Key. 1985. Presence of heat shock mRNAs in field grown soybeans. Plant Physiol. 79: 672–678.

Kloepper, J. W., C. M. Ryu and S. A. Zhang. 2004. Induced systemic resistance and promotion of plant growth by *Bacillus* spp. Phytopathology. 94(11): 1259–1266.

Krause, G. H. 1994. The role of oxygen in photoinhibition in photosynthesis. *In:* C. H. Foyer and P. M. Mullineux [eds.]. Causes of photoxidative stress and amelioration of defense systems in plants. CRC, Boca Raton, FL. pp. 43–76.

Larkindale, J. and M. R. Knight. 2002. Protection against heat induced oxidative damage in *Arabidopsis* involves calcium, abscisic acid, ethylene, and salicylic acid. Plant Physiol. 128: 682–695.

Ledger, S. E., S. C. Deroles and N. K. Given. 1991. Regeneration and *Agrobacterium*-mediated transformation of *Chrysanthemum*. Plant Cell Rep. 10: 195–199.

Lee, Mi-Kyung, Hyoung-Seok Kim, Jung-Sun Kim, Sung-Hoon Kim and Young-Doo Park. 2004. Agrobacterium-mediated transformation system for large-scale producion of transgenic chinese cabbage (*Brassica rapa* L. ssp. pekinensis) plants for insertional mutagenesis. J. Plant Biol. 47: 300–306.

Liu, X. and B. Haung. 2000. Heat stress injury *m* in relation to membrane lipid peroxidation in creeping bentgrass. Crop Sci. 40: 503–510.

Manners, J. M. and H. Way. 1989. Efficient transformation with regeneration of the tropical pasture legume *Stylosanthes humilis* using *Agrobacterium rhizogenes* and a Ti plasmid-binary vector system. Plant Cell Rep. 8: 341–345.

Mathews, H., N. Bharathan, R. E. Litz, K. R. Narayanan, P. S. Rao and C. R. Bhatia. 1990. Transgenic plants of mustard *Brassica juncea* (L.) Czern and Coss. Plant Sci. 72: 245–252.

McCormick, S., J. Niedermeyer, J. Fry, A. Barnson, R. Horsch and R. Fraley. 1986. Leaf disc transformation of cultivated tomato (*Lycopersicon esculentum*) using *Agrobacterium tumefaciens*. Plant Cell Rep. 5: 81–84.

McCue, K. F. and A. D. Hanson. 1990. Drought and salt tolerance: Towards understanding and application. Tibtech. 8: 358.

McGranahan, G. H., C. A. Leslie, S. L. Uratsu, L. A. Martin and A. M. Dandekar. 1988. *Agrobacterium*-mediated transformation of walnut somatic embryos and regeneration of transgenic plants. Bio-Technology. 6: 800–804.

Mendoza, H. A. and R. N. Estrada. 1979. Breeding potatoes for tolerance to stress: heat and frost. *In:* H. Mussell and R. C. Staples [eds.]. Stress Physiology in Crop Plants. Wiley-Interscience, New York. pp. 227–355.

Michelmore, R., E. Marsh, S. Seeley and B. Landry. 1987. Transformation of lettuce (*Lactuca sativa*) mediated by *Agrobacterium tumefaciens*. Plant Cell Rep. 6: 439–442.

Milborrow, B. V. 1981. Abscisic acid and other hormones. *In:* L. G. Paleg and D. Aspinall [eds.]. The Physiology and Biochemistry of Drought Resistance in Plants. Academic Press, Sydney. pp. 347–388.

Mock, J. J. and S. A. Eberhart. 1972. Cold tolerance in adapted maize populations. Crop Sci. 12: 466–471.

Mollel, M. H. N. and E. M. A. Goyvaerts. 2005. Preliminary examination of factors affecting *Agrobacterium tumefaciens*-mediated transformation of marula, *Sclerocarya birrea* subsp. *Caffra* (Anacardiacease). Plant Cell, Tissue & Organ Culture. 79: 321–328.

Moloney, M. M., J. M. Walker and K. K. Sharma. 1989. High efficiency transformation of *Brassica napus* using *Agrobacterium* vectors. Plant Cell Rep. 8: 238–242.

Monk, L. S. and H. V. Davies. 1989. Antioxidant status of the potato tuber and Ca2⁺ deficiency as a physiological stress. Physiol. Plant. 75: 411–416.

Nagao, R. T., J. A. Kimpel and J. L. Key. 1990. Molecular and cellular biology of the heat shock response. Advances in Genet. 28: 235–274.

Naina, N. S., P. K. Gupta and A. F. Mascarenhas. 1989. Genetic transformation and regeneration of transgenic neem (*Azadiarchta indica*) plants using *Agrobacterium tumefaciens*. Curr. Sci. 58: 184–187.

Nehra, N. S., R. N. Chibbar, K. K. Kartha, R. S. Datla, W. L. Crosy and C. Stushnoff. 1990. *Agrobacterium*-mediated transformation of strawberry calli and recovery of transgenic plants. Plant Cell Rep. 9: 10–13.

Nguyen, H. T., C. P. Joshi, N. Klueva, J. Weng, K. L. Hendershot and A. Blum. 1994. The heat-shock response and expression of heat-shock proteins in wheat under diurnal heat stress and field conditions. Austral. J. Plant Physiol. 21: 857–867.

Oh, S. J., S. I. Song, Y. S. Kim, H. J. Jang, S. Y. Kim, M. Kim, Y. K. Kim, B. H. Nahm and J. K. Kim. 2005. Arabidopsis CBF3/DREB1A and ABF3 in transgenic rice increased tolerance to abiotic stress without stunting growth. Plant Physiol. 138: 341–351.

Padmanabhan, P. and S. V. Sahi. 2008. Genetic transformation and regeneration of *Sesbania drummondii* using cotyledonary nodes. Plant Cell Rpt. 28: 31–40.

Patterson, B. D. and L. A. Payne. 1983. Screening for chilling resistance in tomato seedlings. HortScience. 18: 340–347.

Polle, A., K. Chakrabarti, S. Chakrabarti, F. Seifert, P. Schramel and H. Renneberg. 1992. Antioxidants and manganese deficiency in needles of Norway spruce (*Picea abies* L.) trees. Plant Physiol. 99: 1084–1089.

Puonti-Kaerlas, J., T. Eriksson and P. Engstrom. 1990. Production of transgenic pea (*Pisum sativum*) plants by *Agrobacterium tumefaciens*-mediated gene transfer. Theoretical and Applied Genet. 80: 246–252.

Saini, R. and P. K. Jaiwal. 2007. *Agrobacterium tumefaciens*-mediated transformation of blackgram: An assessment of factors influencing the efficiency of *uidA* gene transfer. Biologia Plantarum. 51: 69–74.

Rabinowitch, H. D. and I. Fridovich. 1983. Superoxide radicals, superoxide dismutase, and oxygen toxicity in plants. Photochem. Photobiol. 37: 679–790.

Radin, J. W., Z. Lu, R. G. Percy and E. Zeiger. 1994. Genetic variability for stomata conductance in pima cotton and its relation to improvements in heat adaptation. Proc. Natl. Acad. Sci. USA. 91: 7217–7221.

Rainwater, D. T., D. R. Gossett, E. P. Millhollon, H. Y. Hanna, S. W. Banks and M. C. Lucas. 1996. The relationship between yield and the antioxidant defense system in tomatoes grown under heat stress. Free Radical Res. 25: 421–435.

Rafat, A. M., A. Aziz, A. A. Rashid, S. N. A. Abdullah, H. Kamaladini, M. H. Torabi Sirchi and M.B. Javadi. 2012. Optimization of Agrobacterium tumefaciens-mediated transformation and shoot regeneration after co-cultivation of cabbage (*Brassica oleracea* subsp. capitata) cv. KY Cross with AtHSP101 gene. Scientia Horticulturae 124: 1–8.

Ramagopal, S. 1987. Differential messenger RNA transcription during salinity stress in barley. Proc. of the Natl. Acad. Sci. USA. 84: 94–98.

Redinbaugh, H. D., G. J. Wadsworth and J. G. Scandalios. 1988. Characterization of catalase transcripts and their differential expression in maize. Biophysica. Acta. 951: 104–116.

Reitz, L. P. 1974. Breeding for more efficient wateruse—is it real or a mirage? Agricultural Meteorol. 14: 3–6.

Rhodes, D., P. J. Rich, D. G. Brunk, G. C. Ju, J. C. Rhodes, M. H. Pauly and L. A. Hansen. 1989. Development of two isogenic sweet corn hybrids differing for glycine betaine content. Plant Physiol. 91: 1112–1121.

Richards, R. A. and N. Thurling. 1979. Genetic analysis of drought stress response in rapeseed *(Brassica campestris* and *B. napus).* III. Physiological characters. Euphytica. 28: 755–760.

Rose, A., F. Gindullis and I. Meier. 2003. A novel alpha-helical protein, specific to and highly conserved in plants, is associated with the nuclear matrix fraction. J. Expt. Bot. 54: 1–9.

Sacher, R. F. and R. C. Staples. 1985. Inositol and sugars in adaptation of tomato to salt. Plant Physiol. 77: 206–210.

Saini, A., Y. Li, G. Jagadeeswaran and R. Sunkar. 2012. Role of microRNAs in Plant Adaptation to Environmental Stresses. *In:* R. Sunkar [ed.]. MicroRNAs in Plant Development and Stress Responses—Signaling and Communication in Plants. Springer Heidelberg Dordrecht London New York. pp. 219–232.

Sakai, A. and W. Larcher. 1987. Frost Survival of Plants: Responses and Adaptation to Freezing Stress. Springer Verlag, Berlin.

Schafer, W., A. Gorz and G. Kahl. 1987. T- DNA integration and expression in a monocot crop plant after induction of *Agrobacterium.* Nature. 327: 529–531.

Schoner, S. and G. H. Krause. 1990. Protective systems against active oxygen species in spinach responses to cold acclimation in excess light. Planta. 180: 383–389.

Schrammeijer, B., P. C. Sijmons, P. J. M. van den Elzen and A. Hoekema. 1990. Meristem transformation of sunflower via *Agrobacterium.* Plant Cell Rep. 9: 55–60.

Scott, R. J. and J. Draper. 1987. Transformation of carrot tissues derived from proembryogenic suspension cells: a useful model system for gene expression studies in plants. Plant Molecular Biolog. 8: 265–274.

Shaaltiel, Y., A. Glazer, P. F. Bocion and J. Gressel. 1988. Cross tolerance to herbicidal environmental oxidants of plant biotypes tolerant to paraquat, sulfur dioxide, and ozone. Pest. Chem. Physiol. 13: 13–21.

Shahin, E. A., A. Spielmann, K. Sukhapinda, R. B. Simpson and M. Yasher. 1986. Transformation of cultivated alfalfa using disarmed *Agrobacterium tumefaciens.* Crop Science. 26: 1235–1239.

Sheikholeslam, S. N. and D. P. Weeks. 1987. Acetosysringone promotes high efficiency transformation of *Arabidopsis thaliana* explants by *Agrobacterium tumefaciens.* Plant Molecular Biol. 8: 291–298.

Shinozaki, K. and K. Yamaguchi-Shinozaki. 2000. Gene expression and signal transduction in water-stress response. Curr. Opin. Plant Biol. 3: 217–223.

Shpiler, L. and A. Blum. 1986. Differential reaction of wheat cultivars to hot environ-ments. Euphytica. 35: 483–492.

Soman, P. and J. M. Peacock. 1985. A laboratory technique to screen seedling emergence of sorghum and pearl millet at high soil temperature. Experimental Agri. 21: 335–342.

Spychalla, J. P. and S. L. Desborough. 1990. Superoxide dismutase, catalase, and *a*-tocopherol content of stored potato tubers. Plant Physiol. 94: 1214–1218.

Stamp, P., G. Geisler and R. Thiraporn. 1983. Adaptation to sub and supraoptimal temperatures of inbred maize lines differing in origin with regard to seedling development and photosynthetic traits. Physiologica. Planta. 58: 62–68.

Steponkus, P. L., M. Uemura, R. A. Balsamo, T. Arvinte and D. V. Lynch. 1988. Transformation of the cryobehavior of rye protoplasts by modification of the plasma membrane lipid composition. Proc. Natil. Acad. Sci. USA. 86: 9026–9030.

Stewart, G. R. and I. A. Lee. 1974. The role of proline accumulation in halophytes. Planta. 120: 279–289.

Sundin, G. W. 2003. Stress resistance in Pseudomonas syringae: mechanisms and strategies. *Presentations from the 6th International Conference on Pseudomonas syringae pathovars and related pathogens*, Maratea, Italy, September 15–19. p. 41–49.

Thomashow, M. F. 1990. Molecular genetics of cold acclimation in higher plants. Advances in Gene. 28: 99–131.

Thomashow, M. F. 1999. Plant cold acclimation: freezing tolerance genes and regulatory mechanism. Annu. Rev. Plant Physiol. Plant Mol. Bio. 50: 571–599.

Tian, H., L. Ma, C. Zhao, H. Hao, B. Gong, X. Yu and X. Wang. 2010. Antisense repression of sucrose phosphate synthase in transgenic muskmelon alters plant growth and fruit development. Biochem. Biophys. Res. Commun. 393: 365–70.

Trulson, A., R. Simpson, E. Shahin. 1986. Transformation of cucumber (*Cucumis sativus* L.) plants with *Agrobacterium rhizogenes*. Theoretical and Applied Gene. 73: 11–15.

Uno, Y., T. Furihata, H. Abe, R. Yoshida, K. Shinozaki and K. Yamaguchi-Shimnozuki. 2000. *Arabidopsis* basic leucine zipper transcription factors involved in an abscisic acid-dependent signal transduction pathway under drought and high-salinity condition. Proc. Nati. Sci. USA. 97: 11632–11637.

Vanjildorj, E., T. W. Bae, K. Z. Riu, P. Y. Yun, S. Y. Park, C. H. Lee, S. Y. Kim and H. Y. Lee. 2006. Transgenic *Agrostis mongoloca* Roshev. with enhanced tolerance to drought and heat stresses obtained from *Agrobacterium*-mediated transformation. Plant Cell Tissue Org. 87: 109–120.

Veluthambi, K., W. Ream and S. B. Gelvin. 1988. Virulence genes, borders, and overdrive generate single-stranded T-DNA molecules from the A6 Ti plasmid of *Agrobacterium tumefaciens*. J. Bacteriol. 170: 1523–1532.

Vierling, E. 1991. The role of heat shock proteins in plants. Annu. Rev. Plant Physiol. Plant Mol. Bio. 42: 579–620.

Vierling, E., R. T. Nagao, A. E. DeRocher and L. M. Harris. 1988. A heat shock protein localized to chloroplasts is a member of a eukaryotic super family of heat shock proteins. EMBO Journal. 7: 575–581.

Weretilnyk, E. A. and A. D. Hanson. 1990. Molecular cloning of a plant betaine-aldehyde dehydrogenase, an enzyme implicated in adaptation to salinity and drought. Proc. Natl. Acad. Sci. USA. 87: 2745–2749.

White, D. W. R. and D. Greenwood. 1987. Transformation of the forage legume *Trifolium repens* L. using binary *Agrobacterium* vectors. Plant Molecular Biol. 8: 461–469.

Wise, R. R. and A. W. Naylor. 1987a. Chilling enhanced photooxidation: the peroxidative destruction of lipids during chilling injury to photosynthesis and ultrastructure. Plant Physiol. 83: 272–277.

Wise, R. R. and A. W. Naylor. 1987b. Chilling-enhanced photooxidation: Evidence for the role of singlet oxygen and endogenous antioxidants. Plant Physiol. 83: 278–282.

Wynn Jones, R. G. 1981. Salt tolerance. *In:* C. B. Johnson [ed.]. *Environmental Factors Limiting Plant Productivity*. Butterworths, London. pp. 271–322.

Xiong, L., K. S. Schumaker and J. K. Zhu. 2002. Cell signaling during cold, drought, and salt stress. Plant Cell, 14 (Suppl.): S165–S183.

Zeng, G. W. and A. A. Khan. 1984. Alleviation of high temperature stress by preplant permeation of phthalimide and other growth regulators into lettuce seed via acetone. Journal of the American Society for Horticultural Science. 109: 782–785.

11

The Potent Pharmacological Mushroom *Fomes fomentarius*
Cultivation Processes and Biotechnological Uses

Mohamed Neifar,[1], Atef Jaouani[1] and*
Semia Ellouz Chaabouni[2]

ABSTRACT

Fomes fomentarius is a basidiomycete white-rot fungus belonging to the polyporaceae family. It has been used as a traditional medicine for centuries in treating various diseases such as gastroenteric disorder, hepatocirrhosis, oral ulcer, inflammation, and various cancers. This mushroom possesses a number of therapeutic properties such as antitumour, immunomodulatory, antioxidant, anti-inflammatory, hypocholesterolaemic, antihypertensive, antihyperglycaemic, antimicrobial and antiviral activities. These activities are exhibited by extracts or isolated compounds from *F. fomentarius* fermentation broth, mycelia and fruiting bodies. In particular, polysaccharides appear to be potent antitumour and immuno-enhancing substances, besides possessing other beneficial activities. *F. fomentarius* can be cultivated

[1] Laboratoire Microorganismes et Biomolécules Actives, Faculté des Sciences de Tunis, Université de *Tunis El Manar*, 2092, Tunis, Tunisia.
[2] Unité Enzymes et Bioconversion, Ecole Nationale d'Ingénieurs de Sfax, Université de *Sfax*, 3038, Sfax, Tunisia.
* Corresponding author: mohamed.naifar@gmail.com

on a wide variety of agricultural substrates/byproducts and wastes for the production of feed, enzymes and medicinal compounds, or for waste degradation and detoxification. Many different techniques and substrates have been successfully utilized for mushroom cultivation and biomass production. This chapter focuses on recent advances in the biotechnology of *F. fomentarius*, with emphasis on the production of mycelium and bioactive compounds by solid-state and submerged liquid fermentation. The medicinal properties of this mushroom are also discussed.

Introduction

Mushrooms were originally defined as macrofungi with a distinctive fruiting body, which can either be hypogeous or epigeous, large enough to be seen with the naked eye and to be picked by hand (Chang and Miles 1992). Mushrooms constitute at least 14,000 and perhaps as many as 22,000 known species. Other studies have reported that the number of mushroom species on earth is estimated to be 140,000, indicating that only 10% are known (Lindequist et al. 2005). For millennia, mushrooms have been valued by human kind as an edible and medical resource. Extracts from certain mushrooms could have profound health promoting benefits and, consequently, became essential components in traditional medicine. There are at least 270 species of mushrooms that are known to possess various therapeutic properties (Ying et al. 1987) and the term 'medicinal mushroom' is now increasingly gaining worldwide recognition. Edible mushrooms which demonstrate medicinal or functional properties include species of *Lentinula, Hericium, Grifola, Flammulina, Pleurotus* and *Tremella* while others known only for their medicinal properties, viz. *Ganoderma lucidum, Trametes versicolor,* and *Fomes fomentarius,* are decidedly non-edible because of their coarse texture and bitter taste (Smith et al. 2002). For medicinal purposes, they were almost always prepared either as hot water extracts, concentrates or in powdered form (Gregori et al. 2007, Sanodiya al. 2009). Recent years have seen a surge of commercial interest in medicinal mushroom products, the common market value of which is approximately $13 billion US dollars (Wasser 2005).

Various pharmacological properties have been influenced by the medicinal mushrooms. The biologically active substances are claimed to have profound health promoting benefits such as antibiotics, anti-tumor, antiviral, anti-inflammatory, immunological enhancement, maintenance of homeostasis, regulation of biorhythm, cure of diseases such as cancer, cerebral stroke and heart diseases. It is also being confirmed that mushrooms include effective substances for decreasing blood cholesterol, the improvement of hyperlipidemia, antithrombotic, reduction of blood pressure, hypoglycemic action and various other therapeutic applications

(Wasser and Weis 1999a, b). In fact, recent studies are now confirming their medical efficacy and many of the bioactive molecules have been identified. Medicinal mushrooms accumulate a wide variety of bioactive compounds including terpenoids, steroids, phenols, nucleotides and their derivatives glycoproteins and polysaccharides that display a broad range of biological activities (Smith et al. 2002, Patel and Goyal 2012). These different bioactive compounds have been extracted from the fruiting body, mycelia and culture medium of various medicinal mushrooms (Wasser and Weis 1999a, Wasser 2002).

Historically, most medicinal mushroom species were relatively scarce and were collected from the forests where they grew on dead or living trees and forest litter. They are predominantly lignocelluloses degraders (Stamets 2000). Nowadays almost all of the important medicinal mushrooms have been subjected to large-scale artificial cultivation by solid substrate or low moisture fermentation, thus removing the historical scarcity factor and allowing large commercial operations to develop (Gregori et al. 2007). New technologies and production techniques are being constantly developed as the number of required controllable environment parameters increases (Hölker and Lenz 2004). Currently, solid-state fermentations, other than fruiting body production with medicinal mushrooms, are used either in the transformation of wastes into animal feed or for enzyme production. Submerged liquid fermentation can, on the other hand, provide more uniform and reproducible biomass and can prove interesting for valuable medicinal products or for enzyme production because of uncomplicated downstream processing (Smith et al. 2002, Gregori et al. 2007).

Fomes fomentarius, is a basidiomycete fungus belonging to the genus of *Fomes*. Its fruiting body is called "Mudi", which has been used as a traditional medicine for many centuries for the treatment of various diseases, including oral ulcer, gastroenteric disorder, hepatocirrhosis, inflammation, and various cancers. Recent studies have shown that *F. fomentarius* has the effect of being antioxidant, anti-inflammatory, antidiabetic and antitumor (Park et al. 2004, Lee 2005, Jaszek et al. 2006, Robles-Hernández et al. 2008). As a result of its perceived health benefits, *F. fomentarius* has gained wide popularity as an effective medicine and has become one of the valuable mushrooms. Many different techniques and substrates have been successfully utilized for mushroom cultivation and biomass production by means of solid-state and submerged liquid fermentation (Jaszek et al. 2006, Chen et al. 2008, Neifar et al. 2009, 2010, 2011a, b, c). The purpose of the present chapter is to summarize the pharmacological effects and bioactive compounds of *F. fomentarius* and to describe the development of cultivation techniques of this medicinal mushroom.

General Description of *Fomes fomentarius*

Macroscopic description

Fruit bodies perennial, sessile, mostly solitary, rarely in groups, pileate, dimidiate to ungulate brackets, broadly attached to the substrate, 60–500 x 40–300 mm, 40–250 mm thick, woody hard (Fig. 1). Upper surface smooth, concentrically grooved, hard, glabrous, with 1–3 mm thick blackish crust, which reacts dark red on white paper with KOH. Young regions of pileus ochraceous to red-brown, felty, becoming white, greywhite or grey-black with age depending on altitude and latitude. Lower surface cream to ochraceous later pale brown, bruising brown. Pores round, 2–4 per mm, pale brown, stratified, tubes 2–10 mm turning black with KOH. Trama corklike, tough, pale brown, concentrically zoned. The basidiocarp base has a soft white mottled mycelia core. Smell pleasant, mushroom-like, taste bitter (Schwarze 1994).

Fig. 1. Photograph representing the fruit bodies of the white-rot fungus *F. fomentarius* in its natural habitat (dead plants in the forest) (adapted with permission of Etienne Charles, http:// champignons.moselle.free.fr).

Color image of this figure appears in the color plate section at the end of the book.

Microscopic features

Basidiospores oblong-ellipsoid, smooth, hyaline, partly with drops, 15–22 x 5–7 µm. Basidiocarps sporulating from late spring to early summer. Basidia club-shaped, 20–30 x 7–11 µm with 4 sterigmata, without basal clamp. Hyphal system trimitic, generative hyphae hyaline, thin-walled, 2–4.5 urn in diameter, septa with clamps, skeletal hyphae thick-walled, light brown, up to 6–8 urn. Binding hyphae predominant in trama and tubes, strongly branched, 3–4 µm wide, light brown (Schwarze 1994).

Distribution and Hosts

Fomes fomentarius has a wide distribution and has been reported from Africa, Asia, Europe and North America (Sinclair et al. 1987). In Europe, *F. fomentarius* shows substantial intraspecific morphological variation and preferences in host specificity in different parts of its geographic range. In continental Europe *F. fomentarius* has been recorded from a wide range of angiospermous hosts; e.g., beech, birch, oak, poplar, maple and more rarely alder and hornbeam (Pegler 1973, Breitenbach and Kranzlin 1986). In northern Scotland *F. fomentarius* is restricted to birch, on which it is the principal decay fungus, but, over much of Great Britain the typical decay fungus of birches is *Piptoporus betulinus* (Bull.: Fr.) Karsten. *F. fomentarius* is one of the principal decay fungi on beech in some parts of continental Europe (Butin 1989), but in Great Britain this niche is more often occupied by other fungi, including *Ganoderma* spp., *Stereum* spp. and *Bjerkandera adusta* (Willd.: Fr.) Karsten. Morphologically, populations of *F. fomentarius* in southern England on beech and birch show a resemblance to their counterparts on the continent. In contrast, populations in northern Scotland seem to be rather distinctive in their appearance, particular in respect of their darker colour and the ungulate (hoof-like) form. The tendency for northern Scottish populations to have darker basidiocarps than those from southern England has in the past led to some taxonomic confusion (MacDonald 1938). Until the late 1980s, the taxa of polypores had been described mostly based on their macro- and micromorphological characteristics (e.g., hyphae, spores, basidiomata). In the middle 1990s, more evolutionary based species concepts began to occur, and the use of population tools and DNA-based methods for the identification of taxa emerged (Borneman and Hartin 2000). Analysis of available GenBank data strongly supports the variability in *F. fomentarius*, when ITS sequences could by classified in two genotypes based on the presence or absence of two indels. While genotype A seems to be linked preferably with *Fagus sylvatica* as a host, genotype B is found mainly on other tree species. A rapid method for discrimination of

F. fomentarius genotypes was developed based on ITS-RFLP analysis and applied for analysis of abundance of *F. fomentarius* genotypes in natural forests (Judova et al. 2012).

Decomposer or Wood-Rotting Macrofungus

Fomes fomentarius (Polyporales, Agaricomycetes, Basidiomycota) is economically and ecologically important not only as a source of medicinal and neutraceutical products but also in nutrient cycling in forest ecosystems as decomposers of dead wood (Gilbertson and Ryvarden 1986, Hennon 1995). *F. fomentarius* is a perennial wood-decaying basidiomycete, often found on birch or beech trees (Fig. 1). It most commonly infects stems, causing white rot through the simultaneous degradation of lignin, cellulose and hemicelluloses (Schwarze et al. 2000). There is significant commercial interest in harnessing the power of these wood-degrading enzymes for industrial applications. *F. fomentarius* is widespread; it is found in deciduous forests throughout the northern hemisphere (McDonald 1938, Kotlaba 1997, Schmidt 2006). *F. fomentarius* is both parasitic and saprobic, often infecting live trees and persisting after the tree has died (Schwarze et al. 2000). *F. fomentarius* has been noted as a primary decay wood-rotting polypore (Heilmann-Clausen 2001), and may even stimulate the growth of secondary decay fungi and late stage specialists (Heilmann-Clausen and Boddy 2005).

Medicinal and Pharmacological Potential of *Fomes fomentarius*

Uses of *Fomes fomentarius* in traditional medicine

Except for its use to keep fire embers glowing, its medicinal utilisation goes back to the 5th century BC where it was used for cauterising wounds. This utilisation survived up to the 19th century and perhaps even later in Lapland and Nepal (Delmas 1989). Its name "Agaric of surgeons" is due to its use for stopping light haemorrhages. Its flesh was beaten with a mallet to make it supple and applied as non caustic haemostatic. It was also traditionally used by barbers to stop the bleeding of razor cuts. In Khanty (West Siberia) folk medicine it was used in the same manner: its flesh was pounded in a mortar until soft. This mass was applied on wounds to stop bleeding. They were also using it to make warm compresses for extremities and joints. The aching area was covered with cotton, the mushroom mass, and then tied with cloth. It was left until the pain was gone (Saar 1991). In

China it is used for indigestion and to reduce stasis of digestive vitality, for oesophageal cancer and gastric and uterine carcinomas (Ying et al. 1987, Rogers 2006).

Pharmacological properties of Fomes fomentarius

Increased scientific and medical research in recent years and papers published in peer reviewed journals are increasingly confirming the medicinal efficacy and identifying the bioactive molecules with therapeutic activities. This mushroom modulates the immune system, inhibits tumour growth, has anti-inflammatory, anti-nociceptive activities and hypoglycaemic, lowers blood lipid concentrations, prevents high blood pressure, and has antimicrobial and antioxidant activities (Park et al. 2004, Lee 2005, Jaszek et al. 2006, Rogers 2006, Robles-Hernández et al. 2008, Gao et al. 2009). Recent advances in chemical technology have allowed the isolation and purification of some of the relevant compounds especially polysaccharides which possess strong immunomodulation and anti-cancer activities (Gao et al. 2009, Liu 2009).

Antitumor activity

Tumor diseases are one of the main causes of death worldwide. The current anti-cancer drugs available in market are not target-specific and pose several side-effects and complications in clinical management of various forms of cancer, which highlights the urgent need for novel, effective and less-toxic therapeutic approaches. In this context, some mushrooms with validated anti-cancer properties and their active compounds are of immense interest. Numerous clinical trials have been conducted to assess the benefits of using commercial preparations containing medicinal mushroom extracts in cancer therapy (Patel and Goyal 2012). Their potential uses individually and as adjuncts to cancer therapy have emerged. Mushrooms are known to complement chemotherapy and radiation therapy by countering the side-effects of cancer, such as nausea, bone marrow suppression, anemia, and lowered resistance. A number of bioactive molecules, including anti-tumor agents have been identified from various polypore fungi (Table 1).

Polysaccharides are the best known and most potent mushroom-derived substances with anti-tumor and immunomodulating properties. The polysaccharide, β-glucan is the most versatile metabolite due to its broad spectrum biological activity (Chen and Seviour 2007). Their mechanisms of action involve their being recognized as non-self molecules, so the immune system is stimulated by their presence. *Fomes fomentarius* polysaccharide displayed high homogeneity and was comprised of single polysaccharide,

Table 1. Medicinal uses of some polypore fungi.

Mushroom scientific name	Medicinal uses and therapeutic Effects	Bioactive Compound	Reference
C. volvatus (Peck) Hubb.	Tracheitis, asthma, hemorrhoids, anti-decrepitude, toothache, anti-inflammation	Glucan, sesquiterpenoids	Ying et al. 1987, Hirotani et al. 1991, Kitamura et al. 1994, Huang 1998
F. pinicola (Sw. Ex Fr.) Karst.	Against tumor	Mono-terpenes, triterpene glycosides, steroids, triterpenes, diterpenes	Keller et al. 1996, Huang 1998, Rösecke and König 2000, Kazuko et al. 2001
F. hornodermus Mont.	To calm convulsion, hemostasia, prurigo	No report	Ying et al. 1987
L. betulina (L.) Fr.	Haunch and femora pain, acropathy, apoplexy, cold	Benzoquinones, ergosterol peroxide	Liu 1978, Fujimoto et al. 1994, Lee et al. 1996
P. conchatus (Pers. Ex Fr.) Quel.	To promote blood-circulation and remove blood stasi detoxication, schistosomiasis	No report	Ying et al. 1987
P. coccineus (Fr.) Bond et Sing.	Against tumor, stomatitis, diarrhea, detoxication, rheumatism, prurigo, hemostasis, anti-inflammation	Cinnabarin, laccase	Liu 1978, Ying et al. 1987, Huang 1998, Lomascolo et al. 2002
T. gibbosa (Pers. Ex Fr.) Fr.	Against tumor	Polysaccharides	Huang 1998
T. orientalis (Yasuda) Imaz.	Anti-inflammation, phthisis, bronchitis, rheumatism	No report	Mao 2000
Fomes fomentarius	antioxidant, anti-inflammatory, antidiabetic and antitumor activities, treatment of various diseases, including oral ulcer, gastroenteric disorder, hepatocirrhosis, inflammation, and various cancers	Polysaccharides Laccase Mn Peroxidase Antioxidant enzymes	Ito et al. 1997, Park et al. 2004, Lee 2005, Jaszek et al. 2006, Chen et al. 2008, 2011, Neifar et al. 2009, 2010, 2011a, b, c, 2012

which was formed through the polymerization of galactose. The structure was analyzed by IR suggesting that the polysaccharide was in the form of pyranose containing α-glycosidic bond and C=O group (Liu 2009). Chen et al. (2011) stated that ethanol extract of mycelia biomass and intracellular polysaccharide of *Fomes fomentarius* play crucial roles in gastric cancer intervention. The extracts exhibit anti-proliferative effect on human gastric cancer cell lines SGC-7901 and MKN-45 in a dosedependent manner. In contrast, human normal gastric cell line GES-1 was less susceptible to EEM and IPS. These results suggest that *F. fomentarius* may represent a promising novel approach for gastric cancer intervention. Furthermore, the exopolysaccharide from this mushroom has a direct anti-proliferative effect *in vitro* on SGC-7901 cells in a dose- and time-dependent manner. Also, this exopolysaccharide sensitized doxorubicin (Dox) and induced growth inhibition of SGC-7901 cells at noncytotoxic concentration of 0.25 mg/ml after 24 h treatment (Chen et al. 2008).

Anti-inflammatory activity

In the last years many researchers have studied the possibility that extracts and isolated metabolites from mushrooms stimulate or suppress specific components of the immune system. Immunomodulators can be effective agents for treating and preventing diseases and illnesses that stem from certain immunodeficiencies and other depressed states of immunity (Chirigos 1992). Those metabolites which appear to stimulate the human immune response are being sought for the treatment of cancer, immunodeficiency diseases, or for generalized immunosuppression following drug treatment, for combination therapy with antibiotics, and as adjuvants for vaccines (Zhang et al. 1994). In an attempt to find bioactive natural products with an anti-inflammatory activity, Park et al. (2004) evaluated the effects of the methanol extract of *Fomes fomentarius* (MEFF) on *in vivo* anti-inflammatory and anti-nociceptive activities. MEFF reduced acute paw edema induced by carrageenin in rats, and showed MEFF analgesic activity, as determined by an acetic acid-induced writhing test and a hot plate test in mice. MEFF potently inhibited the production of nitric oxide (NO), prostaglandin E2 (PGE2), and tumor necrosis factor-a (TNF-a) in LPS-stimulated RAW 264.7 macrophages. Inducible NO synthase (iNOS) and cyclooxygenase-2 (COX-2) levels were reduced by MEFF in a dose-dependent manner. Furthermore, MEFF suppressed nuclear factor-k B (NF-k B) activation in LPS-stimulated RAW 264.7 macrophages. These findings suggest that the anti-inflammatory and anti-nociceptive properties of the methanol extract of MEFF may result from the inhibition of iNOS and COX-2 expression through the down-regulation of NF-k B binding activity. Whilst it is known that mushroom extracts have immunomodulatory and/or antitumor activity, the standard approach has been to isolate, characterize, and administer the pure active constituents. However, different components in a mushroom extract may have synergistic activities (Sanodiya et al. 2009). There are several reports of mushrooms containing more than one polysaccharide with antitumor activity. The responses to different polysaccharides are likely to be mediated by different cell surface receptors, which may be present only on specific subsets of cells and may trigger distinct downstream responses. A combination of such responses involving different cell subsets could conceivably provide greater tumor inhibition than could be induced by a single polysaccharide (Tasaka et al. 1988, Sanodiya et al. 2009).

Antibacterial and antiviral activities

Mushrooms need antibacterial compounds to survive in their natural environment. It is therefore not surprising that antimicrobial compounds with more or less strong activities could be isolated from many mushrooms and that they could be of benefit for humans (Lindequist et al. 2005). The woody tinder fungus, *F. fomentarius* inhibited growth of *P. aeruginosa*, *S. marcescens*, *S. aureus*, *B. subtilis*, and *M. smegmatis*, a relative of the pathogenic *M. tuberculosis* (Suay et al. 2000). In contrast to bacterial infectious diseases, viral diseases cannot be treated by common antibiotics and specific drugs are urgently needed. Antiviral effects are described not only for whole extracts of mushrooms but also for isolated compounds. They could be caused directly by inhibition of viral enzymes, synthesis of viral nucleic acids or adsorption and uptake of viruses into mammalian cells. These direct antiviral effects are exhibited especially by smaller molecules. Indirect antiviral effects are the result of the immunostimulating activity of polysaccharides or other complex molecules (Brandt and Piraino 2000, Lindequist et al. 2005). Aoki et al. 1993 reported that the filtrate from the culture of polypore *F. fomentarius*, is highly active against the mechanical transmission of tobacco mosaic virus (TMV) with an IC50 value of 10 µg/ ml, and it has similar effects against the TMV infection on bell pepper and tomato plants.

Antioxidant activities

The antioxidants are important compounds that defend our body against free radicals and mushrooms are rich sources of these compounds. Antioxidant properties of mushrooms are usually related to low-molecular weight compounds, in particular to the phenolic fractions. Therefore, a wide range of these potentially beneficial phenolic compounds could be natural substrates for oxidative enzymes, such as peroxidases or polyphenol oxidases, which are present in high levels in mushrooms. Kalyoncu et al. (2010) investigated the antioxidant properties of 21 wild mushrooms: *Agaricus bresadolanus, Auricularia auricula-judae, Chroogomphus rutilus, Fomes fomentarius, Ganoderma lucidum, Gloeophyllum trabeum, Gymnopus dryophilus, Infundibulicybe geotropa, Inocybe flocculosa* var. *crocifolia, Inocybe catalaunica, Lentinula edodes, Lentinus sajor-caju, Lycoperdon excipuliforme, Macrolepiota excoriata, Morchella esculenta* var. *rigida, Morchella intermedia, Omphalotusolearius, Pleurotus djamor, Postia stiptica, Rhizopogon roseolus* and *Stropharia inuncta*. On the basis of the results, it is suggested that extracts of the wild macrofungi could be of use as an easily accessible source of natural antioxidants. However, at present, the active components in the extracts, responsible for the observed antioxidant activity, are still unknown. The phenolic compounds of mushrooms are the type of antioxidant that possess a strong inhibition effect against lipid oxidation through radical scavenging (Gezer et al. 2006, Gezer et al. 2011). The bioactivity of these phenolics may be related to their ability to chelate metals, inhibit lipoxygenase and scavenge free radicals (Yoon et al. 2011). Further work is necessary on the isolation and purification of the active components from crude extracts of mushrooms to ascertain their mode of action.

Hypoglycemic activity

Diabetes is a disease due to the lack of regulation of the blood sugar level due to the malfunction or non function of specialised pancreatic cells. It is a metabolic disorder affecting 250 million people worldwide. More effective and safer treatment modalities for type 2 diabetes patients need to be investigated, focusing on overcoming peripheral insulin resistance (Konno et al. 2002, Lindequist et al. 2005). Certain mushrooms could have a positive effect on early stages of type 2 diabetes. Many investigators have endeavored to study the hypoglycemic effect either from the fruiting body or mycelia of various edible/medicinal fungi such as *Tremella aurantia, Agrocybe cylindracea*, and *Fomes fomentarius* (Kiho et al. 1994, Lo et al. 2005, Lee 2005, Yang et al. 2008). Two polysaccharides isolated from *Agrocybe cylindracea* were active in normal and streptomicin-induced diabetic mice by i.p. administration (Kiho et al. 1994). According to Lo et al. (2005) *Tremella mesenterica* might be developed as a potential oral hypoglycaemic agent or functional food for diabetic patients and for persons with high risk for diabetes mellitus. Lee et al. (2005) examined the effects of *Fomes fomentarius* supplementation in streptozotocin-induced diabetic rats. In that study *Fomes fomentarius* extract lowered the serum glucose level, increased glutathione peroxidase activity and significantly lowered superoxide dismutase and catalase activities. Yang et al. (2008) investigated the hypoglycemic effects of *F. fomentarius* exo-biopolymers in streptozotocin (STZ)-induced diabetic rats. The rats from each experimental group were orally administered with exo-biopolymers (100 mg/kg BW) daily for 2 weeks. The administration of the exo-biopolymers substantially reduced the plasma glucose level as compared to the saline administered group (control). It also reduced the plasma total cholesterol, triglyceride, aspartate aminotransferase and alanine aminotransferase levels.

Cultivation and Production of *Fomes fomentarius*

World production of mushrooms over the last two decades has shown a phenomenal pattern of growth, with a 5 times increase in tonnage. While *Agaricus bisporus* still retains the highest overall world production, its relative contribution is decreasing due to the dramatic increase in the other species that are both edible and have medicinal properties (*Lentinula* and *Pleurotus* in particular) or are only medicinal (*Ganoderma, Trametes* and *Fomes*). Medicinal mushrooms can be cultivated through a variety of methods. Some methods are extremely simple and demand little or no technical expertise, while cultivations which require aspects of sterile handling technology are much more technically demanding (Stamets 2000, 2005). For production of the mushroom fruit-bodies, various forms of solid substrate or low moisture fermentations are employed whereas, for mycelial biomass production, liquid fermentations are now becoming increasingly important especially for nutriceutical and pharmaceutical productions. Cultivation on solid substrates, stationary liquid medium or by submerged cultivation has become essential to meet the increasing demands of medicinal mushrooms in the international markets.

Submerged Liquid Fermentation with *Fomes fomentarius*

Submerged pure culture fermentation techniques have been widely developed for most of the main medicinal mushrooms and used in the propagation of mycelium for three main applications, viz.: (i) liquid spawn for solid substrate fruit-body production; (ii) biomass that can be used for food and dietary supplements; and (iii) biomass and/or extruded metabolites especially exo-polysaccharides as raw materials for pharmaceutical studies (Smith et al. 2002). In all cases the underlying principle in each approach is to use mycelium in the active physiological state and of known purity. As in any fermentation study, the factors which can affect mycelial growth rate, yield of biomass and metabolic production, include inoculum size, pH, composition of nutrients, aeration and temperature. While many studies have been restricted to shake flask cultures, others have used laboratory and pilot-scale liquid cultivation technology, with dry mass yields of

16–18 g l^{-1} during 4–5 d of cultivation for several medicinal mushroom species (Rozhkova et al. 2001, Solomko 2001, Smith et al. 2002). A recurring problem with the use of Basidiomycete fungi in liquid fermentation conditions has been the low rate of mycelial growth as compared with other microorganisms such as bacteria, yeasts, and filamentous fungi. The vegetative mycelial state of most medicinal mushrooms will be the dikaryon, the binucleate cell containing the opposite sexual nuclei. The vegetative propagation of such cells involves complex clamp connections which may be an impediment to rapid mycelia propagation. The dikaryon is the stable, long-living stage of the Basidiomycete life-cycle whereas the monokaryotic stage is normally shortlived.

SLF for mycelia and polysaccharides production

Chen et al. (2008) optimized the submerged culture conditions and nutritional requirements of exopolysaccharide (EPS) from *F. fomentarius*. Under the optimal culture condition, the maximum EPS concentration reached 3.64 g l^{-1}, which is about four times higher than that at the basal medium. The EPS from *F. fomentarius* has a direct antiproliferative effect *in vitro* on SGC-7901 huaman gastric cancer cells in a dose- and time-dependent manner. Moreover, it was about three times that EPS at noncytocxity concentration of 0.25 mg ml^{-1} could sensitize doxorubicin(Dox)-induced growth inhibition of SGC-7901 cells after 24 h treatment. However, they found that sometimes the properties of EPS were vulnerable to culture conditions in large-scale fermentation, which resulted in poor quality control (Chen et al. 2008, Chen et al. 2011). In comparison, the properties of mycelia and intracellular polysaccharide (IPS) were more stable. As mycelia and IPS belong to two-parameter products, the total amount of IPS cannot reach maximal even if the concentration in mycelium is maximal. A simultaneous higher production of mycelia and IPS in *F. fomentarius* was obtained using desirability functions, which could have a wide application in other microbial fermentation processes. Under the optimal culture condition, the production of mycelia and IPS reached 17.19 and 2.86 g l^{-1} respectively, in a 15l stirred tank bioreactor, which were about twice that of the basal medium (Chen et al. 2011). Exo-biopolymers with hypoglycemic effects were also successfully produced by submerged mycelial cultures of *Fomes fomentarius* as reported by Yang et al. (2008).

SLF for lignocellulolytic enzymes production

The ligninolytic enzymes from *F. fomentarius* has been successfully produced by submerged mycelial culture (Jaszek et al. 2006). The authors found that the application of menadione superoxide generating agent to

F. fomentarius cultures stimulated extracellular laccase and manganese-dependent peroxidase activities in comparison to the control values (without menadione). These results suggest that for *F. fomentarius*, enhancement of the natural ligninolytic metabolism is one of the strategies for adapting to oxidative stress conditions. Elisashvili et al. (2009) investigated the production of lignocellulolytic enzymes by *F. fomentarius* under submerged fermentation of seven lignocellulosic by-products (Mandarin peels, wheat straw, wheat bran, apple pomace, peach pomace, grape pomace and maple leaves). Among the substrates tested, mandarin peels and wheat bran provided the highest laccase and MnP activities (16,590 versus 450 U l^{-1}, respectively) whereas wheat straw induced the highest CMCase and xylanase production (88 versus 195 U ml^{-1}, respectively).

Solid-State Fermentation with *Fomes fomentarius*

It has been reported that the solid state fermentation (SSF) is an attractive alternative process to produce fungal enzymes using lignocellulosic materials from agricultural wastes due to its lower capital investment and lower operating cost (Rodriguez Couto et al. 2003, 2004, 2006, Rodriguez Couto and Sanroman 2005, Gomez et al. 2005, Rodriguez Couto and Toca-Herrera 2007, Neifar et al. 2009). SSF process, for the reasons stated, will be ideal for developing countries. Solid-state fermentations are characterized by the complete or almost complete absence of free liquid. Water, which is essential for microbial activities, is present in an absorbed or in complexed-form with the solid matrix and the substrate (Rodriguez Couto et al. 2003, Rodriguez Couto and Sanroman 2005, Rodriguez Couto and Toca-Herrera 2007). These cultivation conditions are especially suitable for the growth of fungi, known to grow at relatively low water activities. As the microorganisms in SSF grow under conditions closer to their natural habitats they are more capable of producing enzymes and metabolites which will not be produced or will be produced only in low yield in submerged conditions (Rodriguez Couto et al. 2003). SSFs are practical for complex substrates including agricultural, forestry and food-processing residues and wastes which are used as carbon sources for the production of lignocellulolytic enzymes (Haltrich et al. 1996). Like all technologies, SSF has its disadvantages such as heat buildup, bacterial contamination, scale-up, biomass growth estimation and control of substrate content. However, the process has been used for the production of many microbial products and the engineering aspects and the scale-up will depend on the bioreactor design and operation (Rodriguez Couto and Toca-Herrera 2007). Most solid-state fermentations (SSF) with *F. fomentarus* were carried out at small scale, in agar plates, Erlenmeyer flasks or fixed-bed bioreactors. The SSF studies have been focused on the utilization of lignocellulosic organic waste materials for either lignin degradation, use as animal feed, or enzyme production (Neifar et al. 2009, 2010, 2011a, b, c, 2012).

SSF for mycelial biomass and mushroom production

The information about the production of *F. fomentarius* basidiocarp (fruitbody) and mycelium in solid state fermentation is not much established but it can be capitalized for small and large scale industries to meet the increasing demand and supply of present status. *F. fomentarius* MUCL 35117 has been tested for its fruiting ability on different substrates (A, B, C and D) in industrial bags or in polypropylene pots. There was a positive correlation between C:N ratio and fructifying success: the fungus fructified on the substrate with higher nitrogen content (C:N= 75), but the percentage decreased where the substrate had lower Nitrogen content (C:N=265). Mycelial growth was very rapid as it colonised the whole substrate in about 15 days. In all bags, mycelium first took a fawn colour and exuded black liquid, then became dark pigmented and harder. The fruit body on substrate A had two growth phases in 84 d. It had 3 growth phases in 138 d on B1, one growth phase in 100 days in B2 and 2 growth phases in 100 days on B3 (Giovannini 2006).

SSF for laccase production

Laccases (benezenediol:oxygen oxidoreductase, EC 1.10.3.2) are multinuclear copper-containing enzymes that catalyze the oxidation of a variety of phenolic and inorganic compounds, with the concomitant reduction of oxygen to water (Bourbonnais and Paice 1990). Laccases exhibit an extraordinary natural substrate range (phenols, polyphenols, anilines, aryl diamines, methoxysubstituted phenols, hydroxyindols, benzenethiols, inorganic/organic metal compounds and many others) which is the major reason for their attractiveness for dozens of biotechnological applications (Xu 2005, Riva 2006, Kunamneni et al. 2008). Moreover, in the presence of small molecules known as redox mediators, laccases enhance their substrate specificity. Indeed, laccase oxidizes the mediator and the generated radical oxidizes the substrate by mechanisms different from the enzymatic one, enabling the oxidative transformation of substrates with high redox potentials-otherwise not oxidized by the enzyme (Neifar et al. 2011a). The industrial applicability of laccase may therefore be extended by the use of a laccase-mediator system (LMS). Thus, laccase and LMS find potential in environmental applications including pulp delignification, xenobiotics degradation and textile dye bleaching (Yaropolov et al. 1994, Call and Mücke 1997). Different culturing methods, viz, cell immobilization on stainless steel sponges and plastic material and solid-state fermentation (SSF) using wheat bran as substrate were used for laccase production by the white rot fungus *F. fomentarius* MUCL 35117 (Neifar et al. 2009). The SSF study expressed the highest laccase activities, nearly 6400 U l^{-1} after 13 d of laboratory flasks cultivation. When the wheat bran medium was supplemented with 2 mmol l^{-1} copper sulfate, laccase activity increased by threefold in comparison to control cultures, reaching 27,864 U l^{-1}. With the medium thus optimized, further experiments were performed in a 3 l fixed-bed bioreactor leading to a laccase activity of about 6230 U l^{-1} on 13 d (Neifar et al. 2009). Statistical approaches were employed for the optimization of different cultural parameters for the production of laccase by *F. fomentarius* in wheat bran-based solid medium. Screening of production parameters was performed using an asymmetrical design, and the variables with statistically significant effects on laccase production were selected for further optimization studies using a Hoke design. The application of the response surface methodology allows to determine a set of optimal conditions (CaCl$_2$, 5.5 mg/gs, CuSO$_4$, 2.5 mg/gs, inoculum size, 3 fungal discs, and 13 d of static cultivation) leading to a laccase production yield of 150 U/g dry substrate (Neifar et al. 2011b). Laccase produced by *Fomes fomentarius* grown on wheat bran in solid cultures was purified to electrophoretic homogeneity and characterized. The biochemical characteristics of *F. fomentarius* laccase and its high ability to oxidase phenolic compounds and anthraquinonic dyes as well as its relatively easy production and purification, make it very attractive for application in different biotechnological areas, particularly in bioremediation (Neifar et al. 2010, Neifar et al. 2011a, Neifar et al. 2012), in juice clarification (Neifar et al. 2011c) and in chemical synthesis (Neifar et al. 2010). In fact, fungal laccases have been used to synthesize many products of pharmaceutical importance. The first chemical that comes to mind is actinocin, synthesized via a laccase-catalyzed reaction from 4-methyl-3-hydroxyanthranilic acid. This pharmaceutical product has proven effective in the fight against cancer as it blocks transcription of tumor cell DNA (Ossiadacz et al. 1999, Burton 2003). Other examples of the potential application of laccases for organic syntheses include the oxidative coupling of katarantine and vindoline to yield vinblastine. Vinblastine is an important anti-cancer drug, especially useful in the treatment of leukemia (Yaropolov et al. 1994). Laccase coupling has also resulted in the production of several other novel compounds that exhibit beneficial properties, e.g., antibiotic properties (Pilz et al. 2003). The study of new synthetic routes to aminoquinones is of great interest because a number of antineoplast drugs in use, like mitomycin, or under development, like nakijiquinone- derivatives (Stahl et al. 2002) or herbamycin-derivatives (Honma et al. 1992), contain an aminoquinone moiety. Several simple aminoquinones possess activity against a number of cancer cell-lines (Mathew et al. 1986, Zee-Cheng and

Cheng 1970, Pöckel et al. 2006) as well as antiallergic or 5-lipoxygenase inhibiting activity (Pöckel et al. 2006, Timo and Michael 2007). Laccases are also able to oxidize catechins. These molecules are the condensed structural units of tannins, which are considered important antioxidants found in herbs, vegetables and teas. Catechins ability to scavenge free radicals makes them important in preventing cancer, inflammatory and cardiovascular diseases. Oxidation of catechin by laccase has yielded products with enhanced antioxidant capability (Kurisawa et al. 2003, Hosny and Rosazza. 2002). Last but not least, laccase finds applications in the synthesis of hormone derivatives (generating dimers or oligomers by the coupling of the reactive radical intermediates). Nicotra et al. (2004) have recently exploited the laccase capabilities to isolate new dimeric derivatives of the hormone â-estradiol and of the phytoalexin resveratrol, respectively. Similarly, laccase oxidation of totarol, and of isoeugenol or coniferyl alcohol, gave novel dimeric derivatives (Ncanana et al. 2007) and a mixture of dimeric and tetrameric derivatives (Shiba et al. 2000) respectively, whereas an even more complex mixture of products was observed in the oxidation of substituted imidazole (Schäfer et al. 2001). These novel substituted imidazoles or oligomerization products (2–4) are applicable for pharmacological purposes. In another study, derivatization of the natural compound 3-(3,4-dihydroxyphenyl)-propionic acid can be achieved by laccase-catalyzed N-coupling of aromatic and aliphatic amines. The derivatives of this antiviral natural compound 3-(3,4 dihydroxyphenyl)-propionic acid may have interesting pharmaceutical uses.

SSF for waste biodegradation

White-rot basidiomycetes are efficient decomposers of lignocellulose due to their capability to synthesize relevant hydrolytic and oxidative extracellular enzymes. One of the strategies to utilize agro-industrial wastes is to grow these fungi that will not only reduce the toxic nature of the wastes but also help in obtaining protein rich food on cheaper substrate. Olive cake (OC) generated by the oil industries represents a major disposal and potentially severe pollution problem in the Mediterranean countries. Recently, Neifar et al. (2012) studied the bioconversion of OC in solid state fermentation with the medicinal mushroom *F. fomentarius* so as to upgrade its nutritional values and digestibility for its use as ruminants feed. Significant decreases in the values of neutral detergent fiber (hemicellulose, Cellulose and lignin), acid detergent fiber (lignin and cellulose) and acid detergent lignin were detected. The crude protein increased from 6.48% for the control to 22.32% for treated OC. The estimated *in vitro* digestibility improved from 9.31 % (control) to 25.35% for treated OC.

Conclusion

Current investigations demonstrate that *Fomes fomentarius* possesses a number of beneficial medicinal properties such as antitumour, immunomodulatory, antioxidant, anti-inflammatory, anti-allergic, antihypertensive, antihyperglycaemic, antimicrobial and antiviral activities. These activities have been reported for various extracts and isolated compounds, such as polysaccharides, polysaccharide-protein complexes, phenols, organic acids, lactone, steroids, triterpenoids, from mushroom fermentation broth, mycelia or fruiting bodies. In particular, polysaccharides appear to be potent antitumour and immunomodulating substances, besides possessing other beneficial activities. However, the biochemical mechanisms of these therapeutic activities still remain largely unknown. By identifying and modulating intracellular targets of *F. Fomentarius*, this herbal medicine can be developed for clinical usage. While solid substrate fermentations will remain the chosen method for enzyme production and wastes bioconversion purposes, there will be a continued increase in the development of submerged liquid culture to produce a more uniform and reproducible biomass for medicinal products. There are a number of biologically active compounds to be explored in *F. Fomentarius* mycelium and future research may be oriented in that direction.

References

Aoki, M., M. Tan, A. Fukushima, T. Hieda and Y. Mikami. 1993. Antiviral substances with septic effects produced by basidiomycetes such as *Fomes fomentarius*. Biosci. Biotechnol. Biochem. 57: 178–282.

Borneman, J. and R. J. Hartin. 2000. PCR primers that amplify fungal rRNA genes from environmental samples. Appl. Environ. Microbiol. 66: 4356–4360.

Bourbonnais, R. and M. G. Paice. 1990. Oxidation of non-phenolic substrates. An expanded role for laccase in lignin biodegradation. FEBS Lett. 267: 99–102.

Brandt, C. R. and F. Piraino. 2000. Mushroom antivirals. Recent Res. Dev. Antimicrob. Agents Chemother. 4: 11–26.

Breitenbach, J. and F. Kranzlin. 1986. *Pilze der Schweiz*. Band 2: Basidiomyceten. Luzern, Switzerland, Mykologische Gesellschaft. pp. 306–307.

Burton, S. 2003. Laccases and phenol oxidases in organic synthesis. Curr. Org. Chem. 7: 1317–1331.

Butin, H. 1989. *Krankheiten der Wald- und Parkbtiume*. Germany, Stuttgart, Georg Thieme Verlag. pp. 152–153.

Call, H. P. and I. Mücke. 1997. History, overview and applications of mediated ligninolytic systems, especially laccase-mediator-systems (Lignozyme®-process). J. Biotechnol. 53: 163–202.

Chang, S. T. and P. G. Miles. 1992. Mushroom biology—a new discipline. Mycologist. 6: 64–5.

Chen, J. and R. Seviour. 2007. Medicinal importance of fungal b-(1→3) (1→6)-glucans. Mycol. Res. 111: 635–652.

Chen, W., Z. Zhao, S. F. Chen, Y. Q. Li. 2008. Optimization for the production of exopolysaccharide from *Fomes fomentarius* in submerged culture and its antitumor effect *in vitro*. Bioresour. Technol. 99: 3187–3194.

Chen, W., Z. Zhao and Y. Li. 2011. Simultaneous increase of mycelia biomass and intracellular polysaccharide from *Fomes fomentarius* and its biological function of gastric cancer intervention. Carbohydr. Polym. 85: 369–375.

Chirigos, M. A. 1992. Immunomodulators: current and future development and application. Thymus. 19 (suppl. 1): S7–S20.

Delmas, J. 1989. Les champignons et leur culture. Ed.: Flammarion, Paris. p. 689.

Elisashvili, V., E. Kachlishvili, N. Tsiklauri, E. Metreveli, T. Khardziani and S. N. Agathos. 2009. Lignocellulose-degrading enzyme production by white-rot Basidiomycetes isolated from the forests of Georgia. World J. Microbiol. Biotechnol. 25: 331–339.

Fujimoto, H., M. Nakayama, Y. Nakayama and M. Yamazaki. 1994. Isolation and characterization of immunosuppressive components of three mushrooms, *Pisolithus tinctorius, Microporus flabelliformis* and *Lenzites betulina*. Chem. Pharm. Bull. (Tokyo). 42: 694–697.

Gao, H. L., L. S. Lei, C. L. Yu, Z. G. Zhu, N. N. Chen and S. G. Wu. 2009. Immunomodulatory effects of *Fomes fomentarius* polysaccharides: an experimental study in mice. Nan Fang Yi Ke Da Xue Xue Bao. 29: 458–61.

Gezer, K., M. E. Duru, I. Kivrak, A. Turkoglu, N. Mercan, H. Turkoglu and S. Gulcan. 2006. Free-radical scavenging capacity and antimicrobial activity of wild edible mushroom from Turkey. Afr. J. Biotechnol. 5: 1924–1928.

Gezer, K., M. E. Duru, I. Kivrak, A. Turkoglu, N. Mercan, H. Turkoglu and S. Gulcan. 2011. Antioxidant and antityrosinase activities of various extracts from the fruiting bodies of *Lentinus lepideus*. Molecules. 16: 2334–2347.

Gilbertson, R. L. and L. Ryvarden. 1986. North American Polypores. Fungiflora, Oslo. pp. 433.

Giovannini, I. S. 2006. Cultivated basidiomycetes as a source of new products: *In vitro* cultivation development, selection of strains resistant to *Trichoderma viride*, search for new active compounds, factors influencing plasticity in *Grifola frondosa*. Université de Neuchâtel; n°1890.

Gomez, J., M. Pazos, S. Rodrıguez Couto and Ma. A. Sanroman. 2005. Chestnut shell and barley bran as potential substrates for laccase production by *Coriolopsis rigida* under solid-state conditions. J. Food Eng. 68: 315–319.

Gregori, A., M. Svagelj and J. Pohleven. 2007. Cultivation Techniques and Medicinal Properties of *Pleurotus* spp. Food Technol. Biotechnol. 45: 236–247.

Haltrich, D., B. Nidetzky, K. D. Kulbe, W. Steiner and S. Zupancic. 1996. Production of fungal xylanases. Bioresour. Technol. 58: 137–161.

Heilmann-Clausen, J. 2001. A gradient analysis of communities of macrofungi and slime moulds on decaying beech logs. Mycological Res. 105: 575–596.

Heilmann-Clausen, J. and L. Boddy. 2005. Inhibition and stimulation effects in communities of wood decay fungi: exudates from colonized wood influence growth by other species. Microb. Ecol. 49: 399–406.

Hennon, P. E. 1995. Are heart rot fungi major factors of disturbance in gap-dynamic forests? Northwest Sci. 69: 284–293.

Hirotani, M., T. Furuya and M. Shiro. 1991. Cryptoporic acids H and I, drimane sesquiterpenes from *Ganoderma neo-japonicum* and *Cryptoporus volvatus*. Phytochem. 30: 1555–1559.

Hölker, U. and J. Lenz. 2004. Trickle-film processing: An alternative for producing fungal enzymes. BIOforum Europe. 6: 55–57.

Honma, Y., T. Kasukabe, M. Hozumi, K. Shibata, S. Omura. 1992. Effects of herbimycin A derivatives on growth and differentiation of K562 human leukemic cells. Anticancer Res. 12: 189–192.

Hosny, M. and J. P. N. Rosazza. 2002. Novel oxidations of (+)-catechin by horseradish peroxidase and laccase. J. Agric. Food Chem. 50: 5539–5545.

Huang, N. L. 1998. Self-colored illustrated handbook for macrofungi from China. Beijing: Chinese Agriculture Press. p. 61, 76.

Ito, H., K. Shimura, H. Itoh and M. Kawade. 1997. Antitumour effects of a new polysaccharide protein complex (ATOM) prepared from *Agaricus blazei* (Iwade Strain 101) Himematsu-take and its mechanisms in tumour-bearing mice. Anticancer Research. 17: 277–284.

Jaszek, M., J. Zuchowski, E. Dajczak, K. Cimek, M. Graz and K. Grzywnowicz. 2006. Ligninolytic enzymes can participate in a multiple response system to oxidative stress in white-rot basidiomycetes: *Fomes fomentarius* and *Tyromyces pubescens*. Int. Biodeter. Biodegr. 58: 168–175.

Judova, J., K. Dubikova, S. Gaperova, J. Gaper and P. Pristas. 2012. The occurrence and rapid discrimination of *Fomes fomentarius* genotypes by ITS-RFLP analysis. Fungal Biol. 116: 155–160.

Kalyoncu, F., M. Oskay and H. Kayalar. 2010. 'Antioxidant activity of the mycelium of 21 wild mushroom species'. Mycol. 1: 3, 195–199.

Kazuko, Y., I. Mizuho, A. Shigenobu, M. Eiko and K. Satoshi. 2001. Two new steroidal derivatives from the fruit body of *Chlorophyllum molybdites*. Chem. Pharm. Bull. 49: 1030–1032.

Keller, A. C., M. P. Maillard and K. Hostettmann. 1996. Antimicrobial steroids from the fungus Fomitopsis pinicola. Phytochem. 41: 1041–46.

Kiho, T., S. Sobue and S. Udai. 1994. Structural features and hypoglycemic activities of two polysaccharides from a hot-water extract of *Agrocybe cylindracea*. Carbohydr. Res. 3: 81–7.

Kitamura, S., T. Hori, K. Kurita, K. Takeo, C. Hara, W. Itoh, K. Tabata, A. Elgsaeter and B. T. Stokke. 1994. An antitumor, branched (1→3)-beta-D-glucan from a water extract of fruiting bodies of *Cryptoporus volvatus*. Carbohydr. Res. 263: 111–21.

Konno, S., S. Aynehchi, D. J. Dolin, A. M. Schwartz, M. S. Choudhury and H. N. Tazakin. 2002. Anticancer and hypoglycemic effects of polysaccharides in edible and medicinal Maitake mushroom [*Grifola frondosa* (Dicks.:Fr.) S. F. Gray]. Int. J. Med. Mushrooms. 4: 185–95.

Kotlaba, F. 1997. Common polypores (Polyporales s.l.) collected on uncommon hosts. Czech Mycol. 49: 169–188.

Kunamneni, A., S. Camarero, C. García-Burgos, F. J. Plou, A. Ballesteros and M. Alcalde. 2008. Engineering and Applications of fungal laccases for organic synthesis. Microb. Cell Fact. 7: 1–17.

Kurisawa, M., J. E. Chung, H. Uyama and S. Kobayashi. 2003. Laccase-catalyzed synthesis and antioxidant property of poly(catechin). Macromol. Biosci. 3: 758–764.

Lee, I. K., B. S. Yun, S. M. Cho, W. G. Kim, J. P. Kim, I. J. Ryoo, H. Koshino and I. D. Yoo. 1996. Betulinans A and B, two benzoquinone compounds from *Lenzites betulina*. J. Nat. Prod. 59: 1090–1092.

Lee, J. S. 2005. Effects of *Fomes fomentarius* supplementation on antioxidant enzyme activities, blood glucose, and lipid profile in streptozotocin-induced diabetic rats. Nutr. Res. 25: 187–195.

Lindequist, U., T. H. J. Niedermeyer and W. D. Julich. 2005. The Pharmacological Potential of Mushrooms. eCAM. 2: 285–299.

Liu, B. 1978. Medicinal fungi of China. Taiyuan: Sanxi People Press. p. 101, 129.

Liu, J. X. and J. B. JIA. 2009. Isolation, Purification and Identification of Polysaccharides from *Fomes fomentarius*. Food Sci. 30: 80–83.

Lo, H. G., F. A. Tsai, S. P. Wasser, J. G. Yang and B. M. Huang. 2005. Effects of ingested fruiting bodies, submerged culture biomass, and acidic polysaccharide glucuronoxylomannan of *Tremella mesenterica* Retz.:Fr. on glycemic responses in normal and diabetic rats. Life Sci. 78: 1957–66.

Lomascolo, A., J. Cayol, M. Roche, L. Guo, J. L. Robert, E. Record, L. Lesage-Meessen, B. Ollivier, J. Sigoillot and M. Asther. 2002. Molecular clustering of *Pycnoporus* strains from various geographic origins and isolation of monocaryotic strains for laccase hyperproduction. Mycol. Res. 106: 1193–1203.

Mao, X. L. 2000. Macrofungi of China. Zengzhou: Henan Science and Technology Press. p. 447.

Mathew, A. E., K. Y. Zee-Cheng and C. C. Cheng. 1986. Amino-substituted *p*-benzoquinones. J. Med. Chem. 29: 1792–1795.

McDonald, J. A. 1938. *Fomes fomentarius* (Linn.) Gill [Ungulina fomentaria (Linn.) Pat.] on birch in Scotland. Trans. Br. Mycol. Soc. 32: 396–408.

Ncanana, S., L. Baratto, L. Roncaglia, S. Riva and S. G. Burton. 2007. Laccase-mediated oxidation of totarol. Adv. Synth. Catal. 349: 1507–1513.

Neifar, M., A. Jaouani, R. E. Ghorbel, S. E. Chaabouni and M. J. Penninckx. 2009. Effect of culturing processes and copper addition on laccase production by the white-rot fungus *Fomes fomentarius* MUCL 35117. Lett. App. Microbiol. 49: 73–78.

Neifar, M., A. Jaouani, R. E. Ghorbel and S. E. Chaabouni. 2010. Purification, characterization and decolourization ability of *Fomes fomentarius* laccase produced in solid medium. J. Mol. Catal. B: Enzym. 64: 68–74.

Neifar, M., A. Jaouani, A. Kamoun, R. E. Ghorbel and S. E. Chaabouni. 2011a. Decolorization of Solophenyl red 3BL polyazo dye by laccase-mediator system: Optimization through response surface methodology. Enzyme Res. doi:10.4061/2011/179050.

Neifar, M., A. Kamoun, A. Jaouani, R. E. Ghorbel and S. E. Chaabouni. 2011b. Application of Asymetrical and Hoke designs for optimization of laccase production by the white-rot fungus *Fomes fomentarius* in solid-state fermentation. Enzyme Res. doi:10.4061/2011/368525.

Neifar, M., R. E. Ghorbel, A. Kamoun, S. Baklouti, A. Mokni, A. Jaouani and S. E. Chaabouni. 2011c. Effective clarification of pomegranate juice using laccase treatment optimized by response surface methodology followed by ultrafiltration. J. Food Process Eng. 34: 1199–1219.

Neifar, M., A. Ayari, A. Boudabous, A. Cherif and A. Jaouani. 2012. Increasing the feed value of olive oil cake by solid state cultivation of the white-rot fungus *Fomes fomentarius*. J. Environ. Eng. Manage. 11: S69.

Nicotra, S., M. R. Cramarossa, A. Mucci, U. M. Pagnoni, S. Riva and L. Forti. 2004. Biotransformation of resveratrol: synthesis of trans-dehydrodimers catalyzed by laccases from *Myceliophtora thermophyla* and from *Trametes pubescens*. Tetrahedron. 60: 595–600.

Ossiadacz, J., A. J. H. Al-Adhami, D. Bajraszewska, P. Fischer, W. Peczynska-Czoch. 1999. On the use of *Trametes versicolor* laccase for the conversion of 4-methyl-3-hydroxyanthranilic acid to actinocin chromophore. J. Biotechnol. 72: 141–149.

Park, Y. M., I. T. Kim, H. J. Park, J. W. Choi, K. Y. Park, J. D. Lee, B. H. Nam, D. G. Kim, J. Y. Lee and K. T. Lee. 2004. Anti-inflammatory and Anti-nociceptive Effects of the Methanol Extract of *Fomes fomentarius*. Biol. Pharm. Bull. 27: 1588–1593.

Patel, S. and A. Goyal. 2012. Recent developments in mushrooms as anti-cancer therapeutics: a review. 3 Biotech. 2: 1–15.

Pegler, D. N. 1973. Aphyllophorales IV—Poriod Families. *In:* G. C. Ainsworth and F. K. Schwarze F. 1994. Wood rotting fungi: *Fomes fomentarius* (L.: Fr.) Fr. Hoof or Tinder fungus. Mycologist. 8: 32–34. pp. 397–400.

Pilz, R., E. Hammer, F. Schauer and U. Kragl. 2003. Laccase-catalyzed synthesis of coupling products of phenolic substrates in different reactors. Appl. Microbiol. Biotechnol. 60: 708–712.

Puchkova, T. A., V. G. Babitskaya and Z. A. Rozhkova. 2001. Physiology and properties of *Lentinus edodes* (Berk.) Sing. in submerged culture. Int. J. Med. Mush. 3: 206.

Pöckel, D., T. H. J. Niedermeyer, H. T. L. Pham, A. Mikolasch, S. Mundt, U. Lindequist, M. Lalk and O. Werz. 2006. Inhibition of human 5-lipoxygenase and anti-neoplastic effects by 2-amino-1,4-benzoquinones. Med. Chem. 2: 591–595.

Riva, S. 2006. Laccases: blue enzyme for green chemistry. Trends Biotechnol. 24: 219–226.

Robles-Hernández, L., A. Cecilia-González-Franco, J. M. Soto-Parra and F. Montes-Domínguez. 2008. Review of agricultural and medicinal applications of basidiomycete mushrooms. *Tecnociencia Chihuahua*. 2: 95–107.

Rodrıguez Couto, S. and Ma. A. Sanroman. 2005. Application of solid-state fermentation to ligninolytic enzyme production. Biochem. Eng. J. 22: 211–219.

Rodrıguez Couto, S., D. Moldes, A. Liebanas and Ma. A. Sanroman. 2003. Investigations of several bioreactor configurations for laccase production by *Trametes versicolor* operating in solid-state conditions. Biochem. Eng. J. 15: 21–26.

Rodrıguez Couto, S., Ma. A. Sanroman, D. Hofer and G. M. Gubitz. 2004. Stainless steel sponge: a novel carrier for the immobilisation of the white-rot fungus *Trametes hirsuta* for decolourisation of textile dyes. Bioresour. Technol. 95: 67–72.

Rodrıguez Couto, S., E. Lopez and Ma. A. Sanroman. 2006. Utilisation of grape seeds for laccase production in solid state fermentors. J. Food Eng. 74: 263–267.

Rodrıguez Couto, S. and J. L. Toca-Herrera. 2007. Laccase production at reactor scale by filamentous fungi. Biotechnol. Adv. 25: 558–569.

Rogers, R. 2006. The Fungal Pharmacy: Medicinal Mushrooms of Western Canada, Prairie Deva Press, Edmonton, Canada. p. 234.

Rösecke, J. and W. A. König. 2000. Constituents of various wood-rotting basidiomycetes. Phytochem. 54: 603–610.

Rozhkova, Z. A. 2001. Physiology and properties of *Lentinus edodes* (Berk.) Sing. in submerged culture. Int. J. Med. Mush. 3: 206.

Saar, M. 1991. Fungi in Khanty Folk Medicine. J. Ethnopharmacol. 31: 175–179.

Sanodiya, B. S., G. S. Thakur, R. K. Baghel, G. B. K. S. Prasad and P. S. Bisen. 2009. *Ganoderma lucidum*: A Potent Pharmacological Macrofungus. Curr. Pharm. Biotechnol. 10: 717–742.

Schäfer, A., M. Specht, A. Hetzheim, W. Francke and F. Schauer. 2001. Synthesis of substituted imidazoles and dimerization products using cells and laccase from *Trametes versicolor*. Tetrahedron. 57: 7693–7699.

Schmidt, O. 2006. Wood and Tree Fungi. Biology, Damage, Protection, and Use. Springer, Berlin e Heidelberg e New York.

Schwarze, F. W. M. R. 1994. Wood rotting fungi: *Fomes fomentarius* (L.: Fr.) Fr. Mycologist. 8: 131–133.

Schwarze, F. W. M. R., J. Engels and C. Mattheck. 2000. eds. Fungal strategies of wood decay in trees. Springer-Verlag, Berlin, Heidelberg. p. 185.

Shiba, T., L. Xiao, T. Miyakoshi and C. -L. Chen. 2000. Oxidation of isoeugenol and coniferyl alcohol catalyzed by laccase isolated from *Rhus vernicifera* Stokes and *Pycnoporus coccineus*. J. Mol. Catal. B: Enzym. 10: 605–615.

Sinclair, W. A., H. H. Lyon and W. T. Johnson. 1987. *Diseases of Trees and Shrubs*. Cornell University Press. pp. 346–347.

Smith, J. E., N. J. Rowan and R. Sullivan. 2002. Medicinal mushrooms: a rapidly developing area of biotechnology for cancer therapy and other bioactivities. Biotechnol. Lett. 24: 1839–1845.

Solomko, E. F. 2001. Nutritional and medicinal benefits of *Pleurotus ostreatus* (Jacq.:Fr.) Kumm. Submerged cultures. Int. J. Med. Mush. 3: 223.

Stahl, P., L. Kissau, R. Mazitschek, A. Giannis and H. Waldmann. 2002. Natural product derived receptor tyrosin kinase inhibitors: Identification of IGF1R-, Tie-2 and VEGFR3 inhibitors. *Angew.* Chem. Int. Ed. Engl. 41: 1174–1178.

Stamets, P. 2000. Growing gourmet and medicinal mushrooms. 3rd ed. Berkeley (CA): Ten Speed Press.

Stamets, P. 2005. Mycelium running. How mushrooms can help save the world. Berkeley/ Toronto: Ten Speed Press. p. 339.

Suay, I., F. Arenal, F. J. Asensio, A. Basilio, M. A. Cabello, M. T. Díez, J. B. García, A. González del Val, J. Gorrocha-tegui, P. Hernández, F. Peláez and M. F. Vicente. 2000. Screening of basidiomycetes for antimicrobial activities. *Antonie van Leeuwenhoek.* 78: 129–139.

Tasaka, K., M. Mio, K. Izushi, M. Akagi and T. Makino. 1988. Anti-allergic constituents in the culture medium of *Ganoderma lucidum*. (II). The inhibitory effect of cyclooctasulfur on histamine release. Agents Actions. 23: 157–160.

Timo, H. J. N. and L. Michael. 2007. Nuclear amination catalyzed by fungal laccases: Comparison of laccase catalyzed amination with known chemical routes to aminoquinones. J. Mol. Catal. B: Enzym. 45: 113–117.

Wasser, S. P. and A. L. Weis. 1999a. Medicinal properties of substances occurring in higher Basidiomycetes mushrooms: current perspectives (Review). Int. J. Med. Mushrooms. 1: 31–62.

Wasser, S. P. and A. L. Weis. 1999b. General description of the most important medicinal higher Basidiomycetes mushrooms. Int. J. Med. Mushrooms. 1: 351–370.

Wasser, S. P. 2002. Medicinal mushrooms as a source of antitumor and immunomodulating polysaccharides. Appl. Microbiol. Biotechnol. 60: 258–274.

Wasser, S. P. 2005. The Importance of Culinary-Medicinal Mushrooms from Ancient Times to the Present. Int. J. Medicinal Mushrooms. 7: 363–364.

Xu, F. 2005. Applications of oxidoreductases: recent progress. Ind. Biotechnol. 1: 38–50.

Yang, B. K., G. N. Kim, Y. T. Jeong, H. Jeong, P. Mehta and C. H. Song. 2008. Hypoglycemic Effects of Exo-biopolymers Produced by Five Different Medicinal Mushrooms in STZ-induced Diabetic Rats. Mycobiol. 36: 45–49.

Yaropolov, A. I., O. V. Skorobogat'ko, S. S. Vartanov and S. D. Varfolomeyev. 1994. Laccase: Properties, catalytic mechanism, and applicability. Appl. Biochem. Biotechnol. 49: 257–280.

Ying, J. Z., X. L. Mao, Q. M. Ma, Y. C. Zong and H. A. Wen. 1987. Illustrated handbook for medicinal fungi from China. (Transl. Xu, Y.H.), Science Press, Beijing.

Yoon, K. N., N. Alam, K. R. Lee, P. G. Shin, J. C. Cheong, Y. B. Yoo and T. S. Lee. 2011. Antioxidant and Antityrosinase Activities of Various Extracts from the Fruiting Bodies of *Lentinus lepideus*. Molecules. 16: 2334–2347.

Zee-Cheng, K. Y. and C. C. Cheng. 1970. Preparation and the results of antitumor screening of some substituted amino-, azido-, halogeno- and hydroxy-p-benzoquinones. J. Med. Chem. 13: 264–268.

Zhang, J., G. Wang, H. Li, C. Zhuang, T. Mizuno, H. Ito, H. Mayuzumi, H. Okamoto and J. Li. 1994. Antitumor active protein- containing glycans from the Chinese mushroom *songshan lingzhi*, *Ganoderma tsugae* mycelium. Biosci. Biotechnol. Biochem. 58: 1202–1205.

12

Amazonian Microorganisms with Metabolites of Ecological and Economic Importance

Luiz Antonio de Oliveira

ABSTRACT

The Amazonian biodiversity presents great international interest due to the pressing needs of new products for industrial and commercial uses. Products from biological species may result in tens or hundreds of billions of dollars a year. Products like enzymes and antibiotics found in microorganisms are of interest due to its potential for use worldwide. The industrial application of microorganisms, mainly bacteria, fungi and yeast, is extremely diverse, and may provide astronomical yields. Known examples are the fermentations, such as alcoholic beverages and alcohol fuel, dairy products, organic acids and drugs, including antibiotics. Other products of microbial origin are enzymes and polymers, stereo-specific molecules produced by biotransformation, food additives (amino acids, nutraceutic, carbohydrates) and vitamins. Microorganisms are also used in inoculants for agricultural and industrial use, as well as for the biodegradation of toxic compounds in effluent treatments. The number of known enzymes does not reach 5% of the estimated potential, indicating that there is an unused potential involving soil microbiota. Soils of virgin forests are where there is a

National Institute for Amazonian Research (INPA/MCTI), Avenue André Araújo, 2936, Petrópolis, CEP 69060-000, Manaus, Amazonas, Brazil.
Email: luizoli@inpa.gov.br, luizoli51@gmail.com

large number of plant species which we expect to find the greatest diversity of microorganisms and, thus, enzymes. Also, the presence of microorganisms in plants rhizosphere with diverse abilities, such as to degrade pertroleum is highly promising, because the chances of being pathogenic to humans or other animals are extremely rare, since in this environment predominate root exudates excreted by plant roots.

Introduction

The Brazilian biodiversity, particularly the Amazon, has attracted great international interest due to the pressing needs of new products for worldwide commercial and industrial use. The products arising from such biodiversity may result in tens or hundreds of billions of dollars a year, giving to the Brazilians, better living conditions. Investing in research is crucial for these bioproducts to be discovered and marketed in the future. Otherwise, the more developed nations, through biopiracy or not, may be the legal holders of the rights to the use of this biodiversity. Products like enzymes and antibiotics found in microorganisms, are of interest due to their potential for use worldwide. Soil microorganisms produce secondary metabolites such as antibiotics, anticancer agents, immunosuppressants, enzyme inhibitors, antifungal components, antiparasitic agents, herbicides, insecticides and growth promoters (Omura 1992, Doelle 2007).

This work summarizes the knowledge about soil microbes with biotechnological potential metabolisms, emphasizing research on the agronomic and ecological areas with Amazonian microorganisms, one of the most diverse and little-known of the planet.

Microbial Metabolites

The industrial application of microorganisms, primarily bacteria, fungi and yeast, is extremely diverse and may provide astronomical yields. Classic examples are the fermentations, such as alcoholic beverages and alcohol fuel, dairy products, organic acids and drugs, including antibiotics. Other products of microbial origin are enzymes and polymers of industrial application, stereo-specific molecules produced by biotransformation, food additives (amino acids, nutraceutics, carbohydrates) and vitamins. Microorganisms are also used in the formulation of inoculants for agricultural and industrial use, and for the biodegradation of toxic compounds in effluent treatment to environmental remediation.

It is estimated that 40% of the available drugs have been developed from natural sources: 25% from plants, 13% from microorganisms and 3% from animals. Natural products of plants and microbes involve different researches, such as the authenticity of the identity, chemical composition and pharmacological action, as well as the elucidation of active chemical structures (Calixto and Yunes 2001).

Approximately 50,000 secondary metabolites are produced by microorganisms. Of 12,000 antibiotics identified, 22% are from filamentous fungi, where penicillin, cephalosporin C, griseofulvin and fusidic acid are clinically important and biosynthesized by *Cephalosporium*, *Aspergillus* and *Penicillium* (Crueger and Crueger 1990, Pearce 1997, Manfio 2012). The secondary metabolites produced by microorganisms can be synthesized via non-ribosomal and ribosomal (Kolter and Moreno 1992).

The biocompounds, especially those produced by fungi, are potential sources of new chemical substances. These may become promising in bioactive prototypes among which are antibiotics, cholesterol lowering, antitumor, antifungal and immunosuppressive agents, antiprotozoal and enzymes (Göhrt and Zeek 1992, Demain 1999, Newman et al. 2000, Sotero-Martins et al. 2004, Silva Neves et al. 2006).

There is still a large number of other enzymes that should be better researched in the microbiota, as well as antibiotic producers, since there is the appearance of antibiotic resistant pathogens (Fischbach and Walsh 2009). From the discovery of streptomycin, a large number of antibiotics of various structures, such as aminoglycosides, chloramphenicol, tetracycline, and macrolides, were isolated from microorganisms belonging to the genus *Streptomyces*.

However, with the continuous discovery of new antimicrobials and biomolecules in general, the possibility of discovering new 'Streptomyces' which produce various new biomolecules has become increasingly difficult. The frequency to re-isolate organisms producing antibiotics identical to those known is increasing.

The "golden age of antibiotics" came at the middle of the 1960's, after which the discovery of new antimicrobials, from members of the genus *Streptomyces* began to decline rapidly, making the discovery of new pharmacologically interesting molecules increasingly difficult, uncertain and costly. However, it is interesting to note that already during the decade of the 1950's and the early 1960's, and even at the height of the "golden age of streptomicetes as producers by excellence of antimicrobials", occasional publications reporting the discovery of new genera of actinomycetes appeared in the scientific literature, such as members of the genus *Actinoplanes* (Couch 1950), *Streptosporangium* and later, others such as *Microechinospora*, *Microellobosporia* and *Microbispora* (Couch 1950, 1963, Cross et al. 1963, Konev et al. 1967, Nonomura and Ohara 1969).

However, the isolation of representatives of these new genera, also called rare *Actinomycetes* genera, although not only extremely interesting from the standpoint of its novelty but also as a potential producer of new biologically active molecules, their isolation from soil samples with available methods then was totally incidental and hence unsuitable for use as a screening target for industrial scale. The critical element in a research process of microorganisms as producers of new biologically active molecules, is precisely the availability of techniques for isolating members of the genus unexplored and are simultaneously, producers of secondary metabolites.

The importance of rare *Actinomycetes* as producers of new and important biomolecules was evident by the fact that many antibiotics that were successful in the market are produced by representatives of rare genera, such as Riphamycin produced by *Amycalotopsis mediterranei*, Erythromycin by *Saccharopolyspora erythreae*, Teicoplanin by *Actinoplanes teichomyceticus*, Vancomycin by *Amycolatopsis orientalis*, Gentamicin by *Micromonospora purpurea* (Lancini and Lorenzetti 1993).

In order to enhance the isolation of representatives of rare genera, the efforts of many research laboratories of pharmaceutical companies began to direct their efforts to the study of soil samples from exotic niches such as marine environments, sludge ponds, brine, high or very low pH environments, assuming that samples of such places might harbor microorganisms with unique characteristics and thus also with a potential to produce new and pharmacologically interesting molecules (Lazzarini et al. 2000).

But there has not been great success obtained from these exotic, highly selective environments, where representatives of rare genera may occur. This limited success was mainly due to technology adopted:

a) Sample dilution before plating, as is usual to proceed to the isolation of *Streptomyces,* resulting in a near total loss of the rare representatives present in very low concentrations.
b) Generally slower growth of the members of the rare genera species.
c) Preference of representatives of rare genera by poor medium.

To eliminate these drawbacks, a new method was developed for screening (Thiemann 2012) that allows the isolation of a large number of representatives of rare *Actinomycetes*, as well as the isolation of all genera of new Actinomycetales. This isolation method has allowed the isolation of a large number of representatives of rare genera, such as *Microbispora, Microtetraspora, Actinoplanes, Ampulariella, Pilimelia, Dactylosporangium, Streptosporangium,* and also, isolates currently under taxonomic classification, from probably not yet described genera.

Enzymes diversity

Most metabolic compounds secreted by microorganisms are made up of enzymes, which have quite different properties, sometimes acting as antibiotics or antimicrobial agents, or as processers of organic compounds.

Because of these highly desirable characteristics, economically and ecologically, their importance is quite high.

The global market involves about $40 billion per year from products of microbial origin (Beilen and Li 2002), with enzymes representing about $1.3 billion (Panke and Wubbolts 2002, Castro et al. 2004). Lipases have a global market of around $450 million per year (Castro et al. 2004), showing how microbial enzymes are important in today's world.

The International Committee on Nomenclature of Enzymes cataloged in 1992, 3196 enzymes (http://www.chem.qmul.ac.uk/iubmb/enzyme/history.html) (accessed on 20.06.2012). Between 1993 and 2012, 18 new supplements of enzymes were reported by this committee (http://www.chem.qmul.ac.uk/iubmb/enzyme/supplements/). In 2012, more than 120 new enzymes were listed, but the diversity can be much greater than this number, especially in regions of high biological diversity as in the Amazon rainforest.

There is a direct relationship between the number of enzymes present in soil microbes in the litter layer of forests and the number of plant species that are there. The leaves, branches and fruits that fall to the ground, are converted into water and carbon dioxide by the microflora. Theoretically, each species must have at least one chemical component unique in its leaves, branches and trunk, that justifies them as a single species. And this particular chemical compound is converted to a minor compound or into water and carbon dioxide by a specific enzyme. Thus, theoretically it is estimated that the number of enzymes found in the microflora of the soil is very similar to the number of biological species comprising terrestrial ecosystems.

It is estimated that 265,000–280,000 is the number of known plants species in the world, i.e., species formally described and documented in biological collections (for specimens, but also, sometimes by an iconography). Considering species yet to be discovered and identified, this total number may reach 500,000 species, according to botanical experts. Brazil is considered as the country of greatest biological diversity, with about 14% of plant diversity in the world (Shepherd 2002). There is an estimation that in the Brazilian territory between 45 and 50 thousand plant species may be found (Lewinsohn and Prado 2002). Of these, about 20,000 are in the Amazon, although some estimates can reach 75,000 species (15% of 500,000, according to some botanical experts).

Thus, we expect to find in the world's ecosystems, a number of microbial enzymes in the litter layers on the ground, very similar to the number of plant species that compose them. If we also consider the enzymes found inside the plants, these numbers can be much larger. Therefore, we know less than 5% of the enzymes present on the planet.

This high enzyme potential, which can be found in the Brazilian biodiversity, especially in the Amazon, must be better known and explored in order to contribute effectively to the development of mankind.

Microorganisms are also excellent bioindicators in ecosystems. They can serve as a basis for evaluating the effect of human activity on terrestrial and aquatic ecosystems well before macroscopic changes are detected. Studying the diversity of microbial enzymes in soils is a way to evaluate environmental impacts on ecosystems, because there is a direct relationship between enzymatic diversity and biological diversity.

So far, few studies with soil microorganisms and enzyme profiles have been performed in the Amazon (Oliveira and Oliveira 2003a, b, 2005a, b, Oliveira et al. 2003, 2006a, b, 2007, Ferreira et al. 2004, Hara and Oliveira 2004, 2005). Enzymes such as cellulases, hemicellulases, ligninases, phosphatases, nitrogenases, lipases, chitinases and pectinases have been studied in agriculture and forestry systems, with microorganisms being isolated for studies for future economic and ecological uses.

Enzymes of soil microorganisms as biological indicators in the Amazon

Tables 1 and 2 show the populations of microorganisms that produce enzymes found in some gaps opened in the Amazon canopy to explore petroleum and natural gas and those found in the forests around these gaps (Prado 2009), in the Urucu Petroleum Province, Coari, Amazonas. We observed clear differences in the populations of bacteria and fungi found in JAZ 05, compared to those found in its adjacent forest (FN 05), for both the producers of amylases and proteases (Table 1), as well as the cellulase and urease producers (Table 2).

The populations of microorganisms producing amylases (Table 1) and ureases (Table 2) were higher in the JAZ 05, while the producers of proteases (Table 1) and cellulases (Table 2) were higher in the neighbor forest (FN 05).

Table 1. Proteolytic and amylolytic bacteria and fungi population in soils from Urucu Petroleum Province at Urucu, Coari, Amazonas.

Soils	Amylolytic bacteria	Amylolytic fungi	Proteolytic bacteria	Proteolytic fungi
	---------------------------$10^{3} \cdot g^{-1}$ soil -----------------------			
FN 05 (Native Forest)	3.3	0.7	300	180.0
JAZ 05 (gap 05)	14.0	3.3	11	0.3
FN IMT-1 (Native Forest)	0.7	6.6	53	20.0
JAZ IMT-1 (gap IMT-1)	14.3	0.3	15	<0.1

Source: Prado (2009).

When comparing the results obtained in the JAZ IMT-1 with its adjacent forest (FN IMT-1), we observed the same behavior, except for the amylolytic fungi population, which had a lower population in the JAZ IMT-1 than in the forest FN IMT-1 (Table 1) (Prado 2009).

Distinct populations of microorganisms can also be found in the rhizoplane of fruit species (Table 3), as a result of the differences among their exudates released by plant roots (Bais et al. 2006). The number of amylolytic bacteria were reduced on the cupuassu and banana roots and higher on the assai, peach palm and tucuman, while the amylolytic fungi were higher on banana roots (Prado 2009).

As protease-producing microorganisms, the roots of assai palms showed the smallest population when compared to those found in the other species. There were no differences among the populations of proteolytic bacteria found on the roots of these other species, but there were major differences among plant species in terms of proteolytic fungi populations, with the highest being on roots of banana plants (Prado 2009).

Table 2. Cellulolytic and ureolytic bacteria and fungi population in soils from Urucu Petroleum Province at Urucu, Coari, Amazonas.

Soils	Cellulolytic bacteria	Cellulolytic fungi	Ureolytic bacteria	Ureolytic fungi
	----------------------- 10^{3} g^{-1} soil ----------------------			
FN 05 (Native Forest)	257	23	97	60
JAZ 05 (gap 05)	21	1	457	700
FN IMT-1 (Native Forest)	223	20	40	7
JAZ IMT-1 (gap IMT-1)	19	2	2626	1433

Source: Prado (2009).

Table 3. Population of amylolytic and proteolytic bacteria and fungi found in the rhizoplane of fruit species in the rural community of Brasileirinho, Manaus, Amazonas.

Soils/Species	Amylolytic bacteria	Amylolytic fungi	Proteolytic bacteria	Proteolytic Fungi
	-------------------- 10^{3} g^{-1} soil ----------------------			
Assai (*Euterpe oleracea*)	30.0	<0.1	28	<0.1
Peach-palm (*Bactris gasipaes*)	30.0	6.0	302	6.0
Tucuman (*Austrocaryum aculeatum*)	30.0	10.0	300	10.0
Buriti (*Mauritia flexuosa*)	30.0	13.3	301	13.3
Cupuassu (*Theobroma grandiflorum*)	<0.1	2.0	253	2.0
Banana (*Musa* sp.)	<0.1	30.0	300	30.0
Acerola (*Malphigia glabra*)	3.0	<0.1	265	<0.1

Source: Prado (2009).

Degrading microbial metabolites and derivatives of oil in the Amazon

With the intensification of petroleum and natural gas uses in the Amazon Region, it is important to evaluate the ability of soil microorganism population to produce enzymes able to degrade these organic compounds. The emphasis on use of beneficial microorganisms present in the rhizosphere of plants, such as rhizobia and phosphate solubilizing microorganisms may be of great value, since the surface of forests or crop soils are covered with a mantle of roots. Thus, it is the first layer of soil to be affected by contaminants, if there is a spill of oil or its derivatives (gasoline, diesel, etc.). Moreover, non pathogenic microorganisms with these biodegradation characteristics can be used as inoculant without causing adverse environmental impacts as compared to those that are pathogenic to humans, other animals or plants. Some of these pathogenic microorganisms are also capable of producing enzymes for degrading oil, as *Streptococcus aureus, Clostridium faecales, Klebsiella pneumoniae*, etc., but should not be used for such purposes.

The data presented in Tables 4 and 5 illustrate the occurrence of bacteria in the rhizosphere of forest soils of Urucu (city of Coari, Table 4, Mari 2008) and four species of fruit from the Rural Community of Brasileirinho (Manaus, Table 5, Mari 2008), able to degrade the oil extracted by Petrobras in the Urucu Petroleum Province.

When the soils were enriched with diesel oil, populations of these bacteria were higher compared with those of samples where there was no added oil, indicating that this population used both sources of carbon,

Table 4. Bacteria population from rhizosphere of forest soils of Urucu Petroleum Province, able to degrade diesel oil.

Soil / Treatments	*Enriched	7 days	14 days	21 days	28 days
		.... 10^6 g^{-1} soil			
FS1**	Not	25	207	195	1
	Yes	187	96	160	2
FS2**	Not	15	127	33	2
	Yes	200	97	402	1
FS3**	Not	24	152	35	2
	Yes	149	2600	247	5
FS4**	Not	14	21	97	3
	Yes	199	188	163	2

*Enriched or not with diesel oil.
**Forest Soils.
Source: Mari (2008).

Table 5. Bacteria population in the rhizosphere soil from the Rural Community of Brasileirinho (Manaus, Amazonas) able to degrade diesel oil.

Soil/Species	*Enriched	7 days	14 days	21 days	28 days
		\multicolumn{4}{c}{10^6 g^{-1} soil}			
Cupuassu (*Theobroma grandiflorum*)	Not	19	11	2.60	1.00
	Yes	99	667	73.50	39.00
Soursop (*Annona muricata*)	Not	2	13	0.02	≤ 0.01
	Yes	4700	1031	26.30	5.10
Camu-camu (*Myrciaria dubia*)	Not	3	26	12.10	≤ 0.01
	Yes	5	138	64.00	30.00
Arassa-boi (*Eugenia stipitata*)	Not	9	14	8.60	13.80
	Yes	69	131	26.50	40.00

*Enriched or not with diesel oil.
Source: Mari (2008).

soil organic matter and oil (Mari 2008). These results indicate that even the soil that was never contaminated with petroleum, rhizosphere soils of these plants species contain a microorganism population with enzymes that degrade diesel oil.

These kinds of enzymes were also found in another study with grasses, legumes and *Vismia guianensis* in the region of Urucu, where Petrobras explores petroleum and natural gas. These plant species were planted along pipelines in the Urucu Petroleum Province to protect soil and Lima (2010) found the presence of microorganisms capable of degrading petroleum in the rhizosphere of these plant species (Table 6).

Table 6. Bacteria population in the rhizosphere of plants along the pipeline in the Urucu Oil Province, able to degrade petroleum.

Treatments	Rhizosphere	Petroleum	Rainy season	Dry season
			\multicolumn{2}{c}{10^8 g^{-1} soil}	
T1	*Vismia guianensis*	Absent	1.93 h	3.93 f
T2	*Vismia guianensis*	Present	2.33 g	5.83 b
T3	*Mucuna pruriens*	Absent	2,53 f	4.56 e
T4	*Mucuna pruriens*	Present	3.53 c	3.86 g
T5	*Brachiaria decumbens*	Absent	3.70 b	5.40 c
T6	*Brachiaria decumbens*	Present	3.40 d	8.93 a
T7	*Brachiaria humidicola*	Absent	2.70 e	1.63 h
T8	*Brachiaria humidicola*	Present	4.10 a	4.75 d

Different letters in columns indicate different means by Tukey test (5%).
Source: Lima (2010).

It was observed that during the rainy season, the bacteria population with the ablity to degrade petroleum was much higher in the rhizosphere of all the plant species, except for *B. humidicola* without the addition of petroleum (T7). The addition of petroleum resulted in increases of all the bacteria population in all the soil samples, indicating that the soil microbiota used oil organic matter and petroleum as carbon sources. The highest bacteria population during the rainy season was found in the rhizosphere of *B. humidicola* with the addition of petroleum (T8), and at the dry season, *B. decumbens* with petroleum added (T6), which almost reached 9×10^8 CFU.g^{-1} soil, surpassing all other plant species in both periods (Table 6, Lima 2010).

The presence of microorganisms with the ability to degrade pertroleum in the rhizosphere is highly promising, because the chances of being pathogenic to humans or other animals are extremely rare, since in this environment predominate root exudates excreted by plant roots (Bais et al. 2006).

Wood and plant residues degradation by microorganisms

The enzymes capable of degrading plant residues such as cellulases, hemicellulases and ligninases, for example, are very important for recycling nutrients in the forest and agroforestry systems of economic value, as the mineralization of soil organic matter can release nutrients essential for the development of vegetation. But they can also be used to convert plant waste into disposable bioproducts of economic value, such as alcohol. The use of microorganisms for ethanol production has been known for hundreds of years. At the moment, they are being selected to increase the efficiency of biofuel production, using crop residues (Mai et al. 2004), such as sugar cane bagasse, which represents 70% of the weight of the plant or, vegetable waste in general, especially scrap lumber or mobile plants.

Tables 7 and 8 illustrate soil microbiota variations in terms of the ability to convert wood residues (sawdust of assacu—*Hura crepitans*) or industrial residues (bagasse of cane sugar—*Saccharum L.*) in a number of substances (soluble solids) which can later be used as carbon sources in the fermentation process for alcohol production or other products of economic interest. This experiment was carried out using 10 g of soil and 2 g of assacu sawdust or sugar cane bagasse (Braga 2009).

According to Braga (2009), the potential of the soil microbiota to produce soluble solids using assacu as carbon source (Table 7) indicated variation among soil samples. The soil microbiota from soils collected in the Urucu forest (Points 2 and 3) showed the highest values. The highest values were found at 47 and 57 days of incubation (8.3° Brix) provided by the soil microflora of the Point 3. The highest soluble solids percentages found by

Table 7. Production of soluble solids by soil microbiota using assacu sawdust as carbon source.

Soils	Days of incubation										
	8	14	17	22	32	47	57	60	65	77	80
	.... °Brix										
Control	1.0	1.0	1.0	1.0	1.0	1.0	1.0	1.0	1.0	1.0	1.0
Point 2	1.8	2.0	3.6	4.3	5.0	5.1	5.0	5.3	6.0	6.0	3.0
Point 3	1.0	1.5	4.0	4.0	5.6	8.3	8.3	6.3	4.5	4.6	3.0
Inpa 1	1.0	1.5	3.0	3.3	3.0	3.8	3.0	2.8	1.1	4.0	2.0
Inpa 3	1.3	1.5	3.3	3.6	4.6	3.3	4.5	5.1	2.8	3.3	2.0
Inpa 12	1.0	1.1	1.8	1.6	1.0	1.0	1.0	1.0	0.6	0.3	1.0
Inpa 13	1.3	1.5	3.3	3.3	2.0	3.6	2.3	1.5	1.5	2.5	2.0
Averages	1.2	1.4	2.9	3.0	3.2	3.7	3.6	3.3	2.5	3.1	2.0

Source: Braga (2009).

Table 8. Production of soluble solids by soil microbiota using bagasse of sugar cane as carbon source.

Soils	Days of incubation										
	8	14	17	22	32	47	57	60	65	77	80
	°Brix										
Control	1.0	1.0	1.0	1.0	1.0	1.0	1.0	1.0	1.0	1.0	1.0
Point 2	4.0	2.0	1.5	2.7	0.0	1.7	2.0	4.2	2.0	3.1	2.5
Point 3	3.5	1.5	1.5	2.1	3.0	2.3	2.0	3.3	2.6	3.8	3.0
Inpa 1	3.3	1.5	1.5	2.0	1.5	1.0	2.0	1.8	2.6	2.3	2.0
Inpa 3	3.1	1.5	1.5	2.0	2.0	1.7	3.5	3.6	2.1	3.0	2.5
Inpa 12	3.0	1.5	1.5	2.1	1.5	1.1	1.5	1.0	2.2	0.9	1.0
Inpa 13	3.1	1.5	3.0	2.8	0.0	2.0	4.0	3.0	2.1	1.7	2.0
Averages	3.0	1.5	1.6	2.1	1.3	1.5	2.3	2.6	2.1	2.3	2.0

the microbiota of the soil identified as Point 2 were observed at 65 and 77 days of incubation (6.0° Brix). The microbiota of soil collected in the INPA produced less soluble solids, highlighting the sample INPA 3 with 5.1° brix at 60 days and the INPA 1 with 4.0° Brix at 77 days. Soil microbiota of INPA 12 showed an average content of solids soluble equivalent to those observed in the control (no soil), suggesting that it was converting sawdust assacu to soluble products, but which was consuming these soluble elements at the same rate that produced it. This microbial consumption may be observed at 17 and 22 days of incubation (Table 7) when brix values were higher than those observed in the control and those observed at 65 and 77 days, lower than those of the control (Braga 2009).

It was observed that differences in the production of soluble solids are already present on the 8th day of incubation, for the soil microbiota of Point 2, INPA 3 and INPA 13. However, the highest yields of these compounds were observed in the period between 32 and 60 days of incubation, decreasing after that period as can be observed by the averages of each evaluation day.

Thus, the microorganisms found in these soil samples produce enzymes capable of degrading assacu sawdust, such as hemicellulose, cellulose and lignin, the main components of the structure of the plant material (Van Soest 1994), and converting them to soluble solids. However, the differences among the values and period of soluble solids production suggest different microbial enzymatic systems and/or dynamic activities when comparing the different soil samples studied.

When analyzing the activity of the soil microbiota using sugar cane bagasse as carbon source (Table 8), differences among soil microbiota were present and there was also a variation of soluble solids production. In this case, all microbiotas produced significant amounts of these compounds when compared with control (Braga 2009).

However, unlike the results with sawdust assacu, the highest values were observed at 8 days of incubation (Table 8), perhaps due to sugar residues contained in the bagasse and the presence of plant components more easily decomposed at the beginning. Different behaviors were observed, for example, the highest yields of soluble solids caused by soil microbiota collected in Urucu region (Point 2 and Point 3) occurred in the 8th and 60th days, while INPA 3 and INPA 13 microbiotas produced more soluble solids at days 8, 57 and 60. The microbiota of the soil samples INPA 1 and INPA 12 had good production of soluble solids only on the 8th day, suggesting that their enzymes act only on the initial use of sugars residues, and decomposition of the more easily degradable bagasse compounds (Braga 2009).

Whitman et al. (1998) mentions that the soil is an important habitat for microorganisms, where they play a fundamental role in the decomposition of soil organic matter and catalyze unique and indispensable transformations in the global carbon and nitrogen cycles. Torsvik et al. (1990a, b) estimated that 1 g of soil may have 4000 different 'genomic units' of bacteria based on DNA-DNA. In contrast, Paoletti & Bressan (1996) describe that the soil can present more than 10,000 species of organisms per gram of soil, with difficulties for identification, which represent about 85% of the biomass or 90% of CO_2 flux between biotic components responsible for decomposition of the litter layer of deciduous forests.

This diversity of microorganisms could explain the differences in activity observed in this study, when comparing the samples of soil and the addition of different plant materials such as assacu sawdust and sugar cane bagasse.

Solubilization of calcium and aluminum phosphate

Other products to impose economic and ecological importance from the metabolism of microorganisms are responsible for phosphate solubilization. Soil phosphates bind to Aluminum and Iron in very acidic soils (pH<5.0), and to Ca and Mg in soils with a pH higher than 5.5. Some soil microorganisms may present the ability to solubilize these phosphates, increasing their availability to plants (Sylvester-Bradley et al. 1982).

Tables 9 and 10 show the rates of solubility of calcium and aluminum phosphate for strains of rhizobia as illustrative of this capacity and that it is variable (Chagas Jr. 2007). Some behaved early, by starting solubilizing on the 3rd day of incubation, but others started solubilizing later, at 6, 9, 12 or 15 days of incubation. Moreover, the rate of solubilization also varied widely, ranging from 1.08 to 5.15 at the end of the experiment using calcium

Table 9. Calcium phosphate solubilization by rhizobia isolates.

Rhizobia Isolates	Initial Solubilization (days)	SI Index (solubilization)		Final medium pH
		Initial (mm)	Final (mm)	
INPA R806	6	1.16	1.84 (low)	Decreased
INPA R808	6	1.10	1.38 (low)	Not changed
INPA R809	3	1.17	1.35 (low)	Decreased
INPA R813	3	1.52	5.15 (High)	Decreased
INPA R814	15	1.12	1.12 (low)	Decreased
INPA R815	12	1.08	1.08 (low)	Decreased
INPA R839	3	1.47	2.57 (medium)	Decreased
INPA R841	3	1.26	2.06 (medium)	Decreased
INPA R843	3	1.41	2.10 (medium)	Decreased
INPA R851	3	1.40	1.51 (low)	Decreased
INPA R852	6	1.19	1.27 (low)	Decreased
INPA R869	3	1.23	1.93 (low)	Decreased
INPA R871	6	0.20	1.33 (low)	Decreased
INPA R892	3	1.13	1.13 (low)	Decreased
INPA R894	6	1.07	2.36 (medium)	Not changed
INPA R896	9	1.25	1.46 (low)	Not changed
INPA R839	3	1.47	2.57 (medium)	Decreased

Source: Chagas Jr. (2007).

Table 10. Aluminum phosphate solubilization by rhizobia isolates.

Rhizobia isolates	Initial Solubilization (days)	IS Index (solubilization)		Final medium pH
		Initial (mm)	Final (mm)	
INPA R808	3	1.09	1.28 (low)	Increased
INPA R809	9	1.02	1.03 (low)	Increased
INPA R812	15	1.06	1.06 (low)	Increased
INPA R814	3	1.06	1.06 (low)	Not changed
INPA R816	9	0.99	1.00 (low)	Decreased
INPA R817	3	1.00	1.00 (low)	Decreased
INPA R820	15	1.05	1.05 (low)	Not changed
INPA R824	12	1.09	1.09 (low)	Decreased
INPA R825	3	1.06	2.41 (medium)	Decreased
INPA R829	3	1.00	1.00 (low)	Increased
INPA R830	9	1.00	1.00 (low)	Not changed
INPA R841	9	1.35	2.07 (medium)	Decreased
INPA R851	9	1.09	1.09 (low)	Increased
INPA R852	9	1.27	1.47 (low)	Not changed
INPA R871	6	1.19	1.97 (low)	Decreased
INPA R892	9	1.04	1.07 (low)	Decreased
INPA R911	3	1.00	1.23 (low)	Decreased
INPA R914	3	1.00	2.36 (medium)	Not changed

Source: Chagas Jr. (2007).

phosphate (Table 9), and ranging from 1.00 to 2.41 when using aluminum phosphate (Table 10), which according to criteria by Berraquero et al. (1976), are low, medium or high. According to these authors, the solubilization index (SI) is calculated by dividing the diameter of the halo of solubilization by the diameter of the colony of bacteria in a petri dish.

The final medium pH decreased in most of the cases using calcium phosphate in the medium (Table 9), and increased or decreased when using aluminum phosphate in the medium (Table 10), with some of them not changing the medium pH at the final of the experiments. This indicates different products of the bacteria metabolism when growing in these media, and may be useful to understanding final products of their metabolism.

Production of growth hormone

Production of indole acetic acid (IAA) is another very important metabolite, especially by bacteria which occur in the rhizosphere of plants or rhizoplane, such as the rhizobia (Antoun et al. 1998, Hameed et al. 2004). A study of 92 strains of rhizobia indicated that they all produced indole acetic acid (IAA) when supplied with different concentrations of its precursor (tryptophan) in the medium (Chagas Jr. 2007).

However, there was a great variation among them on this capacity. This variation is shown with the data from 16 of these isolates (Table 11, Chagas Jr. 2007). When 10 mg of tryptophan L^{-1} were applied in the medium, isolate INPA R892 produced 581 ug of IAA. mL^{-1}, but INPA R843 produced only 6 ug of IAA. mL^{-1}. Increasing tryptophan concentration to 50 or 100 mg. L^{-1}, the highest IAA production was shown by isolate INPA R815 with 642 ug of IAA. mL^{-1}. These different responses by rhizobia isolates depending on tryptophan concentration suggest that their behavior on the plant root system may be variable, depending on the production rate and concentration of tryptophan in this soil environment. Bais et al. (2006) shows evidence that there is a great variation of chemical compounds in the rhizosphere soil, suggesting that rhizobia isolates may produce different IAA rates when in different plants rhizospheres.

Table 11. Indol Acetic Acid production by rhizobia isolates.

Rhizobia isolates	Tryptophan (mg.L⁻¹)		
	10	50	100
	...............IAA Production (ug mL⁻¹)		
INPA R806	16 f	33 ef	58 e
INPA R808	42 e	46 e	50 e
INPA R809	34 e	25 fg	24 ef
INPA R813	54 e	149 c	139 c
INPA R814	23 f	36 ef	212 b
INPA R815	90 c	642 a	642 a
INPA R839	32 e	40 e	41 e
INPA R841	96 c	256 b	29 ef
INPA R843	6 f	14 g	15 f
INPA R851	54 e	149 c	139 c
INPA R852	48 e	80 d	81 d
INPA R869	37 e	46 e	56 e
INPA R871	36 e	46 e	51 e
INPA R892	581 a	253 b	225 b
INPA R894	83 d	167 c	31 ef
INPA R896	112 b	82 d	117 c

Source: Chagas Jr. (2007).

Functional genes

Another tool that can be used with great success today is the elucidation of the complete genome or part of a particular body, trying to detect the functional genes present in it. An example that can be illustrated is in the publication of the total genome of *Chromobacterium violaceum* (BRAZILIAN NATIONAL CONSORTIUM Genome Project 2003). Based on their functional genes, the bacteria found in Amazonian soils and water of rivers such as the Rio Negro, produce antibiotics such as aztreonam, violacein, aerocyanidina, aerocavina and have antibacterial, antiviral and anticancer activities, and can synthesize polymers capable of replacing synthetic plastic. They also present the ability to hydrolyze plastics and assist in the process of extracting gold from the soil without the use of mercury, reducing the impacts of environmental contamination.

Conclusion

The soil microorganisms have a high biodiversity, whose secondary metabolites are still largely unknown. However, a wide range of microbial products are already used in the pharmaceutical, food and environmental areas of activities. The industrial applications of microorganisms, primarily bacteria, fungi and yeast, are extremely diverse, providing astronomical financial returns. Classic examples are the fermentations, such as alcoholic beverages and alcohol fuel, dairy products, organic acids and drugs, including antibiotics. Other products of microbial enzymes and polymers are of industrial interest, stereo-specific molecules produced by biotransformation, food additives (amino acids) and vitamins. Microorganisms are also used in the formulation of inoculants for agricultural and industrial use, such as probiotics, inoculants for the biodegradation of toxic compounds in effluent treatment, bioremediation inoculants for environmental and agricultural inoculants. Antibiotics, enzymes, nutraceutic, carbohydrates, are the most common. Microorganisms are able to produce biofuels as well as the bioremediation of environments contaminated by fuel and various effluents. The amount of enzymes known yet reaches less than 5% of the estimated potential, indicating that there is an unused potential involving soil microbiota. Soils of virgin forests where there are a large number of plant species are where you expect to find the greatest diversity of microorganisms and thus enzymes.

References

Antoun, H., C. J. Beauchamp, N. Goussard, R. Chabot and R. Lalande. 1998. Potential of *Rhizobium* and *Bradyrhizobium* species as plant growth promoting rhizobacteria on nonlegumes: Effect on radishes (*Raphanus sativus* L.). Plant and Soil. 204: 57–67.

Bais, H. P., T. L. Weir, L. G. Perry, S. Gilroy and J. M. Vivanco. 2006. The role of root exudates in rhizosphere interactions with plants and other organisms. Annu. Rev. Plant Biol. 57: 233–66.

Beilen, J. B. and Z. Li. 2002. Enzyme technology: an overview. Current opinion in Biotechnology. 13: 338–344.

Berraquero, F. R., A. M. Baya and A. R. Cormenzana. 1976. Establecimiento de índices para el estudio de la solubilizacion de fosfatos por bacterias del suelo. Ars. Pharmacéutica. 17: 399–406.

Braga, C. M. S. 2009. Microrganismos de solos amazônicos com potencial na bioconversão de material ligninocelulósico em sólidos solúveis. Dissertação de Mestrado em Biotecnologia e Recursos Naturais, Universidade do Estado do Amazonas, Manaus, A. M. p. 71.

BRAZILIAN NATIONAL GENOME PROJECT CONSORTIUM. 2003. The complete genome sequence of *Chromobacterium violaceum* reveals remarkable and exploitable bacterial adaptability. Proc. Natl. Acad. Sci. USA. 100(20): 11660–5.

Calixto, J. B. and R. A. Yunes. 2001. Plantas Medicinais sob a ótica da moderna química medicinal. Editora Argos, Chapecó.

Castro, H. F., A. A. Mendes, J. C. dos Santos and C. L. de Aguiar. 2004. Modificação de óleos e gorduras por biotransformação. Quim. Nova. 27: 146–156.

Chagas Junior, A. F. 2007. Características agronômicas e ecológicas de rizóbios isolados de solos ácidos e de baixa fertilidade da Amazônia. Tese de Doutorado em Biotecnologia. Universidade Federal do Amazonas. p. 172.

Couch, J. N. 1950. Actinoplanes, a new genus of the Actinomycetales. J. Elisha Mitchell Sc. Soc. 66: 87–92.

Couch, J. N. 1963. Some new genera and species of the Actinoplanaceae. J. Elisha Mitchell Sc. Soc. 79: 53–70.

Cross, T., M. P. Lechevalier and H. Lechevalier. 1963. A new genus of the Actinomycetales: Microellobosporia gen. nov. J. Gen. Microbiol. 31: 421–429.

Crueger, W. and A. Crueger. 1990. Biotechnology: A textbook of industrial microbiology. 2nd ed. Sunderland: Sinauer Associates, USA.

Demain, A. L. 1999. Phamaceutically active secondary metabolites of microorganisms. Appl. Microbiol. Biotechnol. 52: 455–483.

Doelle, H. W. 2007. Microbial metabolism and biotechnology. http://www.twinamasiko.com/IOBB/Publications/Biotechnology_eBook.pdf (Acessed on 24/06/2012). p. 461.

Ferreira, W. A., C. M. Ferreira, A. P. M. Sehettini, J. C. G. Sardinha, A. S. Benzaken, M. A. Garcia, E. G. Garcia and L. A. Oliveira. 2004. *Neisseria gonorrhoeae* produtoras de betalactamase resistentes a azitromicina em Manaus, Amazonas, Brasil. DST- J. Bras. Doenças Sex. Transm. 16(2): 28–32.

Fischbach, M. A. and C. T. Walsh. 2009. Antibiotics for emerging pathogens. Science. 325: 1089–1093.

Hameed, S., S. Yasmin, K. A. Malik, Y. Zafar and F. Y. Hafeez. 2004. *Rhizobium , Bradyrhizobium* and *Agrobacterium* strain isolated from cultivated legumes. Biol. Fertil. Soils. 39: 179–185.

Hara, F. A. S. and L. A. Oliveira. 2004. Características fisiológicas e ecológicas de isolados de rizóbios oriundos de solos ácidos e álicos de Presidente Figueiredo, Amazonas. Acta Amazonica. 34(2): 343–357.

Hara, F. A. S. and L. A. Oliveira. 2005. Características fisiológicas e ecológicas de isolados de rizóbios oriundos de solos ácidos de Iranduba, Amazonas. Pesq. Agropec. bras, Brasília. 40(7): 667–672.

Kolter, R. and F. Moreno. 1992. Genetics of ribosomally synthesized peptide antibiotics. Annual Review of Microbiology. 46: 141–163.

Konev, I. E., V. A. Tsyganov, P. Minbayev and V. M. Morozov. 1967. A new genus of Actinomycetes, Echinospora gen. nov. Mikrobiologiya. 36: 309–317.

Lancini, G. and R. Lorenzetti. 1993. *Biotechnology of antibiotics and other bioactive microbial metabolites*. Plenum Press NY and London. pp. 49–57.

Lazzarini, A., L. Cavaletti, G. Toppo and F. Marinelli. 2000. Rare genera of actinomycetes as potential producers of new antibiotics. Antonie van Leeuwenhoek. 78: 399–405.

Lewinsohn, T. M. and P. I. Prado. 2002. Biodiversidade brasileira. Síntese do estado atual do conhecimento. São Paulo. Contexto. p. 176.

Lima, D. C. R. 2010. Microrganismos degradadores de petróleo isolados de solos rizosféricos da Província Petrolífera de Urucu, Coari, Amazonas. Dissertação de mestrado em Biotecnologia, Universidade do Estado do Amazonas. p. 73.

Manfio, G. P. 2012. Biodiversidade: Perspectivas e oportunidades tecnológicas. Microrganismos e aplicações industriais. Base de Dados Tropical. Availbility in: http://www.bdt.fat.org. br. (Accessed in July 20, 2012).

Mai, C., U. Kiles and H. Militz. 2004. Biotechnology in the wood industry. Appl. Microbiol. Biotechnol. 63: 477–494.

Mari, A. O. 2008. Microrganismos degradadores de gasolina e óleo diesel isolados de solos do Amazonas. Dissertação de Mestrado em Biotecnologia e Recursos Naturais, Universidade do Estado do Amazonas, Manaus, A. M. p. 100.

Newman, D. J., G. M. Cragg and K. M. Snader. 2000. The influence of natural products upon drug discovery. Nat. Prod. Rep. 17: 215–234.

Nonomura, H. and Y. Ohara. 1969. Distribution of Actinomycetes in soil VI. A culture method effective for both isolation and enumeration of *Microbispora* and *Streptosporangium* strains in soil (Part 1). J. Ferment. Technol. 47: 463–468.

Oliveira, A. N. and L. A. Oliveira. 2003a. Características químicas do solo ea esporulação e colonização micorrízica em plantas de cupuaçuzeiro e de pupunheira na Amazônia Central. Revista Ciências Agrárias. 40: 33–44.

Oliveira, A. N. and L. A. Oliveira. 2003b. Sazonalidade, colonização radicular e esporulação de fungos micorrízicos arbusculares em plantas de cupuaçuzeiro e de pupunheira na Amazônia Central. Revista Ciências Agrárias, Belém. 40: 145–154.

Oliveira, A. N. and L. A. Oliveira. 2005a. Seasonal dynamics of arbuscular mycorrhizal fungi in plants of *Theobroma grandiflorum* Schum and *Paullinia cupana* Mart of an Agroforestry system in Central Amazonia, Amazonas State, Brazil. Brazilian Journal of Microbiology. 36: 262–270.

Oliveira, A. N. and L. A. Oliveira. 2005b. Colonização por fungos micorrízicos arbusculares em cinco cultivares de bananeiras em um latossolo da Amazônia. R. Bras. Ci. Solo. 29: 481–488.

Oliveira, A. N., L. A. Oliveira and J. S. Andrade. 2006b. Enzimas hidrolíticas extracelulares de isolados de rizóbia nativos da Amazônia Central, Amazonas, Brasil. Ciênc. Tecnol. Aliment. Campinas. 26(4): 853–860.

Oliveira, A. N., L. A. Oliveira and J. S. Andrade. 2007. Produção de amilase por rizóbios, usando farinha de pupunha como substrato. Ciênc. Tecnol. Aliment. 27: 61–66.

Oliveira, A. N., L. A. Oliveira, J. S. Andrade and A. F. Chagas Junior. 2006a. Atividade enzimática de isolados de rizóbia nativos da Amazônia Central crescendo em diferentes níveis de acidez. Ciênc. Tecnol. Aliment. 26(1): 204–210.

Oliveira, A. N., L. A. Oliveira and A. F. Figueiredo. 2003. Colonização micorrízica e concentração de nutrientes em três cultivares de bananeiras em um latossolo amarelo da Amazônia Central. Acta Amazonica. 33(3): 345–352.

Omura, S. 1992. Trends in the search for bioactive microbial metabolites. Journal of Industrial Microbiology. 10: 135–156.

Panke, S. and M. G. Wubbolts. 2002. Enzyme technology and bioprocess engineering. Current Opinion in Biotechnology. 13: 111–116.

Paoletti, M. G. and M. Bressan. 1996. Soil invertebrates as bioindicators of human disturbance. Critical Review in Plant Sciences. 15: 21–62.

Pearce, C. 1997. Biologically active fungal metabolites. Adv. Appl. Microbiol. 44: 1–80.

Prado, K. L. L. 2009. Microrganismos produtores de amilase, celulase, fosfatase, lípase, protease e urease nos solos amazônicos do ramal do Brasileirinho (Manaus) e de urucu (Coari). Dissertação de mestrado em Biotecnologia, Universidade do Estado do Amazonas. p. 70.

Shepherd, G. 2002. Conhecimento de diversidade de plantas terrestres do Brasil. *In*: T. M. Lewinsohn and P. I. Prado. Biodiversidade brasileira. Síntese do estado atual do conhecimento. São Paulo. Contexto. pp. 155–159.

Silva Neves, K. C., A. L. Porto and M. F. S. Teixeira. 2006. Seleção de Leveduras da Região Amazônica para Produção de protease Extracelular. Acta Amazonica. 36(3): 299–306.

Sotero-Martins. A., P. S. E. Bon and E. Carvajal. 2004. Asparaginase II-Gep fusion as a tool for studying the secretion of the enzyme under nitrogen starvation. J. Microbiology 34: 373–377.

Sylvester-Bradley, R., N. Asakawa, S. LaTorraca, F. M. M. Magalhaes, L. A. Oliveira and R. M. Pereira. 1982. Levantamento quantitativo de microrganismos solubilizadores de fosfato na rizosfera de gramíneas e leguminosas forrageiras na Amazônia. Acta Amazonica. 12(1): 15–22.

Thiemann, J. E. 2012. A direct screening method for the investigation of microbial biodiversity. International Research Journal of Microbiology. 3(1): 36–38.

Torsvik, V., J. Goksoyr and F. L. Daae. 1990a. High diversity in DNA of soil bacteria. Appl. Environ. Microbiol. 56: 782–787.

Torsvik, V., K. Salte, R. Sorheim and J. Goksoyr. 1990b. Comparison of phenotypic diversity and DNA heterogeneity in a population of soil bacteria. Appl. Environ. Microbiol. 56: 776–781.

Van Soest, P. J. 1994. Nutritional ecology of the ruminant. Ithaca: Cornell University Press. p. 476.

Whitman, W. B., D. C. Coleman and W. J. Wiebe. 1998. Prokaryotes: The unseen majority. Proc. Natl. Acad. Sci. USA. 95: 6578–6583.

13

Heavy Metal Niches
Microbes and Their Metabolic Potentials

Chih-Ching Chien

ABSTRACT

Microorganisms play important roles in biogeochemical cycles. Microbe-metal interactions are of key importance in microbial mineralization and transformation of metal ions and related elements. Although many heavy metals and metalloids are toxic to microorganisms, studies on the microbial community in heavy metal contaminated sites still demonstrated a high diversity of microorganisms. Microorganisms residing in heavy metal rich environments have developed strategies to thriving in these niches. Numerous approaches including various culture-independent methods have been proposed to analyze the microbial diversity. Microorganisms possess a variety of mechanisms to deal with high concentrations of heavy metals, including active efflux of toxic metal ions, enzymatic detoxification (usually reduction of the ions), and bioaccumulation or sequestration of the metals. The characteristics of microbe-metal interactions provide a brand new domain of biotechnological applications. This chapter briefly recites microorganisms and their metabolic potential with heavy metals in the above mentioned aspects.

Graduate School of Biotechnology and Bioengineering, Yuan Ze University, Chung-Li, Taiwan.
Email: ccchien@saturn.yzu.edu.tw

Introduction

Heavy metals comprise a group of loosely-defined subset of metallic elements including the transition metals and some metalloids (Duffus 2002). There is no authoritative definition of heavy metals and here we will only focus on metals/metalloids associated with contamination and potential toxicity or ecotoxicity. Heavy metals are ubiquitous and abundant on Earth since the formation of the planet. The evolutions of life on Earth, including microorganisms, have been exposed to these toxic elements since the beginning of history. There is no doubt that the features for toxic metal resistance in microorganisms in this diverse environment have not evolved in response to anthropogenic activities but rather, are an adaptation to the natural environment over billions of years (Silver and Phung 2005). In the natural environment, the amount of biologically available heavy metals can vary greatly in different locations. For example, elevated concentrations of toxic heavy metals can be found in volcanic soils, hot springs and sediments, but are generally low in the average ecosystem (Gadd 2009a). Therefore, the occurrence of microbial resistance systems to toxic metals, albeit widespread in the microbial world, can vary greatly with frequencies ranging from a few percent in pristine environments to nearly all microbial isolates in heavily polluted environments (Silver and Phung 2005, 2009, Gadd 2010). However, the disturbances of environmental niches by anthropogenic activities has led to the redistribution of toxic heavy metal ions in aquatic and terrestrial ecosystems and significantly affected the populations and diversity in different microbial habitats (Haferburg and Kothe 2007).

Trace amount of some metals (e.g., Cu, Zn, and Mn, etc.) in external media are essential for many organisms, but are toxic at high concentration, yet strongly toxic heavy metals (e.g., Cr and Cd) are microcidal in most cases (Nies 1999). The impacts on the shifts and changes of microbial communities are more commonly associated with the redistribution of toxic heavy metals due to anthropogenic activities although the effects can also arise from natural milieus (Ellis et al. 2003, Gadd 2009a). The sources of these anthropogenic stress factors can be present in the atmosphere and deposited into aquatic and terrestrial environments from agricultural practices, industrial activities such as metallurgy, energy production, microelectronics and metalliferous mining and smelting, sewage sludge and wastes disposal (Landa 2005, Haferburg and Kothe 2007, Gilmour and Riedel 2009, Gadd 2010). In spite of the potential toxicity and adverse effects on biota, certain microorganisms can still adapt and survive in toxic metal-polluted environments. Studies on microbial diversity in heavy metal contaminated niches still demonstrated a significant diversity of microorganisms (Dean-Ross and Mills 1989, Hattori 1992, Gelmi et al. 1994, Roane and Kellogg 1996, Filali et al. 2000, Ellis et al. 2003, Chien et al. 2008, Islam and Sar 2011, Margesin et al. 2011). They are indigenous microorganisms that have not only adapted to the new environments but have also flourished under them (Haq and Shakoori 2000, Roane and Pepper 2000). Microorganisms possess a variety of mechanisms, both active and incidental, to deal with high concentrations of heavy metals and often are specific to one or a few metals (Silver and Misra 1988, Silver and Phung 1996, Mejáre and Bulow 2001, Nies 2003, Piddock 2006, Harrison et al. 2007, Gadd 2009a, Chien et al. 2011).

Diversity of microbial communities can be attributed to the degree of heavy metal pollutions in their habitats. Therefore, specific heavy metal resistance determinants can be used to ascertain the impact of the stress caused by anthropogenic activities in natural environments and hence as parameters for environmental forensic as biosensors (Ellis et al. 2003, Zhou 2003, Verma and Singh 2005, Narancic et al. 2012). However, the required approaches to most reliably determine the bacterial community diversity of an environment is often hotly debated. The use of nucleic acid-based culture-independent approaches is believed to be more appropriate for the assessment of the composition of the microbial communities than those based on culturing techniques (Øvreås and Torsvik 1998, Ellis et al. 2003, Hemme et al. 2010, Islam and Sar 2011, Mohapatra et al. 2011), however, the relevance of the microbial activities to the metal pollutants cannot be totally elucidated (Kell et al. 1998, LeBaron et al. 2001). Although it is widely accepted that the enrichment culture approach only identifies a small proportion of microorganisms in any given sample, and culturable microorganisms may only stand for less than 1% of the total population (Rappé and Giovannoni 2003, Sharma et al. 2005, Margesin et al. 2011), the isolation of the microorganisms and the bacterial effects is essential for microbial activities on heavy metals from the application of bioremediation point of view. In this chapter, the approaches for the assessment of microbial diversity in heavy metal contaminated niches and the metabolic potential of microorganisms that reside in these niches will be discussed.

Tracking the Microbial Diversity in Heavy Metal Niches

Although none of the approaches currently available can provide a complete view of the structure and function of microorganisms in any environmental samples, culture-independent methods are still receiving particular attention in the assessment of the diversity of microbial communities and community structure because of their lack of bias inflicted by the limitation of media and isolation techniques in the laboratory. The use of culture-independent approaches has been vigorously developed to study microbial communities from various environments since the last two decades of the 20th century (Woese 1987, Hugenholtz et al. 1998, Ogram 2000, Su et al. 2012). Molecular methods based on analysis of nucleic acids sequences, for example cloning and sequencing small subunit ribosomal RNA gene, denaturing gradient gel electrophoresis (DGGE), real-time polymerase chain reaction (real time PCR), microarray and metagenomics, have been widely applied on survey of microbial population and diversity of heavy metal contaminated niches. Figure 1 summarises the diagram of some major culture-independent approaches frequently applied for assessing the diversity of microbial communities in various environments.

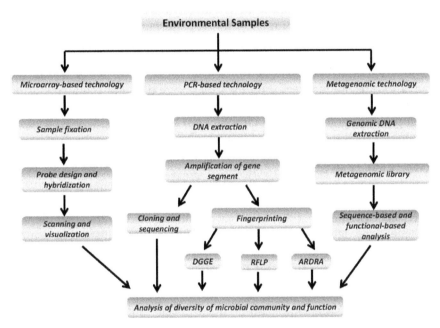

Fig. 1. Major culture-independent approaches frequently applied for assessing the diversity of microbial communities in environments.

PCR-based technology

Microbial diversity in heavy-metal-contaminated soils has been explored by PCR amplification of 16S ribosomal RNA gene with universal primers followed by various analysis assay technologies, including cloning and sequence determination, fingerprinting methods such as DGGE, terminal restriction fragment length polymorphism (T-RFLP) and amplified rDNA restriction analysis (ARDRA), etc. In culture-independent methods, direct cloning and sequence determination of 16S rRNA gene has been extensively applied for phylogenetic analysis of microbial diversity in heavy metal-contaminated bulk and rhizosphere soils, sediments, as well as agricultural fields (Gremion et al. 2003, Chien et al. 2008, Rastogi et al. 2011). There are drawbacks of using PCR to amplify ribosomal RNA (rRNA) genes from extracted environmental DNA, for example, the differential amplification of different DNA templates, sensitivity to rRNA gene copy number and template concentration, primer specificity, amplification of contaminant DNA and formation of chimeric sequences may affect the assessment of microbial diversity (Hugenholtz et al. 1998, Dubey et al. 2006). PCR-based fingerprinting techniques were also employed for the assessment of the composition of microbiota in heavy metal contaminated environments. These included the use of fingerprinting patterns with restriction enzyme digestion such as T-RFLP (Hartmann et al. 2005, Frey et al. 2006, Lazzaro et al. 2008, Said et al. 2010, Iannelli et al. 2012) and ARDRA (Smit et al. 1997, Wenderoth et al. 2001, Pe´rez-de-Mora et al. 2006). In recent years the advanced techniques based on the isolation of different fragments using various electrophoretic separation techniques such as DGGE and TGGE (temperature gradient gel electrophoresis) have been widely applied on evaluation of the structural and functional diversity of microbiota in heavy metal contaminated environments (Li et al. 2006, Hu et al. 2007, Vivas et al. 2008, Khan et al. 2010, Dos Santos et al. 2012). DGGE separates the PCR-amplified DNA fragments of identical or near-identical lengths (ca 200–500 bp) on the basis of differences in nucleotide composition resulting in the differences in their mobility in polyacrylamide gels containing gradient of a denaturing agent (Muyzer et al. 1993, Muyzer 1999). Instead of using a gradient of the denaturant, TGGE relies on temperature dependent changes in structure to separate DNA fragments and has also been extensively applied for characterization of microbial communities in heavy metal contaminated environments (Martins et al. 2010, Vílchez et al. 2011). Study on the relationship between the diversity of the culturable portion of the microbial community and the microbial diversity obtained by direct amplification of 16S rDNA genes and DGGE from soil have been discussed (Ellis et al. 2003, Islam and Sar 2011).

Microarray-based technology

Microarray-based technologies have been introduced as an emerging powerful tool for characterization of complex microbial communities and their structure and function in the environment (Lucchini et al. 2001, Zhou 2003, Bodrossy and Sessitsch et al. 2004). Microarray-based assays not only can analyze the structure, function and dynamic of microbial population in the environment, but also can detect many functional genes simultaneously. The technologies display great potential as specific, sensitive, quantitative, parallel and high throughput tools for the detection, identification and functional characterization of microorganisms in various niches (Zhou and Thompson 2004, Dubey et al. 2006, Mohapatra et al. 2011). For example, functional gene array-based analysis has been used to investigate how different contaminants including heavy metals affect the microbial community diversity, heterogeneity, and functional structure (Waldron et al. 2009). Application of 16S rRNA gene microarray (PhyloChip) was useful in elucidation of the influence of heavy metal contamination on microbial diversity (Field et al. 2010).

Besides this, utilization of heavy metal-resistant bacterium *Cupriavidus metallidurans* CH34 microarray chip proved to be an effective method to determine putative genes involved in microbe–mineral interaction (Olsson-Francis et al. 2010).

Metagenomic technology

Metagenomics and related technologies offers an effective culture independent approach to study microbial diversity in the environment as well as their adaption and evolution in the ecological niches (Handelsman 2004, Streit and Schmitz 2004, Kimura 2006, Schmeisser et al. 2007). This approach involves the DNA extraction from environmental samples, construction of metagenomic libraries and followed by functional-based or sequenced-based analysis (Kimura 2006, Schmeisser et al. 2007). Recent development of next-generation sequencing (NGS) technologies provide practical, vastly parallel sequencing at low cost, making generation of large sequence data sets originated in various environments possible (Hudson 2008, Goel et al. 2011, Simon and Daniel 2011). Analyses of these sequence data sets opened brand new insights into the enormous environmental microbial diversity in structure, function and taxonomy (Simon and Daniel 2011). The employment of metagenomic technologies are being in introduced more and more in studying and unraveling the microbial communities in heavy metal contaminated niches (Hemme et al. 2010, Navarro-Noya et al. 2010, Mohapatra et al. 2011, Daffonchio et al. 2012).

Microbial Metabolic Potential Against Heavy Metal Toxicity

Heavy metal toxicity to microorganisms

Many heavy metal species interact with microbial cells and exert toxic effects on microorganisms through a variety of biochemical mechanisms (Harrison et al. 2007). These biochemical pathways include damage of DNA, RNA and proteins, inhibition of enzymatic activities and disruption of cell membrane. For example, the toxic metal ions may bind to proteins in place of essential metals from their native binding sites or through cellular ligand interactions and consequently change their biological functions (Nies 1999, Bruins et al. 2000, Harrison et al. 2007). Toxic metals can disrupt proteins by binding to sulfhydryl (SH) groups and thus inhibit the activity or the function of enzymes that contain sensitive S groups (Stohs and Bagchi 1995, Nies 1999, Harrison et al. 2007, Zannoni et al. 2007). These reactions frequently involve the production of reactive oxygen species (ROS) and place the cell in oxidative stress state and lead to the damage of DNA, proteins and lipids (Stohs and Bagchi 1995, Harrison et al. 2007). Heavy metals may also affect oxidative phosphorylation, membrane permeability and transport processes (Harrison et al. 2007). Figure 2 illustrate various toxic effects of heavy metals on the microbial cell.

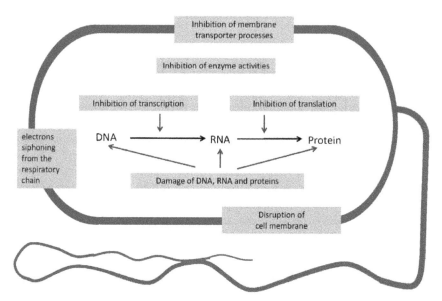

Fig. 2. Various toxic effects of heavy metals on the microbial cell.

Bacterial heavy metal resistance and metabolism

Microorganisms indigenous to the contaminated area usually possess a variety of mechanisms which allows them to tolerate/resist high concentrations of pollutants (e.g., heavy metals and metalloids) (Taylor 1999, Teitzel and Parsek 2003, Haferburg and Kothe 2007, Gadd 2010). These mechanisms include (1) active efflux of toxic metal ions, (2) enzymatic detoxification (usually reduction of the ions), and (3) bioaccumulation or sequestration of the metals (Silver and Phung 1996, Mejáre and Bülow 2001, Nies 2003, Vainshtein et al. 2003). Figure 3 illustrate the general mechanisms involved in detoxification and transformation of heavy metals

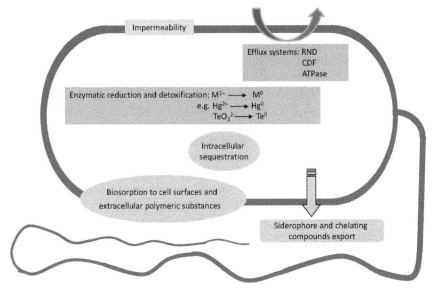

Fig. 3. General mechanisms involved in detoxification and transformation of heavy metals in microorganisms.

in microorganisms. It is unlikely or not sensible that microorganisms utilize a sole mechanism, for example reduction or enzymatic detoxification, to resist heavy metal toxicity. Efflux pumps are the most frequent mechanism of toxic inorganic ion resistance found in microorganisms residing in an environment contaminated with high concentrations of heavy metals, alone, or in combination with other mechanisms (Nies 1999, Silver and Phung 2005, Piddock 2006). The ability of microbial biofilm formation and extracellular polymeric substances production are also associated with their metal resistance, tolerance and bioremediation (Harrison et al. 2007, Pal and Paul 2008).

Impermeability and Efflux systems

Efflux-pump genes and proteins are ubiquitous and present in all microorganisms and active transport or efflux systems represent the major category of metal resistance systems. Families of efflux-pump systems associated with heavy metal resistance include the resistance-nodulation-cell division (RND) protein family, cation diffusion facilitators (CDF) family and P-type ATPase (Nies 2003, Silver and Phung 2005). The RND protein family was first described as a related group of bacterial transport proteins (Saier Jr. et al. 1994, Nies 2003). The RND protein family has become a superfamily that can be divided into seven groups (Tseng et al. 1999). The heavy metal efflux family includes metal transporters, for example

CzcCAB which mediate resistance to Co^{2+}, Zn^{2+} and Cd^{2+} and Cus for copper resistance (Nies 1995, Grass and Rensing 2001, Nies 2003, Kim et al. 2011). Unlike the RND protein families for exporting organic substances which are nonspecific, metal-transporting RND proteins have rather limited substrate spectrum, including the monovalent cation Cu^+ and Ag^+ or divalent cations of the transition metals such as Zn^{2+}, Ni^{2+}, Co^{2+} and Cd^{2+} (Nies 2003, Ma et al. 2009, Conroy et al. 2010, Kim et al. 2011). Families of the RND proteins other than the heavy metal efflux family may also contribute to microbial heavy metal resistance. Several novel RND efflux pumps have been noted in complete genome sequencing and functional genomic analysis in multidrug resistant *Stenotrophomonas maltophilia* (Crossman et al. 2008). *S. maltophilia* is a nosocomial opportunistic pathogen with remarkable multidrug resistance and multimetal tolerance (Alonso and Martinez 1997, Alonso et al. 2000, Minkwitz and Berg 2001, Chien et al. 2007, Pages et al. 2008). The efflux systems that contribute to microbial antibiotic resistance therefore may also be able to participate in microbial metal resistance (Zhang et al. 2000). *Cupriavidus metallidurans* (formally known as *Ralstonia metallidurans* and *Wausteria metallidurans*) possesses a high number of heavy metal resistance genes and is an interesting model organism to study microbial response to heavy metals (Goris et al. 2001, Mergeay et al. 2003, Monsieurs et al. 2011). Metal resistance traits of this soil bacterium are governed by many types of efflux systems including Czc system and P-type ATPases (Legatzki et al. 2003).

The P-type ATPases constitute a large family of cation-transporting proteins (Lutsenko and Kaplan 1995, Rensing et al. 1997). Similar to RND proteins, some families of P-type ATPases have been suggested to be involved in metal homeostasis in microorganisms (Lutsenko and Kaplan 1995, Solioz and Vulpe 1996, Rensing et al. 1997, Chan et al. 2010). The soft metal ATPases such as ZntA from *Escherichia coli* and CadA from *Staphylococcus aureus* have been shown to transport the divalent soft metals, Zn^{2+} and Cd^{2+} (Tsai et al. 1992, Rensing et al. 1997, Sharma et al. 2000, Wu et al. 2004). The ZntA and homologues can be widely found in the prokaryotic cells and were shown to be conferring the resistance to divalent heavy metal cations such as Pb^{2+}, Zn^{2+} and Cd^{2+} in some bacteria (Rensing et al. 1997, 1998, Legatzki et al. 2003, Xiong et al. 2011). Members of this family exist in all three domains of life and can be considered as the basic defense system against heavy metal cations (Nies 2003).

Cation diffusion facilitators (CDF) proteins also occur in all three domains of life and are one of the most ubiquitous classes of metal transporters (Paulsen and Saier Jr. 1997, Nies 2003, Haney et al. 2005, Rosch et al. 2009). The CDF families have been extensively studied in both prokaryotes and eukaryotes, and in most bacterial instances, CDF appear to be involved in metal tolerance and/or homeostasis (Grass et al. 2001,

Haney et al. 2005, Rosch et al. 2009). The mechanisms of the CDF in heavy metal tolerance are via the export of the cations from the cytoplasm to the periplasm or exterior of the cell (Guffanti et al. 2002, Lee et al. 2002, Grass et al. 2005, Haney et al. 2005). Bacterial CDF have been suggested to be capable of zinc efflux (Grass et al. 2001), but some members of this family can also selectively transport divalent cations of Co, Mn, Fe, Cd, and Ni (Delhaize et al. 2003, Anton et al. 2004, Munkelt et al. 2004, Grass et al. 2005, Haney et al. 2005, Moore and Helmann 2005, Rosch et al. 2009). Some examples include ZitB and FieF of *E. coli*, CzcD of *Bacillus subtilis*, DmeF, FieF and CzcD of *C. metallidurans* (Nies 2003, Haney et al. 2005, Rosch et al. 2009).

Enzymatic reduction and detoxification

Enzymatic detoxification of a metal to a less toxic form and reduced metal sensitivity of cellular targets is another important mechanism for microorganisms to resist the toxicity of heavy metals. Heavy metal transformations are usually heavy metal specific and may be directed against one metal or a group of chemically related metals (Nies and Silver 1995, Silver 1996, Bruins et al. 2000, Chien and Han 2009). The reduction/transformation of heavy metals have introduced the detoxification of metal ions such as chromium, mercury, molybdenum, arsenic, lead, copper, uranium, selenium, tellurium (Bautista and Alexander 1972, Lovley 1993).

One of the model examples of enzymatic detoxification of heavy metal in microorganisms is mercury resistance systems and is still the best understood of microbial toxic metal resistance systems (Bruins et al. 2000, Silver and Phung 2005). Mercury resistance exists widely in both Gram positive (e.g., *Staphylococcus aureus*, *Bacillus cereus* and *Clostridium butyricum*, etc.) and Gram negative bacteria (e.g., *E. coli*, *Pseudomonas aeruginosa*, *Serratia marcescens* and *Thiobacillus ferrooxidans*, etc.) (Misra 1992, Barkay et al. 2003). Mercury resistance bacteria can be found in environmental, clinical and industrial isolates and mercury resistance genes are frequently plasmid-encoded, although chromosomal encoded resistance determinants are not uncommon (Misra 1992, Narita et al. 2003). The details of bacterial mercury resistance and the genetic systems of mercury resistance operon (*mer*) have been reviewed thoroughly (Barkay et al. 2003, Mathema et al. 2011).

Together with mercury resistance, arsenic resistance is probably another best studied and most known of microbial metal resistance systems (Mukhopadhyay et al. 2002, Silver and Phung 2009). Arsenite (As^{3+}) resistance and arsenate (As^{5+}) resistance are also widely present in both Gram positive and Gram negative bacteria (Rouch et al. 1995, Sato and Kobayashi 1998, Bruins et al. 2000, Silver and Phung 2009). Arsenic resistance usually is mediated by an *ars* operon consisting of at least three genes, *arsR* (encode an inducer-dependent repressor of transcription), *arsB* (determining an inner membrane protein of the arsenite extrusion pump) and *arsC* (encode a reductase that converts arsenate to arsenite) (Rosenstein et al. 1992, Sato and Kobayashi 1998, Silver and Phung 2005). Two additional genes, *arsA* (encode an intracellular ATPase induced by arsenite and antimonite) and *arsD* (encode a transcriptional regulator), occasionally can be found in *ars* operons in some Gram negative bacteria such as *E. coli* (Tisa and Rosen 1990, Silver and Phung 2005).

Microbial reduction, oxidation and methylation of chalcogen metalloids selenium (Se) and tellurium (Te) affect the bioavailability and toxicity of these metalloids toward microorganisms. Bacterial response to the calgogen metalloids selenium and tellurium has been thoroughly reviewed (Zannoni et al. 2007). The reduction of selenate and selenite to elemental selenium and the reduction of tellurate and tellurite to elemental tellurium can be catalyzed by numerous microorganisms (Taylor 1999, Stolz et al. 2006, Hunter and Manter 2008, Zannoni et al. 2007, Chien and Han 2009, Soudi et al. 2009, Chien et al 2011). Reduction to elemental selenium and tellurium can be considered as detoxification mechanisms of the metalloid oxyanions by microorganisms (Yurkov et al. 1996, Rathgeber et al. 2002, Zannoni et al. 2007, Soudi et al. 2009). In addition to reduction, methylation is another common response of microbial exposure to Se and Te (Chasteen and Bentley 2003).

Chromium (Cr) is a non-beneficial and highly toxic metal for microorganisms (Cervantes et al. 2001). Detoxification mechanisms of Cr compounds by microorganisms include biosorption, diminished accumulation, precipitation, efflux and reduction (Cervantes et al. 2001, Ramírez-Díaz et al. 2008, Viti et al. 2009, Poljsak et al. 2010). Reduction of hexavalent chromium (chromate (CrO_4^{2-}) and di-chromate (Cr_2O7^{2-}) to trivalent chromium ($Cr(OH)_3$) has been noted in a wide range of bacterial species, and the oxidation–reduction reactions of Cr^{6+} to Cr^{3+} can be considered as a chromate resistance mechanism (Cervantes et al. 2001, Francisco et al. 2002, Camargo et al. 2005, Ramírez-Díaz et al. 2008, Morais et al. 2011). Some microorganisms can immobilize several other metals such as hexavalent uranium (U(VI)), hexavalent molybdenum (Mo(VI)), pertechnetate (Tc(VII)), etc., and accordingly reduce their bioavailability and potential toxicity (Lovley 1993, Tebo and Obraztsova 1998, Ramírez-Díaz et al. 2008, Gadd 2009a).

Biosorption, bioaccumulation and sequestration

Biosorption can be defined as the removal of substances from aqueous environments by biological materials (including both living and dead organisms as well as their components) through metabolically mediated or physicochemical process (Volesky and Holan 1995, Kratochvil and Volesky 1998, Vieira and Volesky 2000, Gadd 2010). Due to the high surface to volume ratio, interaction with the mobile metal fraction of the environment is very efficient. The biosorption potential for metal cations is comparable in Gram positive and Gram negative bacteria (Jiang et al. 2004, Haferburg and Kothe 2007). Metals can be immobilized by either intracellular sequestration by protein binding or extracellular sequestration by complexation with released chelating compounds (Bruins et al. 2000). Intracellular sequestration is a cytosolic sequestration mechanism developed by many metal resistance microorganisms (Bruins et al. 2000). Examples of intracellular sequestration include metallothionein production by cyanobacteria (e.g., *Synechococcus* sp.) and cysteine-rich protein of *Pseudomonas* sp. (Belliveau et al. 1987, Bruins et al. 2000, Mejáre and Bülow 2001, Blindauer 2011, Naik et al. 2012). Many metal resistance microorganisms produce internal inclusion bodies such as volutin (polyphosphate) which can bind metal cations (Gonzalez and Jensen 1998, Haferburg and Kothe 2007). Some microorganisms have the capacity to reduce metal/metalloid ions and precipitate the reduced ions (e.g., reduction of silver to metallic silver and tellurite to metallic tellurium) inside the cells (Borghese et al. 2004, Klonowska et al. 2005, Baesman et al. 2007, Chien and Han 2009).

Microbial extracellular polymeric substances (EPS) mainly comprise a mixture of polysaccharides, proteins and some other substances such as nucleic acids, uronic acids, humic substances and lipids, and can bind significant amounts of heavy metals (Pal and Paul 2008, Gadd 2009a). Microbial EPS is crucial to the formation of biofilm and cell aggregates, which contribute to protect cells from hostile environments (Gutnick and Bach 2000, Fang et al. 2002, Pal and Paul 2008). Biofilm and planktonic cells have distinct heavy metal and metalloid susceptibility (Teitzel and Parsek 2003, Harrison et al. 2004a, 2004b, 2007). It is suggested that the complexation or sequestration of heavy metals and retarding their diffusion into the biofilm may be responsible for protecting cells from heavy metal toxicity (Teitzel and Parsek 2003). Microbial EPS are also of particular interest and relevance to the bioremediation process due to their involvement in flocculation and binding of heavy metals from solutions (Salehizadeh and Shojaosadati et al. 2003, Gavrilescu 2004, Pal and Paul 2008, De Philippis et al. 2011, Sundar et al. 2011, Mikutta et al. 2012).

Biotechnology Application

Bioremediation

The potential application in terms of microbes-metal interaction and transformation consists in environmental biotechnology, which can be applied to bioremediation of environmental heavy metal pollution (Mullen et al. 1989, Veglio' and Beolchini 1997, Gavrilescu 2004, Gadd 2010). The processes have been applied to field studies although most of them are still in laboratory scales (Lloyd and Renshaw 2005, Gadd 2010). Microbial processes for metal removal are relevant to mobilization and immobilization of metal ions due to microbial metabolic potentials (Valls and De Lorenzo 2002, Gadd 2004). These processes include bioleaching, bioprecipitation, oxidation-reduction, methylation, biosorption and bioaccumulation (Gadd 2009a, 2010).

Bioleaching is a solubilization mechanism by microorganisms for removal of metals from solid minerals and industrial wastes and is well established process in bioremediation of soil matrices, solid wastes, metal recovery and recycling (Gadd 2010). Microorganisms involved in the bioleaching process include chemolithotrophic (autotrophic) bacteria (e.g., *Acidithiobacillus ferrooxidans, Leptospirillum ferrooxidans*), chemoorganotrophic (heterotrophic) bacteria (e.g., *Sulfolobus* spp., *Bacillus* spp.) and fungi (e.g., *Penicillium simplissimum, Aspergillus niger*) (Jain and Sharma 2004, Brandl and Faramarzi 2006, Lee and Pandey 2012). Bioleaching has been adapted from the mining industry and used in various metal removal operations from soils, sediments, and sludges and proved to be an effective and environmentally friendly approach (Bosecker 1997, Akinci and Guven 2011).

Similar to metal mobilization which can be carried out by many microorganisms, a variety of microorganisms can participate in metal immobilization by biosorption to cell components or EPS, intracellular sequestration or bioprecipitation as organic and inorganic compounds (e.g., sulfides or phosphates) (White et al. 1997). Bioprecipitation by sulfides and phosphates have been extensively studied because of the low solubility of their metal compounds (Macaskie 1990, Gadd 2010, Appukuttan et al. 2011). Precipitation of metals with hydrogen sulfide produced by sulfate reducing bacteria (SRB) has also been applied to bioremediation for the treatment of metal-bearing effluents (Foucher et al. 2001, Pagnanelli et al. 2010). This mechanism consists of the production of hydrogen sulfide by sulfate reducing bacteria and the selective precipitation of metals with biologically produced hydrogen sulfide (Foucher et al. 2001).

Biosorption is a physicochemical process which has been demonstrated as a promising biotechnology for the removal of metal or metalloid species, compounds and particulates from solution by biological material (Gadd 1993, Wang and Chen 2009, Gadd 2010). Biosorption can be carried out by living and dead biomass as well as cellular products such as extracellular polymeric substances or exopolysaccharide (Malik 2004, Sannasi et al. 2006, Pal and Paul 2008, Gadd 2009b). Metal accumulation by microbial biosorption can be metabolism dependent or independent processes (Wang and Chen 2009). Biosorption can also be utilized for the recovery of precious metals such as gold (Reith et al. 2006, Wang and Chen 2009, Gadd 2009b, 2010). Biosorption processes can utilize not only naturally occurring but also genetically engineered microorganisms (Valls and De Lorenzo 2002, Bae et al. 2003, Kiyono and Pan-Hou 2006, Srivastava and Majumder 2008, Vijayaraghavan and Yun 2008, Almaguer-Cantú et al. 2011), and are affected by environmental conditions such as pH, temperature, dissolved oxygen concentration and the presence of other harmless metals and other organic contaminants (Gavrilescu 2004).

Another biotechnological aspect of remediation is the use of plants to remove or detoxify environmental contaminants known as phytoremediation (Raskin et al. 1997, Salt et al. 1998, Abhilash et al. 2012). Symbiotic microorganisms and free-living microorganisms may affect heavy metal phytoremediation by influencing metal bioavailability and interaction and productivity of plants. Studies of phytoremediation integrated with the potential contributory processes from local microorganisms are gaining the attention of researchers working on bioremediation now (Stout and Nüsslein 2010, Gómez-Sagasti et al. 2012, Olguín and Sánchez-Galván 2012).

Novel biomaterials, biocatalysts and biosensors

Microbial–metal interaction has been explored for the production of metal nanoparticles and has the applications in various fields such as biosensors, biocatalysts, antimicrobial activity, bioimaging as well as food industry, etc. (Klaus-Joerger et al. 2001, Verma and Singh 2005, Oh et al. 2009, Krumov et al. 2009, Popescu et al. 2010, Mohammed Fayaz et al. 2011, De Corte et al. 2012). Biosynthesis of nanoparticles has been described in bacteria, yeast, fungi as well as algae (Krumov et al. 2009). Formation of nanoparticles by microorganisms is a result of microbial energy production, detoxification of noxious materials in a hostile environment or use of the nanoparticles for special functions (Mohammed Fayaz et al. 2011, Krumov et al. 2009). Several metal nanoparticles synthesized by microorganisms have been explored including gold, silver, cadmium, zinc, palladium, titanium, magnetite nanoparticles, etc. (Lang et al. 2007, Popescu et al. 2010, Hennebel et al. 2011). A variety of microorganisms are able to interact with metals for the formation of biogenic metal nanoparticles (Hennebel et al. 2009). Precious metals such as gold, silver, palladium, and platinum can be deposited in a reduced form or even in the form of zerovalent nanometals. Manganese (Mn) and iron (Fe) can be precipitated as oxides while magnetotactic bacteria can produce magnetosomes which consist of magnetite (Fe_3O_4) or greigite (Fe_3S_4) (Lovley et al. 2004, Tebo et al. 2004, Fortin and Langley 2005, Schüler 2008, Ash et al. 2011). Using biotechnology for the production of tailor-made biogenic metal nanoparticles is under intensive exploration and is of great interest due to their broad applications. The potential of microorganisms to produce inorganic metal nanoparticles will be a new domain with challenging and new perspectives.

Conclusion

As stated many years ago by Baas Becking "everything is everywhere, but the milieu selects... in nature and in the laboratory" (Baas-Becking 1934). After billions of years' evolution, microorganisms have developed a remarkably wide diversity of metabolic and physiological mechanisms against toxic environments, including heavy metals and metalloids. Microbe-metal interaction plays an important role of biogeochemical cycles for metals, metalloids and related elements. With the new insight of modern biotechnology, the metabolic potentials of microorganisms are being extensively explored and their applications in many facets will be examined with a different approach, both in the laboratory and at a commercial scale.

Acknowledgements

This work was partially supported by grants (contract number NSC100-2221-E-155-002 and NSC101-2221-E-155-031-MY3) from National Science Council, Taiwan.

References

Abhilash, P. C., J. R. Powell, H. B. Singh and B. K. Singh. 2012. Plant-microbe interactions: Novel applications for exploitation in multipurpose remediation technologies. Trends Biotechnol. 30: 416–420.

Akinci, G. and D. E. Guven. 2011. Bioleaching of heavy metals contaminated sediment by pure and mixed cultures of *Acidithiobacillus* spp. Desalination. 268: 221–226.

Almaguer-Cantú, V., L. H. Morales-Ramos and I. Balderas-Rentería. 2011. Biosorption of lead (II) and cadmium (II) using *Escherichia coli* genetically engineered with mice metallothionein I. Water Sci. Technol. 63: 1607–1613.

Alonso, A. and J. L. Martínez. 1997. Multiple antibiotic resistance in *Stenotrophomonas maltophilia*. Antimicrob. Agents Chemother. 41: 1140–1142.

Alonso, A., P. Sanchez and J. L. Martínez. 2000. *Stenotrophomonas maltophilia* D457R contains a cluster of genes from gram-positive bacteria involved in antibiotic and heavy metal resistance. Antimicrob. Agents Chemother. 44: 1778–1782.

Anton, A., A. Weltrowski, C. J. Haney, S. Franke, G. Grass, C. Rensing and D. H. Nies. 2004. Characteristics of zinc transport by two bacterial cation diffusion facilitators from *Ralstonia metallidurans* CH34 and *Escherichia coli*. J. Bacteriol. 186: 7499–7507.

Appukuttan, D., C. Seetharam, N. Padma, A. S. Rao and S. K. Apte. 2011. PhoN-expressing, lyophilized, recombinant *Deinococcus radiodurans* cells for uranium bioprecipitation. J. Biotechnol. 154: 285–290.

Ash, A., K. Revati and B. D. Pandey. 2011. Microbial synthesis of iron-based nanomaterials—A review. Bull. Mat. Sci. 34: 191–198.

Baas-Becking, L. G. M. 1934. Geobiologie of inleiding tot de milieukunde. W.P. van Stockum and Zoon N. V., The Hague, The Netherlands. Cited from: Leadbetter, E. R. 1996. Prokaryotic diversity: from ecophysiology and habitat. In: Manual of Environmental Microbiology. ASM press. pp. 14–24.

Bae, W., C. H. Wu, J. Kostal, A. Mulchandani and W. Chen. 2003. Enhanced mercury biosorption by bacterial cells with surface-displayed MerR. Appl. Environ. Microbiol. 69: 3176–3180.

Baesman, S. M., T. D. Bullen, J. Dewald, D. Zhang, S. Curran, F. S. Islam, T. J. Beveridge and R. S. Oremland. 2007. Formation of tellurium nanocrystals during anaerobic growth of bacteria that use Te oxyanions as respiratory electron acceptors. Appl. Environ. Microbiol. 73: 2135–2143.

Barkay, T., S. M. Miller and A. O. Summers. 2003. Bacterial mercury resistance from atoms to ecosystems. FEMS Microbiol. Rev. 27: 355–384.

Bautista, E. M. and M. Alexander. 1972. Reduction of inorganic compounds by soil microorganisms. Soil Sci. Soc. Am. J. 36: 918–920.

Belliveau, B. H., M. E. Starodub, C. Cotter and J. T. Trevors. 1987. Metal resistance and accumulation in bacteria. Biotechnol. Adv. 5: 101–127.

Blindauer, C. A. 2011. Bacterial metallothioneins: Past, present, and questions for the future. J. Biol. Inorg. Chem. 16: 1011–1024.

Bodrossy, L. and A. Sessitsch. 2004. Oligonucleotide microarrays in microbial diagnostics. Curr. Opin. Microbiol. 7: 245–254.

Borghese, R., F. Borsetti, P. Foladori, G. Ziglio and D. Zannoni. 2004. Effects of the metalloid oxyanion tellurite (TeO_3^{2-}) on growth characteristics of the phototrophic bacterium *Rhodobacter capsulatus*. Appl. Environ. Microbiol. 70: 6595–6602.

Bosecker, K. 1997. Bioleaching: Metal solubilization by microorganisms. FEMS Microbiol. Rev. 20: 591–604.

Brandl, H. and M. A. Faramarzi. 2006. Microbe-metal-interactions for the biotechnological treatment of metal-containing solid waste. China Part. 4: 93–97.

Bruins, M. R., S. Kapil and F. W. Oehme. 2000. Microbial resistance to metals in the environment. Ecotox. Environ. Safe. 45: 198–207.

Camargo, F. A. O., B. C. Okeke, F. M. Bento and W. T. Frankenberger. 2005. Diversity of chromium-resistant bacteria isolated from soils contaminated with dichromate. Appl. Soil Ecol. 29: 193–202.

Cervantes, C., J. Campos-García, S. Devars, F. Gutiérrez-Corona, H. Loza-Tavera, J. C. Torres-Guzmán and R. Moreno-Sánchez. 2001. Interactions of chromium with microorganisms and plants. FEMS Microbiol. Rev. 25: 335–347.

Chan, H., V. Babayan, E. Blyumin, C. Gandhi, K. Hak, D. Harake, K. Kumar, P. Lee, T. T. Li, H. Y. Liu, T. Chung, T. Lo, C. J. Meyer, S. Stanford, K. S. Zamora and M. H. Saier Jr. 2010. The P-Type ATPase superfamily. J. Mol. Microbiol. Biotechnol. 19: 5–104.

Chasteen, T. G. and R. Bentley. 2003. Biomethylation of selenium and tellurium: Microorganisms and plants. Chem. Rev. 103: 1–25.

Chien, C. -C. and C. -T. Han. 2009. Tellurite resistance and reduction by a *Paenibacillus* sp. isolated from heavy metal-contaminated sediment. Environ. Toxicol. Chem. 28: 1627–1632.

Chien, C. -C., C. -W. Hung and C. -T. Han. 2007. Removal of cadmium ions during stationary growth phase by an extremely cadmium-resistant strain of *Stenotrophomonas* sp. Environ. Toxicol. Chem. 26: 664–668.

Chien, C., Y. Kuo, C. Chen, C. Hung, C. Yeh and W. Yeh. 2008. Microbial diversity of soil bacteria in agricultural field contaminated with heavy metals. J. Environ. Sci. 20: 359–363.

Chien, C. -C., M. -H. Jiang, M. -R. Tsai and C. -C. Chien. 2011. Isolation and characterization of an environmental cadmium- and tellurite-resistant *Pseudomonas* strain. Environ. Toxicol. Chem. 30: 2202–2207.

Conroy, O., E. -H. Kim, M. M. McEvoy and C. Rensing. 2010. Differing ability to transport nonmetal substrates by two RND-type metal exporters. FEMS Microbiol. Lett. 308: 115–122.

Crossman, L. C., V. C. Gould, J. M. Dow, G. S. Vernikos, A. Okazaki, M. Sebaihia, D. Saunders, C. Arrowsmith, T. Carver, N. Peters, E. Adlem, A. Kerhornou, A. Lord, L. Murphy, K. Seeger, R. Squares, S. Rutter, M. A. Quail, M. -A. Rajandream, D. Harris, C. Churcher, S. D. Bentley, J. Parkhill, N. R. Thomson and M. B. Avison. 2008. The complete genome, comparative and functional analysis of *Stenotrophomonas maltophilia* reveals an organism heavily shielded by drug resistance determinants. Genome Biol. 9:art. no. R74.

Daffonchio, D., F. Mapelli, A. Cherif, H. I. Malkawi, M. M. Yakimov, Y. R. Abdel-Fattah, M. Blaghen, P. N. Golyshin, M. Ferrer, N. Kalogerakis, N. Boon, M. Magagnini and F. Fava. 2012. ULIXES, unravelling and exploiting Mediterranean Sea microbial diversity and ecology for xenobiotics' and pollutants' clean up. Rev. Environ. Sci. Biotechnol. (in Press) DOI 10.1007/s11157-012-9283-x.

De Corte, S., T. Sabbe, T. Hennebel, L. Vanhaecke, B. De Gusseme, W. Verstraete and N. Boon. 2012. Doping of biogenic Pd catalysts with Au enables dechlorination of diclofenac at environmental conditions. Water Res. 46: 2718–2726.

De Philippis, R., G. Colica and E. Micheletti. 2011. Exopolysaccharide-producing cyanobacteria in heavy metal removal from water: Molecular basis and practical applicability of the biosorption process. Appl. Microbiol. Biotechnol. 92: 697–708.

Dean-Ross, D. and A. L. Mills. 1989. Bacterial community structure and function along a heavy metal gradient. Appl. Environ. Microbiol. 55: 2002–2009.

Delhaize, E., T. Kataoka, D. M. Hebb, R. G. White and P. R. Ryan. 2003. Genes encoding proteins of the cation diffusion facilitator family that confer manganese tolerance. Plant Cell. 15: 1131–1142.

Dos Santos, E. C., I. S. Silva, T. H. N. Simões, K. C. M. Simioni, V. M. Oliveira, M. J. Grossman and L. R. Durrant. 2012. Correlation of soil microbial community responses to contamination with crude oil with and without chromium and copper. Int. Biodeter. Biodegr. 70: 104–110.

Duffus, J. H. 2002. "Heavy metals"—A meaningless term? (IUPAC technical report). Pure Appl. Chem. 74: 793–807.

Dubey, S. K., A. K. Tripathi and S. N. Upadhyay. 2006. Exploration of soil bacterial communities for their potential as bioresource. Bioresour. Technol. 97: 2217–2224.

Ellis, R. J., P. Morgan, A. J. Weightman and J. C. Fry. 2003. Cultivation-dependent and -independent approaches for determining bacterial diversity in heavy-metal-contaminated soil. Appl. Environ. Microbiol. 69: 3223–3230.

Fang, H. H. P., L. -C. Xu and K. -Y. Chan. 2002. Effects of toxic metals and chemicals on biofilm and biocorrosion. Water Res. 36: 4709–4716.

Field, E. K., S. D'Imperio, A. R. Miller, M. R. Vanengelen, R. Gerlach, B. D. Lee, W. A. Apel and B. M. Peyton. 2010. Application of Molecular Techniques to Elucidate the Influence of Cellulosic Waste on the Bacterial Community Structure at a Simulated Low-Level-Radioactive-Waste Site. Appl. Environ. Microbiol. 76: 3106–3115.

Filali, B. K., J. Taoufik, Y. Zeroual, F. Z. Dziri, M. Talbi and M. Blaghen. 2000. Waste water bacterial isolates resistant to heavy metals and antibiotics. Curr. Microbiol. 41: 151–156.

Fortin, D. and S. Langley. 2005. Formation and occurrence of biogenic iron-rich minerals. Earth-Sci. Rev. 72: 1–19.

Foucher, S., F. Battaglia-Brunet, I. Ignatiadis and D. Morin. 2001. Treatment by sulfate-reducing bacteria of Chessy acid-mine drainage and metals recovery. Chem. Eng. Sci. 56: 1639–1645.

Francisco, R., M. C. Alpoim and P. V. Morais. 2002. Diversity of chromium-resistant and -reducing bacteria in a chromium-contaminated activated sludge. J. Appl. Microbiol. 92: 837–843.

Frey, B., M. Stemmer, F. Widmer, J. Luster and C. Sperisen. 2006. Microbial activity and community structure of a soil after heavy metal contamination in a model forest ecosystem. Soil Biol. Biochem. 38: 1745–1756.

Gadd, G. M. 1993. Interactions of fungi with toxic metals. New Phytol. 124: 25–60.

Gadd, G. M. 2004. Microbial influence on metal mobility and application for bioremediation. *Geoderma.* 122: 109–119.

Gadd, G. M. 2009a. Heavy metal pollutants: Environmental and biotechnological aspects. In *Encyclopedia of Microbiology.* Edited by M. Schaechter. Oxford: Elsevier. pp. 321–334.

Gadd, G. M. 2009b. Biosorption: Critical review of scientific rationale, environmental importance and significance for pollution treatment. J. Chem. Technol. Biotechnol. 84: 13–28.

Gadd, G. M. 2010. Metals, minerals and microbes: Geomicrobiology and bioremediation. Microbiol. 156: 609–643.

Gavrilescu, M. 2004. Removal of heavy metals from the environment by biosorption. Eng. Life Sci. 4: 219–232.

Guffanti, A. A., Y. Wei, S. V. Rood and T. A. Krulwich. 2002. An antiport mechanism for a member of the cation diffusion facilitator family: Divalent cations efflux in exchange for K^+ and H^+. Mol. Microbiol. 45: 145–153.

Gelmi, M., P. Apostoli, E. Cabibbo, S. Porru, L. Alessio and A. Turano. 1994. Resistance to cadmium salts and metal absorption by different microbial species. Curr. Microbiol. 29: 335–341.

Gilmour, C. and G. Riedel. 2009. Biogeochemistry of trace metals and metalloids. In *Encyclopedia of Inland Waters.* Edited by G. E. Likens. Amsterdam: Elsevier. pp. 7–15.

Goel, R., S. M. Kotay, C. S. Butler, C. I. Torres and S. Mahendra. 2011. Molecular biological methods in environmental engineering. Water Environ. Res. 83: 927–955.

Gómez-Sagasti, M. T., I. Alkorta, J. M. Becerril, L. Epelde, M. Anza and C. Garbisu. 2012. Microbial monitoring of the recovery of soil quality during heavy metal phytoremediation. Water Air Soil Pollut. 223: 3249–3262.

Gonzalez, H. and T. E. Jensen. 1998. Nickel sequestering by polyphosphate bodies in *Staphylococcus aureus.* Microbios. 93: 179–185.

Goris, J., P. De Vos, T. Coenye, B. Hoste, D. Janssens, H. Brim, L. Diels, M. Mergeay, K. Kersters and P. Vandamme. 2001. Classification of metal-resistant bacteria from industrial biotopes as *Ralstonia campinensis* sp. nov., *Ralstonia metallidurans* sp. nov. and *Ralstonia basilensis* Steinle et al. 1998 emend. Int. J. Syst. Evol. Microbiol. 51: 1773–1782.

Grass, G., B. Fan, B. P. Rosen, S. Franke, D. H. Nies and C. Rensing. 2001. ZitB (YbgR), a member of the cation diffusion facilitator family, is an additional zinc transporter in *Escherichia coli.* J. Bacteriol. 183: 4664–4667.

Grass, G., M. Otto, B. Fricke, C. J. Haney, C. Rensing, D. H. Nies and D. Munkelt. 2005. FieF (YiiP) from *Escherichia coli* mediates decreased cellular accumulation of iron and relieves iron stress. Arch. Microbiol. 183: 9–18.

Grass, G. and C. Rensing. 2001. Genes involved in copper homeostasis in *Escherichia coli.* J. Bacteriol. 183: 2145–2147.

Gremion, F., A. Chatzinotas and H. Harms. 2003. Comparative 16S rDNA and 16S rRNA sequence analysis indicates that *Actinobacteria* might be a dominant part of the metabolically active bacteria in heavy metal-contaminated bulk and rhizosphere soil. Environ. Microbiol. 5: 896–907.

Gutnick, D. L. and H. Bach. 2000. Engineering bacterial biopolymers for the biosorption of heavy metals, new products and novel formulations. Appl. Microbiol. Biotechnol. 54: 451–460.

Haferburg, G. and E. Kothe. 2007. Microbes and metals: Interactions in the environment. J. Basic Microbiol. 47: 453–467.

Handelsman, J. 2004. Metagenomics: Application of genomics to uncultured microorganisms. Microbiol. Mol. Biol. Rev. 68: 669–685.

Haney, C. J., G. Grass, S. Franke and C. Rensing. 2005. New developments in the understanding of the cation diffusion facilitator family. J. Ind. Microbiol. Biotechnol. 32: 215–226.

Haq, R. U. and A. R. Shakoori. 2000. Microorganisms resistant to heavy metals and toxic chemicals as indicators of environmental pollution and their use in bioremediation. Folia Biol. 48: 143–147.

Harrison, J. J., H. Ceri and R. J. Turner. 2007. Multimetal resistance and tolerance in microbial biofilms. Nat. Rev. Microbiol. 5: 928–938.

Harrison, J. J., H. Ceri, C. Stremick and R. J. Turner. 2004a. Differences in biofilm and planktonic cell mediated reduction of metalloid oxyanions. FEMS Microbiol. Lett. 235: 357–362.

Harrison, J. J., H. Ceri, C. A. Stremick and R. J. Turner. 2004b. Biofilm susceptibility to metal toxicity. Environ. Microbiol. 6: 1220–1227.

Hartmann, M., B. Frey, R. Kölliker and F. Widmer. 2005. Semi-automated genetic analyses of soil microbial communities: Comparison of T-RFLP and RISA based on descriptive and discriminative statistical approaches. J. Microbiol. Methods. 61: 349–360.

Hattori, H. 1992. Influence of heavy metals on soil microbial activities. Soil Sci. Plant Nutr. 38: 93–100.

Hemme, C. L., Y. Deng, T. J. Gentry, M. W. Fields, L. Wu, S. Barua, K. Barry, S. G. Tringe, D. B. Watson, Z. He, T. C. Hazen, J. M. Tiedje, E. M. Rubin and J. Zhou. 2010. Metagenomic insights into evolution of a heavy metal-contaminated groundwater microbial community. ISME J. 4: 660–672.

Hennebel, T., B. De Gusseme, N. Boon and W. Verstraete. 2009. Biogenic metals in advanced water treatment. Trends Biotechnol. 27: 90–98.

Hennebel, T., S. Van Nevel, S. Verschuere, S. De Corte, B. De Gusseme, C. Cuvelier, J. P. Fitts, D. van der Lelie, N. Boon and W. Verstraete. 2012. Palladium nanoparticles produced by fermentatively cultivated bacteria as catalyst for diatrizoate removal with biogenic hydrogen. Appl. Microbiol. Biotechnol. 91: 1435–1445.

Hu, Q., M. Dou, H. Qi, X. Xie, G. Zhuang and M. Yang. 2007. Detection, isolation, and identification of cadmium-resistant bacteria based on PCR-DGGE. J. Environ. Sci. 19: 1114–1119.

Hudson, M. E. 2008. Sequencing breakthroughs for genomic ecology and evolutionary biology. Mol. Ecol. Resour. 8: 3–17.

Hugenholtz, P., B. M. Goebel and N. R. Pace. 1998. Impact of culture-independent studies on the emerging phylogenetic view of bacterial diversity. J. Bacteriol. 180: 4765–4774.

Hunter, W. J. and D. K. Manter. 2008. Bio-reduction of selenite to elemental red selenium by *Tetrathiobacter kashmirensis*. Curr. Microbiol. 57: 83–88.

Iannelli, R., V. Bianchi, C. Macci, E. Peruzzi, C. Chiellini, G. Petroni and G. Masciandaro. 2012. Assessment of pollution impact on biological activity and structure of seabed bacterial communities in the Port of Livorno (Italy). Sci. Total Environ. 426: 56–64.

Islam, E. and P. Sar. 2011. Culture-dependent and -independent molecular analysis of the bacterial community within uranium ore. J. Basic Microbiol. 51: 372–384.

Jain, N. and D. K. Sharma. 2004. Biohydrometallurgy for nonsulfidic minerals—A review. Geomicrobiol. J. 21: 135–144.

Jiang, W., A. Saxena, B. Song, B. B. Ward, T. J. Beveridge and S. C. B. Myneni. 2004. Elucidation of functional groups on Gram-positive and Gram-negative bacterial surfaces using infrared spectroscopy. Langmuir. 20: 11433–11442.

Kell, D. B., A. S. Kaprelyants, D. H. Weichart, C. R. Harwood and M. R. Barer. 1998. Viability and activity in readily culturable bacteria: A review and discussion of the practical issues. *Antonie van Leeuwenhoek.* 73: 169–187.

Khan, S., A. E. -L. Hesham, M. Qiao, S. Rehman and J. -Z. He. 2010. Effects of Cd and Pb on soil microbial community structure and activities. Environ. Sci. Pollut. Res. 17: 288–296.

Kim, E. -H., D. H. Nies, M. M. McEvoy and C. Rensing. 2011. Switch or funnel: How RND-type transport systems control periplasmic metal homeostasis. J. Bacteriol. 193: 2381–2387.

Kimura, N. 2006. Metagenomics: Access to unculturable microbes in the environment. Microbes Environ. 21: 201–215.

Kiyono, M. and H. Pan-Hou. 2006. Genetic engineering of bacteria for environmental remediation of mercury. J. Health Sci. 52: 199–204.

Klaus-Joerger, T., R. Joerger, E. Olsson and C. -G. Granqvist. 2001. Bacteria as workers in the living factory: Metal-accumulating bacteria and their potential for materials science. Trends Biotechnol. 19: 15–20.

Klonowska, A., T. Heulin and A. Vermeglio. 2005. Selenite and tellurite reduction by *Shewanella oneidensis*. Appl. Environ. Microbiol. 71: 5607–5609.

Kratochvil, D. and B. Volesky. 1998. Advances in the biosorption of heavy metals. Trends Biotechnol. 16: 291–300.

Krumov, N., I. Perner-Nochta, S. Oder, V. Gotcheva, A. Angelov and C. Posten. 2009. Production of inorganic nanoparticles by microorganisms. Chem. Eng. Technol. 32: 1026–1035.

Landa, E. R. 2005. Microbial biogeochemistry of uranium mill tailings. Adv. Appl. Microbiol. 57: 113–130.

Lang, C., D. Schüler and D. Faivre. 2007. Synthesis of magnetite nanoparticles for bio- and nanotechnology: Genetic engineering and biomimetics of bacterial magnetosomes. Macromol. Biosci. 7: 144–151.

Lazzaro, A., F. Widmer, C. Sperisen and B. Frey. 2008. Identification of dominant bacterial phylotypes in a cadmium-treated forest soil. FEMS Microbiol. Ecol. 63: 143–155.

LeBaron, P., P. Servais, H. Agogué, C. Courties and F. Joux. 2001. Does the High Nucleic Acid Content of Individual Bacterial Cells Allow Us to Discriminate between Active Cells and Inactive Cells in Aquatic Systems? Appl. Environ. Microbiol. 67: 1775–1782.

Lee, S. M., G. Grass, C. J. Haney, B. Fan, B. P. Rosen, A. Anton, D. H. Nies and C. Rensing. 2002. Functional analysis of the *Escherichia coli* zinc transporter ZitB. FEMS Microbiol. Lett. 215: 273–278.

Lee, J. -C. and B. D. Pandey. 2012. Bio-processing of solid wastes and secondary resources for metal extraction—A review. Waste Manage. 32: 3–18.

Legatzki, A., G. Grass, A. Anton, C. Rensing and D. H. Nies. 2003. Interplay of the Czc system and two P-type ATPases in conferring metal resistance to *Ralstonia metallidurans*. J. Bacteriol. 185: 4354–4361.

Li, Z., J. Xu, C. Tang, J. Wu, A. Muhammad and H. Wang. 2006. Application of 16S rDNA-PCR amplification and DGGE fingerprinting for detection of shift in microbial community diversity in Cu-, Zn-, and Cd-contaminated paddy soils. Chemosphere. 62: 1374–1380.

Lloyd, J. R. and J. C. Renshaw. 2005. Bioremediation of radioactive waste: Radionuclide-microbe interactions in laboratory and field-scale studies. Curr. Opin. Biotechnol. 16: 254–260.

Lovley, D. R. 1993. Dissimilatory metal reduction. Annu. Rev. Microbiol. 47: 263–290.

Lovley, D. R., D. E. Holmes and K. P. Nevin. 2004. Dissimilatory Fe(III) and Mn(IV) reduction. Adv. Microb. Physiol. 49: 219–286.

Lucchini, S., A. Thompson and J. C. D. Hinton. 2001. Microarrays for microbiologists. Microbiol. 147: 1403–1414.

Lutsenko, S. and J. H. Kaplan. 1995. Organization of P-type ATPases: Significance of structural diversity. Biochem. 34: 15607–15613.

Ma, Z., F. E. Jacobsen and D. P. Giedroc. 2009. Coordination chemistry of bacterial metal transport and sensing. Chem. Rev. 109: 4644–4681.

Macaskie, L. E. 1990. An immobilized cell bioprocess for the removal of heavy metals from aqueous flows. J. Chem. Technol. Biotechnol. 49: 357–379.

Malik, A. 2004. Metal bioremediation through growing cells. Environ. Int. 30: 261–278.

Margesin, R., G. A. Płaza and S. Kasenbacher. 2011. Characterization of bacterial communities at heavy-metal-contaminated sites. Chemosphere. 82: 1583–1588.

Martins, M., M. L. Faleiro, S. Chaves, R. Tenreiro, E. Santos and M. C. Costa. 2010. Anaerobic bio-removal of uranium (VI) and chromium (VI): Comparison of microbial community structure. J. Hazard. Mater. 176: 1065–1072.

Mathema, V. B., B. C. Thakuri and M. Sillanpää. 2011. Bacterial mer operon-mediated detoxification of mercurial compounds: A short review. Arch. Microbiol. 193: 837–844.

Mejáre, M. and L. Bülow. 2001. Metal-binding proteins and peptides in bioremediation and phytoremediation of heavy metals. Trends Biotechnol. 19: 67–73.

Mergeay, M., S. Monchy, T. Vallaeys, V. Auquier, A. Benotmane, P. Bertin, S. Taghavi, J. Dunn, D. van der Lelie and R. Wattiez. 2003. *Ralstonia metallidurans*, a bacterium specifically adapted to toxic metals: towards a catalogue of metal responsive genes. FEMS Microbiol. Rev. 27: 385–410.

Mikutta, R., A. Baumgärtner, A. Schippers, L. Haumaier and G. Guggenberger. 2012. Extracellular polymeric substances from *Bacillus subtilis* associated with minerals modify the extent and rate of heavy metal sorption. Environ. Sci. Technol. 46: 3866–3873.

Minkwitz, A. and G. Berg. 2001. Comparison of antifungal activities and 16S ribosomal DNA sequences of clinical and environmental isolates of *Stenotrophomonas maltophilia*. J. Clin. Microbiol. 39: 139–145.

Misra, T. K. 1992. Bacterial resistances to inorganic mercury salts and organomercurials. Plasmid. 27: 4–16.

Mohammed Fayaz, A., M. Girilal, M. Rahman, R. Venkatesan and P. T. Kalaichelvan. 2011. Biosynthesis of silver and gold nanoparticles using thermophilic bacterium *Geobacillus stearothermophilus*. Process Biochem. 46: 1958–1962.

Mohapatra, B. R., W. Douglas Gould, O. Dinardo and D. W. Koren. 2011. Tracking the prokaryotic diversity in acid mine drainage-contaminated environments: A review of molecular methods. Miner. Eng. 24: 709–718.

Monsieurs, P., H. Moors, R. Van Houdt, P. J. Janssen, A. Janssen, I. Coninx, M. Mergeay and N. Leys. 2011. Heavy metal resistance in *Cupriavidus metallidurans* CH34 is governed by an intricate transcriptional network. BioMetals. 24: 1133–1151.

Moore, C. M. and J. D. Helmann. 2005. Metal ion homeostasis in *Bacillus subtilis*. Curr. Opin. Microbiol. 8: 188–195.

Morais, P. V., R. Branco and R. Francisco. 2011. Chromium resistance strategies and toxicity: What makes *Ochrobactrum tritici* 5bvl1 a strain highly resistant? BioMetals. 24: 401–410.

Mukhopadhyay, R., B. P. Rosen, L. T. Phung and S. Silver. 2002. Microbial arsenic: From geocycles to genes and enzymes. FEMS Microbiol. Rev. 26: 311–325.

Mullen, M. D., D. C. Wolf, F. G. Ferris, T. J. Beveridge, C. A. Flemming and G. W. Bailey. 1989. Bacterial sorption of heavy metals. Appl. Environ. Microbiol. 55: 3143–3149.

Munkelt, D., G. Grass and D. H. Nies. 2004. The chromosomally encoded cation diffusion facilitator proteins DmeF and FieF from *Wautersia metallidurans* CH34 are transporters of broad metal specificity. J. Bacteriol. 186: 8036–8043.

Muyzer, G. 1999. DGGE/TGGE a method for identifying genes from natural ecosystems. Curr. Opin. Microbiol. 2: 317–322.

Muyzer, G., E. C. De Waal and A. G. Uitterlinden. 1993. Profiling of complex microbial populations by denaturing gradient gel electrophoresis analysis of polymerase chain reaction-amplified genes coding for 16S rRNA. Appl. Environ. Microbiol. 59: 695–700.

Naik, M. M., A. Pandey and S. K. Dubey. 2012. *Pseudomonas aeruginosa* strain WI-1 from Mandovi estuary possesses metallothionein to alleviate lead toxicity and promotes plant growth. Ecotox. Environ. Safe. 79: 129–133.

Narancic, T., L. Djokic, S. T. Kenny, K. E. O'Connor, V. Radulovic, J. Nikodinovic-Runic and B. Vasiljevic. 2012. Metabolic versatility of Gram-positive microbial isolates from contaminated river sediments. J. Hazard. Mater. 215–216: 243–251.

Narita, M., K. Chiba, H. Nishizawa,H. Ishii, C. -C. Huang, Z. Kawabata, S. Silver and G. Endo. 2003. Diversity of mercury resistance determinants among Bacillus strains isolated from sediment of Minamata Bay. FEMS Microbiol. Lett. 223: 73–82.

Navarro-Noya, Y. E., J. Jan-Roblero, D. C. González-Chávez, R. Hernández-Gama and C. Hernández-Rodríguez. 2010. Bacterial communities associated with the rhizosphere of pioneer plants (Bahia xylopoda and Viguiera linearis) growing on heavy metals-contaminated soils. *Antonie van Leeuwenhoek*. 97: 335–349.

Nies, D. H. 1995. The cobalt, zinc, and cadmium efflux system CzcABC from *Alcaligenes eutrophus* functions as a cation-proton antiporter in *Escherichia coli*. J. Bacteriol. 177: 2707–2712.

Nies, D. H. 1999. Microbial heavy-metal resistance. Appl. Microbiol. Biotechnol. 51: 730–750.

Nies, D. H. 2003. Efflux-mediated heavy metal resistance in prokaryotes. FEMS Microbiol. Rev. 27: 313–339.

Nies, D. H. and S. Silver. 1995. Ion efflux systems involved in bacterial metal resistances. J. Ind. Microbiol. 14: 186–199.

Ogram, A. 2000. Soil molecular microbial ecology at age 20: Methodological challenges for the future. Soil Biol. Biochem. 32: 1499–1504.

Oh, K. S., R. S. Kim, J. Lee, D. Kim, S. H. Cho and S. H. Yuk. 2009. Gold/chitosan/pluronic composite nanoparticles for drug delivery. J. Appl. Polym. Sci. 108: 3239–3244.

Olguín, E. J. and G. Sánchez-Galván. 2012. Heavy metal removal in phytofiltration and phycoremediation: the need to differentiate between bioadsorption and bioaccumulation. New Biotech. (In Press).

Olsson-Francis, K., R. Van Houdt, M. Mergeay, N. Leys and C. S. Cockell. 2010. Microarray analysis of a microbe-mineral interaction. Geobiol. 8: 446–456.

Øvreås, L. and V. Torsvik. 1998. Microbial diversity and community structure in two different agricultural soil communities. Microb. Ecol. 36: 303–315.

Pagnanelli, F., C. Cruz Viggi and L. Toro. 2010. Isolation and quantification of cadmium removal mechanisms in batch reactors inoculated by sulphate reducing bacteria: Biosorption versus bioprecipitation. Bioresour. Technol. 101: 2981–2987.

Pages, D., J. Rose, S. Conrod, S. Cuine, P. Carrier, T. Heulin and W. Achouak. 2008. Heavy metal tolerance in *Stenotrophomonas maltophilia*. PLoS ONE. 3:art. no. e1539.

Pal, A. and A. K. Paul. 2008. Microbial extracellular polymeric substances: Central elements in heavy metal bioremediation. Indian J. Microbiol. 48: 49–64.

Paulsen, I. T. and M. H. Saier Jr. 1997. A novel family of ubiquitous heavy metal ion transport proteins. J. Membr. Biol. 156: 99–103.

Pérez-De-Mora, A., P. Burgos, E. Madejón, F. Cabrera, P. Jaeckel and M. Schloter. 2006. Microbial community structure and function in a soil contaminated by heavy metals: Effects of plant growth and different amendments. Soil Biol. Biochem. 38: 327–341.

Piddock, L. J. V. 2006. Multidrug-resistance efflux pumps—Not just for resistance. Nat. Rev. Microbiol. 4: 629–636.

Poljsak, B., I. Pócsi, P. Raspor and M. Pesti. 2010. Interference of chromium with biological systems in yeasts and fungi: A review. J. Basic Microbiol. 50: 21–36.

Popescu, M., A. Velea and A. Lorinczi. 2010. Biogenic production of nanoparticles. Dig. J. Nanomater. Biostruct. 5: 1035–1040.

Ramírez-Díaz, M. I., C. Díaz-Pérez, E. Vargas, H. Riveros-Rosas, J. Campos-García and C. Cervantes. 2008. Mechanisms of bacterial resistance to chromium compounds. BioMetals. 21: 321–332.

Rappé, M. S. and S. J. Giovannoni. 2003. The Uncultured Microbial Majority. Annu. Rev. Microbiol. 57: 369–394.

Raskin, I., R. D. Smith and D. E. Salt. 1997. Phytoremediation of metals: Using plants to remove pollutants from the environment. Curr. Opin. Biotechnol. 8: 221–226.

Rastogi, G., S. Barua, R. K. Sani and B. M. Peyton. 2011. Investigation of Microbial Populations in the Extremely Metal-Contaminated Coeur d'Alene River Sediments. Microb. Ecol. 62: 1–13.

Rathgeber, C., N. Yurkova, E. Stackebrandt, J. T. Beatty and V. Yurkov. 2002. Isolation of tellurite- and selenite-resistant bacteria from hydrothermal vents of the Juan de Fuca Ridge in the Pacific Ocean. Appl. Environ. Microbiol. 68: 4613–4622.

Reith, F., S. L. Rogers, D. C. McPhail and D. Webb. 2006. Biomineralization of gold: Biofilms on bacterioform gold. Science. 313: 233–236.

Rensing, C., B. Mitra and B. P. Rosen. 1997. The *zntA* gene of *Escherichia coli* encodes a Zn(II)-translocating P-type ATPase. Proc. Natl. Acad. Sci. USA. 94: 14326–14331.

Rensing, C., Y. Sun, B. Mitra and B. P. Rosen. 1998. Pb(II)-translocating P-type ATPases. J. Biol. Chem. 273: 32614–32617.

Roane, T. M. and S. T. Kellogg. 1996. Characterization of bacterial communities in heavy metal contaminated soils. Can. J. Microbiol. 42: 593–603.

Roane, T. M. and I. L. Pepper. 2000. Microbial responses to environmentally toxic cadmium. Microb. Ecol. 38: 358–364.

Rosch, J. W., G. Gao, G. Ridout, Y. -D. Wang and E. I. Tuomanen. 2009. Role of the manganese efflux system *mntE* for signalling and pathogenesis in *Streptococcus pneumonia*. Mol. Microbiol. 72: 12–25.

Rosenstein, R., A. Peschel, B. Wieland and F. Gotz. 1992. Expression and regulation of the antimonite, arsenite, and arsenate resistance operon of *Staphylococcus xylosus* plasmid pSX267. J. Bacteriol. 174: 3676–3683.

Rouch, D. A., B. T. O. Lee and A. P. Morby. 1995. Understanding cellular responses to toxic agents: A model for mechanism-choice in bacterial metal resistance. J. Ind. Microbiol. 14: 132–141.

Said, O. B., M. Goñi-Urriza, M. E. Bour, P. Aissa and R. Duran. 2010. Bacterial community structure of sediments of the bizerte lagoon (Tunisia), a southern mediterranean coastal anthropized lagoon. Microb. Ecol. 59: 445–456.

Saier Jr., M. H., R. Tam, A. Reizer and J. Reizer. 1994. Two novel families of bacterial membrane proteins concerned with nodulation, cell division and transport. Mol. Microbiol. 11: 841–847.

Salehizadeh, H. and S. A. Shojaosadati. 2003. Removal of metal ions from aqueous solution by polysaccharide produced from *Bacillus firmus*. Water Res. 37: 4231–4235.

Salt, D. E., R. D. Smith and I. Raskin. 1998. Phytoremediation. Annu. Rev. Plant Biol. 49: 643–668.

Sannasi, P., J. Kader, B. S. Ismail and S. Salmijah. 2006. Sorption of Cr(VI), Cu(II) and Pb(II) by growing and non-growing cells of a bacterial consortium. Bioresour. Technol. 97: 740–747.

Sato, T. and Y. Kobayashi. 1998. The *ars* operon in the skin element of *Bacillus subtilis* confers resistance to arsenate and arsenite. J. Bacteriol. 180: 1655–1661.

Schmeisser, C., H. Steele and W. R. Streit. 2007. Metagenomics, biotechnology with non-culturable microbes. Appl. Microbiol. Biotechnol. 75: 955–962.

Schüler, D. 2008. Genetics and cell biology of magnetosome formation in magnetotactic bacteria. FEMS Microbiol. Rev. 32: 654–672.

Sharma, R., C. Rensing, B. P. Rosen and B. Mitra. 2000. The ATP hydrolytic activity of purified ZntA, a Pb(II)/Cd(II)/Zn(II)- translocating ATPase from *Escherichia coli*. J. Biol. Chem. 275: 3873–3878.

Sharma, R., R. Ranjan, R. K. Kapardar and A. Grover. 2005. 'Unculturable' bacterial diversity: An untapped resource. Curr. Sci. 89: 72–77.

Silver, S. 1996. Bacterial resistances to toxic metal ions - A review. Gene. 179: 9–19.

Silver, S. and T. K. Misra. 1988. Plasmid-mediated heavy metal resistances. Annu. Rev. Microbiol. 42: 717–743.

Silver, S. and L. T. Phung. 1996. Bacterial heavy metal resistance: New surprises. Annu. Rev. Microbiol. 50: 753–789.

Silver, S. and L. T. Phung. 2005. A bacterial view of the periodic table: Genes and proteins for toxic inorganic ions. J. Ind. Microbiol. Biotechnol. 32: 587–605.

Silver, S. and L. T. Phung. 2009. Heavy metals, bacterial resistance. In *Encyclopedia of Microbiology*. Edited by M. Schaechter. Oxford: Elsevier. pp. 220–227.

Simon, C. and R. Daniel. 2011. Metagenomic analyses: Past and future trends. Appl. Environ. Microbiol. 77: 1153–1161.

Smit, E., P. Leeflang and K. Wernars. 1997. Detection of shifts in microbial community structure and diversity in soil caused by copper contamination using amplified ribosomal DNA restriction analysis. FEMS Microbiol. Ecol. 23: 249–261.

Solioz, M. and C. Vulpe. 1996. CPx-type ATPases: A class of P-type ATPases that pump heavy metals. Trends Biochem. Sci. 21: 237.

Soudi, M. R., P. T. M. Ghazvini, K. Khajeh and S. Gharavi. 2009. Bioprocessing of seleno-oxyanions and tellurite in a novel *Bacillus* sp. strain STG-83: A solution to removal of toxic oxyanions in presence of nitrate. J. Hazard. Mater. 165: 71–77.

Srivastava, N. K. and C. B. Majumder. 2008. Novel biofiltration methods for the treatment of heavy metals from industrial wastewater. J. Hazard. Mater. 151: 1–8.

Stohs, S. J. and D. Bagchi. 1995. Oxidative mechanisms in the toxicity of metal ions. Free Radic. Biol. Med. 18: 321–336.

Stolz, J. F., P. Basu, J. M. Santini and R. S. Oremland. 2006. Arsenic and selenium in microbial metabolism. Annu. Rev. Microbiol. 60: 107–130.

Stout, L. and K. Nüsslein. 2010. Biotechnological potential of aquatic plant-microbe interactions. Curr. Opin. Biotechnol. 21: 339–345.

Streit, W. R. and R. A. Schmitz. 2004. Metagenomics—The key to the uncultured microbes. Curr. Opin. Microbiol. 7: 492–498.

Su, C., L. Lei, Y. Duan, K. -Q. Zhang and J. Yang. 2012. Culture-independent methods for studying environmental microorganisms: Methods, application, and perspective. Appl. Microbiol. Biotechnol. 93: 993–1003.

Sundar, K., A. Mukherjee, M. Sadiq and N. Chandrasekaran. 2011. Cr (III) bioremoval capacities of indigenous and adapted bacterial strains from Palar river basin. J. Hazard. Mater. 187: 553–561.

Taylor, D. E. 1999. Bacterial tellurite resistance. Trends Microbiol. 7: 111–115.

Tebo, B. M. and A. Y. Obraztsova. 1998. Sulfate-reducing bacterium grows with Cr(VI), U(VI), Mn(IV), and Fe(III) as electron acceptors. FEMS Microbiol. Lett. 162: 193–198.

Tebo, B. M., J. R. Bargar, B. G. Clement, G. J. Dick, K. J. Murray, D. Parker, R. Verity and S. M. Webb. 2004. Biogenic manganese oxides: Properties and mechanisms of formation. Annu. Rev. Earth Planet. Sci. 32: 287–328.

Teitzel, G. M. and M. R. Parsek. 2003. Heavy metal resistance of biofilm and planktonic *Pseudomonas aeruginosa*. Appl. Environ. Microbiol. 69: 2313–2320.

Tisa, L. S. and B. P. Rosen. 1990. Molecular characterization of an anion pump. The ArsB protein is the membrane anchor for the ArsA protein. J. Biol. Chem. 265: 190–194.

Tsai, K. -J., K. P. Yoon and A. R. Lynn. 1992. ATP-dependent cadmium transport by the *cad*A cadmium resistance determinant in everted membrane vesicles of *Bacillus subtilis*. J. Bacteriol. 174: 116–121.

Tseng, T. -T., K. S. Gratwick, J. Kollman, D. Park, D. H. Nies, A. Goffeau and M. H. Saier Jr. 1999. The RND permease superfamily: An ancient, ubiquitous and diverse family that includes human disease and development proteins. J. Mol. Microbiol. Biotechnol. 1: 107–125.

Vainshtein, M., P. Kuschk, J. Mattusch, A. Vatsourina and A. Wiessner. 2003. Model experiments on the microbial removal of chromium from contaminated groundwater. Water Res. 37: 1401–1405.

Valls, M. and V. De Lorenzo. 2002. Exploiting the genetic and biochemical capacities of bacteria for the remediation of heavy metal pollution. FEMS Microbiol. Rev. 26: 327–338.

Veglio', F. and F. Beolchini. 1997. Removal of metals by biosorption: A review. Hydrometallurgy. 44: 301–316.

Verma, N. and M. Singh. 2005. Biosensors for heavy metals. BioMetals. 18: 121–129.

Vieira, R. H. S. F. and B. Volesky. 2000. Biosorption: a solution to pollution? Int. Microbiol. 3: 17–24.

Vijayaraghavan, K. and Y. -S. Yun. 2008. Bacterial biosorbents and biosorption. Biotechnol. Adv. 26: 266–291.

Vílchez, R., C. Gómez-Silván, J. Purswani, J. González-López and B. Rodelas. 2011. Characterization of bacterial communities exposed to Cr(III) and Pb(II) in submerged fixed-bed biofilms for groundwater treatment. Ecotoxicology. 20: 779–792.

Viti, C., F. Decorosi, A. Mini, E. Tatti and L. Giovannetti. 2009. Involvement of the *oscA* gene in the sulphur starvation response and in Cr(VI) resistance in *Pseudomonas corrugata* 28. Microbiol. 155: 95–105.

Vivas, A., B. Moreno, C. Del Val, C. MacCi, G. Masciandaro and E. Benitez. 2008. Metabolic and bacterial diversity in soils historically contaminated by heavy metals and hydrocarbons. J. Environ. Monit. 10: 1287–1296.

Volesky, B. and Z. R. Holan. 1995. Biosorption of heavy metals. Biotechnol. Prog. 11: 235–250.

Waldron, P. J., L. Wu, J. D. Van Nostrand, C. W. Schadt, Z. He, D. B. Watson, P. M. Jardine, A. V. Palumbo, T. C. Hazen and J. Zhou. 2009. Functional gene array-based analysis of microbial community structure in groundwaters with a gradient of contaminant levels. Environ. Sci. Technol. 43: 3529–3534.

Wang, J. and C. Chen. 2009. Biosorbents for heavy metals removal and their future. Biotechnol. Adv. 27: 195–226.

Wenderoth, D. F., E. Stackebrandt and H. H. Reber. 2001. Metal stress selects for bacterial ARDRA-types with a reduced catabolic versatility. Soil Biol. Biochem. 33: 667–670.

White, C., J. A. Sayer and G. M. Gadd. 1997. Microbial solubilization and immobilization of toxic metals: Key biogeochemical processes for treatment of contamination. FEMS Microbiol. Rev. 20: 503–516.

Woese, C. R. 1987. Bacterial evolution. Microbiol. Rev. 51: 221–271.

Wu, C. -C., N. Bal, J. Perard, J. Lowe, C. Boscheron, E. Mintz and P. Catty. 2004. A cloned prokaryotic Cd^{2+} P-type ATPase increases yeast sensitivity to Cd^{2+}. Biochem. Biophys. Res. Commun. 324: 1034–1040.

Xiong, J., D. Li, H. Li, M. He, S. J. Miller, L. Yu, C. Rensing and G. Wang. 2011. Genome analysis and characterization of zinc efflux systems of a highly zinc-resistant bacterium, *Comamonas testosteroni* S44. Res. Microbiol. 162: 671–679.

Yurkov, V., J. Jappé and A. Verméglio. 1996. Tellurite resistance and reduction by obligately aerobic photosynthetic bacteria. Appl. Environ. Microbiol. 62: 4195–4198.

Zannoni, D., F. Borsetti, J. J. Harrison and R. J. Turner. 2007. The Bacterial Response to the Chalcogen Metalloids Se and Te. Adv. Microb. Physiol. 53: 1–72.

Zhang, L., X. Z. Li and K. Poole. 2000. Multiple antibiotic resistance in *Stenotrophomonas maltophilia*: Involvement of a multidrug efflux system. Antimicrob. Agents Chemother. 44: 287–293.

Zhou, J. 2003. Microarrays for bacterial detection and microbial community analysis. Curr. Opin. Microbiol. 6: 288–294.

Zhou, J. and D. K. Thompson. 2004. Microarray technology and applications in environmental microbiology. Adv. Agron. 82: 183–270.

14

Sophorolipids: Production, Characterization and Biologic Activity

Isabel A. Ribeiro, Matilde F. Castro and
*Maria H. Ribeiro**

ABSTRACT

Sophorolipids (SLs) are one of the most promising glycolipid biosurfactants. They are biodegradable, show low ecotoxicity and besides surfactant properties show other diverse biological activities. SLs are produced by non-pathogenic yeasts, in high yields, as mixture of lactonic and acidic compounds mainly composed of a fatty acid with C16 or C18 chain length, different unsaturation grade and terminal (ω) or sub-terminal (ω-1) hydroxylation linked to a sophorose unit that can also show different acetylation grade. *Candida bombicola* or the teleomorph *Starmerella bombicola* are larger producers of SLs and consequently selected for several studies. Phylogenetically different yeast, the basidomycete *Rhodotorula bogoriensis*, can also produce SLs, different from the previously mentioned compounds. *Rhodotorula*

Research Institute for Medicines and Pharmaceutical Sciences (i-Med-UL), Faculty of Pharmacy, University of Lisbon, Av. Prof. Gama Pinto, 1649-003 Lisboa, Portugal.
* Corresponding author: mhribeiro@ff.ul.pt

bogoriensis produces only acidic SLs with a longer fatty acid chain mainly with 22 carbon atoms. The sophorose moiety can also present different acetylation grades, and is linked to the fatty acid at C13 position. SLs are produced as a mixture of compounds and one of the problems is the separation of each compound. Depending on SLs structure different properties can be obtained, therefore the individualization of each SL can assume particular significance in the properties characterization and biological activity evaluation.

Introduction

Glycolipids (GLs) are amphiphilic compounds which contain a hydrophilic head of one or more monosaccharide residues bound by a glycosidic linkage to a hydrophobic lipid tail. GLs can be found nearly in all vertebrate cells and body fluids and above all on cellular membranes (Yu and Yanagisawa 2010) (Fig. 1). They are ubiquitously distributed in eukaryotic cell surface membranes usually associated to phospholipids and the most common are glycosphingolipids. More than 400 glycosphingolipids varieties have been identified according to their sugar moiety structure. Since the lipid moiety is also extremely variable the number of existing glycosphingolipids is at least 10 times higher (Sonnino et al. 2009, Iwabuchi et al. 2010). Besides being structural components of cell membranes, glycolipids play an important role in cellular processes such as cell-cell interaction, cell signal transduction, cell proliferation, and cell recognition (Im et al. 2001).

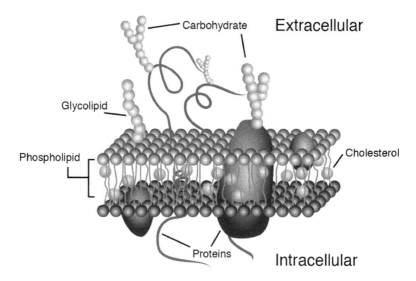

Fig. 1. Eukaryotic cell membrane.

At cell surfaces are receptors for bidding to cells, which may interact with virus, bacteria, and bacterial toxins, as the initial step of cell invasion, establishement of infection in tissues and production of toxic effects (Souza et al. 2010, Bektas and Spiegel 2004). Also malignant transformation in cancer progression is often related with changes in the hydrophilic structures of glycolipids. Examples are the glycolipid GM3/GD3, a melanoma associated antigen that is related to metastasis and the expression of the glycolipid glycosylceramide associated with multidrug resistance in many cancer cells (Bektas and Spiegel 2004). These changes in glycolipid profile outcome are predominantly from altered levels of glycosyltransferase activities involved in biosynthesis (Schnaar et al. 2009).

Analysis of glycolipid composition profile of cellular membrane of the most studied yeast, *Saccharomyces cerevisiae*, showed that major glycolipids are glycosphingolipids (cerebrosides, ceramide polyhexosides and sulfatides). The sugar moiety is mainly constituted of hexoses (mannose, galactose, glucose) while the lipidic moiety comprises mostly C16:0, C18:0, C18:1, C18:2, C18:3 fatty acids (Malhotra and Singh 2006).

Glycolipid Biosurfactants

Due to their amphiphilic characteristics some glycolipids synthesized by microorganisms show surfactant properties and are especially classified as "biosurfactant". Biosurfactants (BS) can be divided into four categories: (i) glycolipids, (ii) fatty acids, (iii) lipopeptides and (iv) polymers. BS has gained attention as "alternative surfactants" for being environmentally friendly compounds. The large majority of the presently used surfactants are petroleum-based compounds produced by chemical synthesis, often toxic to the environment. Therefore BS appears as a viable alternative concerning the ecotoxicity, bioaccumulation, and biodegradability issues. Other advantages of these compounds are their stability at different temperatures, ionic strength, pH and enhanced foaming properties (Mukherjee et al. 2006). Moreover waste or cheaper agro-based substrates can be used by the microorganisms in the fermentation process to produce BS (Mukherjee et al. 2006). Despite the high potential for several applications, the reduction of production costs or generation of higher yield mutant strains is required in order to make the BS production process profitable on a commercial scale (Mukherjee et al. 2006, Williams 2009).

Glycolipids are the most promising biosurfactants, concerning industrial production and applications. Glycolipids BS are classified in accordance with their carbohydrate structure as: (i) sophorolipids, (ii) mannosylerythritol lipids, (iii) rhamnolipids, (iv) trehalose lipids, (v) succinoyl-trehalose lipids, (vi) cellobiose lipids and (vii) oligosaccharide lipids (Fig. 2) (Kitamoto et al. 2002, Kitamoto et al. 2009).

Sophorolipids (SLs) group is one of the most studied and promising biosurfactants produced by non pathogenic microorganisms in high yields and easy recovery (Kitamoto et al. 2002, Bogaert et al. 2007).

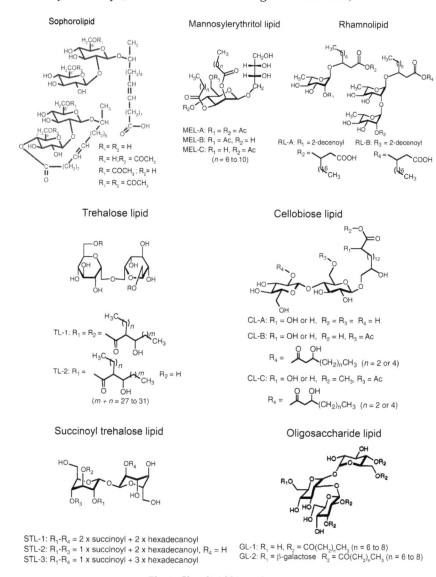

Fig. 2. Glycolipid biosurfactants.

SLs were identified in 1961 (Gorin et al. 1961), and because they are environmentally friendly surfactants (e.g., biodegradable, low ecotoxicity) and present different biological activities (not observed in conventional chemical synthesized surfactants) the interest in these compounds as multifunctional materials has been attracting attention in diverse fields. Among the different biological activities shown by SLs are the bactericide, fungicide, virucide, anticancer and anti-inflammatory that will further be discussed (Bogaert et al. 2007).

Sophorolipids

Molecular structure

The backbone of sophorolipids molecule consists of the glucose dimmer, sophorose (2-O-β-D-glucopyranosyl-β-D-glucopyranose) linked through a glycosidic bond to a hydroxy fatty acid. SLs are produced as mixture of compounds that individually show some variations concerning the acetylation grade of sophorose moiety, the hydroxylation position, the unsaturation grade and the chain length of the hydroxy fatty acid, and the molecule configuration, that can be in the acidic or in the lactonized form. Sophorose moiety acetylation can be observed or not at its 6' or/and 6" carbon (Solaiman et al. 2007, Ashby et al. 2006 and 2008, Bogaert et al. 2007, Nunez et al. 2001). The hydroxy fatty acid hydroxylation is mainly at a terminal (ω) or sub-terminal (ω-1) position (Nunez et al. 2001), however there are some exceptions like in SLs produced by *Rhodotorula bogoriensis* with hydroxylation at carbon 13 (Nunez et al. 2004). The internal esterefication between the carboxylic end and the sophorose moiety (lactonized form), normally occurs at 4" position of the disaccharide but can also occur at 6' or 6" position (Bogaert et al. 2007). Examples of SLs produced by *Candida bombicola* (A1 and A2) and a SL produced by *Rhodotorula bogoriensis* are shown on Fig. 3.

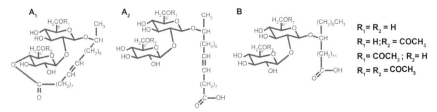

Fig. 3. Examples of SLs produced by *Candida bombicola* in the lactonized form (A₁) and acid form (A₂) and acid form SLs produced by *Rhodotorula bogoriensis* (B). Different grade of acetylation on sophorose unit can occur: $R_1 = R_2 = H$; $R_1 = H$ and $R_2 = COCH_3$; $R_1 = COCH_3$ and $R_2 = H$; $R_1 = R_2 = COCH_3$.

Ten different possibilities of SLs may occur by combination of different lactonization and acetylation profiles (Davila et al. 1997), when considering compounds with equal hydroxy fatty acid chain.

Producing microorganisms

SLs were first discovered in 1961 by Gorin and colleagues (Gorin et al. 1961) who identified the producing species as the yeast *Torulopsis magnolia* and later on, in 1968 Tulloch and Spencer (1968a) verified that the microorganism was in fact *Torulopsis apicola* known nowadays as *Candida apicola*. Also, in 1968 Tulloch and Spencer (1968b) discovered a new SL produced by the yeast *Candida bogoriensis*, today *Rhodotorula bogoriensis*, and two years later the same authors (Spencer et al. 1970) identified a new SLs producer, *Torulopsis bombicola* known today as *Candida bombicola*. *Candida bombicola* is many times referred as *Starmerella bombicola* since the last one includes the teleomorph (sexual stage) of the first one (Rosa and Lachance 1998). In the later years *Wickerhamiella domercqiae* (Chen et al. 2006a), *Pichia anomala* (Thaniyavarn et al. 2008), *Candida batistae* (Konishi et al. 2008) have also been mentioned as SLs producers. More recently Kurtzman and co-workers (2010) screened 19 species of the *Starmerella* yeast clade and found that *Candida riodocensis, Candida stellata* and *Candida* sp. NRRL Y-27208 (a new species) showed a significant production of SLs. A resume of the evolution of SLs main producing microorganisms is shown in Fig. 4.

Despite *Rhodotorula bogoriensis*, a basydomycete, the other known SLs producer species are ascomycetes and majority belong to the *Starmerella* yeast clade.

SLs are secondary metabolites, therefore not essential for the yeast growth, development and reproduction. Their function is not entirely understood but regarding sophorolipids properties some assumptions can be made. SLs can be considered extracellular carbon storage material and a way for adaptation to high osmotic pressure ensuing from the elevated sugar concentrations (Davila et al. 1997, Kitamoto et al. 2002). In addition

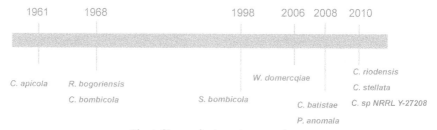

Fig. 4. SLs producing microorganisms.

their biocide activity against other microorganisms can act as a survival interspecies competition mechanism (Desai et al. 1997, Kitamoto et al. 2002).

Different SLs or different acid/lactonic SLs ratios can be obtained depending on the microorganism used. *Candida bombicola* produces mainly C16 and C18 hydroxy fatty acid chain SLs with different unsaturation grade and ω or ω-1 hydroxylation. Qualitatively, a higher number of different acidic S can generally be found, however quantitatively lactonic SLs represent the majority of compounds present in the mixture. The major compound produced by *Candida bombicola* is 17-L-[(2-O-β-glucopyranosyl-β-D-glucopyranosyl)-oxy]-9-octadecenoic acid 1,4-lactone 6′,6″-diacetate SL (Solaiman et al. 2004). Longer chain SLs such as C20 can also be produced but in trace quantity. Other producing yeasts from *Starmerella* clade such as *Candida apicola*, *Candida batistae*, *Candida riodocensis*, *Candida stellata* and *Candida* sp. NRRL Y-27208 also produce equivalent SLs and mostly C18 hydroxy fatty acid chain SLs (Kurtzman et al. 2010). Some small variations can occur and an example is *Candida batistae* that according to Konishi and co-workers (2008) produced a higher fraction of acidic SLs (more than 60%) than lactonic contrary to *Candida bombicola*. The same authors also mentioned that in SLs mixture produced by *Candida batistae* the main compound produced was the C18:1 (ω) diacetylated acidic SL, 18-L-[(2′-O-β-D-glucopiranosyl-β-D-glucopiranosyl)-oxy]-octadecenoic acid, 6′,6″-di-O-acetylate (Konishi et al. 2008). Moreover Thaniyavarn and colleagues (2008) studied SLs mixture composition of *Pichia anomala* and concluded that SLs had a fatty acid chain length of 18 (C18:1) and 20 carbon atoms under the acidic, lactonic and diacetylated forms. Although *Wickerhamiella domercqiae* is less closely related to *Starmerella* clade, SLs produced are also equivalent to *Candida bombicola* (Shao et al. 2012). The *Rhodotorula bogoriensis*, a phylogenetically different SLs producer (basydomycete) is known for producing acidic SLs with a longer hydroxy fatty acid chain, where the sophorose moiety is linked to 13-hydroxydocosanoic acid (Nunez et al. 2004). A new C24:0 monoacetylated and diacetylated acidic SLs were produced by *Rhodotorula bogoriensis* (Ribeiro et al. 2012a).

Biosynthetic pathway

The SLs biosynthetic pathway includes data collected from observations with different producing species and assumes that production process is equal or similar among every producing microorganism. Several reactions are involved in the SLs biosynthetic pathway (Fig. 5).

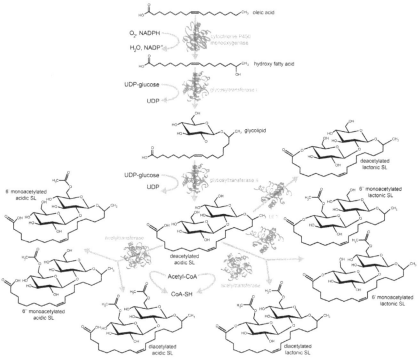

Fig. 5. SLs biosynthesis pathway.

Production can start from a fatty acid supplemented to the fermentation medium or from the fatty acid synthesis. The fatty acid synthesis can occur from n-alkanes, alcohols, aldehydes or triglycerides supplementation to the media or from de novo synthesis from acetyl-coenzyme A (CoA) (Albrecht et al. 1996, Brakemeier et al. 1998, Bogaert et al. 2007). Initially, through the action of a membrane bound nicotinamide adenine dinucleotide phosphate (NADPH reduced form) dependent monooxygenase enzyme cytochrome P450 the fatty acid will be converted into a hydroxylated fatty acid (Hommel et al. 1994, Bogaert et al. 2007, 2009, 2011a). Hereafter, according to the mechanism proposed for *Rhodotorula bogoriensis* (Esders and Light 1972, Breithaupt and Light 1982) and confirmed for *Candida bombicola* (Bogaert et al. 2011), a specific glycosyltransferase I will mediate the linkage between glucose at C1′ position with hydroxyl group of the hydroxy fatty acid. Afterwards a glycosyltransferase II will link a second glucose molecule at C1″ position to the first at C2′ position. For accomplishment of these two steps a nucleotide activated glucose (uridine diphosphate (UDP)-glucose) is needed as glycosyl donor for both glycosyltransferases activity (Esders and Light 1972, Breithaupt and Light 1982, Bogaert et al. 2011). At this stage is formed an acidic non acetylated SL. Further acetylation of the sophorose molecule can occur at 6' and/or 6" position by the action of acetyl-coenzyme A (CoA) dependent acetyl transferase (Esders and Light 1972, Bucholtz and Light 1976, Bogaert et al. 2011a). The SL can also become cyclic (lactone) by internal esterification of one hydroxyl group mainly at 4" position of sophorose moiety by the action of specific lipases (L) or esterase (E) that till now have not been identified (Bogaert et al. 2011a).

Factors affecting SLs production

SLs production can be influenced by different factors such as the microorganisms as previously mentioned, the media composition (e.g., carbon and nitrogen sources and inorganic compounds), bioreactor and environmental conditions.

Media composition

SLs are produced at the end of the exponential growth and stationary phase. (Casas et al. 1997, Rau et al. 2001), usually under nitrogen limiting conditions (Davila et al. 1997, Rau et al. 2001). Some authors consider that the nitrogen limiting conditions block growth and SLs production might be an overflow metabolism that regulates yeast intracellular energy level (Davila et al. 1997).

Candida bombicola is by far the most studied SLs producer organism and different fermentation media have been used for the biosurfactants production. A resume of some of the media employed during the past years is present in Table 1 and 2.

Carbon Source

The carbon substrate has a dual role in fermentation processes. It will be consumed to assure energy generation and also product biosynthesis. SLs can be produced from a single hydrophilic carbon source (carbohydrates), however production yield will rise considerably if a second hydrophobic carbon source (lipids) is simultaneously added to the medium (Davila et al. 1994, 1997). SLs producer species can synthesise the fatty acid moiety but if these compounds are added to the medium, direct incorporation can occur, with an increase in process efficiency (Cavalero and Cooper 2003). Therefore a lipidic carbon source is normally supplemented to the fermentation medium as co-substrate.

Among hydrophilic sources glucose is the one used for excellence. Besides being the primary energy source for the yeast it is incorporated

Table 1. Examples of fermentation media supplemented of a single hydrophobic carbon source used for SLs production by *Candida bombicola*.

Carbon source II (g L⁻¹)	Carbon source I (g L⁻¹)				Nitrogen source (g L⁻¹)							Minerals (g L⁻¹)							References
	Glucose	DW	SMp	SM	Yeast extract	Urea	Peptone	NH_4Cl	NH_4NO_3 or $(NH_4)_2SO_4$	Malt extract	Cornsteep liquor	KH_2PO_4	$MgSO_4 \cdot 7H_2O$	NaCl	$CaCl_2$	$FeCl_3 \cdot 6H_2O$	$Zn(AcO)_2 \cdot 2H_2O$	H_3BO_3	
OA	100	-	-	-	1.5	-	-	4	-	-	-	1	0.5	0.1	-	-	-	-	Kurtzman et al. 2010
OA (100)	10	90	-	-	2	-	-	-	-	-	-	-	-	-	-	-	-	-	Daverey and Pakshirajan 2010
OA (40)	100	-	-	-	10	1	-	-	-	-	-	-	-	-	-	-	-	-	Fu et al. 2008, Hardin et al. 2007, Shah et al. 2005
OA (90)	-	-	√ᵃ	√ᵇ	10	1	-	-	-	-	-	-	-	-	-	-	-	-	Solaiman et al. 2007, 2004
OA (90)	100	-	-	-	10	1	-	-	-	-	-	-	-	-	-	-	-	-	Shah and Prabhune 2007
AA (2)	20	-	-	-	3	-	10	-	-	3	-	-	-	-	-	-	-	-	Shah et al. 2007b
OA (100)	100	-	-	-	10	1	-	-	-	-	-	-	-	-	-	-	-	-	Felse et al. 2007
MA; PA; SA; OA; LA; LoA; EA (40)	100	-	-	-	10	1	-	-	-	-	-	-	-	-	-	-	-	-	Felse et al. 2007
SA OA (var)	100	-	-	-	10	1	-	-	-	-	-	-	-	-	-	-	-	-	Felse et al. 2007
OA (40)	100	-	-	-	10	1	-	-	-	-	-	-	-	-	-	-	-	-	Azim et al. 2006
Alk; FA; (5)	10	-	-	-	1	0.1	-	-	-	-	-	-	-	-	-	-	-	-	Cavalero and Cooper 2003
OA(100)	100	-	-	-	5	-	-	4	-	-	-	1	0.5	-	-	-	-	-	Rau et al. 2001
none	12	-	-	-	1	-	-	1.9	-	-	-	7.4	0.2	0.1	0.1	0.35	0.16	0.06	Casas et al. 1997
none	120	-	-	-	10	1	-	-	-	-	-	-	-	-	-	-	-	-	Albrecht et al. 1996
$C_{16}H_{34}$	10	-	-	-	1.5	-	-	-	√	-	-	1	5	0.1	0.1	-	-	-	McCaffrey and Cooper 1995
None*; Alk	50	-	-	-	-	-	-	-	-	-	5	-	-	-	-	-	-	-	Davila et al. 1994

OA-Oleic acid; AA-Araquidonic acid; MA-myristic acid; PA-palmitic acid; SA-Stearic acid; LA-linoleic acid; LoA-Linolenic acid; EA-eicosanoic acid; Alk-Alkanes; FA-fatty acids; DW-Deproteinized whey; SM-Soy molasses; SMp-Soy molasses pretreated; (var)-variable concentration along the assay

Table 2. Examples of fermentation media supplemented of complex hydrophobic carbon source used for SLs production by *Candida bombicola*.

Carbon source II (g L⁻¹)	Carbon Source I (g L⁻¹)					Nitrogen Source I (g L⁻¹)						Minerals (g L⁻¹)				References
	Glucose	DW	Sugar cane molasses[a]	Glycerol	Honey	Yeast extract	Urea	Peptone	NH_4NO_3	$(NH4)_2SO_4$	Cornsteep liquor	KH_2PO_4	$MgSO_4 \cdot 7H_2O$	NaCl	$CaCl_2$	
SbO (100)	100	-	-	-	-	10	1	-	-	-	-	-	-	-	-	Daverey et al. 2009
SbO; SfO; OO (100)	-	-	100	-	-	10	1	-	-	-	-	-	-	-	-	Daverey et al. 2009
SbO; SfO; OO (100)	-	-	100	-	-	-	-	-	-	-	-	-	-	-	-	Daverey et al. 2009
SbDO; SbO; CO; RgO; RsO (100)	100	-	-	-	-	5	-	0.7	-	-	-	1	0.5	0.1	0.1	Kim et al. 2009
RsO (10)	50	-	-	-	-	5	-	0.7	-	-	-	1	0.5	0.1	0.1	Kim et al. 2009
Restaurant waste oil (100)	100	-	-	-	-	10	1	-	-	-	-	-	-	-	-	Shah et al. 2007b
TfA; CfA (40); (var)	100	-	-	-	-	10	1	-	-	-	-	-	-	-	-	Felse et al. 2007
Me-Soy; Et-Soy; Prop-Soy	-	-	-	√	-	10	1	-	-	-	-	-	-	-	-	Ashby et al. 2006
Turkish Corn Oil (100)	100	-	-	-	-	10	1	-	-	-	-	-	-	-	-	Pekin et al. 2005
Turkish Corn Oil (100)	-	-	-	-	-	10	1	-	-	-	-	-	-	-	-	Pekin et al. 2005
Turkish Corn Oil (100)	100	-	-	-	20	10	1	-	-	-	-	-	-	-	-	Pekin et al. 2005
Sunflower Oil (5)	10	-	-	-	-	1	0.1	-	-	-	-	-	-	-	-	Cavalero et al. 2003

Table 2. contd....

Table 2. contd.

Carbon source II (g L⁻¹)	Carbon Source I (g L⁻¹)					Nitrogen Source I (g L⁻¹)						Minerals (g L⁻¹)				References
	Glucose	DW	Sugar cane molasses[a]	Glycerol	Honey	Yeast extract	Urea	Peptone	NH_4NO_3	$(NH4)_2SO_4$	Cornsteep liquor	KH_2PO_4	$MgSO_4 \cdot 7H_2O$	NaCl	$CaCl_2$	
RsO (100)	100	-	-	-	-	-	-	-	-	4	5	1	0.5	-	-	Rau et al. 2001
CcScO (triglycerides)	80	-	-	-	-	-	-	-	-	-	-	-	-	-	-	Daniel et al. 1999
Rapeseed oil (100)	-	20L	-	-	-	4	-	-	-	-	-	-	-	-	-	Otto et al. 1999
CcScO + RsO	-	20	-	-	-	4	-	-	-	-	-	-	-	-	-	Otto et al. 1999
Soybean oil (100)	100	-	-	-	-	5	-	-	-	3.3	-	1	5	5	0.1	Kim et al. 1997
Rapeseed ethyl esters	100	-	-	-	-	-	-	-	-	-	-	-	-	-	-	Davila et al. 1997
Sunflower oil	10	-	-	-	-	1.5	-	-	√	√	-	1	5	0.1	0.1	McCaffrey and Cooper 1995
RsO; SfO; PmO; FO	50	-	-	-	-	-	-	-	-	-	-	-	-	0.1	-	Davila et al. 1994
RsO; SfO; PmO; LsO esters	50	-	-	-	-	-	-	-	-	-	5	-	-	-	-	Davila et al. 1994

SbO–soybean oil; SfO–sunflower oil; OO–olive oil; SbDO–Soybean dark oil; CO–corn oil; RgO–Rice germ Oil; RsO–Rapeseed Oil; TfA–Tallow fatty acid; CfA–Coconut fatty acid; PmO–Palm oil; LsO–Linseed oil; FO–Fish oil; CcScO–*Cryptococcus curvatus* Single cell Oil–DW–Deproteineized whey (although not directly) (Fig. 5) into the sophorolipid structure assuring higher production yields. Other hydrophilic sources have also been used on the study of SLs production and examples are lactose, fructose, sucrose and glycerol (Zhou and Kosaric 1993, Rispoli et al. 2010). Instead of pure carbohydrates, molasses have also been supplemented to fermentation medium. Molasses cost is very competitive when comparing with pure carbohydrates, however they contain several impurities that generally will turn the extraction/purification process more difficult and expensive (Stanbury and Whitaker 1995). In SLs production sugar cane molasses (Davery and Pakshirajan 2009a) and soy molasses (Solaiman et al. 2004 and 2007) have been used. Similar yields were obtained when comparing, within the same experimental conditions, sugar cane molasses (23.3 g L^{-1}) with glucose (29.3 g L^{-1}) (Davery and Pakshirajan 2009a) and pre-treated soy molasses (75 g L^{-1}) with glucose (79 g L^{-1}) (Solaiman et al. 2007). Deproteinized whey in combination with glucose has also been supplemented to fermentation medium in order to produce SLs (Daniel et al. 1999, Otto et al. 1999).

Among the hydrophobic carbon sources tested in SLs production are alkanes (with a chain of 12, 14, 15, 16, 17, 18 or 20 carbon atoms), saturated (e.g., C18:0) and unsaturated (e.g., C16:1; C18:1, C20:4) fatty acids, alcohols (e.g., 1, 12-dodecanediol) and cellular triglycerides (e.g., *Cryptococcus curvatus* single cell oil) (Cavalero and Cooper 2003, Davila et al. 1994, Felse et al. 2007, Bogaert 2011b, Daniel et al. 1999). Vegetable (e.g., sunflower, soybean, rapeseed, linseed, olive and corn oil) and animal oils (e.g., tallow and fish oil) that include several lipidic compounds in their composition have also been used as well as their esters (Kim et al. 2009, Davery et al. 2009, Davila et al. 1994, Felse et al. 2007). Moreover SLs are also produced if waste oils (e.g., restaurant waste oil) (Shah et al. 2007) processing bioproducts oil (e.g., soybean dark oil) (Kim et al. 2009), dairy industry wastewater (Davery et al. 2009, Davery and Pakshirajan 2011) and oil refinery waste (Bednarski et al. 2004) are added as co-substrate, making the fermentation process environmentally friendly.

Moreover hydrophobic sources can also influence final SLs mixture composition. The supplementation of production media with different co-substrates C15, C16, C17 or C18 alkanes, allowed an increase on SLs production (Cavalero and Cooper 2003). However, the addition of saturated fatty acids resulted in low SLs yield (Cavalero and Cooper 2003). Davila and colleagues (1994) had already verified that production increased with the number of carbons of the alkane chain supplemented to the fermentation media by the following order C18> C16> C14> C12. In this experiment C14 and C12 SLs were produced from respective co-substrates in very low yields. According to Brakemeier and co-workers (1995) short chain precursors (e.g., C10, C12, C14) fatty acids, esters or primary alcohols are used in *de novo* synthesis of C16 and C18 fatty acids instead of incorporated into the sophorolipid structure. The use of shorter or longer fatty acids included in coconut (C8:0, C10:0, C12:0, C14:0) and meadowfoam oil (C20:0, C20:1, C22:2) respectively did not enhance SLs production (Felse et al. 2007, Bogaert et al. 2010). According to Bogaert and colleagues coconut oil had a toxic effect towards *Candida bombicola*, which was attributed to medium chain fatty acids suggesting interference with cell membrane. However, in Felse and co-workers work (Felse et al. 2007) an enhanced biomass production was verified when coconut fatty acid residue was added with other lipid percursor. Moreover, according to Davila and colleagues (1994) the supplementation of esters or vegetable oils SLs production revealed direct incorporation of fatty acids present in the lipidic co-substrate and a higher incorporation of C18 and C16 fatty acids. Therefore, improvement of the production of specific SLs structural classes can be obtained by supplementation of different co-substrates. The supplementation of esters instead of oils increases SLs production yield. SLs produced from oils or stearic or oleic acid present higher content in diacetylated lactones than SLs obtained from esters. In contrast a higher acid fraction can be obtained with co-substrates composed of polyunsaturated fatty acids (Davila et al. 1994). Felse and co-workers (2007) observed that C18:1 and C18:0 lactone formation seems to occur preferably with ω-1 hydoxy fatty acids.

Not naturally occurring SLs can be obtained if specific co-substrates are added to the fermentation media.

Longer chain SLs can also be produced by *Candida bombicola* if the adequate precursor substrate is added. Example is the production of non conventional C22 SLs by the supplementation of erucic acid (C22:1) or respective ester and supplementation of rapeseed oil supplemented of docosanoic (C22:0) and docosadienoic acid (C22:2) or respective ester (Shin et al. 2010). Similar results were obtained with the addition of arachidonic acid (C20:4) to the fermentation media originating SLs with the same fatty acid chain length as the precursor substrate (Shah and Prabhune 2007).

Furthermore, *Starmerella bombicola* was able to produce C18:3 diacetylated acidic and lactonic SLs with the supplementation of borage oil. The existence of new glycolipid composed of a C18:1 diacetylated acidic SL linked to a second sophorose unit was detected in the SLs mixture produced by *Starmerella bombicola* meaning that this yeast can also produce these new glycolipids (Ribeiro et al. 2012b).

Brakemeier and co-workers (1995, 1998) added to the fermentation media secondary alcohols with a carbon chain from C12 to C16 and obtained SLs with a lipophilic moiety correspondent to the unmodified alcohol substrate. With the addition of 2-dodecanol 88% of SL composition included direct incorporation of the co-substrate (Brakemeier et al. 1998). Moreover, the addition of hydroxylated substrates such as 12-hydroxydodecanoic acid or 1,12-dodecanediol led to the synthesis of medium chain (C12) fatty acid SLs by substrate direct incorporation contrary to the use of dodecanoic acid that is not incorporated (Bogaert et al. 2011b). Carboxylated esters (e.g., dodecyl glutarate) and alkyl esters (e.g., pentenyldodecanoate) substrates can also be incorporated into the SLs structure, with the first allowing a better yield (Bogaert et al. 2011b). When the incorporated molecule presents an internal ester bond like dodecyl glutarate an alkali hydrolysis can originate medium chain SLs (Bogaert et al. 2011b).

Other producing microorganisms like *Rhodotorula bogoriensis* have not received the same attention as *Candida bombicola*, so few carbon sources have been tested. Glucose at a concentration of 30 g L^{-1} (Esders and Light 1972), 50 g L^{-1} (Cutler and Light 1979), and 80 g L^{-1} (Tulloch and Spencer 1968b, Nunez et al. 2004) had been added to fermentation media as single carbon source. More recently Zhang and co-workers (2011) used rapeseed and meadowfoam oil (35 g L^{-1}) as co-substrate and observed a 70% increase in production yield (1.21 g L^{-1}) when rapeseed oil was added.

The culture media of *Wickerhamiella domercqiae* for SLs production have included glucose at 50 g L^{-1} or 80 g L^{-1} with rapeseed oil in a concentration of 20 g L^{-1} or 60 g L^{-1} respectively, or glucose at 50 g L^{-1} with oleic acid at 60 g L^{-1} (Chen et al. 2006a, Shao et al. 2012).

Nitrogen Source

The SLs producer species can metabolize either organic or inorganic nitrogen sources. Organic nitrogen can be supplemented as amino acids, proteins, urea or as complex sources, like peptone and yeast extract. Yeast extract is one of the most important for microbial growth and SLs production (Casas et al. 1997), therefore most of fermentation media contain yeast extract in concentrations between 1.5 and 10 g L^{-1} (Table 1). The replacement of yeast extract by peptone or urea influenced negatively microbial growth and SLs production (Bogaert et al. 2007). Several fermentation media for SLs production combine 2 nitrogen sources, namely yeast extract used in association with urea or peptone. According to Casas and Garcia-Ochoa (1999) yeast extract can also modulate SLs production since in concentrations lower than 5 g L^{-1} the lactonic SLs fraction increased and in concentrations higher than 5 g L^{-1} the acidic SLs fraction rose. Some multi-component sources also contain nitrogen sources besides carbon sources and example is malt extract or corn steep liquor. Corn steep liquor is a by-product after starch extraction from maize that contains small amounts of reducing sugars and complex polysaccharides and was primarily used as nitrogen source (Stanbury and Whitaker 1995). Corn steep liquor (Davila et al. 1994) and malt extract (Shah and Prabhume 2007) have also been supplemented for SLs production.

Other possible nitrogen sources are inorganic such as ammonium salts or nitrates. Ammonium salts like ammonium sulphate, normally originates acidic conditions as the ammonium ion is consumed while nitrates will usually raise the medium pH as it is utilized (Stanbury and Whitaker 1995). Either NH_4NO_3 (McCaffrey and Cooper 1995), $NaNO_3$ (Cooper and Paddock 1984), $(NH_4)_2SO_4$ (Rau et al. 2001) or NH_4Cl (Kurtzman et al. 2010) have also been included in *Starmerella bombicola* fermentation media.

The culture media of *Rhodotorula bogoriensis* for SLs production have included yeast extract as solo nitrogen source at concentrations of 1.5 g L^{-1} (Cutler and Light 1979, Esders and Light 1972) or 4 g L^{-1} (Tulloch and Spencer 1968b, Nunez et al. 2004).

The effect of different nitrogen sources on SLs production by *Wickerhamiella domercqiae* was recently studied by Ma and colleagues (2011). They used yeast extract, peptone, urea, L-glutamic acid and CH_3COONH_4 as organic nitrogen sources and $(NH_4)_2SO4$, NH_4HCO_3, NH_4Cl, NH_4NO_3, and $NaNO_3$ as inorganic nitrogen sources. It was observed that nitrogen source affected SLs production. The lactonic SLs production was reduced in the presence of ammonium ion included in different nitrogen sources (except NH_4HCO_3) (Ma et al. 2011).

Inorganic compounds

Certain mineral elements are essential for microorganism's growth and metabolism. Some SLs culture media were added of mineral components in specific concentrations. Among the mineral components frequently supplemented are KH_2PO_4, $MgSO_4$, $MgCl_2$, NaCl and $CaCl_2$. In semi-synthetic culture media phosphorous supplementation was important to Candida bombicola growth since it influenced maximum growth (Casas et al. 1997).

The inorganic compounds added to culture media for SLs production by *Rhodotorula bogoriensis* included KH_2PO_4 (1g L^{-1}) and $MgSO_4.7H_2O$ (0.2 g L^{-1}) (Tulloch and Spencer 1968b, Nunez et al. 2004) while by *Wickerhamiella domercqiae* included KH_2PO_4 (1g L^{-1}), $Na_2HPO_4.12H_2O$ (1g L^{-1}), $MgSO_4.7H_2O$ (0.5 g L^{-1}) with or without $(NH_4)_2SO_4$ (4g L^{-1}) (Chen et al. 2006a, Ma et al. 2011, Shao et al. 2012).

Strategies for microbial growth and SLs production

For the production of SLs three major strategies have been applied to grow microorganisms (i) batch, (ii) fed-batch and (iii) continuous culture (Fig. 6).

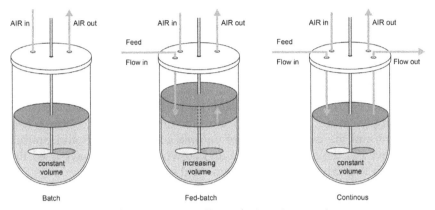

Fig. 6. Strategies to grow SLs producing microorganisms.

In batch culture (i) there is no additional supplementation of nutrients besides the ones initially included into the culture media. It is a closed system and products are only removed at the end of fermentation when the essential nutrients have been depleted (Stanbury and Whitaker 1995, Baltz et al. 2010). The batch culture technique is easy to put into practice and has been carried out in several SLs production studies. Batch culture strategies for SLs production have been applied in smaller laboratory scale approaches such as shake flasks (SF) (Shah et al. 2007b, Felse et al. 2007, Kim et al. 2009, Davery and Pakshirajan 2009, 2010) or in larger scale approach, in bioreactors (Albrecht et al. 1996, Daniel et al. 1999, Shah et al. 2007b, Davery and Pakshirajan 2010). The SF approach is ideal to proceed with initial basic studies for SLs production. SF ideally filled with medium culture of about 20% of the nominal flask volume and operated at temperature controlled shaking devices act as simple bioreactors and have been employed in initial SLs screenings, production optimization and simple SLs production. An alternative viable batch approach, especially for SLs screening production with different experimental conditions, can be the use of microtiter plates (MTPs) as minireactors. This approach applied to SLs production studies allowed the use of smaller volumes, larger number of experiments, costs reduction and time consumption (Ribeiro et al. 2012a, b).

To attain higher SLs productivities, fed-batch cultures had been implemented. In the fed-batch culture the hydrophilic carbon source essential for growth (e.g., glucose) can be added constantly into the bioreactor simultaneously with the addition of the hydrophobic carbon source, in order to increase SLs production. In this type of culture the volume increases with time and at the end total or partial recovery of culture medium can be performed to allow repeating the process (Baltz et al. 2010). Different fed-batch cultivation consists of linear or exponential feeding of one or more nutrients, being the supplementation constant or exponential per unit of time (Baltz et al. 2010). Examples of fed-batch approach used for production of SLs include additions of lipid source (Pekin et al. 2005, Shah et al. 2007b) or daily carbohydrate feedings simultaneously with constant lipid substrate supplementation at constant rate (Solaiman et al. 2004, 2007, Davila et al. 1994).

In the continuous culture where the carbon source is added to the microorganism culture at a specific rate, the total volume of culture is kept constant and the excess of culture volume containing cells and product is removed at the same rate. Kim and co-workers (1997) have used this approach to produce SLs from glucose and soybean oil.

Environmental factors

Environmental factors like temperature, pH, agitation and oxygen availability influence SLs production. Examples of different factors affecting SLs production by *Candida bombicola* are presented in Table 3. Fermentation media oxygenation plays a relevant role in SLs production. Growth profile of *Candida bombicola* was similar when cultured in shake flasks and in an aerated bioreactor, but after reaching stationary phase a decrease in biomass was verified in shake flasks, while in the bioreactor stationary phase remains constant (Casas et al. 1997). Guilmanov and colleagues (2002) studied the effect of oxygen transfer rate on SLs production in a fed-batch shake flask and concluded that optimal aeration was between 50 and 80 mM O_2 L^{-1} h^{-1}. Oxygenation level can also influence individual SLs production. In fact lower aeration conditions decreased unsaturated fatty acid SLs and increased saturated fatty acid SLs. The conditions of substrate feeding influence SLs mixture composition (Davila et al. 1997), namely the distribution of acidic/lactonic SLs and the acetylation grade. If hydrophilic and hydrophobic carbon sources were simultaneously supplied at the beginning of growth phase, SLs production was improved, although hydrophobic sources were not consumed in this phase, they seemed to potentiate SLs production. If only one substrate was supplied (hydrophilic or hydrophobic) lower SLs production was obtained. The type of growth kinetics (e.g., batch, fed-batch, continuous) selected also influences SLs production. In Table 3 are resumed some experimental conditions and yields observed in SLs production. Production yields of 365 g L^{-1} (Kim et al. 2009) and 340 g L^{-1} (Davila et al. 1994) with supplementation of rapeseed oil or 300 g L^{-1} with supplementation of oleic acid (Rau et al. 2001) were obtained in bioreactors working in fed-batch mode. The optimum temperature for SLs production by *Candida bombicola* had been determined to be 21°C by Gobbert and colleagues (1984) and around 22°C and 25°C by Spencer and co-workers cited in Vadescateele (2008). Moreover Davila and colleagues

Table 3. Examples of fermentation conditions for production of SLs by *Candida bombicola*.

Vessel (L)	Fermentation Process	Medium volume (L)	Temp. (°C)	pH	Time (d)	Yield (g L⁻¹)	References
SF (0.05)	-	0.01	25	4.5	4-7	-	Kurtzman et al. 2010
SF (1)	Batch	0.2	30	-	8	23.3	Daverey and Pakshirajan 2010
Bioreactor (3)	Batch	1	30	-/3.5	8	25.5/33.3	
SF (0.25)	Batch	0.05	30	-	5	17.5-29.4	Daverey et al. 2009
SF (0.25)	Batch	0.05	25	-	7	65-120	Kim et al. 2009
Bioreactor (2.5)	Fed-batch	1	30	3.5	8	365	
SF (0.5)	-	0.1	30	-	7-8	-	Fu et al. 2008, Hardin et al. 2007, Shah et al. 2005
Bioreactor (12)	Fed-batch	4	26	3.5	7	75 [a] / 21 [b]/79**	Solaiman et al. 2007, 2004
SF (1)	-	0.25	30	-	4	1.44	Shah and Prabhune 2007
SF (0.125)	Batch and Fed-batch	0.025	30	-	10	30-38 and 21-22	Shah et al. 2007b
Bioreactor (3.3)	Batch and Fed-batch	1	30	-		30-40 and 15-20	
SF (0.5)	Batch and Fed-batch	0.1	-	-	7 / 10	20-60 / 40-120	Felse et al. 2007
Bioreactor (3)	Fed-batch	2	25	3.5	18.2	>400	Perkin et al. 2005
SF	Batch	0.01	-	-	11	-	Cavalero et al. 2003
SF (2)	Batch	0.5	-	-	3	-	Rau et al. 2001
Bioreactor	Fed-batch	-	-	3.5	5.2	300	
Bioreactor (100)	Batch	70	30	-	-	-	Daniel et al. 1999
Bioreactor (100)	Fed-batch	70	30	-	-	422	Otto et al. 1999
Bioreactor (5)	Continuous	3	25	4	-	-	Kim et al. 1997
SF (0.25)	Batch	0.05	30	5.6	-	-	Casas et al. 1997
Bioreactor (50)	-	-	27	-	-	-	Albrecht et al. 1996
Bioreactor	Fed-batch	-	30	-	6-7	-	McCaffey et al. 1995
Bioreactor (4)	Fed-batch	-	25	3.5	-	17-340	Davila et al. 1994

(1997) verified that production of SLs decreased in the following order 25 > 28.5 > 32.5 > 37.0°C. Studies about the pH influence on SLs production by *Candida bombicola* showed that optimum pH was 3.5 (Gobbert et al. 1984, Davila et al. 1997). At pH 3.5 acidic SLs precipitate preventing lower yields due to the product inhibition since SLs solubility increased with pH (Davila et al. 1997).

Some yeasts, producers of SLs, showed slight differences when compared with *Candida bombicola* production conditions. For instance, optimum pH for SLs production by *Pichia anomala* has been mentioned to be pH 5.5 (Thaniyavarn et al. 2008).

Derivatives of natural SLs

The modification of natural SLs allowed the production of several different compounds, which can be an advantage on the research for new biological activities. Some of possible reactions are addressed and summarized in Fig. 7. As mentioned before, SLs are produced as mixture of compounds with different acetylation grade on the sophorose moiety and different unsaturation grade, chain length and hydroxylation position of the hydroxy fatty acid. The hydrolysis of SLs ester positions is a way to synthetize a homogeneous mixture of acidic compounds.

Alkaline or acidic (Fig. 7 *i*) hydrolysis can convert lactonic SLs into acidic non acetylated SLs (Develter and Flurackers 2010). At more aggressive acidic conditions the glycosydic bond can breake, the sugar moiety removed and as final product a hydroxy fatty acid is obtained (Develter and Flurackers 2010) (Fig. 7 *ii*). These reactions can be advantageous in the synthesis of hydroxy fatty acids or to obtain a higher yield in acidic SLs that present higher water solubility and different applications from the native mixture. Moreover the enzymes acetylesterase (E.C.3.1.1.6) (Fig. 7 *iii*) or cutinase (Fig. 7 *iv*) from *Fusarium solani* can also be used *in vitro* to obtain deacetylated SLs, the first promotes 6' and 6" deacetylation while the last only 6' deacetylation (Bogaert et al. 2007, Koster et al. 1995). Another possible approach is the *in vitro* enzymatic modification of SLs with glycosidases or lipases. Several glycosidases (e.g., hesperidinase, pectolyase, naringinase) (Fig. 7 *v*) have been used in SLs modification with the intention of removing one hexose molecule and although the properties were similar, the water solubility decreased (Rau et al. 1999). Lipases, such as porcine pancreatic lipase, can be used to promote deacetylation and polymerization of lactonic SLs (Hu and Ju 2003) (Fig. 7 *vi*). Alternatively Novozyme 435 lipase can re-acetylate 6' and 6" positions of SLs esters and promote lactonization of acidic SL into lactonic SL at 6" position of sophorose moiety (Bisht et al. 1999)

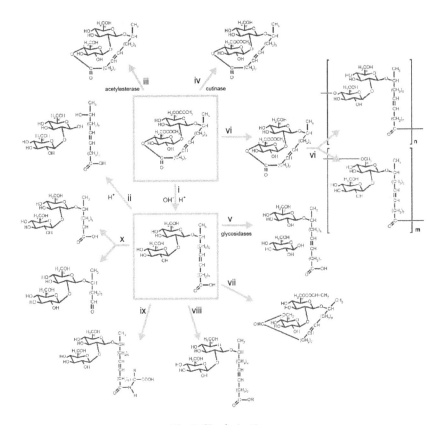

Fig. 7. SLs derivatives.

(Fig. 7 *vii*). Other common modification is in the carboxylic end of the fatty acid, an example is the modification to originate SLs esters (Shah et al. 2005, Hardin et al. 2007) (Fig. 7 *viii*) or to form amino acid conjugated SLs (Azim et al. 2006) (Fig. 7 *ix*). If a shorter hydrophobic chain is desired, that can be obtained by ozonolysis that cleaves double bonds existent on the fatty acid (Bogaert et al. 2011a) (Fig. 7 *x*).

Isolation, analysis and characterization of SLs

Separation and Purification

Few studies have been performed on SLs downstream processing. SLs are generally recovered from fermentation media by solvent extraction. A first extraction is usually performed with ethyl acetate and when a lipidic source is supplemented as co-substrate, a second extraction is necessary usually with n-hexane to remove the remaining co-substrate. Other solvents like diethyl ether (Bogaert el al. 2011b) or n-pentane (Daniel et al. 1999) can also be used in the final extraction. Some authors state that when produced in large amount SLs can be obtained by decantation (Bogaert et al. 2007) while according to others (Rau et al. 2001) when oil is totally consumed, to heat the suspension till 60°C and let it cool down to room temperature, will make SLs precipitate. Hu and Ju (2001b) pointed out a method for purifying lactonic SLs with phosphate buffer. The methodology used comprised a first wash at room temperature with phosphate buffer pH 6.5 to remove acidic SLs followed by a second wash with the same buffer at 65°C to dissolve lactonic SLs. After cooling, crystals of lactonic SLs are obtained with low content of acidic SLs. If separation and purification of specific compounds is required column chromatography separation or High Performance Liquid Chromatography (HPLC) preparative chromatography are typically performed (Deshpande and Daniels 1995, Azim et al. 2006, Konishi et al. 2008, Chen et al. 2006a). Column chromatography is normally used for separation and purification of SLs after the extraction process. Most frequent methods employed utilize a glass column packed with silica gel loaded with SLs previously dissolved in appropriate solvent (e.g., ethyl acetate). The elution was carried out with a chloroform/methanol gradient system and fractions of purified compounds were collected along time (Deshpande and Daniels 1995, Bisht et al. 1999, Davery and Pakshirajan 2010, Gross and Shah 2004, 2007, 2012). A possible variant approach of this method is the Medium Performance Liquid Chromatography (MPLC) technique. The same stationary phase and eluents have been used but the chromatographic separation is performed with a solvent delivery system (Asmer et al. 1988, Rau et al. 2001, Chen et al. 2011). According to Chen and colleagues (2011) lactonic SLs were isolated by MPLC using a gradient of chloroform/methanol from 98:2 to 96:4 while acidic SLs were eluted from 80:20 to 50:50. Shah and Prabhune (2007) have proposed an alternative purification method that included a dialyse membrane loaded with silica gel F_{254} as packing material instead of a glass column. The chloroform/methanol (9:1) solvent system was allowed to migrate till it reached the end of the column. Bands of separated compounds were observed under UV-visible light at 254 nm, cut out and extracted with ethyl acetate.

Moreover, preparative HPLC has also been performed by some authors to separate SLs produced by *Wickerhamiella domercqiae* (Chen et al. 2006a, Ma et al. 2011, Shao et al. 2012). In these studies preparative C18 columns were used in combination with a gradient elution of water/acetonitrile. Besides the use of preparative HPLC, preparative TLC can be used for compound purification by loading sample solutions onto a higher adsorbent thickness (normally 1–2 mm) plate. After development in the solvent system, the dried plate was submitted to a non destructive method for bands visualization. From the scrapped bands dissolved in appropriate solvent compounds were recovered. Examples of the use of this technique for SLs initial purification are published in Ogawa and Ota (2000) and Davila and colleagues (1993) work.

A possible alternative method for simultaneously extraction of SLs from fermentation broth and purification is the adsorption into polymeric adsorbents. As previously mentioned for purification of SLs, silica gel columns have been used in association with a chloroform/methanol gradient. As a more environmentally friendly approach polymeric adsorbents could be tested and experiments could be performed to evaluate SLs recovery and purification. Besides being more environmentally friendly if used as in-situ separation approach for removal of SLs from fermentation broth it can avoid product inhibition phenomena and avoid SL conversion along time, increasing the desired recovery of products (SLs). This methodology can especially be useful in laboratory scale; however at industrial level an in-stream SLs recovery with adsorbents can be more adequate such as the continuous pumping of the fermentation broth through a column that contains the adsorbent with a micro-filtration unit to prevent biomass contamination.

Adsorption technique requires adsorbents to retain the solutes of interest reversibly and the two major groups used to hold biological molecules are activated carbons and synthetic resins (Belter et al. 1988). The carbons can be from vegetable or mineral origin while other adsorbents such as ion exchange resins, neutral polymeric resins (styrene and divinylbenzene), polyacrilamide hydrogels, zeolites, silica gel or alumina are, mainly synthetically produced (Belter et al. 1988). Non polar adsorbents (e.g., neutral resins) have a relative high affinity for organic compounds and low affinity for water, therefore those materials were tested for SLs removal from fermentation broth (Ribeiro 2012).

The physical adsorption involves relative weak forces between the adsorbent and the adsorbate (solute) and consequently is reversible. For non polar adsorbents van der Walls forces are predominant and the relative affinity is determined largely by the polarizability and the size of the adsorbent molecule and the dimension of the adsorbent pores (Ruthven, 2008). Various factors can affect physical adsorption such as temperature, surface area of the adsorbent, solvent, nature of the solute, pH of the solution, presence of competing solutes (Cooney 1999).

Qualitative and quantitative analysis

For identification of SLs present in mixtures a frequently used method is thin layer chromatography (TLC). Other methods used comprise HPLC that allows simultaneously identification and quantification of SLs and the "anthrone assay", a spectrophotometric method that allows an approximate quantification of SLs.

TLC has been used as an initial method for the detection and compositional analysis of SLs present in fermentation broth extracts. It can allow an initial screening of the produced compounds, follow the production along time and assess purity after purification steps. Since SLs discovery, TLC has been employed in several studies (Deshpande and Daniel 1995, Pekin et al. 2005, Daniel et al. 1999, Casas and Garcia-Ochoa 1999, Rau et al. 2001, Cavalero and Cooper 2003, Chen et al. 2006a, Davery and Pakshirajan 2010, Ashby et al. 2011) and a widely stated method is the one used by Asmer and co-workers (1988) for analysing sophorolipids produced by *Candida bombicola*. Majority of methods combine a stationary phase of silica gel 60 with a mobile phase of chloroform/methanol/water. To locate sample components after separation non specific detection of compounds can be obtained by iodine vapours, sulphuric acid spraying or choosing a stationary phase incorporated of a fluorescent material (e.g., silica Gel 60 F_{254}). However a specific reagent ought to be used to detect SLs such as glycolipid revealing reagent, colouring only the compounds of interest.

Colorimetric methods used to quantify carbohydrates such as anthrone (Scott and Melvin 1953, Helbert and Brown 1957, Smith et al. 2010) can be carried out to detect and approximately quantify glycolipids presence in fermentation broth (Smith et al. 2010). The anthrone assay can indirectly quantify SLs by assessment of the glucose units included into the sophorolipids structure. Some chemicals and carbohydrates can interfere, so the obtained results can be imprecise if measurements are performed in a complex matrix such as fermentation broth. In this assay the heat (100°C) and strong acidic (H_2SO_4, 75% $_{(v/v)}$) environment hydrolyses glycosidic bonds, promotes dehydratation of monomers and produces furfuraldehyde derivatives (Smith et al. 2010). These compounds react with anthrone (9, 10-dihydro-9-oxoanthracene) and originate a green coloured product which absorbance can be measured at 620 nm. Some published material has mentioned the use of anthrone assay to assess SLs concentration (Asmer et al. 1988, Davila et al. 1992, Schippers et al. 2000, Hu and Ju 2001a).

The HPLC has been used in SLs analysis since it allows separation, identification and quantification of SLs, when coupled to an evaporative light scattering (ELSD) or a mass spectrometry (MS) detector (further discussed). The HPLC coupled to a ELSD detector allows analysis of compounds that lack chromophore groups such SLs and has been used by some authors (Davila et al. 1992, 1994, Casas and Garcia-Ochoa 1999, Bogaert et al. 2008, Ribeiro et al. 2012c). Isocratic elutions with water/methanol (Casas and Garcia-Ochoa 1999) or water/acetonitrile (Davila et al. 1992, 1993) mobile phases have been used as well as gradient elution with a water/acetonitrile (Davila et al. 1994) mobile phase. In the isocratic methods two columns were used (RP8+RP18) while in gradient methods the only column used was a RP18.

Structural Characterization analysis

For structural characterization of SLs MS, Fourrier-transform infrared spectroscopy (FTIR), ^{1}H and ^{13}C Nuclear Magnetic Resonance (NMR) have frequently been used for SLs structural identification.

MS represents a powerful tool for analysing SLs providing data of the molecular mass and valuable structural information of the compound under investigation. An example is the use of a MS analyser such as Matrix-Assisted Laser Desorption/Ionisation-Time of Flight (MALDI-TOF) MS for identifying SLs and used in the work of Bisht and co-workers (1999) and Kurtznam and colleagues (2010).

Advantages such as sensitivity and selectivity can be obtained by coupling HPLC with MS. This methodology has been preferred for SLs analysis principally for allowing individual separation and characterization, and has assumed special importance in SLs identification since they are produced as a mixture of compounds and no standards are available. Atmospheric pressure chemical ionization (APCI) and electrospray ionization (ESI) are the common HPLC-MS interfaces used in the analysis of SLs or SLs derivatives. The use of the HPLC-APCI-MS operating in the positive mode of ionization has been mentioned in several published reports (Nunez et al. 2001, Nunez et al. 2004, Ashby et al. 2006, 2008, Solaiman et al. 2004, 2007). Moreover the APCI probe was also used coupled to an ultra performance liquid chromatography (UPLC) that uses the same principles as HPLC with faster runs and smaller solvent volumes (Ratsep and Shah 2009). The HPLC-ESI-MS interface has been used either in the positive mode (Davery and Pakshirajan 2009, Ribeiro et al. 2012c) or in the negative mode (Bogaert et al. 2011b).

The use of Tandem MS (MS/MS) or multi-stage MS (MS^n) in mixtures provide unequivocal evidence for the presence of known compounds and valuable structural information of unknown compounds. With this approach a precursor ("parent") ion from the initial fragmentation is selected and allowed or induced to further fragmentation originating "daughter" ions (Silverstein et al. 2005). The obtained "daughter" ions will allow a more consistent identification or structural characterization of the SL under study. Few MS/MS and MS^n studies on SLs fragmentation have been realized. Koster and colleagues (1995) addressed SLs analysis by fast atom bombardment (FAB)-MS and tandem MS. Analysis were operated in the positive and negative mode and collision induced dissociation (CID) MS/MS data for [M+H]+, [M+Na]+ and [M-H]- were presented. Moreover

Hu and Ju (2001a) have also identified and quantity produced SLs by HPLC-ESI-MS[n].

Ribeiro et al. 2012c developed a correlation between TLC, HPLC-ELSD and HPLC-ESI-MS SLs analysis methods. It was important for compound identification of within each methodology. HPLC-ESI-MS/MS method was crucial for SLs identification/characterization and a SLs fragmentation pattern could be established from [M+Na]+ collision-induced dissociation (CID). The product ions obtained were essentially hexose moiety fragments, fatty acid portion fragments and SLs fragments after losing one hexose molecule. With this approach it was also possible to identify new SLs produced by *Starmerella bombicola* and *Rhodotorula bogoriensis* (Ribeiro et al. 2012c). The HPLC-ESI-MS/MS method allowed the identification for the first time of new SLs produced by *Rhodotorula bogoriensis*. These new compounds are longer chain SLs with a fatty acid chain of 24 carbon atoms and were identified as C24:0 monoacetylated and diacetylated acidic SLs (Ribeiro et al. 2012c).

Gaseous chromatography coupled with mass spectrometry (GC-MS) has also been used for identification of hydroxy fatty acid moiety present in the SL structure. SLs show several variations in the lipid backbone therefore lipid portion analysis can provide structural information essential for identification of the SL structure. To accomplish that information it is necessary to promote hydrolysis of the glycosidic bond that links the sophorose moiety to the fatty acid and subsequent esterification of the fatty acid. Davila and colleagues (1993) added to SLs methanol and H_2SO_4 to obtain methyl esters of hydroxy fatty acids. After extraction the hydroxy group was silynated to obtain a good reproducibility in the analysis and the obtained silynated hydroxy fatty acids methyl esters were analysed by GC-MS. The same methodology with few variations was carried out by Cavalero and Cooper (2003) and Ma and co-workers (2011).

The use of MS or tandem MS allows the identification of the SL molecular structure to relatively high degree but not entirely. The use of NMR guarantees full structural determination by allowing the identification of functional groups and position of linkages within the SLs molecules. With ^{1}H and ^{13}C NMR analysis one-dimensional (1D) spectra or two-dimensional (2D) spectra have been obtained. NMR analysis has been done with pure SLs or pure SLs derivatives and results can be observed in the published studies from several group researchers, such as Asmer (1988), Koster (1995), Bisht (1999), Carr (2003), Zhang (2004), Azim (2006), Shin (2010).

Analyses of SLs performed with FTIR are published in Hu and Ju (2001a) and Davery and Pakshirajan (2009b, 2010) works, showing SLs functional groups and allowing confirmation of acidic and lactonic structures.

Surface properties analysis

As surface active compounds, there is a great interest on SLs surface properties evaluation. The critical micelle concentration (CMC) and surface tension at CMC (γ CMC) can be easily obtained by measuring surface tension of different concentration solutions of surfactant. SLs surface properties have been determined by the pendant drop (Konishi et al. 2008), Wilhelmy plate (Solaiman et al. 2004, 2007) and Du Nouy ring method (Otto et al. 1999). The pendant drop method uses a tensiometer where the surface tension is determined by the geometry of the drop, analyzed optically. With the two last mentioned methods a tensiometer with a wettable probe suspended from a precision balance is necessary to determine surface or interfacial tension. This probe is usually a platinum plate or platinum-iridium alloy ring, respectively in the Wilhelmy plate method and the Du Nouy ring method. As the probe touches the liquid surface a force acts on the balance and this force is measured and used to calculate interfacial or surface tension. The tension read in the equipment is calculated using the following expression: $\gamma = F/L \times \cos\theta$ where γ is the surface or interfacial tension, F the force acting on the balance, L the wetted length and θ the contact angle.

Current interest in sophorolipids

Sophorolipids have gathered a prominent interest in commercial and academic fields being the subject of several publications in recent years. Main attribute of SLs is their ability to act as surface active agents since they are non ionic surfactants. Surfactants are surface-active compounds, capable of reducing surface and interfacial tension respectively at the surface between a condensed phase (liquid or solid) and a gaseous phase or at the interface between non miscible condensed phases (liquid or solid). As result, phases will mix or disperse freely and a stable suspension or emulsion will be attained. BS activity includes wetting capacity, foaming ability and detergency power but typical desirable properties are surface tension reduction, solubility enhancement and low critical micelle concentration (Mulligan 2005, Kitamoto et al. 2009).

As BS, SLs show potential application in the industrial field, like pharmaceutical, cosmetic, food, and cleaning products industries (Banat et al. 2000).

SLs are presently being used in industry for their surfactant properties. SLs surfactant properties are suitable for cleaning health products and cosmetics but besides those properties SLs are also effective against dandruff and body odors, as depigmentation and desquamating agents. Additionally, SLs are suitable for cellulite treatments (by reducing subcutaneous fat) and show inhibitory effects against oxygen reactive species (Bogaert et al. 2007).

SLs are presently being used in the cleaning products industry by Ecover Belgium NV and Saraya Co., Ltd., and in the cosmetic industry by Soliance and more recently MG Intobio Co., Ltd. The Belgium Company Ecover is dedicated to the development and production of effective and ecological cleaning and washing products and was the first to use sophorolipids in hard surface cleaning products. Saraya based in Japan, commercializes several products with sophorolipids as active ingredient in dish washer detergent, automated jet washers detergent, laundry soap and fresh wash products. A good example of sophorolipids potential is the active ingredient for cosmetics named *Sopholiance S* (sophorolipid biosurfactant) produced and sold to the cosmetic industry by Soliance Group in France. This product possesses anti-bacterial and sebo-regulation activity, and can be used in acne prone skin treatment, deodorant, face cleaner, shower gel and make up remover formulations. More recently the South Korean MG Intobio Company introduced into the market a few cosmetic products (functional soaps) with sophorolipids in their composition under the brand name *Sopholine*. Moreover surfactants SLs exhibit lower cytotoxicity than the lipopeptide surfactin, a commercially used surfactant in cosmetic products, obtained from *Bacillus subtilis* (Williams 2009). Besides surfactant properties SLs also present diverse biological activities and examples are their biocide activity towards bacteria (Kitamoto et al. 2002, Shah et al. 2007a), virus (Shah et al. 2005a, Gross and Shah, 2004, 2007, 2012), fungi (Yoo et al. 2005) and algae (Lee et al. 2008).

In general the effect of SLs on bacteria appears to be more evident against gram positive bacteria than gram negative most likely because of their differences in cell wall (Kitamoto et al. 2002, Kim et al. 2002, Shah et al. 2007a). SLs, produced by *Candida bombicola*, have shown activity against gram positive bacteria that can become pathogenic or involved in disease (even though some of those are commensal bacteria). Among those it is possible to mention *Staphylococcus xylosus*, *Staphylococcus epidermidis*, *Streptococcus faecium*, *Streptococcus mutans*, *Rhodococcus erythropolis*, *Streptococcus agalactiae* (Kitamoto et al. 2002, Kim et al. 2002, Shah et al. 2007a). SLs produced by *Candida bombicola* have also revealed effect on the actinobacteria *Propionibacterium acne* responsible for the acne inflammation (Kim et al. 2002) and on the gram positive bacteria *Corynebacterium xerosis* responsible for dandruff (Bogaert et al. 2011a). In the group of gram negative bacteria susceptible to *Candida apicola* lactonic SLs are *Pseudomonas aeruginosa*, *Escherichia coli* and *Serratia marcescens* with a minimum inhibitory concentration (MIC) of 7.8, 7.8 and 1.95 mg L^{-1}, respectively (Kitamoto et al. 2002). More recently the effect of SLs from *Candida bombicola* and SLs derivatives on clinically relevant bacteria was studied, namely *Escherichia coli, Staphylococcus aureus, Klebsiella pneumoniae, Pseudomonas aeruginosa* and *Proteus mirabilis* were tested and the best results for minimum inhibitory concentrations were higher than 0.128 µg L^{-1} with ethyl ester diacetate SL derivative (Sleiman et al. 2009).

As mentioned before, SLs also present antifungal activity towards some fungus and water molds. That activity was verified in *Glomerella cingulata* [conidia germination inhibition (50 mg L^{-1})] (Kitamoto et al. 2002), *Phytophthora* sp. *and Pythium* sp. (plant pathogens) [mycelial growth inhibition in 8% (500 mg L^{-1} of SL) for *phytophthora* sp. and zoospore motility decrease of 80% (100 mg L^{-1} SL)] (Yoo et al. 2005) and *Botrytis cineria* (plant pathogen) [cellular growth inhibition in 50% (100 mg L^{-1})] (Kim et al. 2002).

The antiviral activity of SLs from *Candida bombicola* and SLs derivatives was tested and revealed effects against Human Immunodeficiency Virus (HIV) and Herpes virus (Shah et al. 2005, Gross and Shah, 2007, 2012). The anti-HIV activity was evident with acidic SL but superior with the ethyl ester SL derivative (Shah et al. 2005). The anti-herpes activity was evident with the diacetylated lactonic SL, methyl ester derivative and ethyl ester derivative showing a 50% effective concentration (EC$_{50}$) of 25.8, 18.4 and inferior to 0.03 µM respectively (Gross and Shah 2007).

Moreover inhibition of harmful algae bloom has also been verified with sophorolipid/yellow clay mixture (5 mg L^{-1}/1 g L^{-1}) that efficiently mitigated the *Cochlodinium* bloom (with 95% removal efficiency after 30 min) and presented fewer negative effects on the pelagic ecosystem (Lee et al. 2008).

Another important feature of SLs is their activity towards cancerous cell lines. Isoda and co-workers (1997) found that SL (from *Candida bombicola*) promoted human promyelocytic leukemia cell line HL60 differentiation instead of cell proliferation. The authors also verified that SLs inhibited cell line HL60 protein kinases family (PKC) responsible for cell regulation, proliferation and differentiation. In addition, SLs from *Candida bombicola* and SLs derivatives, demonstrated cytotoxic effect against human pancreatic carcinoma cells (Fu et al. 2008). More recently studies with SLs from *Wickerhamiella domercqiae* revealed cytotoxic activity towards human leukemia HL60 and K562 (Chen et al. 2006a), human liver cancer H7402, human lung cancer A549 (Chen et al. 2006a, b) and human esophageal cancer KYSE109 and KYSE450 (Shao et al. 2012) cell lines. According to Chen and co-workers (2006b) apoptosis seems to be involved in the anti-cancer activity observed with the supplementation of SLs to human liver cancer cells H7402.

Additionally SLs from *Candida bombicola* and SLs derivatives demonstrated to mediate anti-inflammatory effect by decreasing nitric oxide and inflammatory cytokine production *in vitro* and decreasing sepsis-related mortality with an *in vivo* rat model of sepsis peritonitis (Hardin et al. 2007). Other studies referred SLs as immunomodelators and comprise IgE regulation leading to its decrease (Bluth et al. 2006, Hagler et al. 2007) and decreasing asma severity in experimental asma model (Bluth et al. 2008).

SLs have also been used in nanoparticle tecnologies. Acidic SLs were used as water soluble capping agents for Co nanoparticles making the nanoparticle stable and water dispersible (Kasture et al. 2007). Additionally sophorolipid-conjugated gellan gum reduced/capped gold nanoparticles, revealed activity by killing the glioma cell lines and glioma stem cell lines. The effects observed were enhanced when doxorubicin hydrochloride was also conjugated to these gold nanoparticles (Dhar et al. 2011).

Other possible applications of SLs were similar to other BS such as environmental protection applications in soils decontamination by organochloride compounds like PCBs, polyaromatic hydrocarbons (PAHs) and heavy metals (Shippers et al. 2000, Kitamoto et al. 2002, Mulligan 2005); energy saving technologies by controlling ice particle agglomeration; in the pharmaceutical field as drug or gene delivery carriers (Kitamoto et al. 2002, Kasture et al. 2007); as biofilm adhesion and formation preventing strategy in medical devices (Banat et al. 2010).

Structure-activity relationship (SAR)

SAR and surface properties

Hydrophilic/hydrophobic balance of SLs molecule plays an important role on surface properties of SLs molecule. Consequently, acetylation grade, chain length of SLs and SLs configuration profile will influence CMC and surface tension at CMC. On the subject of acetylation grade Otto and colleagues (1999) observed a significant difference in the γCMC as result of the addition of a second acetyl group to a SL. The authors verified that the C18:1 monoacetylated lactonic SL presented a γCMC of 40 mN m^{-1} while in the C18:1 diacetylated lactonic SL it was 36 mN m^{-1}. Additionally from surface properties measurements of alkyl (e.g., methyl, propyl, butyl) esters SLs derivatives it was observed that CMC and γCMC decreased as the alkyl chain length increased (Zhang et al. 2004). In this condition the increase on chain length resulted in higher hydrophobic profile and consequently decreased γCMC. Furthermore SLs lactonization also enhances surface activity of molecules when comparing to the respective acidic SL structure (Glenns and Cooper 2006).

Most of published data about SLs surface properties are relative to SLs mixtures obtained as final products. SLs mixture surface properties are dependent on co-substrate supplemented to fermentation media since it will influence SL mixture composition. Ashby et al. (2008) noticed that SLs mixtures obtained from palmitic and linoleic acid presented higher water solubility and higher CMC than SLs obtained from stearic or oleic acid. In those experiments γCMC obtained was around 30–35 mN m^{-1}.

It must also be noticed that depending on the method used different CMC and different γCMC or minimal surface tention (γ_{min}) can be observed. Most of published data about SLs surface properties parameters are from *Candida bombicola* SLs and a resume is presented in Table 4.

Modulation of the hydrophilic/lipophilic balance in naturally occurring SLs, in SLs derivatives or in not naturally occurring SLs, can assume special importance on surfactants with enhanced specific properties. Example is the SLs mixture obtained when adding a different co-substrate such as lauryl alcohol where the CMC and γCMC were 0.68 mg L^{-1} and 24 mN m^{-1}, respectively (Pulate et al. 2012).

Acidic SLs can act as moisturizing agents (Davila et al. 1994) and their derivatives can be used as solubilising, wetting, emulsifying and cleaning agents. Improvement of SLs properties can also be obtained by structure modifications after SLs production. An example is propylene glycol SL derivative commercialized as skin moisturizer and softener in cosmetics (Kitamoto et al. 2002). Also an improvement on CMC was verified by increasing the carbon chain of alkyl ester SL derivatives (Shah et al. 2005). Alternatively shorter chain acid SLs obtained from ozonolysis of double bonds of the fatty acid moiety can result in good wetting agents (Bogaert et al. 2011a).

SAR and biological activity

SLs molecule structure also influences their biological activity. Concerning SLs configuration, lactonic SLs are referred to present a more evident biological activity as antibacterial (Yoo et al. 2005, Shah et al. 2007a), fungicide (Yoo et al. 2005), spermicidal (Shah et al. 2005) and antitumor (Shao et al. 2012) agents while acidic SLs are described to present a slightly higher antiviral activity (Shah et al. 2005). The fatty acid chain length also influences SLs activity which was verified by Shah et al. (2005) when studying SLs spermicidal activity. The SL alkyl derivative with a chain length of 20 carbon atoms revealed lower spermicidal activity than the correspondent SL derivative with a chain length of 24 carbon atoms. Nevertheless in the virucidal activity towards HIV the contrary was verified since SLs with a shorter carbon chain length present higher effect (Shah et al. 2005). Moreover the acetylation grade of SL molecule can also influence SLs activity. It was also observed that lactonic diacetylated SLs or SLs derivatives presented higher toxicity towards virus and cancer cells than monoacetylated and deacetylated (Shah et al. 2005, Shao et al. 2012). In the evaluation of human esophageal cancer cells viability when supplemented

Table 4. CMC, γ_{min} of SLs and mixtures of SLs.

Microorganism	Co-substrate	SLs	Method	Solvent	CMC (mg L⁻¹)	γCMC (mN m⁻¹)	γmin (mN m⁻¹)	Reference
Candida bombicola	RsO	mixture	Wilhelmy plate	-	89	36.3	-	Bogaert 2011b
Starmerella bombicola	OO	mixture	Pendant drop	H₂O	16.6	36.4	-	Konishi et al. 2008
Candida bombicola	Soy m/OA	mixture	Wilhelmy plate	-	-	-	36–38	Solaiman et al. 2007
Candida bombicola	Soy m/OA	mixture	Wilhelmy plate	Acetate buffer pH6 TrisHCl pH9	6 / 13	- / -	37 / 38	Solaiman et al. 2004
Candida bombicola	SfO	mixture	Du Nouy ring	Phosphate buffer pH 7.4	11	-	37.7	Otto et al. 1999
Starmerella bombicola	OA	L-C18: 1diacetylated	Wilhelmy plate	H₂O	27.17	-	34.18	Davery 2010
Starmerella bombicola	OO	A-C18:1	Pendant drop	H₂O	94.7	43.2	-	Konishi et al. 2008
Candida bombicola	ScO +RsO	L-C18:1 monoacetylated L-C18:1 diacet	Du Nouy ring	Phosphate buffer pH 7.4	130 / 10	- / -	40.4 / 36.1	Otto et al. 1999
Pichia anomala	SbO	mixture	Du Nouy ring	-	-	-	28–31	Thaniyavarm et al. 2008
Candida batistae	OO	mixture	Pendant drop	H₂O	366	40.2	-	Konishi et al. 2008
Candida batistae	OO	A-C18:1	Pendant drop	H₂O	138	39.3	-	Konishi et al. 2008

RsO-Rapeseed oil; OA - Oleic acid; Soy m-Soy molasses; SfO-sun flower oil; OO-Olive Oil; SbO-Soybean; ScO-C. curvatus single cell oil. a -acidic A-C18:1 (ω-1) 6' 6''Ac ; b-acidic A-C18:1 (ω) 6'6''Ac with SLs produced by *Wickerhamiella domercqiae* C18:1 diacetylatedlactonic SLs cytotoxic effect was higher than C18:2 diacetylated lactonic SL effect (Shao et al. 2012). Moreover, in those studies mono- and diacetylated acidic SLs presented lower cytotoxic activity towards esophageal cancer cells than lactonic SLs (Shao et al. 2012).

In some cases SLs derivatives show higher biological activity than natural SLs. For instance acetylated ethyl ester SLs presented a higher spermicidal activity than lactonic SLs. Also the diacetylated ethyl ester SL derivative and the etlyl ester derivative SL were more effective against HIV and herpes virus respectively than natural SLs (Shah et al. 2005, Gross and Shah 2007, 2012). Amino acid conjugated SLs have also presented antibacterial and anti HIV activity and the leucine conjugated SL ethyl ester derivative was the most effective among the derivatives studied (Azim et al. 2006).

The cytotoxic effect of SLs towards cancer cells are summarized in Table 5. The path by which SLs act on the cancer cell has not yet been deeply studied. It was verified that acetylation and lactonization of SLs lead to a stronger effect against esophageal cancer cells (Shao et al. 2012). Moreover, C18:1 diacetylated lactonic SL was cytotoxic towards lung cancer cells and additional studies pointed towards SLs ability to trigger apoptosis (Chen et al. 2006b). Alternatively, SLs were also able to promote cell differentiation of HL 60 leukemia cells (Isoda et al. 1997). In this study it was also verified that SLs inhibited a family of protein kinases (PKC) implicated in regulation, differentiation and proliferation of cells (Isoda et al. 1997). Additionally, SLs activity towards cancer cells might also be related to their surface properties and membrane disturbing activity.

Conclusions

This work gathers achievements on sophorolipids production, extraction/ separation and properties as surface active compounds and biologic activities. The conjugation of fermentation time with different lipid co-substrate or nitrogen sources can lead to more selective production of SLs. Production with or without lipid co-substrates originates different composition in SLs mixtures, which can be an advantage in SLs purification.

For structural characterization of SLs MS, Fourrier-transform infrared spectroscopy, ^{1}H and ^{13}C Nuclear Magnetic Resonance have frequently been used for SLs structural identification.

SLs molecular structure influenced their biological activity. Lactonic SLs configuration was referred to present a higher biological activity as antibacterial, fungicide, spermicidal and antitumor agents while acidic SLs are described to present a slightly higher antiviral activity. The fatty acid chain length also influenced SLs activity. The acetylation grade of SL

Table 5. Cytotoxic activity studies of SLs towards cancer cells.

Microorganism	SLs	Human Cells	Tests	Results observed	Reference
Wickerhamiella domercqiae	L-C18:0 diacet. L-C18:1 diacet. L-C18:2 diacet. L-C18:1 monoacet. A-C18:1 monoacet A-C18:1 diacet. A-C18:2 diacet.	Esophageal cancer KYSE 109 KYSE 450	Viability (0–60 mg L^{-1})	Anticancer activity. L-C18:1 diacet. SL showed strongest effect	Shao et al. 2012
Candida bombicola	SL-gellan gum-gold nanoparticle	Glioma LN-229 HNGC-2	Viability (1–12.5 mgL^{-1})	Killed glioma cells	Dhar et al. 2011
Candida bombicola	Mixture A-SL L-SL diacetate SLs ester	pancreatic carcinoma	Viability (0–200 mgL^{-1})	Methyl ester SL showed strongest effect	Fu et al. 2008
Wickerhamiella domercqiae	L-C18:1 diacet.	Liver cancer-H7402 Lung cancer-A549 Leukimia-HL60; K562	Viability (0–120 mg L^{-1})	Anticancer activity	Chen et al. 2006a
Wickerhamiella domercqiae	L-C18:1 diacet.	Liver cancer- H7402 Lung cancer-A549	Viability (0–50 mg L^{-1}) Morphology TEM analysis TUNEL assay Caspase 3 assay Intracellular Ca^{2+} assay	Triggered apoptosis	Chen et al. 2006b
Candida bombicola	–	Leukemia HL60	Viability Cell differentiation Kinase activity	Cell differentiation PKC kinase activity inhibition	Isoda et al. 1997

A-SL acidic SL; L-SL lactonic SL

molecule can also affect SLs activity. In fact, it was observed that lactonic diacetylated SLs or SLs derivatives presented higher toxicity towards virus and cancer cells than monoacetylated and deacetylated.

References

Albrecht, A., U. Rau and F. Wagner. 1996. Initial steps of sophoroselipid biosynthesis by *Candida bombicola* ATCC 22214 grown on glucose. Appl. Microbiol. Biotechnol. 46: 67–73.

Ashby, R. D., D. K. Y. Solaiman and T. A. Foglia. 2008. Property control of sophorolipids: influence of fatty acid substrate and blending. Biotechnol. Lett. 30: 1093–1100.

Ashby, R. D., J. A. Zerkowski, D. K. Y. Solaiman and L. S. Liu. 2011. Biopolymer scaffolds for use in delivering antimicrobial sophorolipids to the acne-causing bacterium Propionibacterium acnes. New Biotechnol. 28: 24–30.

Ashby, R. D., Solaiman D. K. Y. and T. A. Foglia. 2006. The use of fatty acid esters to enhance free acid sophorolipid synthesis. Biotechnol. Lett. 28: 253–260.

Asmer, H. J., S. Lang, F. Wagner and V. Wray. 1988. Microbial Production, Structure Elucidation and Bioconversion. JAOCS. 65: 1460–1466.

Azim, A., V. Shah, G. F. Doncel, N. Peterson, W. Gao and R. Gross. 2006. Amino Acid Conjugated Sophorolipids: A New Family of Biologically Active Functionalized Glycolipids. Bioconjugate Chem. 17: 1523–1529.

Baltz, R. H., A. L. Demain and J. E. Davies. 2010. Manual of Industrial Microbiology and Biotechnology, 3rd ed., ASM Press, United States of America.

Banat, I. M., A. Franzetti, I. Gandolfi, G. Bestetti, M. G. Martinotti, L. Fracchia, T. J. Smyth and R. Marchant. 2010. Microbial biosurfactants production, applications and future potential. Appl. Microbiol. Biotechnol. 87: 427–444.

Banat, I. M., R. S. Makkar and S. S. Cameotra. 2000. Potential commercial applications of microbial surfactants. Appl. Microbiol. Biotechnol. 53: 495–508.

Bednarski, W., M. Adamczak, J. Tomasik and M. Płaszczyk. 2004. Application of oil refinery waste in the biosynthesis of glycolipids by yeast. Bioresource Technol. 95: 15–18.

Bektas, M. and S. Spiegel. 2004. Glycosphingolipids and cell death. Glyco. J. 20: 39–47.

Belter, P. A., E. L. Cussler and W. S. Hu. 1988. Bioseparations: Downstream Processing for Biotechnology. Wiley-Interscience Publiation, United States of America.

Bisht, K. S., R. A. Gross and D. L. Kaplan. 1999. Enzyme-Mediated Regioselective Acylations of Sophorolipids. J. Org. Chem. 64: 780–789.

Bluth, M. H., S. L. Fu, A. Fu, A. Stanek, T. A. Smith-Norowitz, S. R. Wallner, R. A. Gross, M. Nowakowski and M. E. Zenilman. 2008. Sophorolipids Decrease Asthma Severity And Ova-specific IgE Production In A Mouse Asthma Model. J. Allergy Clin. Immunol. 121: S2.

Bluth, M. H., T. A. Smith-Norowitz, M. Hagler, R. Beckford, S. Chice, V. Shah, R. Gross, M. Nowakowski, R. Schulze and M. Zenilman. 2006. Sophorolipids Decrease IgE Production in U266 Cells. J. Allergy Clin. Immunol. 117: S220.

Bogaert, I. N. A. V., D. Develter, W. Soetaert and E. J. Vandamme. 2008. Cerulenin inhibits de novo sophorolipid synthesis of Candida bombicola. Biotechnol. Lett. 30: 1829–32.

Bogaert, I. N. A. V., J. Zhang and W. Soetaert. 2011a. Microbial synthesis of Sophorolipids. Process Biochem. 46: 821–833.

Bogaert, I. N. A. V., K. Saerens, C. Muynck, D. M. Develter, W. Soetaert and E. J. Vandamme. 2007. Microbial production and application of sophorolipids. Appl. Microbiol. Biotechnol. 76: 23–34.

Bogaert, I. N. A. V., M. Demey, D. Develter, W. Soetaert and E. J. Vandamme. 2009. Importance of the cytochrome P450 monooxygenase CYP52 family for the sophorolipid-producing yeast *Candida bombicola*. FEMS Yeast Res. 9: 87–94.

Bogaert, I. N. V., S. Roelants, D. Develter and W. Soetaert. 2010. Sophorolipid production by *Candida bombicola* on oils with a special fatty acid composition and their consequences on cell viability. Biotechnol. Lett. 32: 1509–1514.

Bogaert, I. V., S. Fleurackers, S. V. Kerrebroeck, D. Develter and W. Soetaert. 2011b. Production of New-to-Nature Sophorolipids by Cultivating the Yeast *Candida bombicola* on Unconventional Hydrophobic Substrates. Biotechnol. Bioeng. 108: 734–741.

Brakemeier, A., D. Wullbrandt and S. Lang. 1998. *Candida bombicola*: production of novel alkyl glycosides based on glucose/2-dodecanol. Appl. Microbiol. Biotechnol. 50: 161–166.

Brakemeier, A., S. Lang, D. Wullbrandt, L. Merschel, A. Benninghoven, N. Buschmann and F. Wagner. 1995. Novel sophorose lipids from microbial conversion of 2-alkanols. Biotechnol. Lett. 17: 1183–1188.

Breithaupt, T. B. and R. J. Light. 1982. Affinity Chromatography and Further Characterization of the Glucosyltransferases Involved in Hydroxydocosanoic Acid Sophoroside Production in *Candida bogoriensis*. J. Biol. Chem. 257: 9622–9628.

Bucholtz, M. L. and R. J. Lights. 1976. Acetylation of 13-Sophorosyloxydocosanoic Acid by an Acetyltransferase Purified from *Candida bogoriensis*. J. Biol. Chem. 251: 424–430.

Carr, J. A. and K. S. Bisht. 2003. Enzyme-catalyzed regioselective transesterification of peracylated sophorolipids. Tetrahedron. 59: 7713–7724.

Casas, J. A. and F. García-Ochoa. 1999. Sophorolipid production by *Candida bombicola*: medium composition and culture methods. J. Biosci. Bioeng. 88: 488–94.

Casas, J. A., S. Garcia de Lara and F. Garcia-Ochoa. 1997. Optimization of a synthetic medium for *Candida bombicola* growth using factorial design of experiments. Enz. Microb. Technol. 21: 221–229.

Cavalero, D. A. and D. G. Cooper. 2003. The effect of medium composition on the structure and physical state of sophorolipids produced by *Candida bombicola* ATCC 22214. J. Biotechnol. 103: 31–41.

Chen, J., X. Song, H. Zhang and Y. Qu. 2006a. Production, structure elucidation and anticancer properties of sophorolipid from *Wickerhamiella domercqiae*. Enz. Microb. Technol. 39: 501–506.

Chen, J., X. Song, H. Zhang, Y. Qu and J. Miao. 2006b. Sophorolipid produced from the new yeast strain Wickerhamiella domercqiae induces apoptosis in H7402 human liver cancer cells. Appl. Microbiol. Biotechnol. 72: 52–59.

Chen, M., C. Dong, J. Penfold, R. K. Thomas, T. J. P. Smyth, A. Perfumo, R. Marchant, I. M. Banat, P. l. Stevenson, A. Parry, I. Tucker and R. A. Campbell. 2011. Adsorption of Sophorolipid Biosurfactants on Their Own and Mixed with Sodium Dodecyl Benzene Sulfonate, at the Air/Water Interface. Langmuir. 27: 8854–8866.

Cooney, D. O. 1999. Adsorption design for wastewater treatment. CRC Press LLC.

Cooper, D. G. and D. A. Paddock. 1984. Production of a Biosurfactant from *Torulopsis bombicola*. Appl. Environ. Microbiol. 47: 173–176.

Cutler, A. J. and R. J. Light. 1979. Regulation of Hydroxydocosanoic Acid Sophoroside Production in *Candida bogoriensis* by the Levels of Glucose and Yeast Extract in the Growth Medium. J. Biol. Chem. 254: 1946–1950.

Daniel, H. J., R. T. Otto, M. Binder, M. Reuss and C. Syldatk. 1999. Production of sophorolipids from whey: development of a two-stage process with *Cryptococcus curvatus* ATCC 20509 and *Candida bombicola* ATCC 22214 using deproteinized whey concentrates as substrates. Appl. Microbiol. Biotechnol. 51: 40–45.

Daverey, A. and K. Pakshirajan. 2009a. Production of sophorolipids by the yeast *Candida bombicola* using simple and low cost fermentative media. Food Res. Inter. 42: 499–504.

Daverey, A. and K. Pakshirajan. 2009b. Production, characterization, and properties of sophorolipids from the yeast *Candida bombicola* using a low-cost fermentative medium, Appl. Biochem. Biotechnol. 158: 663–74.

Daverey, A. and K. Pakshirajan. 2010. Sophorolipids from *Candida bombicola* using mixed hydrophilic substrates: Production, purification and characterization. Colloids Surf. B. Biointerfaces. 79: 246–253.

Daverey, A. and K. Pakshirajan. 2011. Pretreatment of synthetic dairy wastewater using the sophorolipid-producing yeast *Candida bombicola*. Appl. Biochem. Biotechnol. 163: 720–728.

Daverey, A., K. Pakshirajan and P. Sangeetha. 2009. Sophorolipids Production by *Candida bombicola* using Synthetic Dairy Wastewater. World Academy of Science, Engineering and Technology. 27: 497–499.

Davila, A. M., R. Marchal and J. P. Vandecasteele. 1994. Sophorose lipid production from lipidic precursors: Predictive evaluation of industrial substrates. J. Ind. Microbiol. 13: 249–257.

Davila, A. M., R. Marchal and J. P. Vandecasteele. 1997. Sophorose lipid fermentation with differentiated substrate supply for growth and production phases. Appl. Microbiol. Biotechnol. 47: 496–501.

Davila, A. -M., R. Marchal and J. -P. Vandecasteele. 1992. Kinetics and balance of a fermentation free from product inhibition: sophorose lipid production by *Candida bombicola*. Appl. Microbiol. Biotechnol. 38: 6–11.

Davila, A. -M., R. Marchal, N. Monin and J. -P. Vandecasteele. 1993. Identification and determination of individual sophorolipids in fermentation products by gradient elution high-performance liquid chromatography with evaporative light-scattering detection. J. Chromatogr. 448: 139–149.

Desai, J. D. and I. M. Banat. 1997. Microbial Production of Surfactants and Their Commercial Potential. Microbiol. Mol. Biol. Rev. 61: 47–64.

Deshpande, M. and L. Daniels. 1995. Evaluation of sophorolipid biosurfactant production by *Candida bombicola* using animal fat. Bioresource Technol. 54: 143–150.

Develter, D. W. G. and S. J. J. Fleurackers. 2010. Sophorolipids and Rhamnolipids. *In*: M. Kjelin and I. Johanssob [eds.]. Surfactants from Renewable Resources, John Wiley and Sons Ltd., UK. pp. 213–238.

Dhar, S., E. M. Reddy, A. Prabhune, V. Pokharkar, A. Shiras and B. L. V. Prasad. 2011. Cytotoxicity of sophorolipid-gellan gum-gold nanoparticle conjugates and their doxorubicin loaded derivatives towards human glioma and human glioma stem cell lines. Nanoscale. 2: 575–580.

Esders, T. W. and R. J. Light. 1972. Glucosyl and Acetyltransferases Involved in the Biosynthesis of Glycolipids from *Candida bogoriensis*. J. Biol. Chem. 247: 1375–1386.

Felse, P. A., V. Shah, J. Chan, J. Kandula, K. J. Rao and R. A. Gross. 2007. Sophorolipid biosynthesis by *Candida bombicola* from industrial fatty acid residues. Enz. Microb. Technol. 40: 316–323.

Fu, S. L., S. R. Wallner, W. B. Bowne, M. D. Hagler, M. E. Zenilman, R. Gross and M. H. Bluth. 2008. Sophorolipids and Their Derivatives Are Lethal Against Human Pancreatic Cancer Cells. J. Surg. Res. 148: 77–82.

Glenns, R. N. and D. G. Cooper. 2006. Effect of Substrate on Sophorolipid Properties. JAOCS. 83: 137–145.

Gobbert, U., S. Lang and F. Wagner. 1984. Sophorose lipid formation by resting cells of *Torulopsis bombicola*. Biotechnol. Lett. 6: 225–230.

Gorin, P. A. J., J. F. T. Spencer and A. P. Tulloch. 1961. Hydroxy fatty acid glycosides of sophorose from *Torulopsis magnolia*. Can. J. Chem. 39: 846–855.

Gross, R. A. and V. Shah. 2004. Antimicrobial properties of various forms of sophorolipids. Patent application World patent W2004/011216 A1.

Gross, R. A. and V. Shah. 2007. Anti-herpes virus properties of various forms of sophorolipids. World patent WO2007US63701.

Gross, R. A. and V. Shah. 2012. Virucidal properties of various forms of sophorolipids. Patent application 20120231068.

Guilmanov, V., A. Ballistreri, G. Impallomeni and R. A. Gross. 2002. Oxygen transfer rate and sophorose lipid production by *Candida bombicola*. Biotechnol. Bioeng. 77: 489–494.

Hagler, M., T. A. Smith-Norowitz, S. Chice, S. Wallner, D. Viterbo, C. M. Mueller, R. Gross, M. Nowakowski, R. Schulze, M. E. Zenilman and M. H. Bluth. 2007. Sophorolipids decrease IgE production in U266 cells by downregulation of BSAP (Pax5), TLR2, STAT3 and IL-6. J. Allergy Clin. Immunol. 119: S263.

Hardin, R., J. Pierre, R. Schulze, C. M. Mueller, S. L. Fu, S. R. Wallner, A. Stanek, V. Shah, R. A. Gross, J. Weedon, M. Nowakowski, M. E. Zenilman and M. H. Bluth. 2007. Sophorolipids Improve Sepsis Survival: Effects of Dosing and Derivatives. J. Surg. Res. 142: 314–319.

Helbert, J. R. and K. D. Brown. 1957. Color reaction of anthrone with monosaccharide mixtures and oligo and polysaccharides containing hexuronic acids. Anal. Chem. 29: 1464–1466.

Hommel, R. K., S. Stegner, K. Huse and H. -P. Kleber. 1994. Cytochrome P-450 in the sophorose-lipid-producing yeast *Candida (Torulopsis) apicola*. Appl. Microbiol. Biotechnol. 40: 724–728.

Hu, Y. and L. K. Ju. 2001a. Sophorolipid production from different lipid precursors observed with LC-MS. Enz. Microb. Technol. 29: 593–601.

Hu, Y. and L. K. Ju. 2003. Lipase-Mediated Deacetylation and Oligomerization of Lactonic Sophorolipids. Biotechnol. Prog. 19: 303–311.

Hu, Y. and L. -K. Ju. 2001b. Purification of lactonic sophorolipids by crystallization. J. Biotechnol. 87: 263–272.

Im, J. H., T. Nakane, H. Yanagishita, T. Ikegami and D. Kitamoto. 2001. Mannosylerythritol lipid, a yeast extracellular glycolipid, shows high binding affinity towards human immunoglobulin G. BMC Biotechnol. 1: 5.

Isoda, H., D. Kitamoto, H. Shinmoto, M. Matsumura and T. Nakahara. 1997. Microbial extracellular glycolipid induction of differentiation and inhibition of the protein kinase C activity of human promyelocytic leukemia cell line HL60. Biosci. Biotechnol. Biochem. 61: 609–14.

Iwabuchi, K., H. Nakayama, C. Iwahara and K. Takamori. 2010. Significance of glycosphingolipid fatty acid chain length on membrane microdomain-mediated signal transduction, FEBS Lett. 584: 1642–1652.

Kasture, M., S. Singh, P. Patel, P. A. Joy, A. A. Prabhune, C. V. Ramana and B. L. V. Prasad. 2007. Multiutility Sophorolipids as Nanoparticle Capping Agents: Synthesis of Stable and Water Dispersible Co Nanoparticles. Langmuir. 23: 11409–11412.

Kim, K., Y. Dalsoo, K. Youngbum, L. Baekseok, S. Doonhoon and K. Eun-Ki. 2002. Characteristics of Sophorolipid as an Antimicrobial Agent. J. Microbiol. Biotechnol. 12: 235–241.

Kim, S. Y., D. K. Oh, K. H. Lee and J. H. Kim. 1997. Effect of soybean oil and glucose on sophorose lipid fermentation by *Torulopsis bombicola* in continuous culture. Appl. Microbiol. Biotechnol. 48: 23–26.

Kim, Y. B., H. S. Yun and E. K. Kim. 2009. Enhanced sophorolipid production by feeding-rate-controlled fed-batch culture. Bioresour. Technol. 100: 6028–6032.

Kitamoto, D., H. Isoda and T. Nakahara. 2002. Functions and potential applications of glycolipid biosurfactants-from energy saving materials to gene delivery carries. J. Biosci. Bioeng. 94: 187–201.

Kitamoto, D., T. Morita, T. Fukuoka, M. Konishi and T. Imura. 2009. Self-assembling properties of glycolipid biosurfactants and their potential applications. Current Opinion in Colloid & Interface Science. 14: 315–328.

Konishi, M., T. Fukuoka, T. Morita, T. Imura and D. Kitamoto. 2008. Production of New Types of Sophorolipids by *Candida batistae*. J. Oleo Sci. 57: 359–369.

Koster, C. G., W. Heerma, H. A. M. Pepermans, A. Groenewegen, H. Peters and J. Haverkamp. 1995. Tandem Mass Spectrometry and Nuclear Magnetic Resonance Spectroscopy and Product Formed on Hydrolysis by Cutinase. Anal. Biochem. 230: 135–148.

Kurtzman, C. P., N. P. J. Price, K. J. Ray and T. M. Kuo. 2010. Production of sophorolipid biosurfacants by multiple species of the *Starmerella (Candida) bombicola* yeast clade. FEMS Microbiol. Lett. 6: 311–140.

Lee, Y. -J., J. -K. Choi, E. -K. Kim, S. -H. Youn and E. -J. Yang. 2008. Field experiments on mitigation of harmful algal blooms using a Sophorolipid—Yellow clay mixture and effects on marine plankton. Harmful Algae. 7: 154–162.

Ma, X. -J., H. Li, L. -J. Shao, J. Shen and X. Song. 2011. Effects of nitrogen sources on production and composition of sophorolipids by *Wickerhamiella domercqiae var.* sophorolipid CGMCC 1576. Appl. Microbiol. Biotechnol. 91: 1623–32.

Malhotra, R. and B. Singh. 2006. Ethanol-Induced Changes in Glycolipids of *Saccharomyces cerevisiae*. Appl. Biochem. Biotechnol. 128: 205–213.

McCaffrey, W. C. and D. G. Cooper. 1995. Sophorolipids Production by *Candida bombicola* Using Self-Cycling Fermentation. J. Ferment. Bioeng. 79: 146–151.

Mukherjee, S., P. Das and R. Sen. 2006. Towards commercial production of microbial surfactants. Trends Biotechnol. 24: 509–515.

Mulligan, C. N. 2005. Environmental applications for biosurfactants. Environ. Pollution. 133: 183–198.

Nuñez, A., R. Ashby, T. A. Foglia and D. K. Y. Solaiman. 2001. Analysis and Characterization of Sophorolipids by Liquid Chromatography with Atmospheric Pressure Chemical Ionization. Chromatographia. 53: 673–677.

Nuñez, A., R. Ashby, T. A. Foglia and D. K. Y. Solaiman. 2004. LC/MS analysis and lipase modification of the sophorolipids produced by *Rhodotorula bogoriensis*. Biotechnol. Lett. 26: 1087–1093.

Ogawa, S. and Y. Ota. 2000. Influence of exogenous natural oils on the omega-1 and omega-2 hydroxy fatty acid moiety of sophorose lipid produced by *Candida bombicola*. Biosci. Biotechnol. Biochem. 64: 2466–2468.

Otto, R. T., H. J. Daniel, G. Pekin, K. Muller-Decker, G. Furstenberger, M. Reuss and C. Syldatk. 1999. Production of sophorolipids from whey II. Product composition, surface active properties, cytotoxicity and stability against hydrolases by enzymatic treatment. Appl. Microbiol. Biotechnol. 52: 495–501.

Pekin, G., F. Vardar-Sukan and N. Kosaric. 2005. Production of Sophorolipids from *Candida bombicola* ATCC 22214 Using Turkish Corn Oil and Honey. Eng. Life Sci. 5: 357–362.

Pulate, V. D., S. Bhagwat and A. Prabhune. 2012. Microbial Oxidation of Medium Chain Fatty Alcohol in the Synthesis of Sophorolipids by *Candida bombicola* and its Physicochemical Characterization. J. Surfact. Deterg. DOI 10.1007/s11743-012-1378-4.

Ratsep, P. and V. Shah. 2009. Identification and quantification of sophorolipid analogs using ultra-fast liquid chromatography-mass spectrometry. J. Microbiol. Meth. 78: 354–356.

Rau, U., R. Heckmann, V. Wray and S. Lang. 1999. Enzymatic conversion of a sophorolipid into a glucose lipid. Biotechnol. Lett. 21: 973–977.

Rau, U., S. Hammen, R. Heckmann, V. Wray and S. Lang. 2001. Sophorolipids: a source for novel compounds. Ind. Crops Prod. 13: 85–92.

Ribeiro, I. A. 2012. Sophorolipids: from biosynthesis to biologic activity evaluation. Ph.D thesis, Faculdade Farmácia, Universidade Lisboa.

Ribeiro, I. A., M. R. Bronze, M. F. Castro and Maria H. L. Ribeiro. 2012a. Design of selective production of sophorolipids by *Rhodotorula bogoriensis* through nutritional requirements. J. Mol. Recogn. DOI: 10.1002/jmr.2188.

Ribeiro, I. A., M. R. Bronze, M. F. Castro and Maria H. L. Ribeiro. 2012b. Sophorolipids: improvement of the selective production by *Starmerella bombicola* through the design of nutritional requirements. Appl. Microb. Biotechnol. DOI 10.1007/s00253-012-4437-x.

Ribeiro, I. A., M. R. Bronze, M. F. Castro and Maria H. L. Ribeiro. 2012c. Optimization and correlation of HPLC-ELSD and HPLC–MS/MS methods for identification and characterization of sophorolipids J. Chromatography B Anal. Technol. in the Biomed. and Life Sciences. 899: 72–80.

Rispoli, F. J., D. Badia and V. Shah. 2010. Optimization of the fermentation media for sophorolipid production from *Candida bombicola* ATCC 22214 using a simplex centroid design. Biotechnol. Prog. 26: 938–944.

Rosa, C. A. and M. -A. Lachance. 1998. The yeast genus *Starmerella gen. nov.* and *Starmerella bombicola sp. nov.*, the teleomorph of *Candida bombicola* (Spencer, Gorin & Tullock) Meyer & Yarrow. Int. J. Syst. Bacteriol. 48: 1413–1417.

Ruthven, D. M. 2008. Fundamentals of Adsorption Equilibrium and Kinetics in Microporous Solids. Mol. Sieve. 7: 1–43.

Schippers, C., K. Gessner, T. Muller and T. Scheper. 2000. Microbial degradation of phenanthrene by addition of a sophorolipid mixture. J. Biotechnol. 83: 189–198.

Schnaar, R. L., A. Suzuki and P. Stanley. 2009. Glycosphingolipids. Chapter 10. *In*: A. Varki, R. D. Cummings and J. D. Esko 2nd [ed.]. Essentials of Glycobiology. Cold Spring Harbor, NY, United States of America.

Scott, T. A. Jr. and E. H. Melvin. 1953. Determination of dextran with anthrone. Anal. Chem. 25: 1656–1661.

Shah, S. and A. Prabhune. 2007. Purification by silica gel chromatography using dialysis tubing and characterization of sophorolipids produced from *Candida bombicola* grown on glucose and arachidonic acid. Biotechnol. Lett. 29: 267–72.

Shah, V., D. Badia and P. Ratsep. 2007a. Sophorolipids Having Enhanced Antibacterial Activity. Antimicrob. Agents Chemother. 51: 397–400.

Shah, V., G. F. Doncel, T. Seyoum, K. M. Eaton, I. Zalenskaya, R. Hagver, A. Azim and R. Gross. 2005. Sophorolipids, Microbial Glycolipids with Anti-Human Immunodeficiency Virus and Sperm-Immobilizing Activities. Antimicrob. Agents Chemother. 49: 4093–4100.

Shah, V., M. Jurjevic and D. Badia. 2007b. Utilization of Restaurant Waste Oil as a Precursor for Sophorolipid Production, Biotechnol. Prog. 23: 512–515.

Shao, L., X. Song, X. Ma, H. Li and Y. Qu. 2012. Bioactivties of Sophorolipid with Different Structures Against Human Esophageal Cancer Cells. J. Surg. Res. 173: 286–291.

Shin, J. D., J. Lee, Y. B. Kim, I. Han and E. K. Kim. 2010. Production and characterization of methyl ester sophorolipids with 22-carbon-fatty acids. Bioresour. Technol. 101: 3170–3174.

Silverstein, R. M., F. X. Webster and D. J. Kiemle. 2005. Spectrometric Identification of Organic Compounds, 7th ed. John Wiley and Sons, Inc., United States of America.

Sleiman, J. N., S. A. Kohlhoff, P. M. Roblin, S. Wallner, R. Gross, M. R. Hammerschlag, M. E. Zenilman and M. H. Bluth. 2009. Sophorolipids as antibacterial agents. Ann. Clin. Lab. Sci. 39: 60–3.

Smyth, T. J. P., A. Perfumo, R. Marchant and I. M. 2010. Banat. Isolation and Analysis of Low Molecular Weight Microbial Glycolipids. *In*: K. N. Timmis [Ed.]. Handbook of Hydrocarbon and Lipid Microbiology. Springer-Verlag, Berlin, Heidelberg, Germany. pp. 3705–3722.

Solaiman, D. K. Y., R. D. A. Nuñez and T. A. Foglia. 2004. Production of sophorolipids by *Candida bombicola* grown on soy molasses as substrate. Biotechnol. Lett. 26: 1241–1245.

Solaiman, D. K. Y., R. D. Ashby, J. A. Zerkowski and T. A. Foglia. 2007. Simplified soy molasses-based medium for reduced-cost production of sophorolipids by *Candida bombicola*. Biotechnol. Lett. 29: 1341–1347.

Sonnino, S., A. Prinetti, H. Nakayama, M. Yangida, H. Ogawa and K. Iwabuchi. 2009. Role of very long fatty acid-containing glycosphingolipids in membrane organization and cell signaling: the model of lactosylceramide in neutrophils. Glyco. J. 26: 615–621.

Souza, A. M. M. S., E. S. Trindade, M. C. Jamur and C. Oliver. 2010. Gangliosides Are Important for the Preservation of the Structure and Organization of RBL-2H 3 Mast Cells. J. Histochem. Cytochem. 58: 83–93.

Spencer, J. F. T., P. A. J. Gorin and A. P. Tulloch. 1970. *Torulopsis bombicola* sp. *n.*, Antonie Van Leeuwenhoek. 36: 129–133.

Stanbury, P. F. and A. Whitaker. 1995. S. J. Hall (ed.). Principles of fermentation Technology, 2nd edition. Butterworth-Heinemann, Oxford, UK.

Thaniyavarn, J., T. Chianguthai, P. Sangvanich, N. Roongsawang, K. Washio, M. Morikawa and S. Thaniyavarn. 2008. Production of sophorolipid biosurfactant by *Pichia anomala*. Biosci. Biotechnol. Biochem. 72: 2061–2068.

Tulloch, A. P. and J. F. T. Spencer. 1968a. Fermentation of long-chain compounds by *Torulopsis apicola*. IV. Products from esters and hydrocarbons with 14 and 15 carbon atoms and from methyl palmitoleate. Can. J. Chem. 46: 1523–1528.

Tulloch, A. P. and J. F. T. Spencer. 1968b. A new hydroxy fatty acid sophoroside from *Candida bogoriensis*. Can. J. Chem. 46: 345–348.

Williams, K. 2009. Biosurfactants for cosmetic application: Overcoming production challenges, MMG 445 basic. Biotechnol. 5: 78–83.

Yoo, D. -S., B. -S. Lee and E. -K. Kim. 2005. Characteristics of microbial biosurfactant as an antifungal agent against plant pathogenic fungus. J. Microbiol. Biotechnol. 15: 1164–1169.

Yu, R. K. and M. Yanagisawa. 2010. Membrane glycolipids in stem cells. FEBS Lett. 584: 1694–1699.

Zhang, J., K. M. J. Saerens, I. N. A. V. Bogaert and W. Soetaert. 2011. Vegetable oil enhances sophorolipid production by *Rhodotorula bogoriensis*. Biotechnol. Lett. 33: 2417–2423.

Zhang, L., P. Somasundaran, S. K. Singh, A. P. Felse and R. Gross. 2004. Synthesis and interfacial properties of sophorolipid derivatives. Colloids and Surfaces A: Physicochem. Eng. Aspects. 240: 75–82.

Zhou, Q. H. and N. Kosaric. 1993. Effect of lactose and olive oil on intra- and extracellular lipids of *Torulopsis bombicola*. Biotechnol. Lett. 15: 477–482.

15

Issues in Algal Biofuels for Fuelling the Future

Ravichandra Potumarthi[1], and *Rama Raju Baadhe[2]*

ABSTRACT

Decreasing fossil fuel resources, competition between bioenergy crops and food crops and environmental concerns drives more attention towards algal biofuels research. Algae have more ability to trap CO_2 than most terrestrial plants. Therefore algae biomass could be used for producing different feed stocks to generate biofuels as well as other value added co-products while potentially mitigating the release of green house gas (GHG) into the atmosphere. Even though CO_2 is abundantly available in nature and algae is an autotroph; biofuel production from algae is not viable commercially due to technology limitations. This chapter highlights the existing technologies for the biodiesel production and discusses some of the engineering, scientific opportunities and challenges of algae biodiesel production.

Introduction

Fast depleting fossil fuels, energy security and environmental concerns are promoting conventional renewable energies, wind, tidal, solar and biofuels as an alternative (IPCC 1991, UK Royal Commission 2000). In

---
[1] Department of Chemical Engineering, Monash University, Clayton, 3800, Australia.
  Email: ravichandra.potumarthi@monash.edu, pravichandra@gmail.com
[2] Department of Biotechnology, National Institute of Technology, Warangal, AP, India.
* Corresponding author

comparison with wind, tidal, and solar forms of renewable energy, biomass based biofuels allow solar energy to be converted into lignocellulosics and stored (Hill et al. 2006) which further can be converted into biofuels by physio, chemico, thermal and biological methods. Lignocellulosic biomass, an important feedstock to alleviate greenhouse gas emissions, to replace fossil fuels and to provide energy security (Goldemberg 2000, Yang et al. 2011, Prasad et al. 2007, Singh et al. 2010). Diverse and abundant biomass is available in the form of agricultural residues, forest residues and they have been investigated as the feedstock for the production of different class of second generation biofuels (Yanqun et al. 2008). While second generation biofuels answers concerns about the intricacy generated by the first generation biofuels such as the food security, overall savings energy, greenhouse gas emissions, etc. it fails to answer the competition between bioenergy crops (switch grass) and food crops for arable land which affects food security (Kenneth et al. 2007). In addition, lignocellulosic based biofuels has a significant environmental impact in comparion with fossil fuels by breaking the carbon cycle—conversion of fixed carbon into CO_2. Hence, without proper replantation it may result in massive biomass deficit which ultimately leads to deforestation (Yanqun et al. 2008). Due to techno-economic non viability of biomass pretreatment technologies, cost of enzymes (ex. cellulases) and lack of viable technology for co-fermentation of glucose and xylose the commercialization of lignocellulosic ethanol (Jay and Timilsina 2010) at industrial level of operation is limited. In addition to these factors nitrogen fertilizers used for crops releases extra N_2O into atmosphere contributing more green house gases than by fossil fuel savings (Crutzen et al. 2007). Issues and concerns raised from lignocellulosic based second generation biofuels has directed scientific and engineering attention towards research and development in algal biofuels. Although research is carried out on micro and macro algal biofuels, further discussions in this chapter are restricted to micro algae.

Micro algae are simple and efficient photosynthetic microorganisms with high growth rates. It is estimated that the biomass productivity of microalgae (traps solar energy 10–20%) could be 50 times more than that of switchgrass (traps 0.5% solar energy), which is the fastest growing terrestrial plants used for bioenergy generation (Lewis and Nocera 2006, Demirbas 2006, Richmond 2000, Huntley and Redalje 2007). In comparison with terrestrial bioenergy crops, microalgae live in diverse ecological habitats such as freshwater, brackish, marine and hyper-saline, with a range of temperatures and pH, and unique nutrient availabilities (Falkowski and Raven 1997). Under favourable growth conditions, algae synthesizes fatty acids which are converted to membrane lipids and constitute about 5 to 20% of their dry cell weight (DCW). Under un-favourable environmental and/ or stress conditions, algae alter their lipid biosynthetic pathways towards the formation and accumulation of neutral lipids (20–50% DCW), mainly in the form of triacylglycerol (TAG) in lipid bodies located in the cytoplasm of the algal cell (Hu et al. 2008). Many algal species have been found to grow rapidly and produce extensive quantities of TAG or lipids, and are thus referred to as oleaginous algae: potential cell factories that produce oils, other lipids for biofuels and other biomaterials (Hu et al. 2008).

Biofuel production using microalgae offers the following advantages (Yanqun et al. 2008, Hu et al. 2008, Rawat et al. 2011):

1) High growth rate: Satisfy the high demand for biofuels using limited land resources without causing potential biomass deficit.
2) Diverse ecological habitats: Consumes less water and tolerates different conditions (e.g., fresh water, brackish, marine and hyper-saline desert, arid- and semi-arid lands) that are not suitable for conventional agriculture.
3) The tolerance of microalgae to high CO_2: Sequesters CO_2 from flue gases emitted from fossil fuels burning and emits less greenhouse gases upon burning.
4) NO_2 release could be minimized when microalgae are used for biofuel production.
5) Synthesizes and accumulates large quantities of neutral lipids/oil (20–50% DCW) compared with traditional oil crops used for biofuel production.
6) Produces value-added co/by-products (e.g., biopolymers, proteins, polysaccharides, pigments, animal feed, fertilizer and H_2).

On the other hand, few major difficulties involved in microalgae biofuel production are (Yanqun et al. 2008, Scott et al. 2010):

1) Low algal biomass concentration due to limited light penetration.
2) Harvesting of algal biomass: Small size of algal cell makes harvesting relatively costly.
3) Dewatering: Removal of large quantities of water makes drying as an energy-consuming process.
4) Lipid/Oil extraction: Lack of effective technology for oil extraction from algal biomass.

Apart from above mentioned issues, higher capital costs and intensive care required during microalgal cultivation compared to conventional agricultural farm is another factor that obstructs the commercialization of microalgal biofuels. However, these problems are expected to be minimized by engineering the existing technologies. Consequently, it makes algal biofuel as one of the most important alternative energy sources to fuel the future. In this chapter we have discussed some of the challenges and engineering aspects of microalgae biofuel production.

Algae Cultivation Methods

Two potential and very common methods in practice till today for large-scale production of microalgae are discussed here.

Open Ponds/Raceway Ponds

Conservative open pond or high rate algal ponds (HRAPs). Algal production systems are age old systems for algal biomass production as open ponds can be built and operated very economically. The vast bulk of microalgae cultivated during early days are in open ponds (Weissman et al. 1988). Open ponds are built in a raceway pond design, a closed loop of the rectangular grid with recirculation channel. Generally, raceway ponds operate at water depths of 15–20 cm, this results in biomass concentrations about 1 g dry weight/L and productivities of 60–100 mg/L/day can be obtained (Pulz 2001). Paddlewheel is used to mix and circulate algal biomass and nutrients as shown in Fig. 1. Continuous operation of paddlewheel prevents sedimentation of biomass which affects growth of the algae. The raceways are relatively inexpensive to build and operate; this feature may cut down the algal biofuel production cost.

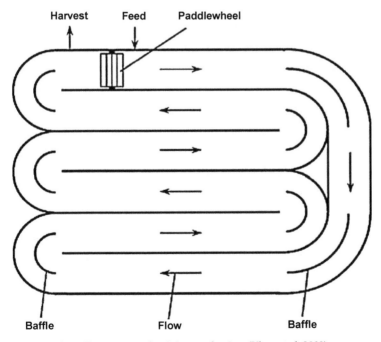

Fig. 1. Raceway pond: mixing mechanism (Khan et al. 2009).

However open ponds often suffer low productivity due to contamination, poor mixing, insufficient light, and inefficient use of CO_2 (Chisti 2008, Mata et al. 2010). Due to open system there are chances of loss of water through evaporation. Theoretical biomass yield does not match with practical yield levels (Shen et al. 2009) in open pond systems. Maintenance of monoculture of algae is very difficult, which is commonly decimated by predators and bacteria species (Schenk et al. 2008). Usage of commercial fertilizers can significantly increase production costs and leads to hike in algal derived biofuel production prices (US DOE 2010).

Engineering algal cultures to sustain extreme culture environments, such as high salinity (*Dunaliella*), high alkalinity (*Spirulina*) and high nutritional status (*Chlorella*) (Lee 2001) allowed a limited range of microalgae to be maintained as a monoculture in open ponds in long term operation. In addition acetate prevents the excessive bacteria growth (Lee 2001) but does not limit the growth algal biomass. Alternatively, cultivation of algae on wastewater in open pond/raceway ponds are economically viable compared with high costs of solo treatment of wastewater. This process simultaneously treats wastewater and produces biofuels as outcomes from the waste (Christenson and Sims 2011). Due to high nutritional quality of the waste water, maintaining a monoculture is very difficult as dominant algal strains will grow in the wastewater. However, prior treatment of wastewater for removing some compounds to make it selective for mono algal culture cultivation is a better alternative although it is difficult to perform practically. Open ponds can potentially be carbon limited due to mass transfer limitations; artificially bubbling carbon dioxide concentration may increase algal biomass which in turn may increase lipid content. But this approach has limitations due to less residence times of CO_2 bubbles (Malta et al. 2010, Christenson and Sims 2011). Optimizing CO_2 delivery through direct bubbling or other means remains an engineering challenge.

Closed Photobioreactors

Due to the limitations of open pond systems for algal biomass cultivation and the necessity of monoculture, that grow in mild and more controlled culture conditions has led to the development of enclosed photobioreactors (Lee 1986). Closed systems include tubular, flat plate and stirred tank photobioreactors. But photosynthetic efficiency in tubular reactors was higher than those in other systems due to uniform light penetration by the geometry of tubular photobioreactors.

Tubular Photobioreactors

Tubular photobioreactors are potentially attractive and most suitable type for outdoor mass culturing of algal biomass (Molina et al. 2001, Ugwu et al. 2008). Tubular photobioreactors consist of an array of straight/vertical (airlift/bubble column), horizontal coiled, or looped transparent tubes that are usually made of transparent plastic or glass as shown in Fig. 2 (Torzillo et al. 1986, Richmond 1987, Zittelli et al. 1999). An airlift device offers exchange of CO_2 and O_2 gases between the liquid medium and aeration by gas minimizes mechanical cell damage.

Surface to volume ratio plays vital role during algae culturing in tubular photobioreactors. Increasing tube diameter results in a decrease in the surface/volume ratio which in turn affects light penetration to the core of the reactor. As the algal culture grows and cell density increases, light intensity decreases at the core which leads to volumetric reduction in biomass per unit of incident light. Scale up of a tubular photobioreactor can be achieved by either increasing the length or diameter of the tube. Hence it influences the circulation by increase in residence time (Molina et al. 2000, Molina et al. 2001, Converti et al. 2006). Apart from above issues, increase in tube length allows accumulation of O_2 produced by photosynthesis which inhibits the photosynthesis. Simultaneously increase in tube length decreases CO_2 gradient leading to starvation of some algae for carbon (Ling et al. 2009) hence scale up is not economical. Another important disadvantage is large angle relative to the direction of sunlight, which causes a high fraction of incident energy to be reflected back and thus lost in terms of biomass growth purposes. These issues could be solved by engineering better mixing, aeration, handling large volumes of cultures, improving the penetration of light into the reactor (Ling et al. 2009).

Flat Plate Reactors

Flat plate reactors (FPR) are conceptually designed to make efficient use of sunlight; hence, narrow panels are usually built so as to attain high area to volume ratios (Carvalho et al. 2006); compared to tubular bioreactors their narrow U-turns may use less space than coiled tubes, and their wall thickness can be thinner than tubular bioreactor (Ling et al. 2009) Fig. 3. In this system algal culture circulate from an open gas exchange unit through several parallel panels placed horizontally. The great advantage of this system is overcoming the problem of oxygen-buildup, because of its provision of an open gas transfer unit. However, there are few disadvantages, algal cells adhesion to walls, high stress damage by aeration, systems are not amenable

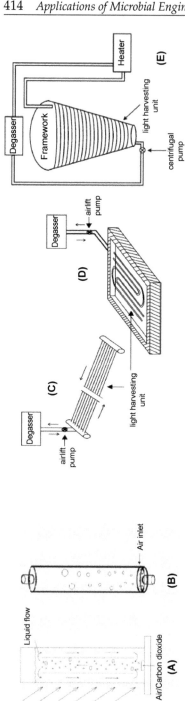

Fig. 2. Schematic representation of different tubular reactors A) Airlift B) Bubble column; horizontal tubular reactor with C) sets of tubes or a (D) loop tube E) helical tubular reactors (Carvalho et al. 2006).

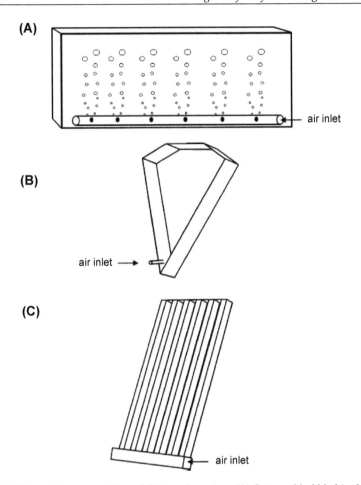

Fig. 3. Schematic representation of flat panel reactors: (A) flat panel bubbled in the bottom (B) V-shaped panel and (C) alveolar panel (Carvalho et al. 2006).

to sterilization and incompatible with off the shelf fermentation equipment (Ugwu et al. 2008) and also temperature control is a problem and sprinkler systems are often used for evaporative cooling.

Stirred Tank Reactors (STR)

STRs were originally proposed to grow microalgae photoautotrophically using artificial light sources or sunlight. STR was built (Pohl et al. 1988) with stainless steel and illuminated internally by fluorescent lamps placed inside a narrow glass (or Plexiglass) tube CO_2-enriched air bubbled at the bottom, through a V-shaped (i.e., low shear stress) stirrer. Such a system operated batch-, semicontinuous- and continuous-wise (Fig. 4). The main advantage

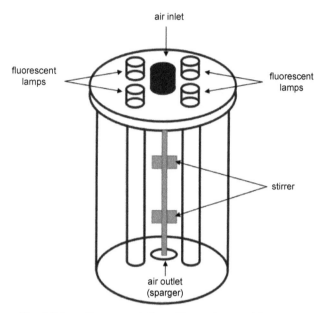

air inlet

fluorescent lamps

fluorescent lamps

stirrer

air outlet (sparger)

Fig. 4. Schematic representation of fermenter-type bioreactor.

of these types of systems is of course accurate control of processing parameters, including light, coupled with vast experience available from food and pharmaceutical industries in terms of scale-up. Therefore, STR would certainly become a competitive alternative for microalgae mass production. This system was recently used to establish a deterministic model to predict microalgal productivity in order to establish practical feasibility for large scale application; however, subsequent implementation has not yet been reported (Carvalho et al. 2006).

All these issues related to reactors need to be solved or engineered in order to improve the productivity. Recently there have been some efforts for improving the productivity three times higher than that of the control photobioreactor utilizing spatial light dilution thus indicating that photobioreactor engineering is promising (Dan et al. 2011).

Algae Harvesting/Dewatering of Microalgal Cultures

A large amount of studies have been completed and are still on going on harvesting and recovery techniques of microalgal cells in dilute suspensions. Due to the differences in the biology of microalgal species it is difficult to name one method as superior. The choice of harvesting technique depends on the species of microalgae and the final product desired. Desired microalgal properties that simplify harvesting are large cell size, high specific gravity compared to medium and reliable auto flocculation. In addition, optimum harvesting method chosen for a particular microalgal species should have minimum energy requirements and should be as economical as possible. Few important harvesting techniques are summarized in Table 1. Centrifugation is seen as most efficient recovery technique; however energy and capital costs associated are unappealing.

Flocculation is an efficient technique, which is currently used in industries for harvesting and dewatering microalgae. Freshwater microalgae flocculation has been immensely studied throughout the years. However, there is a dearth of information and comparative studies for marine microalgae. With growing concern about freshwater resources in many parts of the world, further research on the processing of marine microalgae would be very beneficial.

However, separating algae from water remains a major hurdle to industrial scale processing partly because of the small size of the algal cells. In addition, relatively dilute cultures of 200–600 mg/l are common (Nyomi et al. 2010), and require large volumes of water to be processed. Recovery has been estimated to contribute 20–30% of the total cost of biomass production (Christenson and Sims 2011). The initial harvesting step not only affects cost of dewatering but also affects any later downstream processing. Lowering the cost of harvesting and dewatering algae in a way to control the cost of algal biofuels remains a challenge.

Drying

Drying is one of the general pre-processing steps required to achieve high biomass concentrations. The harvested biomass slurry (5–15% dry solid) must be processed rapidly to prevent spoiling of biomass due to hot environment. Removal of 1 kg H_2O by drying requires more than 800 kcal of energy. Generally two types of methods are adopted for drying, Indirect and direct heating. In indirect heating rotating disks to accelerate heat exchange are used and direct heat uses an open flame to create steam. This technique is efficient but energy intensive. Another method is flash drying for rapid removal of moisture by spraying or injecting a mixture of dried and undried material in a hot gas. In fluid Bed method of drying biomass must have low moisture content for better drying efficiencies. Microwave drying is another method in which energy is delivered electromagnetically, rather than as heat. This method can reduce drying time but has a potential to dry unevenly and it is also energy intensive. Conventional sun drying is one of the oldest methods that work well in low humid climates. Sun drying is the cheapest dehydration method; but the main disadvantages include long drying times, requirement for large drying surfaces/area, and the risk of material loss (Desmorieux and Decaen 2006, Brennan and Owende 2010).

Table 1. Summary of the performances of different harvesting and dewatering techniques (Nyomi et al. 2010).

Dewatering process	Highest possible yield	Highest possible water removal concentration factor	Energy usage	Reliability	Limitations
Flocculation	Flocculation 95% removal of microalgae	200–800, efficiencies of 80%	Low energy requirement for slow mixing; varies largely	Very good	Flocculants can be expensive, especially at high dose, and may cause contamination issues
Centrifugation	12% TSS	Self-cleaning disk stack centrifuge, 120	Very high, 8 kWh/m^3	Very good	High energy input
Gravity sedimentation	0.5%–1.5% TSSc	Lamella separator, 16g	Low lamella separator, 0.1 kWh/m^3 g	Poor	Process is slow
Filtration and screening (natural filtration)	1%–6% TSS	15–60	Low vibrating screen filter, 0.4 kWh/m^3 g	Good	Filters and screens need to be replaced periodically
Filtration and screening (pressure filtration)	5%–27%	50–245	Moderate chamber filter press, 0.88 kWhg	Very Good	Filters and screens need to be replaced periodically
Tangential flow filtration	70%–89% microalgal recovery	5–40	High, 2.06 kWh/m^3	Good to very good	The high energy input; filters needs to be replaced periodically Good to very good
Flocculation–flotation	Dissolved air flotation, 1%–6% TSSc, dispersed air flotation, 90% microalgae removal	N/A	High, dissolved air flotation, 10–20 kWh/m3	Good to very good	Electrodes need to be replaced periodically
Electrocoagulation	95% microalgae removal 99.5% TSS	N/A	Medium-high, 0.8–1.5 kWh/m3 k	Very good	The electrodes need to be replaced periodically
Electroflotation	3%–5% TSSc	300–600l	Very high, N/A	Very good	The electrodes need to be replaced periodically
Electrolytic flocculation	90% microalgae removal	N/A	Low-medium, 0.331 kWh/m3	Very good	The electrodes need to be replaced periodically

Lipids or Oil Extraction

Cell disruption

Before extraction of lipids from algae, cell disruption is often required. The cell wall is the main barrier for extraction process (Heger 2009). Most cell disruption methods applicable to microalgae have been adapted from applications on intracellular non-photosynthetic byproducts (Middelberg 1994). Cell disruption methods that have been used successfully include high-pressure homogenisers, autoclaving, and the addition of chemical based hydrochloric acid, sodium hydroxide, or alkaline lysis (Liam and Owende 2009).

Methods for extraction of lipid

Numerous methods have been applied for extraction of lipids from microalgae; but most common methods are expeller/oil press, liquid–liquid extraction (solvent extraction), supercritical fluid extraction (SFE) and ultrasound techniques (Harun et al. 2010).

Expeller/oil pressing is a mechanical method for extracting oil from raw materials such as nuts and seeds. Press uses high pressure to squeeze and break cells. For effective extraction algae first needs to be dried. Although this method can recover 75% of oil and no special skill is required, it was reported less effective due to comparatively longer extraction time (Harun et al. 2010). Solvent extraction proved to be successful in order to extract lipids from microalgae. In this approach, organic solvents, such as benzene, cyclo-hexane, hexane, acetone, chloroform are added to algae paste. Solvents destroy the algal cell wall, and extract oil from aqueous medium because of their higher solubility in organic solvents than water. Solvent extract can then be subjected to distillation process to separate oil from solvent. The latter can be recovered for further use. Hexane is reported to be the most efficient solvent for extraction based on its highest extraction capability and low cost (Harun et al. 2010). Supercritical extraction makes use of high pressures and temperatures to rupture the cells. This particular method of extraction has proved to be extremely time-efficient and is commonly employed (Harun et al. 2010). Another promising method to be used in extraction of microalgae is the application of ultrasound. This method exposes algae to a high intensity ultrasonic wave, which creates tiny cavitation bubbles around cells. Collapse of bubbles emits shock waves, shattering the cell wall and releasing the desired compounds into the solution. Although the extraction of oil from microalgae using ultrasound is already in extensive use at laboratory scale, sufficient information on the feasibility or cost for a commercial-scale operation is unavailable. This approach seems to have a high potential, but more research is needed. In a recent comparative study on autoclaving, bead-beating, microwaves, sonication methods; that microwave method is more effective and released more lipids and it is simple to operate (Lee et al. 2010).

Engineering and Technology Challenges

There are three primary challenges to cost-effective algae production:

1. Concentrating the algal culture: Before oil extraction from each cell they should follow an energy-intensive de-watering stage.
2. Breaking algal cell wall: The challenge is to maximize the number of algae cells to be cracked with the smallest amount of energy.
3. Harvesting the by-products: In order to make algae cultivation cheap and economical and to achieve the best possible energy balance.

Recently new technologies like single step extraction, live Extraction (Origin Oil 2012), enzyme-assisted aqueous extraction (Hong et al. 2012) have been developed for effective extraction of lipids and by products (hydrogen) from algae (Origin Oil 2012). Immobilization of algal cells also offers advantages for reduced downstream processing and operation costs for effective nutritional removal from medium (Zeng et al. 2012a, Zeng et al. 2012b).

Transesterification/Biodiesel Production

After lipids extraction, resulting microalgal oil can be converted into biodiesel or FAME (Fatty Acid Methyl Esters) through a process called transesterification (Fig. 5).

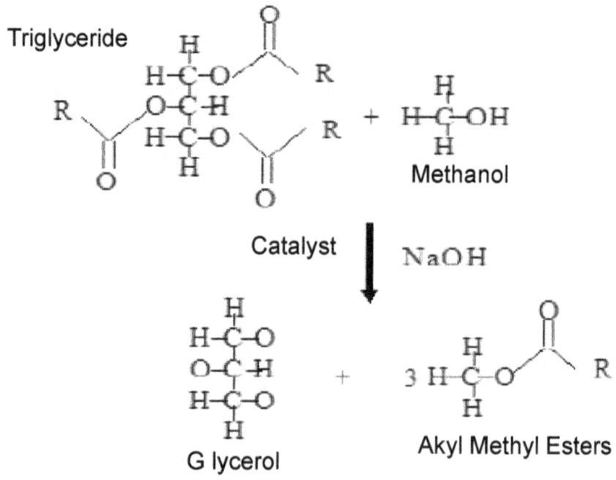

Fig. 5. Schematic representation of the transesterification.

In this reaction triglycerides transformed into fatty acid alkyl esters, in the presence of an alcohol, such as methanol or ethanol, and a catalyst, such as an alkali or acid, with glycerol as a byproduct. High levels of algae FFAs (20–50% of total fatty acids) under alkaline catalyst-based transesterification forms soap. Hence it is not suitable for making fuel-grade biodiesel as per the ASTM D6751 standards (Kosaric and Velikonja 1995, Mansour et al. 2005). Acid, enzyme-based catalysts are recommended for making algal biodiesel because of they do not form soap during the transesterification. Lipase enzyme from different origins (*Pseudomonas fluorescens, Mucor javanicus, Burkholderia cepacia, Peniciliumcyclopium, Thermomyces lanuginose, Bacillus subtilis, Enterobacter aerogenes, Burkholderia Cepacia, Candida* sp.) are used for the transesterification reaction (Nevena et al. 2011, Swetha et al. 2003).

However, these two types of the catalysts have inherent limitations. For example, the acid-based transesterification rates are very low, while the enzymes are expensive and unable to complete the transesterification that meet the ASTM standard (Mutanda et al. 2011). Recently a new method was introduced for transesterifcation which is termed as *in situ* transesterification, which does not require the extraction and purification of the oil prior to FAME synthesis (Michael et al. 2007). It may be applicable to algal culture because of their cell wall nature. There are also some efforts for biodiesel production in a single step by combining extraction and conversion process to produce biodiesel in high yield from mixed microalgae culture (Bradley et al. 2011).

Apart from the above issues there is a key attention on the genetic engineering and metabolic engineering of algae for improved production of lipids (Radakovits et al. 2010, Yu et al. 2011). Due to antibiotic resistant marine algae, maintenance of algal monoculture makes genetic engineering and metabolic engineering prospects tedious. But a lot of scope and challenges exist which will enable an improvement in process technologies that will facilitate the widespread commercialization of algal biofuels to fuel the future.

Conclusions

Promise of algae based biofuels creates motivation for research and industrial investment. However, reduction in capital costs, increases in yield, and increase in product value, all in combination, are critical for the extensive application of algae as a feedstock for biofuel production. Furthermore, demonstrating the technology at different scales of operation and integration is necessary for commercialization. Based on this, focusing on high value products and byproducts is crucial for commercial success of the process. In order to make the micro algae base biofuel generation engineering of various aspects during process development is crucial.

As domain knowledge improves, it can be expected that it will enable an improvement in process technologies that will facilitate the widespread commercialization of algal biofuels to fuel the future.

Acknowledgements

Dr. Potumarthi is thankful to Monash University and Department of Chemical Engineering for supporting present work.

References

Bradley, D. Wahlen, Robert M. Willis and Lance C. Seefeldt. 2011. Biodiesel production by simultaneous extraction and conversion of total lipids from microalgae, cyanobacteria, and wild mixed-cultures. Bioresour. Technol. 102: 2724–2730.

Brennan, L. and P. Owende. 2010. Biofuels from microalgae—A review of technologies for production, processing, and extractions of biofuels and co-products. Renew. Sust. Energ. Rev. 14: 557–577.

Carvalho, A. P., L. A. Meireles and F. X. Malcata. 2006. Microalgal reactors: a review of enclosed system designs and performances. Biotechnol. Prog. 22: 1490–1506.

Chisti, Y. 2008. Biodiesel from microalgae beats bioethanol. Trends. Biotechnol. 26: 126–31.

Christenson, L. and R. Sims. 2011. Production and harvesting of microalgae for wastewater treatment, biofuels, and bioproducts. Biotechnol. Adv. 29: 686–702.

Converti, A., A. Lodi, D. Borghi and C. Solisio. 2006. Cultivation of Spirulina platensis in a combined airlift-tubular reactor system. Biochem. Eng. J. 32: 13–18.

Crutzen, P. J., A. R. Mosier, K. A. Smith and W. Winiwarter. 2007. N_2O release from agro-biofuel production negates global warming reduction by replacing fossil fuels. Atmos. Chem. Phys. 7: 1191–11205.

Dan, D., J. Muhs, B. Wood and R. Sims. 2011. Design and performance of a solar photobioreactor utilizing spatial light dilution. J. Sol. Energ-T Asme. 133: 1067–1076.

Demirbas, A. 2006. Oily products from mosses and algae via pyrolysis. Energ. Source. Part A. 28: 933–940.

Desmorieux, H. and N. Decaen. 2006. Convective drying of spirulina in thin layer. J. Food. Eng. 66: 497–503.

Falkowski, P. G. and J. A. Raven. 1997. Aquatic Photosynthesis. Blackwell Science, Malden, MA, USA.

Goldemberg, J. 2000. World Energy Assessment, Preface. United Nations Development Programme, New York, NY, USA.

Harun, R., M. Singh, G. M. Forde and M. K. Danquah. 2010. Bioprocess engineering of microalgae to produce a variety of consumer products. Renew. Sust. Energ. Rev. 14: 1037–1047.

Heger, M. 2009. New processing scheme for algae biofuels. MIT Technology review. Available on http://www.technologyreview.com/news/413325/a-new-processing-scheme-for-algae-biofuels/

Hill, J., E. Nelson, D. Tilman, S. Polasky and D. Tiffany. 2006. Environmental economic and energetic costs and benefits of biodiesel and ethanol biofuels. Proc. Natl. Acad. Sci. USA. 103: 11206–11210.

Hong, L. K., Q. Zhang and W. Cong. 2012. Enzyme-assisted aqueous extraction of lipid from microalgae. J. Agric. Food Chem. 10.1021/jf302836v (In press).

Hu, Q., M. Sommerfeld, E. Jarvis, M. Ghirardi, M. Posewitz, M. Seibert and A. Darzins. 2008. Microalgal triacylglycerols as feedstocks for biofuel production: perspectives and advances. Plant. J. 54: 621–639.

Huntley, M. E. and D. G. Redalje. 2007. CO_2 mitigation and renewable oil from photosynthetic microbes: a new appraisal. Mitig. Adapt. Strat. Global Change. 12: 573–608.

[IPCC] Intergovernmental Panel on Climate Change 1991. Climate change: the IPCC 1990 and 1992 assessments.Washington DC: Island Press. Available on http://www.ipcc.ch/ publications_and_data/publications_and_data_reports.shtml#1.

Jay, J. C. and G. R. Timilsina. 2010. Advanced Biofuel Technologies: Status and Barriers. Policy Research Working Paper http://www-wds.worldbank.org/external/default/ WDSContentServer/WDSP/IB/2010/09/01/000158349_20100901162217/Rendered/ PDF/WPS5411.pdf.

Kenneth, T., Gillingham, S. J. Smith and D. Ronald. 2007. Sands Impact of bioenergy crops in a carbon dioxide constrained world: an application of the MiniCAM energy-agriculture and land use model. Mitig. Adapt. Strat. Global Change. 13: 675–701.

Kosaric, N. and J. Velikonja. 1995. Liquid and gaseous fuels from biotechnology: Challenge and opportunities. FEMS. Microbiol. Rev. 16: 111–142.

Lee, J. Y., C. Yoo, S. Y. Jun, C. Y. Ahn and H. M. Oh. 2010. Comparison of several methods for effective lipid extraction from microalgae. Bioresour. Technol. 101: S75–S77.

Lee, Y. K. 1986. Enclosed bioreactors for the mass cultivation of photosynthetic microorganisms: the future trend. Trends. Biotechnol. 4: 186–189.

Lee, Y. K. 2001. Microalgal mass culture systems and methods: Their limitation and potential. J. Appl. Phycol. 13: 307–315.

Lewis, N. S. and D. G. Nocera. 2006. Powering the planet: chemical challenges in solar energy utilization. Proc. Natl. Acad. Sci. USA. 103: 15729–15735.

Ling, X., P. J. Weathers, X. R. Xiong and C. Z. Liu. 2009. Microalgal bioreactors: Challenges and opportunities. Eng. Life Sci. 9: 178–189.

Mansour, M. P., D. M. F. Frampton, P. D. Nichols, J. K. Volkman and S. I. Blackburn. 2005. Lipid and fatty acid yield of nine stationary-phase microalgae: Applications and unusual C24-C28 polyunsaturated fatty acids. J. Appl. Phycol. 17: 287–300.

Mata, T. M., A. A. Martins and N. S. Caetano. 2010. Microalgae for biodiesel production and other applications: a review. Renew. Sust. Energ. Rev. 14: 217–32.

Michael, J. H., K. M. Scott, T. A. Foglia and W. N. Marmer. 2007. The general applicability of in situ transesterification for the production of fatty acid esters from a variety of feedstocks. J. Am. Oil. Chem. Soc. 84: 963–970.

Middelberg, A. P. J. 1994. The release of intracellular bioproducts. *In*: G. Subramanian [ed.]. Bioseparation and bioprocessing: a handbook. Wiley-Verlag GmbH, Newyork. pp. 131–164.

Molina, E., F. G. Fernandez Acien and Y. Chisti. 2001. Tubular photobioreactor design for algal cultures. J. Biotechnol. 92: 113–131.

Molina, E., F. G. Acien Fernandez, F. G. Camacho, F. C. Rubio and Y. Chisti. 2000. Scale-up of tubular photobioreactors. J. Appl. Phycol. 12: 355–368.

Mutanda, T., D. Ramesh and S. Karthikeyan. 2011. Bioprospecting for hyper-lipid producing microalgal strains for sustainable biofuel production. Bioresour. Technol. 102: 57–70.

Nevena, L., K. J. Zorica and D. Bezbradica. 2011. Biodiesel Fuel Production by Enzymatic Transesterification of Oils: Recent Trends, Challenges and Future Perspectives. *In*: Maximino Manzanera [eds.]. Alternative Fuel. In Tech Europe, Croatia. pp. 47–72.

Nyomi, U., Q. Ying, M. K. Danquah, G. M. Forde and A. Hoadley. 2010. Dewatering of microalgal cultures: A major bottleneck to algae-based fuels. J. Renew. Sustain. Energ. 2: 012701 doi: 10.1063/1.3294480.

Origin Oil. Next Generation Technology 2012. Available on http://www.originoil.com/ technology/overview.html.

Pohl, P., M. Kohlhase and M. Martin. 1988. Photobioreactors for the axenic mass cultivation of microalgae. *In:* T. Stadler, J. Mollion, M. C. Verdus, Y. Karamanos, H. Morvan and D. Christiaen [eds.]. In Algal Biotechnology. Elsevier, New York, USA. pp. 209–218.

Prasad, S., A. Singh and H. C. Joshi. 2007. Ethanol as an alternative fuel from agricultural, industrial and urban residues. Resour. Conserv. Recy. 50: 1–39.

Pulz, O. 2001. Photobioreactors: production systems for phototrophic microorganisms. Appl. Microbiol. Biot. 57: 287–93.

Radakovits, R., R. E. Jinkerson, A. Darzinsandre and M. C. Posewitz. 2010. Genetic Engineering of Algae for Enhanced Biofuel Production. Eukaryot. Cell. 9: 486–501.

Rawat, I., R. R. Kumar, T. Mutanda and F. Bux. 2011. Dual role of microalgae: Phycoremediation of domestic wastewater and biomass production for sustainable biofuels production. Appl. Energ. 88: 3411–3424.

[RCEP] Royal Commission on Environmental Pollution Sponsorship. UK Royal Commission on Environmental Protection. Energy—the changing climate. 2000. *In:* 22nd Report of the RCEP.

Richmond, A. 1987. The challenge confronting industrial microagriculture: high photosynthetic efficiency in large-scale reactors. Hydrobiologia. 151/152: 117–121.

Richmond, A. 2000. Microalgal biotechnology at the turn of the millennium: a personal view. J. Appl. Phycol. 12: 441–451.

Schenk, P. M., S. R. Thomas-Hall, E. Stephens, U. C. Marx, J. H. Mussgnug, C. Posten, O. Kruse and B. Hankamer. 2008. Second generation biofuels: high-efficiency microalgae for biodiesel production. Bioenerg. Res. 1: 20–43.

Scott, S. A., M. P. Davey, J. S. Dennis, I. Horst, C. J. Howe, D. J. Lea-Smith and A. G. Smith. 2010. Biodiesel from algae: challenges and prospects. Curr. Opin. Biotechnol. 21: 277–286.

Shakeel, A. K., Rashmi, M. Z. Hussain, S. Prasad and U. C. Banerjee. 2009. Prospects of biodiesel production from microalgae in India. Renew. Sust. Energ. Rev. 13: 2361–2372.

Shen, Y., W. Yuan, Z. J., Q. Wu and E. Mao. 2009. Microalgae mass production methods. Trans. ASABE. 52: 1275–1287.

Singh, A., D. Pant, N. E. Korres, A. S. Nizami, S. Prasad and J. D. Murphy. 2010. Key issues in life cycle assessment of ethanol production from lignocellulosic biomass: challenges and perspectives. Bioresour Technol. 101: 5003–5012.

Swetha, S., S. sharma and M. N. Gupta. 2003. Enzyme transesterification for biodiesel production. Ind. Biochem. Biophy. 40: 392–399.

Torzillo, G., B. Puspararaj, F. Bocci, W. Balloni, R. Materassi and G. Florenzano. 1986. Production of Spirulina biomass in closed photobioreactors. Biomass. 11: 61–74.

Ugwu, C. U., H. Aoyagi and H. Uchiyama. 2008. Photobioreactors for mass cultivation of algae. Bioresour. Technol. 99: 4021–4028.

[U.S. DOE] USA Department of Energy. National algal biofuels technology roadmap. 2010. Report no.: DOE/EE-0332. U.S. Department of Energy, Office of Energy Efficiency and Renewable Energy, Biomass Program.

Weissman, J. C., R. P. Goebe and J. R. Benemann. 1988. Photobioreactor design: mixing, carbon utilization, and oxygen accumulation. Biotechnol. Bioeng. 31: 336–44.

Yang, J., M. Xu, X. Zhang, Q. Hu, M. Sommerfeld and Y. Chen. 2011. Life-cycle analysis on biodiesel production from microalgae: water footprint and nutrients balance. Bioresour. Technol. 102: 159–65.

Yanqun, L., M. Horsman, N. Wu, C. Q. Lan and N. D. Calero. 2008. Biofuels from Microalgae. Biotechnol. Prog. 24: 815–820.

Yu, W. L., W. Ansari, N. G. Schoepp, M. J. Hannon, S. P. Mayfield and M. D. Burkart. 2011. Modifications of the metabolic pathways of lipid and triacylglycerol production in microalgae. Microb. Cell. Fact. 2, 10: 91.

Zeng, X., M. Danquah, R. Potumarthi, J. Cao, X. D. Chen and Y. Lu. 2012a. Characterization of sodium cellulose sulphate/poly-dimethyldiallyl-ammonium chloride biological capsules for immobilized cultivation of microalgae. J. Chem. Tech. Biotech. In press. DOI: 10.1002/jctb.3869.

Zeng, X., M. Danquah, C. Zheng, R. Potumarthi, X. D. Chen and Y. Lu. 2012b. NaCS-PDMDAAC immobilized autotrophic cultivation of *Chlorella* sp. for wastewater nitrogen and phosphate removal. Chem. Eng. J. 187: 185–192.

Zittelli, G. C., F. Lavista, A. Bastianini, L. Rodolfi, M. Vincenzini and M. R. Tredici. 1999. Production of eicosapentaenoic acid by *Nannochloropsis* sp. cultures in outdoor tubular photobioreactors. J. Biotechnol. 70: 299–312.

16

The Forefront of Low-cost and High-volume Open Microalgae Biofuel Production

Navid R. Moheimani,[1],* Mark P. McHenry[2] and Karne de Boer[3]

ABSTRACT

This review summarises the status of seven selected companies at the cutting edge of liquid fuel production from microalgae cultured in open systems. Open microalgae production technologies and companies have fostered commercial collaborations and technological progress towards producing low-cost, high-volume microalgae biofuels, although significant large scale-demonstration is required. This work provides a concise summary of each respective company's technology, background, and where available, productivity and economics. Based on the review findings, an unprecedented level of collaboration between entities

---

[1] Algae R&D Centre, School of Biological Sciences & Biotechnology, Murdoch University, Western Australia.
   Email: n.moheimani@murdoch.edu.au
[2] School of Engineering and Energy, Murdoch University, Western Australia.
   Email: mpmchenry@gmail.com
[3] Regenerate Industries, Maddington, Perth, Western Australia.
   Email: karne@regenerateindustries.com
* Corresponding author

is needed across the production chain to achieve cost-competitive production systems to displace current liquid fuels. In addition to microalgae mass production and culturing challenges, there are also notable challenges surrounding harvesting/dewatering, extraction, and conversion to a final product. The microalgae biofuel industry will thus require a 'cross-pollination' of industries historically unrelated to biological production systems, and there is likely to be a consolidation of current microalgal biofuel capability and expertise.

Introduction

The two fundamental microalgae biomass production technologies aiming to supply the global demand for renewable biofuels are known as 'open' ponds and 'closed' bioreactor/fermentor systems (Borowitzka 1999, Ugwu et al. 2008, Vasudevan and Briggs 2008, McHenry 2010). Whilst many microalgae species must be cultured in highly controlled conditions, species that exhibit a selective advantage in particular environments are able to be cultured in open-air systems such as ponds (Borowitzka 1999). The majority of microalgae biofuel companies are designing commercial production systems based on open cultivation of particular species/cultivars due to the advantages of relatively lower capital and production costs of open systems compared to the currently available closed systems, and a historically proven record of production (Borowitzka 1999, Spolaore et al. 2006). At present, the most common open system designs include the use of natural water bodies, constructed ponds of a variety of shapes, depths, and sizes, and also tanks (Ugwu et al. 2008). Each system exhibits unique advantages and disadvantages. For example, while commonly used raceway ponds improve yields and reliability relative to simple shallow ponds, they can be an expensive design to construct and operate (Borowitzka 1992, Kunjapur and Eldridge 2010).

There are numerous technical components to this analysis, including biological, engineering, economic, and environmental challenges. While open systems may exhibit relatively low capital requirements compared to closed systems (Huntley and Redalje 2006, Chisti 2007), the microalgae yield is often decreased with relatively poor water body mixing, contamination, and microalgal predatory organism issues (Huntley and Redalje 2006, Chisti 2007, Ugwu et al. 2008). Further production limitations can include low solar resource utilisation, low atmospheric CO_2 diffusion, and high land and water demands (Ugwu et al. 2008, Clarens et al. 2010, Borines et al. 2011b, McHenry 2012, Moheimani et al. in press), and extreme events such as high rainfalls or low temperatures can lead to total pond culture losses (Moheimani and Borowitzka 2006, Amin 2009, Borowitzka and Moheimani 2010). Therefore, industrial-scale microalgae developments will likely occur in dry to arid regions with good solar resources on land unsuitable for conventional agriculture (Gross 2007, Hankamer et al. 2007, Borowitzka and Moheimani 2010, McHenry 2010). However, the use of arid regions may present problems related to suitable water access and availability to replace the extremely large evaporative losses from open ponds in such areas (Chisti 2007, Clarens et al. 2010). Further water-related challenges involve the resultant wastewater treatment. Hence, significant collaborative research and development is required to reduce microalgae production demand for freshwater, output effluent wastewater issues, and also relatively high energy demands, which all require significant cross-disciplinary investment (Wyman and Goodman 1993, Xiong et al. 2008, Charcosset 2009, Borowitzka and Moheimani 2010, Clarens et al. 2010).

Highly productive microalgal industrial developments in non-agricultural areas may reduce arable land competition between conventional biofuel and agricultural food production (Sheehan et al. 1998, Huntley and Redalje 2006, Chisti 2007, Gross 2007, Hankamer et al. 2007, Cantrell et al. 2008, Borowitzka and Moheimani 2010). However, cost-effective microalgal biofuel production will reflect conventional biology-based primary industries, and remain fundamentally tied to total factor productivity of biological systems. Thus, industrial-scale commercial production will be dependent on microalgal biology, the climate, and inputs costs such as energy, water, land, capital, labour, transport, and nutrients (Borowitzka 1999, Kunjapur and Eldridge 2010). As such, productive regions and production systems will be selected according to technological, economic, infrastructure, and environmental conditions suitable for particular species and strains (Borowitzka 1992, Kunjapur and Eldridge 2010). While microalgal biomass production in open ponds may seem a relatively trivial technical challenge, the complexity of downstream biofuel processing imposes additional production costs that will make it difficult for the final fuel to compete with existing energy supply prices (Lee 2001, Borowitzka and Moheimani 2010). For microalgae producers to achieve and sustain a competitive advantage will require capital, culturing operations, harvesting, drying and processing equipment costs to all remain low, whilst achieving a sustained high-level productivity over the entire year (Borowitzka 1999).

This review's primary focus is the technical research and development challenge to open pond production, specifically in terms of existing partnerships and capabilities of each company, and the development needs of the industry as a whole. This review companies primarily using open systems, (in reverse-alphabetical order): Synthetic Genomics Inc.; Seambiotic; Sapphire Energy Inc.; Muradel Pty. Ltd.; General Atomics; Aquaflow Bionomic Corporation, and; Aurora Algae Inc. Due to the limited peer-reviewed literature available on the forefront of commercial microalgal production, the authors contacted each company to update the publically available information presented in this review. The authors have made every attempt to provide only facts and commentary, without undue criticism of strategies or opinions on commercialisation potential. However, this review aims to collate and clarify the often rudimentary unqualified information publically available regarding microalgae biofuel production developments, and also facilitate sustained progress in the expansion of the microalgae industrial endeavour.

Synthetic Genomics Inc. www.syntheticgenomics.com

Synthetic Genomics Inc. (SGI) is a privately owned company based in La Jolla, California, USA, with co-founders Craig Venter (as the Chairman, CEO, and Co-Chief Scientific Officer), and Hamilton O. Smith (Co-Chief Scientific Officer). The company focuses on commercialising genomics and policy research capability at the J. Craig Venter Institute. SGI development scope encompasses biochemistry, bioinformatics, climate change, environmental genomics, genome engineering, microbiology, plant genomics, and synthetic biology.

SGI partnerships encompass a range of sectors and companies, including the Asiatic Centre for Genome Technology, Biotechonomy LLC, BP, Draper Fisher Juvetson, Exxon Mobil, Meteor Group, and Plenus. In 2009, Exxon Mobil agreed to provide up to USD300 million to SGI for research and development and production of biofuels from photosynthetic microalgae on a milestone basis. SGI's broad scope includes developing metabolic pathways for producing biochemicals and biofuels from a number of feedstocks, microbial methods to increase conversion and recovery of subsurface hydrocarbons, and the development of novel plant feedstocks and microbial agents for agriculture.

SGI's Technology Review

SGI have planned to develop a 4 ha test site incorporating both ponds and photobioreactors, although are not expected to become a major fuel producer in their own right. SGI's wide research and development scope is focussing on primary and enabling technology, including harvesting microalgae, increasing strain lipid contents, agricultural organism GM for disease resistance, vaccine development, and gene synthesis. SGI is well known for being the first to synthesise a genome using oligomers, and using polymerase cycle assembly, and SGI have developed cultivation and monitoring technology to assess as yet undiscovered wild microorganisms with novel genes for commercial biofuel production. SGI aim to produce commercial products from these new organisms, and are developing methods to engineer organisms to increase productivity.

SGI Summary

Currently there are no publicised production claims for SGI relating to large-scale production facilities. Synthetic Genomics' commercial scope, technology, and capability remain at relatively primary stages of development, and in terms of microorganisms, or modification, seem to be resisting convergence into technology types, or specific commercial markets. At this stage it is unclear if SGI will specialise into modification for fuel producing organisms, or organisms that produce refining precursors. A large uncertainty for the microalgae industry at present is the regulation and standards required for GM technology applied at the industrial scale. Nonetheless, the staged Exxon Mobil funding will add enormous capability to SGI, and the considerable intellectual property available to SGI promise to generate much innovation,[1] and progress towards developing low-cost, high-volume, sustainable biofuels.

Seambiotic www.seambiotic.com

Seambiotic was the first company to utilise flue gas from coal burning power stations for microalgae cultivation. In 2003, Seambiotic was founded as a private company in Ashkelon, Israel, with an objective to grow and process marine microalgae to produce biofuel and food additives (primarily omega-3 fatty acids). Seambiotic with their major partner, Nature Beta Technologies Ltd. (NBT), a subsidiary of Nikken Sohonsha, currently operate a 10 ha plant in Eilat, Israel, culturing a low-lipid, high-carbohydrate *Dunaliella* strain for beta-carotene production in salty water pumped from the Red Sea. NBT is an Eilat-based nutraceutical-grade microalgae company producing since 1988, and Seambiotic's Chief Advisor, Ami Ben-Amotz, a scientific director of NBT, initiated the biofuel focus. Seambiotic have several partnerships, including with Seattle-based Inventure Chemical. Seambiotic supply microalgae paste for Inventure's thermochemical process where catalysts convert the paste into biodiesel, a sugar solution, and a protein solution. Seambiotic and Rosetta Green, a company specialising in unique gene identification in plants, announced collaborations in 2010 for strain improvement aimed at improving contamination resistance and oil yield. Seambiotic also have a non-commercial computational collaboration with the US National Aeronautics and Space Administration (NASA) to optimise microalgae-based aviation biofuel feedstock. Seambiotic signed a licence

---

[1]WO2010068821, WO2009048971, WO2009076559, WO2009064910, O2009144192, and WO2009140701. In addition to SGI intellectual property, the company has access to the considerable intellectual property and capabilities of the hundreds of researchers at the J. Craig Venter Institute.

agreement and a joint venture with Yantai Hairong Electricity Technology Ltd. and Penglai Weiyuan Science & Trading Ltd for construction of a 10 ha, USD10 million facility in Penglai power station China, owned by China Guodian Corporation, China's fifth largest power company.

Seambiotic Review

Seambiotic's research and development pilot study over the last five years has occurred at the Ashkelon Rutenberg coal-fired flue gas desulphurisation (FGD) power station of the Israeli Electric Corporation in Ashkelon. The following alga have been grown in Seambiotic's 0.1 ha open raceway pond pilot plant: *Amphora* sp.; *Dunaliella* sp.; *Chlorococcum* sp.; *Nannochloris* sp.; *Nannochloropsis* sp.; *Navicula* sp.; *Phaeodactylum tricornutum*; *Skeletonema* sp.; *Tetraselmis* sp. (etc.). Seambiotic's impressive productivity results in their 0.15 cm deep raceway ponds include both seasonal and annual studies over several years (2005–2007, and ongoing). Average yearly productivities of 20 g m^{-2} day^{-1} (~700 t ha^{-1} yr^{-1}) using *Nannochloropsis* 'fed' with FGD inputs, in addition to conventional inputs. In contrast to many microalgae companies, Seambiotic's production parameters and results are often publically available (see Table 1).

Seambiotic were the first microalgae company to successfully grow microalgae outdoors at a medium scale for several years, and the company posses gas transfer, cleaning, and various other production control technologies.[2] In terms of economics, Seambiotic's own microalgal oil cost estimates are around USD50 L^{-1}, based on 25% lipid content and 0.8 kg L^{-1} of dry microalgae. As the company uniquely holds a high resolution commercial understanding of current microalgae production system costs, these costs estimates have high reliability. Seambiotic aim to produce microalgal oil at USD0.34 kg^{-1}, equivalent to approximately USD2.25 L^{-1} of biodiesel assuming a 12% microalgae lipid content. The authors note that this estimate is simply based on the *Dunaliella* facility costs in Eliat.

Table 1. Growth conditions and parameters for the period March 2005 to November 2006.

Species	*Skeletonema costatum*	*Nannochloropsis*
Microalgal Biomass (g L^{-1})	0.5–1.5	0.5–1.0
Turbine seawater at max, (m^3 h^{-1}, °C)	450 000, 12–35°C	450 000, 12–35°C
Flue Gas after FGD at max, (tCO$_2$ h^{-1})	431	431

---

[2]WO2008107896 Method for growing photosynthetic organisms,

Seambiotic Summary

Seambiotic's successful history of commercial microalgae production and demonstration of the full production chain is unique among microalgae companies focussing on biofuels. Furthermore, Seambiotic are one of very few microalgae companies that are able to produce both biodiesel and bioethanol from their own product. The transition from a high-cost, high-return model of *Dunaliella* production into high-volumetric production at low-cost will require significant innovation, partnerships, investment, and likely new algal strains. Therefore the company's biofuel productivity data will change with the culturing of other microalgae other than *Dunaliella*. Nonetheless, Seambiotic believe that high value co-production can subsidise biofuels production to successfully compete with mineral oil, a distinct difference from simply a low-cost, high-volume model.

Sapphire Energy Inc. www.sapphireenergy.com

Sapphire is a privately owned company operating a pilot facility (40 ha) in Las Cruces, New Mexico, and an integrated microalgal biorefinery (using hydroprocessing) and production facility (120 ha of ponds) in Columbus, New Mexico, designed to produce biofuels. Sapphire Energy is supported by ARCH Venture Partners, Cascade Investment LLC, the Wellcome Trust, and Venrock Partners. This impressive list of partners have provided around USD100 million. Sapphire received a further USD104 million through the US Department of Energy (DoE), and a loan guarantee from the US Department of Agriculture for the integrated microalgal biorefinery in Columbus. Sapphire's large number of collaborators also include: Amec; Amex Geomatrix; Brown and Caldwell; the DoE's Joint Genome Institute; Deutsche Bank; Dynamic Fuels; Harris Group; New Mexico State University; San Diego Center for Algal Biotechnology; Sandia National Laboratories; Scandia National Institute; the Scripps Research Institute; Shaw; Square 1 Bank; Praxair; University of California; University of Kentucky; and the University of Tulsa.

Sapphire Review

The company uses GM microalgae to achieve high productivity, by creating strains through both selective breeding and synthetic biology. Sapphire Energy is looking at the entire microalgae production value chain, and the company states it has over 230 patents or applications spanning the entire microalgae-to-fuel process,[3] including systems for cultivation, CO_2 utilisation, refining, and harvesting (reportedly having used ultrasonic, microwave, and dissolved air flotation harvesting technology).

Sapphire's estimates the Columbus integrated microalgal biorefinery project's productivity at around 25 g m^{-2} day^{-1}, the equivalent of 30 t day^{-1} total area output. The expanded Columbus facility is expected to produce 100 BPD of microalgal oil, yielding a minimum of 60 BPD of fuels, including jet fuel, petrol, and diesel. The inputs include 35,000 t yr^{-1} of CO_2 for microalgal culture, and 210 t yr^{-1} of H_2 for the refining process, in addition to conventional inputs, such as water, nutrient, energy, etc. The logistical challenge for on-site delivery of these input gases will necessitate careful site selection, and will require large reticulation systems to maintain high productivity over the entire facility area. Sapphire's expansion plans include a 1,200 ha plant in New Mexico, proposed to be completed in 2013, supplying around 400 ML of microalgal oil by 2018. A simple calculation finds the company's stated expected output of around 100 BPD at the 120 ha facility is roughly 6.5 ML of microalgal oil, or roughly 54,000 L ha^{-1} yr^{-1}. In the authors' opinions, this is an optimistic scenario.

Nonetheless, Sapphire has demonstrated it has the capability to produce infrastructure-compliant ASTM (American Society of Testing and Materials) certified fuels. In 2008, Sapphire produced 91-octane petrol from microalgae that met ASTM standards, and in 2009, Sapphire supplied microalgae jet fuel to a Boeing 737–800 twin-engine test flight with Continental and JAL airlines. The jet fuel trial reduced fuel consumption by around 4% due to higher energy density of the microalgal fuel. In 2009, Sapphire partnered with Algaeus, FUEL, and The Veggie Van Organisation to demonstrate an unmodified plugin-hybrid Toyota Prius consuming microalgae-based fuels. Sapphire provided 190 L for the Prius trial. The authors note that much of the microalgal biomass to produce Sapphire's fuels in these trials were not

---

[3]The authors were able to locate only the following patents: WO2010042138 System for capturing and modifying large pieces of genomic DNA and constructing organisms with synthetic chloroplasts; WO2010051489 Animal feedstock comprising genetically modified algae; WO2010019813 Production of fatty acids by genetically modified photosynthetic organisms; WO2009158658 Induction of flocculation in photosynthetic organisms; WO2009036067 Molecule production by photosynthetic organisms; WO2009036087 Methods of producing organic products with photosynthetic organisms and products and compositions thereof; US20090246766 High throughput screening of genetically modified photosynthetic organisms; US20090123977; System for capturing and modifying large pieces of genomic DNA and constructing organisms with synthetic chloroplasts; US20090126260 Methods of refining hydrocarbon feedstocks; US20090087890 Methods of producing organic products with photosynthetic organisms and products and compositions thereof; WO2008321074 Production of Fc-fusion polypeptides in eukaryotic algae; CN101544918; M15-M85 vehicle alcohol ether fuel and preparation method thereof; GB2464264 Vector comprising chloroplast replicating sequence, and; GB2460351 Vectors comprising essential chloroplast genes.

produced by Sapphire, and were sourced from other microalgal biomass suppliers. In terms of economics, Sapphire's goal is to be able to produce microalgal oil at between USD60 to USD80 per barrel by 2025. Additional costs of refining to meet ASTM standards are estimated by the authors at between 5–12 US cents L^{-1} (using hydrogenation), which is likely to decrease with the investment from Sapphire's biorefinery project.

Sapphire Energy Summary

The combination of the high profile vehicle and jet trials, impressive intellectual property, alongside the noteworthy group of partners and investors give Sapphire an extremely competitive commercial advantage over many microalgae biofuel producers in progressing towards low-cost, high-volume biofuels. However, at present Sapphire are yet to demonstrate their capability for medium-scale primary microalgae production. The authors are aware that Sapphire representatives approached both Seambiotic in Israel, and Cognis (a *Dunaliella salina* microalgae production company based in Western Australia), to purchase microalgal biomass for their fuel trials. In the author's opinions, Sapphire aim at maintaining productivities of around 25 g m^{-2} day^{-1} over an entire year is relatively optimistic for open pond systems, but is potentially feasible. Maintaining these productivity values over large areas is dependent of the species characteristics and the production system design, which will in turn determine the final cost per barrel of microalgae oil. The authors are unaware of the microalgae species/cultivars used by Sapphire, and the company's open pond designs and cultivation trials are still in the early stages. In terms of the productivity and use of Sapphires GM strains, the author's suspect no GM microalgae will be used in the open pond trials in the near future due to significant regulatory barriers.

Muradel Pty Ltd. muradel.com

Muradel was established to commercialise Murdoch University's and University of Adelaide's microalgal technology intellectual property developed over a 30 year period. The in-house intellectual property incorporates the full microalgal production chain, including species selection, culture management, harvesting, oil extraction, media recycling, biomass disposal/reuse, and culture system design, construction, and operation. Muradel possesses extensive experience in commercial microalgae production and engineering, most notably through the work of Prof. Borowitzka from Murdoch University, who played a vital role in the development of the worlds' largest microalgae production facility of around

750 ha of ponds producing *Dunaliella salina* for commercial beta-carotene production in Hutt Lagoon, Western Australia, operated by Cognis.

Murdoch University and the University of Adelaide have received over AUD2 million from REDGTF (Renewable Energy and Distributed Generation Task Force) to develop a small pilot plant (one 20 m^2 and four 200 m^2 raceway ponds) at Karratha. Murdoch University and the University of Adelaide have also received $1.89 million funding from the Commonwealth Government of Australia as part of the Asia-Pacific Partnership on Clean Development and Climate to develop a commercial biodiesel and aviation fuel feedstock production process. Muradel project investors and partners include Rio Tinto, Laing O'Rourke, Parry Nutraceuticals, South China University of Technology, and Z-Filter. Muradel's proprietary microalgae strains isolated in Western Australia will be used to produce biodiesel and aviation fuel feedstock. The operational locations of Muradel include Perth and Karratha in Western Australia, and Adelaide in South Australia.

Muradel Review

Murdoch University in Perth specialise in microalgae species and cultivar selection, culture optimisation, process design, and economic modelling, while University of Adelaide specialise in culture dewatering, microalgae extraction, conversion and process design. Laboratory-scale low-cost, high-efficiency harvesting for the microalgal strains are undertaken in Adelaide, while laboratory-scale extraction optimisation for the strains occur at both Perth and Adelaide. Continuous small-scale outdoor laboratory trials (using 1 m^2 raceway ponds) in Perth and Adelaide were used to determine productive strains, average productivities, and effective monitoring and management protocols. The long-term (greater than 3 years) continuous trial results from Perth found reliable average microalgal biomass productivities of greater than 20 g m^{-2} day^{-1} with an average lipid content of at least 30% (equivalent to around 36 t of microalgal oil ha^{-1} yr^{-1}). Based on the Perth results, Muradel expect that the Karratha site will be able to fix 60 tCO$_2$-e ha^{-1} yr^{-1}, although the productivity results transferability will be determined in the pilot plant raceway ponds located at the Rio Tinto Yurralyi Maya Power Station in Karratha, while the harvested microalgae will be sent to Adelaide for analysis, lipid extraction, and biofuel production. Muradel is the only company known to the authors that is using a strain of microalgae that can tolerate salinity levels from seawater to around three-times seawater concentration. This strain reduces the flushing, or 'blow-down' requirement of ponds, decreasing effluent and contamination risks, and in combination with a unique harvesting and dewatering technology, as well as lipid extraction methods optimised for the unique strains. In terms of economics, Muradel state that their cost of production of microalgal biomass contracted from AUD12 kg^{-1} in 2008 to below AUD3 kg^{-1} in 2010, with the pilot plant aiming to reduce costs to less than AUD1 kg^{-1} for final biofuels.

Muradel Summary

With several decades of experience in microalgae isolation, culture optimisation, and commercial-scale experience, Muradel are a unique entity in the microalgal biofuels space. The two universities are using proven production systems (raceway ponds), and are focussing on capital expenditure minimisation and supply chain logistics to achieve high-volume, low-cost, high-reliability culture systems for production of biodiesel and aviation fuel feedstock. If the Karratha pilot plant research maintains reasonable productivities, a 20 ha demonstration plant by 2012 at Karratha will require an investment of between AUD10–15 million, a significant barrier to achieving commercial production.

General Atomics www.ga.com

General Atomics (GA), based in San Diego, has a 50-year history, and since 1986 has been owned by Neal Blue and Linden Blue. GA has a long history in environmental, energy, and defence contracts, and affiliated companies include GA Aeronautical Systems, Inc., which manufactures the Predator® unmanned aircraft. In 2009 GA won a staged USD43 million Defence Advanced Research Projects Agency (DARPA) contract to develop cost-effective scalable microalgae production for Jet A biofuel production. GA microalgae partnerships include: Center of Excellence for Hazardous Materials Management (CEHMM), New Mexico State University, the State of New Mexico, Algaeventure Systems Inc., Beach Energy Ltd, Carlsbad City, and AgriLife Research. GA's microalgae operational facilities include the Pecos Valley at New Mexico State University's Agriculture Science Centre, and also potential site near Moomba, in the Cooper basin, South Australia.

General Atomics Review

The 3 year DARPA contract focuses on identifying key cost drivers for increasing productivity and reducing capital and operating expenditures, including strain selection, water, CO_2, and nutrient supply, harvesting, extraction, and conversion, etc. The AgriLife Research partnership received USD4 million from the Texas Emerging Technology Fund (matched with DARPA funds), for high-oil strain evaluation, the development and testing of both open pond production systems and microalgal oil separation systems.

GA's partnership with AgriLife aim to initially demonstrate small-scale (<1 ha) open pond production, increasing to a commercial-scales (20–45 ha) in the final stages. Also under the DARPA program, GA's partnership with Algaeventure Systems from Maysville, Ohio, includes a purchase order and evaluation of Algaeventure's microalgae "Harvesting, Dewatering, and Drying" technology (AVS HDD) which reportedly reduces dewatering costs considerably using unique centrifugal technology. Algaeventure Systems, Inc. is a spinoff company from Univenture, Inc., an environmentally conscious plastics packaging manufacturer, which may be a future higher value output stream which subsidises biofuel production.

GA's joint partnership with CEHMM, New Mexico State University, the State of New Mexico, and the city of Carlsbad, focuses on reducing hazardous wastes from open pond microalgae production. The partnership aims to culture microalgae in high temperature, high saline open pond systems on unused non-arable lands in the southwest of New Mexico. Two small (95 kL) capacity ponds have been constructed with a high density polyethylene liner to prevent groundwater contamination, and at least two years of data has been recorded. Since these initial ponds were built, at least three additional ponds under 1 ha and processing facilities have been constructed, with a solvent-based extraction plant. In early 2010, CEHMM became the world's first fully integrated microalgal biomass biorefinery with a full capacity of 3,800 L day^{-1} of microalgal oil. CEHMM's Executive Director, Doug Lynn, states that the current system has the potential to produce 47,000 L of oil ha^{-1} yr^{-1}, although no specific productivity data has been released to confirm this extremely high productivity. Nonetheless, the joint partnership is now in a unique position of being able to follow the production chain from cultivation to extraction at the New Mexico State University Agriculture Science Centre. New Mexico State University was awarded USD2.36 million by the US Air Force for jet fuel production and refining research with the University of Central Florida.

In 2010, Beach Energy Ltd., an Australian oil and gas company, and GA announced a joint project researching production of biofuels in the Cooper Basin, South Australia. On June 30, 2009, Beach Energy's estimated oil and gas reserves were 66 million barrels of oil equivalent, with production in the 2008/09 financial year of 9.6 million barrels of oil equivalent. The project aims to culture microalgae using effluent and CO_2 emissions from gas and oil production sites. The preliminary project scope includes microalgal biofuels market demand, input power, transport, and other required infrastructure evaluations, in addition to water, land, solar, and CO_2 resource assessments. Depending on the outcomes of preliminary work, between 1,000 and 2,000 ha open ponds could be constructed in the Cooper Basin around 2015.

General Atomic Summary

The leveraged DARPA funding and the existing infrastructure and intellectual property[4] from culture to refining, alongside unique applications in the oil and gas industry, launch GA's considerable technical capability into a strong position in microalgae biofuels sector. The authors are unable to verify the reliability or productivity of the very high productivity claims (especially without microalgae strain/species information), and recommend prudence as the DARPA funded program is at the strain selection and system testing phase. Similarly, the small-scale biorefinery and production ponds, while a significant development, will require larger systems to determine the economics of the production systems. Nonetheless, the formidable technical capability of GA may be able to overcome many of these cost challenges, and the authors would suggest that GA's economic and microalgal strain selection data would add significant value to this review of their potential progress towards supplying low-cost, high-volume sustainable biofuels.

Aquaflow Bionomic Corporation www.aquaflowgroup.com

Aquaflow, based in Marlborough, Blenheim District, in New Zealand was established in 2005. The private company has uniquely specialised in harvesting wild microalgae from municipal sewage ponds, agricultural effluents, and even rivers. Aquaflow has an estimated 100 million shares held by mostly New Zealand investors, with Pure Power having an 18% stake in the company. Aquaflow Bionomics expected to raise an estimated AUD10–15 million to develop continuous flow microalgae production technology with parallel water remediation, fuel production, and high-value co-product capabilities. In 2008 Aquaflow partnered with Honeywell UOP (a world leading fuel refining technology company) to provide microalgal oil to optimise the refining of micralagae oil from Aquaflow's wild microalgae. The agreement gives Aquaflow access to UOP/Eni Ecofining™ processes to produce Honeywell Green Diesel™ and Honeywell Green Jet™ fuels, which uses catalysts and thermal energy for output separation. The cooperative agreement project between Aquaflow and Honeywell's UOP have received USD1.5 million from the U.S. DoE to demonstrate exhaust stack CO_2 microalgal capture with Honeywell's automated control systems. In 2010 a USD3.1 million co-funded collaboration between Aquaflow and the US Gas Technology Institute (GTI), funded in part by the U.S. DoE on

---

[4]WO2008070281 Photosynthetic oil production with high carbon dioxide utilisation, WO2008070280 Photosynthetic carbon dioxide sequestration and pollution abatement, and WO2008048861 Photosynthetic oil production in a two-stage reactor.

microalgal biomass conversion technology aimed to develop hydropyrolysis and hydroconversion technology to produce gasoline and diesel directly from microalgae. The Illinois-based GTI has over 65 years of environmental expertise resulting in around 750 licenses and 1,200 patents.

Aquaflow is also collaborating with Impulse Devices Inc. (IDI) from California specialising in acoustic inertial confinement nuclear fusion using high frequency sound waves to create cavitation in liquids. In terms of microalgae research, IDI's acoustic cavitation technology is able to control algal growth and modify water characteristsics without the use of chemicals. Aquaflow has also partnered with Solray Energy, which is a joint venture between Solvent Rescue (a solvent recovery company), and Rayners (a heating equipment manufacturer). Other Aquaflow partners include NZ Capital Strategies, Milestone Capital (Rutherford Fund) and collaborations with New Zealand Trade and Enterprise.

Aquaflow Bionomic Corporation Review

Aquaflow currently operate with around 60 ha of open sewage oxidation ponds. Aquaflow's pilot plant operates in Nelson, New Zealand, with wild microalgae harvested using dissolved air flotation and flocculation, followed by a belt press for extraction of solids (8 to 10% microalgae by volume, the rest is primarily water). This method is commonly used in wastewater treatment, although uniquely, the Aquaflow harvesting technology has been in operation at the Blenheim Municipal Wastewater site for more than 2 years harvesting microalgae. Aquaflow's continuous harvester processes 35 m^3 of water each hour, and achieves between 70 and 90% recovery of microalgae with minimal human oversight. Each harvester is built inside a 40 ft sea container, enabling modular networks, built with standard components, exhibiting low electricity consumption. The company's approach is unique as it avoids challenges of most other microalgae biofuel system strategies based around microalgal monocultures, ponds, or bioreactor technologies. Aquaflow's 60 ha of open oxidation ponds serves a population of 27,000, with a mix of sewage, municipal, and agro-industrial waste, with a water flow of 5 GL yr^{-1}. Aquaflow claim that whilst wild microalgae populations change seasonally, an average equilibrium of the total population remains suitable for biofuel production. Despite the relative oil fractions of each species varying considerably, the company utilises an integrated pyrolysis and hydrogenation process using the total microalgae biomass. Therefore, the process is not reliant on direct oil extract from the microalgae.

The economics of the biofuel conversion processes are unavailable to the authors at this stage. However, the company omit sales of microalgae biofuels and potential carbon sequestration value from their business model, and state that they solely rely on water remediation and equipment sales.

Aquaflow would be able to sell refined microalgae biofuel feedstock and have also recently focussed on high value co-products, but may also receive water processing royalties. In addition, as wastewater oxidation ponds culture microalgae as a by-product of the process, the focus on productivity is somewhat depreciated, as long as sufficient biomass is available over the majority of the year.

Aquaflow Bionomic Corporation Summary

Aquaflow's commercial advantage in producing microalgae biofuels by the use of existing sewage oxidation ponds is likely to produce a relatively inexpensive source of microalgae biofuel feedstock. To the extent that this option is reproducible will depend on oxidation sewage pond designs, associated inflow rates, and environmental conditions suitable for microalgal cultures. Therefore, the availability of suitable sites will determine the maximum output of macroalgae biofuels using Aquaflow commercial model. Unfortunately, little data are available to verify the seasonal variance of productivity in the existing pilot plant ponds in Blenheim. However, the notable interest of Aquaflow partners suggests that annual average microalgae production and associated downstream processing economics are attractive. As a comparison with productivity figures from dedicated microalgae ponds, the average annual productivity from wild microalgae is around $5 \text{ g m}^{-2} \text{ day}^{-1}$, or the equivalent of $18.25 \text{ t ha}^{-1} \text{ yr}^{-1}$. For example, if the 70–90% efficient harvester used in the 60 ha pilot facility ponds, the annual microalgal biomass collected would be around 766–985 t yr^{-1}. This relatively small annual primary biomass production illustrates why Aquaflow's commercial focus remains water remediation, and cost-effective sustainable biofuels, carbon sequestration, and other co-production streams to create an additional benefit, rather than simply a low-cost, high-volume model.

Aurora Algae Inc. www.aurorainc.com

Aurora Algae (formerly Aurora Biofuels) was founded in December 2006 in Alameda, California as an owner-operated biofuel company. The company specialises in open pond microalgae culturing developed by Prof. Tasios Melis at Berkeley University, and conversion technology based on transesterification of microalgae oils. Aurora's operational facilities are in Vero Beach, Florida, USA (<1 ha), and Karratha in Western Australia (~2.5 ha), the site of the "Algae Biofuel Demonstration Facility". Aurora Algae has attracted several investors, including Gabriel Venture Partners® (based in Silicon Valley with over USD260 million in ventures under management),

Noventi Ventures (also from Silicon Valley with past investments including Bitfone, EasyMarket, Sygate and M7), and Oak Investment Partners (over USD8 billion in committed capital with more than 481 companies). Aurora Biofuels has raised over USD40 million to fund construction of demonstration open pond systems.

Aurora Algae Review

Aurora completed an 18 month pilot project in early 2009 at the Vero Beach pilot facility. Whilst the capacity of the facility is less than 3,500 L microalgal biofuel yr^{-1}, the company states that it has produced biodiesel to the American Society of Testing and Materials (ASTM) standard. Aurora aims to improve pond productivity primarily through enhanced genetics, and is using proprietary non-transgenic methods to develop unique strains (directed mutagenesis).[5] Aurora is focussing on low-cost production systems and use CO_2 injection to both provide mixing and deliver gas to eliminate the need for conventional aeration components. The company uses floatation technology to harvest, and state they achieve a relatively high microalgae-to-water concentration (20%). Wet extraction of microalgal oil is used, which reduces energy consumption and cost by avoiding the drying process entirely. Aurora states their production process is able to produce fuels other than biodiesel, although little detail is available.

Aurora Algae is in the process of building 20 ha "Algae Biofuel Demonstration Facility" in Karratha, Western Australia, which is expected to use industrial CO_2 emissions for culturing photosynthetic microalgae to produce biodiesel, protein-rich feed for fish and human consumption, including omega-3 oils. This small facility is expected to be the forerunner to a commercial production facility, and produces between 12–15 tonnes of algal biomass per month (Aurora Algae 2012). In terms of economics, Aurora expect to produce biofuel at USD0.5 L^{-1} by 2012, assuming average productivities of around 33,000 $L\,ha^{-1}\,yr^{-1}$ (around 9 $g\,m^{-2}\,day^{-1}$ of fuel) with microalgal biomass sold at USD300 t^{-1}, all produced in an 800 ha production facility. This also assumes a zero-cost CO_2 source. In terms of yield, most company's best average productivity is around 20 g of microalgal biomass $m^{-2}\,day^{-1}$, and oil fractions of oil species are generally between 40 and 60%, excluding associated conversion losses. Therefore Aurora's productivity estimates are reasonable, in the opinion of the authors, and is based on

---

[5]Patents of Aurora include, WO2010011335 Glyphosate applications in aquaculture, WO2010008490 The use of 2-hydroxy-5-oxoproline in conjunction with algae, WO2009149470 VCP-based vectors for algal cell transformation, WO2009082696 Methods for concentrating microalgae, and WO2008060571 Methods and compositions for production and purification of biofuel from plants and microalgae.

actual production at the small scale. However, CO_2 extraction, pumping, and reticulation is not in reality 'free CO_2', and it appears that Aurora are using biomass sales to effectively subsidise biofuel production, and further processing may be required on site. For example, the biomass will exhibit excessive salt concentrations for use as an animal fodder without washing. There is also limited availability of basic cost data for Aurora's Karratha facility including conversion processes, energy inputs, fertiliser requirements, capital and operating costs, etc. Based on the knowledge of the authors of the area, we believe the actual cost of fuel production at this location is likely to be at least 25% higher than the USD0.5 L^{-1} value. In the author's opinions, while good progress is being made, the USD0.5 L^{-1} value is optimistic and speculative, as the full 20 ha facility in Karratha remains unbuilt and the final microalgae strains are being developed.

Aurora Algae Summary

Aurora's relative competitiveness is hinged on their production technology, microalgae productivity, and processing capability. In general, the microalgal concentrating and wet extraction technology, alongside any potential multiple output production makes Aurora an attractive commercial competitor in microalgae low-cost, high-volume biofuel space, although future developments will provide more information of yields and associated costs. Aurora's non-transgenic GM microalga strain development may be extremely successful in terms of high productivity. However, there will be concerns over the high-yielding non-transgenic strains resilience to competition, and questions of selective propagation in open ponds over time. The completed 18 month project is likely to have produced data on the magnitude of the oil yield decline, although this information is unavailable to the authors.

Harvesting, Dewatering, Extraction, & Conversion Considerations

All seven companies considered in this chapter aim to produce large quantities of microalgae biomass using open cultivation systems, and have chosen or will have to choose a process to convert this biomass into saleable products. The focus of many of these companies is biofuel, although some are focussing on alternative value chains such as neutraceuticals, chemicals, or remediation approaches. Figure 1 outlines that cultivation of microalgae is only the first step in a multistage system required to convert microalgae to fuels. Although the focus of this review is on the current commercial

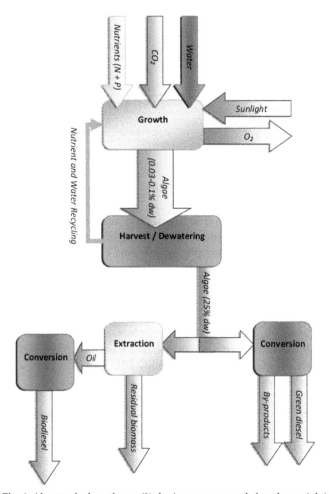

Fig. 1. Algae to fuels pathway (% dw is percentage ash free dry weight).

status of open systems for microalgae cultivation, it is fundamental that the growth system is considered in the context of the entire process, including harvesting/dewatering, extraction, and conversion.

The concentration of microalgae in open pond systems are typically between 0.5 and 2 g L^{-1} (Fon Sing et al. 2011) which is up to 10 times less than the cell density in closed (photobioreactor) systems. This low density requires moving and harvesting massive quantities of water in multiple stages as demonstrated in Fig. 2. Primary dewatering is typically achieved through flocculation, and once flocculated, the concentrated algae solution

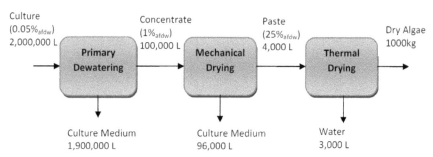

Fig. 2. Dewatering metrics.

is separated from the water via settling or floatation (e.g., dissolved air flotation). Various methods of flocculation can be used with some algae naturally flocculating while other species requiring the addition of chemicals or flocculants like *chittosan* to induce the agglomeration of cells (In addition to these methods, some companies such as Origin Oil and Diversified Technology have used pulsed electric fields with great success to enhance flocculation and lyse the microalgae cells).

Secondary dewatering or mechanical drying is typically achieved through physical separation of the water and the algae on the basis of mass difference via centrifuges or decanters. These machines are energy intensive, however, recent developments by Evodos have seen the power consumption drop significantly (Evodos 2011). In addition, a membrane recently developed by Pall Corporation is able to separate microalgae from water, requiring much lower energy consumption then standard centrifuges. After mechanical drying, the microalgae solution resembles toothpaste or a thick sludge. Removing the remaining water is undertaken by drying with heat. Although a range of dryer technologies exist, the fundamental need to vaporise the water results in a very high energy burden. Consideration of the required downstream harvesting and dewatering operations indicates the need for large microalgae developments to optimise upstream strain selection, water/nutrient recycling, and other downstream processes to be competitive. The wide range of methods proposed for converting microalgae to fuel can be categorised into one of two groups:

- Processing the whole algae for conversion into fuel or other products;
- Extraction of oil or other components from algae for conversion into fuel or other products.

Within these two groupings, there are methods that require dried microalgae biomass and methods that are effective in the presence of water, and Fig. 3 provides a useful guide to characterising the wide range of methods under research or being pursued in commercial application (de

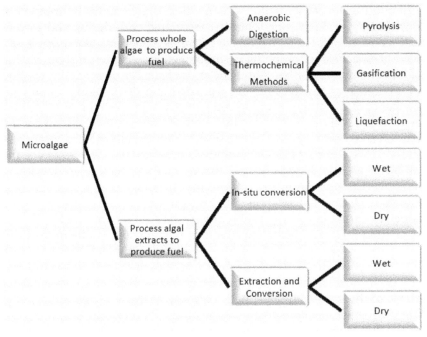

Fig. 3. Microalgae to fuel conversion methods (de Boer et al. 2012).

Boer et al. 2012). It is also possible for microalgae to produce fuel via direct secretion, like the ethanol producing microalgae cultivated by Algenol, and the hydrocarbon secreting chlorophyte *Botryococcus braunii* (Metzger and Largeau 2005). Despite this potential, most companies pursuing large scale open pond cultivation are considering conversion methods that fit in to the two categories listed above.

Although energetically feasible, relatively simple and proven anaerobic digestion yields a low value product, methane, and is typically not pursued commercially except as a means of converting byproducts to energy. Out of the three thermochemical methods, liquefaction is the most promising as it produces a crude oil replacement from microalgae in the presence of water at relatively mild conditions (Biller and Ross 2011). Furthermore, this process was shown to be the most energetically feasible process in a comparison of the four most promising methods available for conversion of microalgae into fuel (de Boer et al. 2012). Instead of processing the whole microalgae cell, it is possible to extract components of interest (e.g., lipids) and convert these compounds to desired fuel products. The most familiar method is some form of solvent extraction (e.g.: soxhlet) to remove the oil from the dried biomass followed by transesterification (Lardon et al. 2009).

Although technically feasible, this method is economically impractical due to the very high energy burden associated with drying and solvent extraction. In response, methods to convert compounds like lipids *in situ* have been developed resulting in simultaneous extraction and conversion of the compounds of interest from the microalgae. The most feasible method in this category involves mixing the dried and disrupted biomass with methanol and a strong acid to produce fatty acid methyl esters (Wahlen et al. 2011). Although effective, this method still results in a high energy burden and associated cost due to the need for drying and evaporation of large excess volumes of methanol (de Boer et al. 2012). To overcome the problems associated with drying, further methods have been developed to convert the lipids directly to fuel products in-situ in wet microalgae biomass. One recent method showing promise in this category is the supercritical ethanolysis and subsequent esterification proposed by Levine et al. (2010). Despite being effective this method still results in the need to purify (evaporate) large volumes of ethanol ultimately resulting in a high energy burden. Unfortunately, it is at present not possible to say that any of the numerous methods in development will become commercially viable at the industrial scale, with very few of them energetically viable (producing more energy than they consume) (Borines et al. 2011a, de Boer et al. 2012, McHenry 2012). With these challenges in mind, it is necessary that companies pursuing high-volumetric production of fuels, chemicals, or other compounds of interest from microalgae carefully consider the effect their cultivation system has on the entire process.

Conclusion

Historically, commercial open microalgae production has been limited to relatively small-scale volumetric production of high-value products (Pulz and Gross 2004). Irrespective of the current range of competing companies using open systems, industrial-scale microalgae production of cost-competitive biofuels will require detailed biotechnical and system control capability to achieve production reliability, and aim to lower production costs (Jamers et al. 2009, Kovacevic and Wesseler 2010). In the recent past, in an attempt to produce high-volumetric microalgal energy products, the industry has witnessed collectively recurring issues associated with biological system unpredictability, variable photosynthetic efficiencies, technological challenges with poor CO_2 addition, high energy inputs, and fundamental difficulties in microalgae harvesting and extraction (Brennan and Owende 2010, Clarens et al. 2010, Borines et al. 2011a). These issues are fundamental to progressing towards sustainable supplies of low-cost and high-volume algal biofuels.

This review suggests that the current suite of open pond microalgae technical developments are progressing towards a low-cost, high-volume sustainable supply of liquid fuel, yet is highly conditional on the sum of the technical resources and commercial strategies of each major player. Competing in low-cost, high-volume energy markets against conventional fuels will require enormous further technological development to achieve commercially viable and sustainable microalgal biofuel production systems (Brennan and Owende 2010, Kovacevic and Wesseler 2010). The authors would like emphasise that these challenges are not insurmountable, but will likely require an unprecedented level of collaboration between several industries historically unrelated with biotechnology or the biological sciences generally, and long-term support from both governments and non-government institutions.

References

Amin, S. 2009. Review on biofuel oil and gas production processes from microalgae. Energ. Convers. Manage. 50: 1834–1840.

Aurora Algae. 2012. Aurora Algae secures full $2 million LEED grant for successful production of algae-based platform. 18 October, 2012 www.aurorainc.com/aurora-algae-secures-full-2-million-leed-grant-for-successful-production-of-algae-based-platform/

Biller, P. and A. B. Ross. 2011. Potential yields and properties of oil from the hydrothermal liquefaction of microalgae with different biochemical content. Bioresource Technol. 102(1): 215–225.

Borines, M. G., R. L. De Leon and M. P. McHenry. 2011a. Bioethanol production from farming non-food macroalgae in Pacific island nations—Chemical constituents, bioethanol yields, and prospective species in the Philippines. Renew. Sust. Energ. Rev. 15(9): 4432–4435.

Borines, M. G., M. P. McHenry and R. L. de Leon. 2011b. Integrated macroalgae production for sustainable bioethanol, aquaculture and agriculture in Pacific island nations. Biofuels, Bioprod. Biorefin. 5(6): 599–608.

Borowitzka, M. 1992. Algal biotechnology products and processes - matching science and economics. J. Appl. Phycol. 4: 267–279.

Borowitzka, M. 1999. Commercial production of microalgae: ponds, tanks, tubes and fermenters. J. Biotechnol. 70: 313–321.

Borowitzka, M. A. and N. R. Moheimani. 2010. Sustainable biofuels from algae. Mitig. Adapt. Strat. Global Change. DOI: 10.1007/s11027-010-9271-9.

Brennan, L. and P. Owende. 2010. Biofuels from microalgae—a review of technologies for production, processing, and extractions of biofuels and co-products. Renew. Sust. Energ. Rev. 14: 557–577.

Cantrell, K. B., T. Ducey, K. S. Ro and P. G. Hunt. 2008. Livestock water-to-bioenergy generation opportunities. Bioresource Technol. 99: 7941–7953.

Charcosset, C. 2009. A review of membrane processes and renewable energies for desalination. Desalination. 245: 214–231.

Chisti, Y. 2007. Biodiesel from microalgae. Biotechnol. Adv. 25: 294–306.

Clarens, A. F., E. P. Resurreccion, M. A. White and L. M. Colosi. 2010. Environmental life cycle comparison of algae to other bioenergy feedstocks. Environ. Sci. Technol. 44: 1813–1819.

de Boer, K., N. Moheimani, M. Borowitzka and P. Bahri. 2012. Extraction and conversion pathways for microalgae to biodiesel: a review focused on energy consumption. J. Appl. Phycol. DOI10.1007/s10811-012-9835-z.

Evodos. 2011. Totally, dewatering algae, alive. 18 October 2012 www.evodos.eu/fileadmin/PDF/evodos_brochure_algae.pdf.

Fon Sing, S., A. Isdepsky, M. A. Borrowitzka and N. R. Moheimani. 2011. Production of biofuels from microalgae. Mitig. Adapt. Strat. Global Change. DOI 10.1007/s11027-011-9294-x.

Gross, M. 2007. Algal biofuel hopes. Curr. Biol. 18(2): R46–R47.

Hankamer, B., F. Lehr, J. Rupprecht, J. H. Mussgnug, C. Posten and O. Kruse. 2007. Photosynthetic biomass and H_2 production by green algae: from bioengineering to bioreactor scale-up. Physiol. Plantarum. 131: 10–21.

Huntley, M. E. and D. G. Redalje. 2006. CO_2 mitigation and renewable oil from photosynthetic microbes: a new appraisal. Mitig. Adapt. Strat. Global Change. 12: 573–608.

Jamers, A., R. Blust and W. De Coen. 2009. Omics in algae: paving the way for a systems biological understanding of algal stress phenomena? Aquat. Toxicol. 92: 114–121.

Kovacevic, V. and J. Wesseler. 2010. Cost-effectiveness analysis of algae energy production in the EU. Energ. Policy. 38: 5749–5757.

Kunjapur, A. M. and R. B. Eldridge. 2010. Photobioreactor design for commercial biofuel production from microalgae. Ind. Eng. Chem. Res. 49: 3516–3526.

Lardon, L., A. Hélias, B. Sialve, J. -P. Steyer and O. Bernard. 2009. Life-Cycle assessment of biodiesel production from microalgae. Environ. Sci. Technol. 43(17): 6475–6481.

Lee, Y. -K. 2001. Microalgal mass culture systems and methods: their limitation and potential. J. Appl. Phycol. 13: 307–315.

Levine, R. B., T. Pinnarat and P. E. Savage. 2010. Biodiesel production from wet algal biomass through *in situ* lipid hydrolysis and supercritical transesterification. Energ. Fuel. 24: 5235–5243.

McHenry, M. P. 2010. Microalgal bioenergy, biosequestration, and water use efficiency for remote resource industries in Western Australia. In: A. M. Harris [Ed.]. Clean energy: resources, production and developments. Nova Science Publishers, Hauppauge, New York.

McHenry, M. P. 2012. Hybrid microalgal biofuel, desalination, and solution mining systems: increased industrial waste energy, carbon, and water use efficiencies. Mitig. Adapt. Strat. Global Change. DOI: 10.1007/s11027-012-9361-y.

Metzger, P. and C. Largeau. 2005. *Botryococcus braunii*: a rich source for hydrocarbons and related ether lipids. Appl. Microbiol. Biot. 66(5): 486–496.

Moheimani, N. R. and M. Borowitzka. 2006. Limits to productivity of the alga *Pleurochrysis carterae* (Haptopphyta) grown in outdoor raceway ponds. Biotechnol. Bioeng. 96(1): 27–36.

Moheimani, N. R., M. P. McHenry, P. Mehrani and Microalgae biodiesel and macroalgae bioethanol: the solar conversion challenge for industrial renewable fuels. In: R. Razeghifard [Ed.] (in press). Natural and artificial photosynthesis: solar power as an energy source. John Wiley & Sons Inc., Hoboken, New Jersey, USA.

Pulz, O. and W. Gross. 2004. Valuable products from biotechnology of microalgae. Appl. Microbiol. Biot. 65: 635–648.

Sheehan, J., T. Dunahay, C. E. Hanson and P. Roessler. 1998. A look back at the US Department of Energy's Aquatic Species Program—biodiesel from algae. National Renewable Energy Laboratory, Golden, USA.

Spolaore, P., C. Joannis-Cassan, E. Duran and A. Isambert. 2006. Commercial applications of microalgae. J. Biosci. Bioeng. 101(2): 87–96.

Ugwu, C. U., H. Aoyagi and H. Uchiyama. 2008. Photobioreactors for mass cultivation of algae. Bioresource Technol. 99: 4021–4028.

Vasudevan, P. T. and M. Briggs. 2008. Biodiesel production—current state of the art and challenges. J. Ind. Microbiol. Biot. 35: 421–430.

Wahlen, B. D., R. M. Willis and L. C. Seefeldt. 2011. Biodiesel production by simultaneous extraction and conversion of total lipids from microalgae, cyanobacteria, and wild mixed-cultures. Bioresource Technol. 102(3): 2724–2730.

Wyman, C. E. and B. J. Goodman. 1993. Biotechnology for production of fuels, chemicals, and materials from biomass. Appl. Biochem. Biotech. 39(40): 41–59.

Xiong, W., X. Li, J. Xiang and Q. Wu. 2008. High-density fermentation of microalga *Chlorella protothecoides* in bioreactor for microbio-diesel production. Appl. Microbiol. Biot. 78: 29–36.

17

Biobutanol: An Important Biofuel and Bio-Product

Jamie A. Hestekin,* Alexander M. Lopez,
Edgar C. Clausen and *Thomas M. Potts*

ABSTRACT

Biobutanol as a fuel has been around since the late 19th century. Since its inception many technologies have improved its technical and economic viability including genetic modification of organisms, separation advancement, and new feedstocks for production. However, 100 years later, problems still remain in economic viability caused by need for new organisms and advanced separations. This chapter will discuss the important advances as well as looking at what still needs to be done in order to realize biobutanol as a large scale fuel replacement strategy.

Introduction

History of biobutanol

Butanol production via bacterial fermentation has taken place for over 100 years. In 1862, Pasteur first recorded the production of butanol by a microorganism he called *Vibrion butyrique*. In 1905, Schardinger isolated a

Ralph E. Martin Department of Chemical Engineering, University of Arkansas, 3202 Bell Engineering, Fayetteville, AR 72701, USA.
* Corresponding author: jhesteki@uark.edu

bacterium that produced acetone, butanol, and ethanol. Within 5 years, a British company, Strange and Graham Ltd., began research on the biological production of solvents for the manufacture of synthetic rubber. One of the Strange and Graham employees, Chaim Weizmann, left the company and while at Manchester University isolated a new bacterium, *Clostridium acetobutylicum*, which produced quantities of acetone, butanol, and ethanol in the ratio of 3:6:1 from potato starch. He was awarded a patent on this process in 1915 (Weizmann 1915).

During the course of World War I, British supplies of acetone were depredated by Axis naval and aerial forces. The Weizmann process was selected for production of acetone, and several plants were constructed in Britain, Canada, and the US for the conversion of corn mash and liquor to acetone. The conclusion of World War I left very large inventories of the byproduct butanol. These stockpiles were used by DuPont as feedstock for the production of butyl acetate, a solvent for nitrocellulose acetate lacquer. Existing acetone plants were converted to butanol production to support this solvent production.

Fermentation production of butanol proliferated and feedstock shifted from corn liquor to molasses, which supported greater solvent productivity in the 1930's. Many new microorganisms were found (and patented) during this time. At the height of World War II, acetone was once again in short supply, and many new production facilities were implemented worldwide for acetone production via fermentation of molasses, corn mash, and other feedstocks (Koepke and Durre 2011). After the war, advances in petrochemical technology and cheap oil yielded processes that greatly lowered the cost of manufacturing of acetone and butanol, and fermentation of these solvents could no longer compete economically. Few fermentation facilities survived and by 1970, the ABE industry in the US was nearly ended (Ni and Sun 2009). Foreign facilities lasted another 20 years, and the last large scale ABE plant in China was closed in 2004. With the increased worldwide demands for crude oil and the increase in price, interest in ABE fermentation has renewed in recent years (Koepke and Durre 2011). China has resumed ABE production in 11 plants, one of which is a 30,000 ton/year facility in Jilin China, operated by Cathay Industrial Biotech (Ni and Sun 2009). In the US, Gevo is operating a 10 million gallon/year of iso-butanol facility at Luverne, MN. The Gevo butanol is being sold to Sasol for solvent applications.

Butanol Uses and Markets

Currently, butanol is produced from petroleum derivatives to be used as a solvent and chemical intermediate for many important products (Ezeji et al. 2005c). As a solvent, butanol is used in the production of paints, dyes, and chemical stabilizers. In the chemical industry, butanol is used in the production of various plastics and polymers such as safety glass, hydraulic fluid, and detergents (Mariano et al. 2012). The worldwide chemical market for butanol is approximately 950 million gallons produced mainly by Dow Chemical Company, DuPont, BASF, and Oxea Group (Yuan and Hui-feng 2012). In Brazil butanol is produced internally by the company Elekeiroz and in China 50% of the butanol consumed is imported with the remainder produced by ABE plants operated by small solvent companies (Mariano et al. 2012). The worldwide demand for butanol is expected to increase by 3.2% per year with concentrated demand in North America, Europe, and Asia (Green 2011). Costs for producing butanol from fossil fuels are at $3.30/ gal with wholesale at $3.80/gal (Ramey and Yang 2004).

Many companies are looking at producing butanol as a biofuel. This would allow greater production of renewable bioenergy in the automotive market and present an opportunity for butanol to enter the energy sector with specialty fuels such as jet fuel. However, the main issue keeping biobutanol from becoming a large scale commercial success as a fuel is that it must compete with existing biofuels such as ethanol. Current estimates for the cost of producing fuel grade biobutanol lie between $3.50–$4.00/ gallon while bioethanol is currently produced at $2.50 (Ramey and Yang 2004). On an energy basis (using LHV) the cost comparison results in butanol outputting 25.6–29.2 MJ/$ while ethanol provides 32.3 MJ/$ (Weast 1978). The main reasons for the higher price of butanol are the low yields from fermentation due to the high toxicity of butanol and costly separations involved in removing water and other non-desirables in fuel grade butanol (Tashiro et al. 2005). When the cost of producing butanol approaches $2.50/ gallon, the butanol market can expand into the fuel additive market and compete for the greater than 22 billion gallon market currently monopolized by ethanol. Figure 1 shows the current market of butanol, mainly as a solvent and additive, and also demonstrates the potential if butanol becomes an additive or ideally a drop-in fuel. As shown in the figure, the current market is a small sliver of the potential market if the price was reduced.

In order for butanol to have the greatest impact in the liquid fuel industry it must become a drop-in fuel. This would require that butanol be produced cost competitively with liquid petroleum on a MJ/$ basis. Today gasoline is produced with an energy output of 35.9MJ/$. However cheap gasoline at $2/gallon has an energy output of 59.9 MJ/$ setting the target even higher for a potential drop-in fuel. When butanol approaches this value, the market for butanol production can expand out of the fuel additive market and begin to alleviate the 135 billion gallons of gasoline currently being consumed in the Unites States annually. Implementation of renewable butanol as a pure energy fuel would require little or no automotive redesign

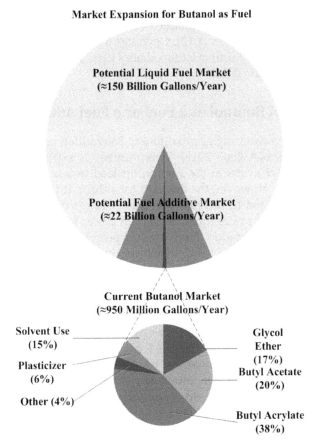

Fig. 1. Current Butanol market and expansion potential to fuel additive and direct fuel competitor.

(Szulczyk 2010). Further, most of the transportation and delivery methods for the fuel industry would remain the same, limiting the time required for complete conversion of U.S. infrastructure for butanol fuel.

In order to facilitate the penetration of butanol into vehicular use, many researchers are looking at developing new production techniques using wild and genetically engineered bacteria in order to boost butanol yield and bacterial tolerance to butanol (Rosgaard et al. 2012). Companies such as Silicon Valley based Cobalt and Brazil's Green Biologics are currently designing and running pilot-scale butanol facilities to develop the technologies needed for large scale butanol production (Herndon 2012, Nielsen 2012). In China companies such as Ji-An Biochemical, Guiping Jinyuan, and Jilin Cathy have ramped up production of butanol from ABE methods to meet the high demand for butanol in the solvent market (Ni and

Sun 2009). BP and DuPont are also developing a butanol production facility through a company named Butamax™ Advance Biofuels with expected production in 2014 (Martin 2012). Gevo has a similar biobutanol making process and plans to retrofit existing ethanol plants for butanol production as early as 2014 (Lane 2012).

Advantages of Butanol as a Fuel or a Fuel Additive

The US is now consuming approximately 360 million gallons of gasoline per year (DOE/EIA-0383 2012). This number is expected to decrease slightly, as higher prices at the fuel pump lead to less driving and thus smaller demand. However, the demand for vehicular fuels in developing countries is expected to increase and off-set lower consumptions in the US. The higher fuel prices translate to high prices for goods and services dependent on transportation. Thus, the European Union issued the biofuels directive, officially 2003/30/EC, which set the goal of replacing 5.75% of all transport fossil fuels (gasoline and diesel) with biofuels by 2010. Some years later, the US Congress passed the Energy Independence and Security Act of 2007, which among other regulations and goals, expands the national renewable fuels standard to 9 billion gallons in 2008, with a phased increase to 36 billion gallons by 2022.

A by-product of using fossil fuels is carbon dioxide, a greenhouse gas. Rise of average global temperatures has been linked to increasing carbon dioxide concentration in the atmosphere. Approximately one fourth of human generated carbon dioxide emissions are from use of fossil derived vehicular fuels for cars and light trucks (Jaffe et al. 2011). Thus, replacement of fossil derived fuel with a (nearly) carbon dioxide-neutral fuel such as butanol could have major implications to the environment and world economy.

The proven oil reserves available to the world as of 2011 were estimated at 1350 billion barrels, with just over 200 billion barrels in North America (Jaffe et al. 2011). In addition to these proven reserves, there are estimates that there are an additional unconventional, mostly in oil shale and shale oil, 1300 billion available in North America and 2129 billion barrels available worldwide. At the current rate of consumption, and using Jaffe's numbers, the US and Canada could supply their own demands for about 24 years with proven reserves and about 178 years with the proven and unconventional, technically recoverable reserves. For the world as a whole, proven supplies would meet current demand for about 43 years while the proven plus unconventional would meet current demand for about 112 years. Of course, these numbers are based on several admittedly flawed assumptions: the demand will not change; there will be no shift in energy consumption from fossil fuels to other forms of energy from nuclear, coal, natural gas, wind power, solar power, or alternative fuels from biomass. It is highly probable that demand for oil in highly industrialized countries with stable populations will continue to decrease slightly, but that demand in developing countries will increase at rates commensurate with availability and price. It is also probable that as oil production shifts from easy to attain proven resources to more difficult to process unconventional reserves, that oil prices will rise sharply and demand will be reduced. Higher oil prices will make the transition to the use of electric energy for many applications currently met with oil more attractive, but will drive cost of living higher and lower living standards for the bulk of the world's population. It is also highly probable that because of the very large consumption numbers, no single alternative energy source can supplant the use of oil, and each potential source will play an important role in niche markets. However, with proven oil reserves being >40 years with unconventional sources, it is likely that in order to have biobutanol as a large scale fuel replacement that the economics and the environmental implications must be verified and improved.

Current biofuels production efforts focus primarily on bioethanol and biodiesel, with other potential fuels, such as biobutanol, poised on the wings to become major players if certain technological advances can be realized. Table 1 lists certain fuel properties of ethanol and butanol. As can be seen from the table, 1-butanol exhibits some characteristics that make it a good candidate for vehicular fuel or fuel additive when compared to ethanol. Butanol has a higher energy density with an LHV (Lower Heating Value) of 99,800 btu/gal as contrasted to ethanol at 76,000 btu/gal. Ramey found in his test of a 1992 Buick driven across the country on B100 (100% butanol) that the gas mileage was 8% better (based on volume of butanol used) than on gasoline (Szulczyk 2010). Edmunds tested several 2007, flex-fuel Chevrolet Tahoes operated on both E85 (85% ethanol, 15% gasoline) and gasoline and found that the gas mileage was 26% less when operated on E85 (Edmunds 2007). Additionally, the Ramey Buick's engine had no modifications whereas the Edmunds Tahoe required an engine specifically designed to run on either gasoline or E85. This is another advantage for butanol as contrasted to ethanol: butanol will run in most vehicles without engine modifications, whereas ethanol at concentrations higher than 10% requires extensive modifications. This is because butanol is much less corrosive than is ethanol. Ethanol has high miscibility with water and will absorb moisture from the air. If the water content of the ethanol/gasoline mixture becomes too high, the fuel will phase separate with a water/ethanol layer in the bottom of the tank. This layer will render the vehicle inoperable. To return the vehicle to service, the tank must be drained, cleaned, and thoroughly dried. Butanol has much less miscibility and is not hygroscopic and thus does not present this problem. As can also be

Table 1. Properties of Various Fuels.

Property	Butanol	Ethanol	Gasoline	No. 2 Diesel	Jet Fuel	Hydrogen, liquid
Chemical Formula	C4H9OH	C2H5OH	C4–C12	C3–C25	C5-C12	H2
Molecular Weight	74.1	46.1	100–105	≈200	~140	2.0
Wt% Carbon	64.8	52.2	85–88	84–87	~84	0.0
Wt% Hydrogen	27.0	13.1	12–15	33–16	~16	100.0
Wt%Oxygen	21.6	34.7	0.0	0.0	0.0	0.0
Specific gravity, 60° F	0.8	0.8	0.72–0.78	0.81–0.89	0.8	0.1
Density, lb/gal @ 60° F	6.8	6.6	6.0–6.5	6.7–7.4	6.8	0.6
Boiling temperature, °F	244.0	172.0	80–437	370–650	349.0	–423.0
Reid vapor pressure, psi	0.3	2.3	8–15	0.2	0.4	na
Heat of vaporization, btu/lb	256	396	150	100	150	198
Research octane no.	96	108	90–100	na	na	na
LHV Energy Density, btu/gal	99,800	76,000	115,000	130,500	135,000	30,000
Freezing point, °F	–130.0	–173.2	–40	–40–30	-54	na
Viscosity @ 60° F, Cp	3.0	1.19	0.37–0.44	2.6–4.1	2	na
Flash point, closed cup, °F	84	55	-45	165	100	na
Autoignition temperature, °F	650	793	495	≈600	410	1058
LFL, wt%	1.4	4.3	1.4	1.0	0.6	4.0
HFL, wt%	11.2	19.0	7.6	6.0	5.0	75.0
Specific heat, Btu/lb °F	0.55	0.57	0.48	0.43	0.51	0.002

seen in Table 1, the Reed vapor pressure of butanol is much lower than that for ethanol or gasoline. This gives some advantages to butanol but also some disadvantages. The lower vapor pressure means that butanol vapor emissions from fuel tanks, storage facilities, etc. are much less than corresponding emissions from ethanol or gasoline. Additionally, the lower vapor pressure and higher flash point of butanol means it is somewhat less of a fire hazard when spilled than is ethanol, although when blended with gasoline, this is a moot point. One disadvantage of the lower vapor pressure of butanol is that the low temperature starting capabilities of the vehicle is less than with higher vapor pressure fuels. Again, when in a gasoline blend, this is a moot point due to the very high vapor pressure of gasoline. A second disadvantage of butanol as contrasted with ethanol is that the butanol octane number is lower, about the same as winter gasoline. Ethanol, with a much higher octane number, can serve as an anti-ping additive or octane booster in gasoline. On the other hand, butanol, with its lower octane number, can be blended with diesel as well as gasoline. Butanol/diesel blends have been found to increase diesel mileage, reduce emissions of hydrocarbons and non-hydrocarbon pollutants, and enhance cold starting properties of diesel engines (Altun et al. 2011, Dogan 2011, Chen and Chen 2011, Yao et al. 2010). Still to be determined is whether butanol added to diesel fuel will adversely affect engine lifetime. Both ethanol and butanol are oxygenates, with butanol at 21.6% and ethanol at 34.7%. EPA requirements for oxygenates in fuels favor the use of ethanol for its higher oxygen content. Butanol is somewhat more toxic to many forms of life (including humans) than is ethanol, which may favor the continuing use of ethanol instead of switching to butanol (Szulczyk 2010). However, butanol is much less toxic than many compounds commonly found in gasoline and from a purely technical viewpoint; the advantage to ethanol is minimal.

However, the primary reason that butanol has not become a major biofuel is that it costs much more to produce than does ethanol. To supplant ethanol as a biofuel, the manufacturing cost will need to be reduced by half. This topic is discussed in greater detail elsewhere in this paper.

Production of Butanol Biofuels

There are three different schemes for the production of butanol that will be talked about in this paper. Isobutanol, 1-butanol production via a one step process, and 1-butanol production via a two-step process. All three of these schemes will be discussed in the paper below and the advantages\ disadvantges of using isobutanol over 1-butanol will also be discussed.

Isobutanol production

An additional form of butanol currently being researched for biofuel production is isobutanol. Isobutanol differs from 1-butanol in a variety of ways. Some key differences are presented in Table 2 and the molecular structures of the two molecules are given in Fig. 2. The main benefit of isobutanol over 1-butanol is that the octane number is much higher, resulting in greater fuel efficiency. Isobutanol has also been shown to reduce stress corrosion cracking in engines and has lower hygroscopicity than bioethanol (Gevo 2012). Currently isobutanol is synthesized using syngas. However, this method is expensive due to high temperature and catalysts

Table 2. Property comparison between N-Butanol and Isobutanol (Glassner 2009).

	Isobutanol	N-Butanol
Octane Number	98–102	87
Oxygen Content (%)	21.6	22
Reid Vapor Pressure (psi)	4–5	0.33
Energy Content (MJ/kg)	32.6	33.4
Viscosity (cP)	3.95	3.00
Cost ($/gal)	3.75–4.25	3.50–4.00
Density	802	810

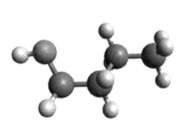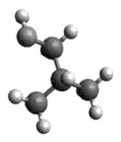

Fig. 2. Molecular representations of (left) 1-butanol and (right) isobutanol. Generated from Avogadro modeling software.

requirements, so economical and renewable methods for isobutanol are attractive (Li et al. 2011a). Renewable isobutanol can be produced from a variety of bacterial fermentation methods. Many are specifically designed bacteria genetically modified to resist isobutanol poisoning and maximize productivity (Atsumi et al. 2010). A few processes focus on the development of isobutanol from non-fermentative pathways, eliminating the need for a source feedstock. The main approaches of each production pathway will be discussed below.

Isobutanol is produced in bacteria following a specific enzymatic process within the bacterial glucose consumption pathway. Glucose or a similar sugar is consumed via the glycolysis pathway to produce pyruvate. Then pyruvate is catalyzed by a series of enzymes and reaction intermediates to form isobutanol. Figure 3 outlines this reaction pathway. Understanding this pathway has allowed researchers to improve the activity of enzymes that lead to isobutanol formation while inhibiting additional enzymes in bacteria that result in unwanted products.

The isobutanol fermentation began in the mid 2000's following interest in bioethanol and other biofuels. Since then a large number of bacteria have been identified naturally and designed for isobutanol production. Currently the main producers of isobutanol are *E. Coli, Bacillus subtilis,* and *Corynebacterium glutamicum* (Blombach and Eikmanns 2011). However, each of these bacterial strains has their limitations with the main issue being that

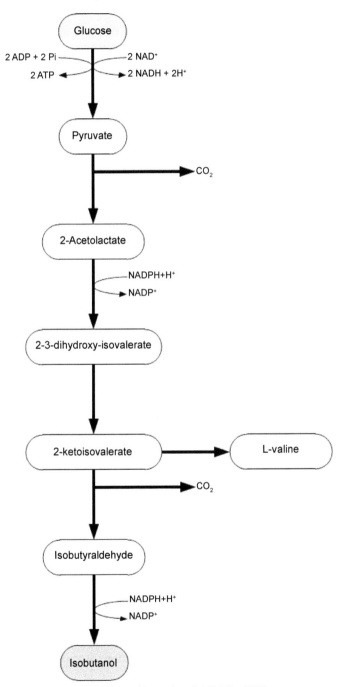

Fig. 3. Isobutanol Enzymatic Pathway (Atsumi et al. 2010, Lu 2012).

most fermentation systems are intolerant of isobutanol, dying off when concentrations of isobutanol exceed 8 g/L (Atsumi et al. 2010). To combat this issue, many researchers have begun to identify methods to extract the isobutanol as it is produced or improve bacterial tolerance to isobutanol. A successful method would involve utilizing of both mitigation regimes.

Escherichia coli (*E. coli*) has been studied over the past decade for isobutanol production. Research has been focused on using the 2-keto acid-based pathway for isobutanol production from glucose (Smith and Liao 2011). Many key enzymes have been isolated via protein purification and verified using SDS-PAGE in order to conduct *in vitro* studies. Some enzymes and genes of study include aldehyde reductase and alcohol dehydrogenase (Atsumi et al. 2009). Another method developed to improve isobutanol production in *E. coli* involves *in situ* product removal (gas sparging). Isobutanol produced by engineered *E. coli* is constantly stripped out of the fermentation system with air and is subsequently condensed and absorbed in chilled water. This method allows the bacteria to continue to produce isobutanol without the issue of product inhibition. With *in situ* product removal, isobutanol productivity reached 50 g/L in 72 hours with an isobutanol yield of 0.29 g isobutanol/g sugar (Baez et al. 2011). This corresponds to 68% of the theoretical maximum yield. Other efforts in *E. coli* isobutanol production involve using elementary mode (EM) analysis to determine the theoretical maximum production of isobutanol and 1-butanol in *E. coli* (Trinh 2012). This method seeks to analyze current work done on engineering *E. coli.* in order to determine the ideal mechanisms for anaerobic biobutanol production. Using EM analysis, a significant understanding of the metabolic pathways that *E. coli* employs was obtained and applied for strain optimization for biofuel production (Trinh et al. 2011).

Bacillus subtilis is another promising isobutanol producing bacteria. *Bacillus subtilis* possesses a much higher isobutanol tolerance than that of *E. coil* and *C. glutamicum* (Jia et al. 2012). However, its production capabilities are much less. The highest reported yield of isobutanol from *B. subtilis* is approximately 0.08 g isobutanol/g sugar (19% theoretical yield). Main issues associated with *B. subtilits* lie in the overproduction of acetate and lactate (Li et al. 2011a). Significant work needs to be done to direct the carbon flux to isobutanol production and improve isobutanol yields.

Corynebacterium glutamicum has been researched as a candidate for isobutanol production. Major advantages of using *C. glutamicum* is that it is more robust compared to *E. coli*. However, many side products are formed with *C. glutamicum* fermentation. A significant genetic modification procedure is required for extractable concentrations of isobutanol. Current research progress shows isobutanol concentrations at approximately 4.0 g/L in 96 hrs. This results in a theoretical yield of 19% (Smith et al. 2010). Other researchers have been able to improve upon this to reach a theoretical yield of 77% by allowing cell growth under aerobic conditions and then depriving the bacteria of oxygen to boost isobutanol production (Blombach et al. 2011). Final isobutanol concentrations also showed promise with >10g/L as a final concentration and a maximum productivity of approximately 0.9g/Lh (Blombach et al. 2011).

Saccharomyces cerevisiae (brewer's yeast) has also been studied for isobutanol production potential. The idea behind using *S. cerevisiae* is that the metabolic pathways for ethanol production can be deleted and the pathways leading to isobutanol production can be overexpressed. Current results are still in the initial phase with concentrations <1g/L being produced and yields less than 1% (Lee et al. 2012, X. Chen et al. 2011). Significant work is still needed to direct the carbon flux toward isobutanol production and match productivities gained by other bacteria.

Recent interesting approaches for isobutanol production include photosynetic pathways. Li et al. (2012) report a method for producing isobutanol in a fermentation system that relies solely on carbon dioxide and energy generated from man-made photovoltaic cells (solar panels). This method is best described as a type of pseudo-photosynthesis that can have a higher efficiency than current fuel biological systems. However, this report only lists biofuel concentrations of 140 mg/L after 100 hours of bacterial activity. The bacteria, *Ralstonia eutropha*, were also reported to have a low contamination tolerance, implying that expensive sterilization procedures will need to be in place for sufficient large scale production (Li et al. 2012). Other research with *Ralpha Eutropha* show that the maximum yield is approsimately 78% with concentrations of 4.5 g/L (Atsumi et al. 2008, Lu 2012). Atsumi et al. have begun looking at producing biofuels from cyanobacteria. Currently they are focusing on the production of 1-butanol, isopropene, and isobutanol using *S. elongates* and *Synechocystis* sp. (Atsumi et al. 2008, Machado and Atsumi 2012). The results show that fuel production is possible, but much more research is needed to match the current yields of other biofuel production methods. A summary of the maximum isobutanol yields obtained is given in Table 3. Much work has been accomplished in this field. However, the main issue with isobutanol production is the low yields from fermentation. Additionally, high costs for genetic modification and maintenance of bacterial strain purity puts a significant financial strain on the process. In order for isobutanol production to reach large scale implementation, these issues must fully be addressed. As such, current 1-butanol production technology possesses a significant advantage for industrial implementation and will become a major player in second generation biofuel production.

Table 3. Summary of Fermentation Bacteria and Maximum Yields.

Microorganism	Maximum Isobutanol Yield g isobutanol/g sugar	Study
Escherichia coli	0.42	Bastian et al. 2011
Bacillum subtilis	0.08	Li et al. 2011a
Saccharomyces cerevisiae	0.006	Kondo et al. 2012
Corynebacterium glutamicum	0.32	Blombach et al. 2011
Ralph eutropha	0.33	Atsumi et al. 2008

Classical ABE process

Butanol has been explored as a transportation fuel for many years, dating back to the First World War. A representative fermentation of 6C sugars with *C. acetobutylicum* can produce acetone, butanol and ethanol in the ratio of 3:6:1 acetone:butanol:ethanol. This fermentation has become known as ABE fermentation. Since that time, other bacteria have been employed for ABE fermentation with slightly different product distributions, nutrient requirements, and carbon source preference. Several bacterial strains developed from *C. acetobutylicum*, including *C. Beijerinckii, C. butyricum*, and *C. Saccharoperbutylacetonicum*, have been found to be productive with glucose, arabinose, and xylose (Chin et al. 1991, Ounine 1983). During batch processing, ABE solvents are generated in two distinct time-based phases. After a stable cell mass has been achieved, the bacteria will typically produce organic acids such as lactic acid, acetic acid, and butyric acid. This is the first phase of production and is characterized by a drop of pH from about 6.5 to less than 4.5, due to the formation of the organic acids. The first phase is termed the acidogenesis phase. The acidogenesis phase is also characterized by the evolution of gases, predominantly hydrogen. During the second metabolic phase, termed the solventogenesis phase, the bacteria will employ additional pathways, which convert the organic acids to the corresponding solvents. This phase is characterized by a gradual rise in pH as the acids are consumed and by increased carbon dioxide evolution. The transition from acidogensis to solventogenesis is probably driven by a predominance of the non-disassociated form of the organic acids, which readily traverse the cell membrane (Awang et al. 1988). The metabolism process for *C. acetobutylicum* is shown in Fig. 4. This metabolic scheme was developed in the mid-1970s by Doelle and Stanier, and has undergone some refinement over the years (Doelle 1975, Stanier et al. 1976). During batch fermentation processes, it was found that the initial glucose concentration needed to be above 60 g/L to ensure enough glucose remained during the second phase to support solvent production (Monot et al. 1982). Glucose limited broths are restricted to acid production. Limited organic nitrogen availability during the solventogenesis

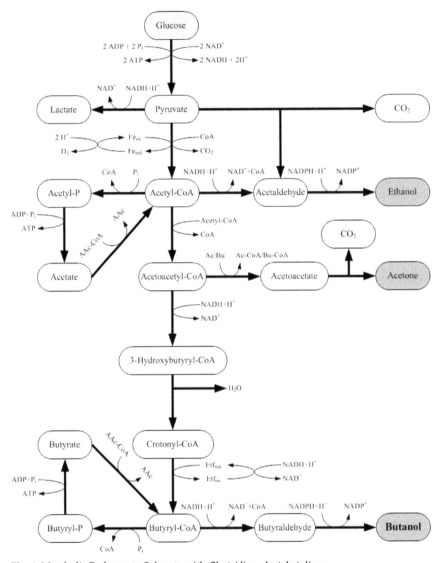

Fig. 4. Metabolic Pathway to Solvents with *Clostridium Acetobutylicum*.

phase leads to greater butanol productivity (Monot and Engasser 1983, Andersch et al. 1982). Organic nitrogen present during the first phase of fermentation promotes cell growth and acid generation. As the pH drops, the utilization of the organic nitrogen is inhibited and solventogenesis is enhanced. On the other hand, the presence of inorganic nitrogen appears to reduce cell growth, acid generation, and solvent production. In a like manner, sulfate and phosphate limitation enhances solventogensis (Bahl and

Gottschalk 1984). Higher levels of sulfate and phosphate shift production from hydrogen to lactic acid and limit solvent productivity and butanol selectivity. When hydrogen partial pressures were maintained at 3 to 5 atmospheres, butanol production increased by 18% (Yerushalmi et al. 1985). During solventgenesis, hydrogen production ceases as the excess reducing agent is used for solvent production (Awang et al. 1988).

Cultures of the various clostridial strains used for ABE fermentation undergo time-dependent degeneration with respect to solvent production. Strain degradation of *C. acetobutylicum* can be mitigated by limiting phosphate (Ezeji et al. 2005a). *C. beijerinckii* benefits from the addition of dilute acetate (Chen and Blaschek 1999). Several bacterial strains have been used for the ABE fermentation of 1-butanol. Table 4 lists several strains, and the carbohydrate feedstocks used by several researchers.

Table 4. Examples of ABE fermentation.

Feedstock	Bacterial strain	Reference
Algal biomass	*Pasteurianum* *Saccharoperbutylacetonicum*	Nakas et al. 1983 Potts et al. 2012
Apple Pomace	*Acetobutylicum, butyricum*	Voget et al. 1985a
Cassava	*Saccharoperbutylacetonicum*	Thang et al. 2010
Cheese Whey	*Acetobutylicum* Unspecified	Maddox et al. 1993 Stoeberl et al. 2011
Corn	*Acetobutylicum* *Beijerinckii* *Acetobutylicum* *Acetobutylicum*	Chiao and Sun 2007 Ezeji et al. 2005b Killeffer 1927 Weizmann 1915
Jerusalem artichokes	*Acetobutylicum*	Marchal et al. 1985
Molasses	*Beijerinckii* Various *Acetobutylicum*	Ezeji et al. 2005b D.T. Jones 2001 Dong et al. 2012
Potatoes	*Beijerinckii, acetobutylicum* *Acetobutylicum* *Acetobutylicum* *Acetobutylicum* *Saccharoperbutylacetonicum* *Acetobutylicum*	Gutierrez et al. 1998 Grobben et al. 1993 Weizmann 1915 Fernbach and Strange 1911 Al-Shorgani et al. 2011 Dong, 2011
Sweet potatoes	*Acetobutylicum* *Beijerinckii*	Chiao and Sun 2007 Ezeji et al. 2005b
Sago	*Saccharoperbutylacetonicum*	Al-Shorgani et al. 2012b
Argi-hydrolysate	*Acetobutylicum* *Beijerinckii*	Dong et al. 2010 Qusheri et al. 2008
Food Waste	*Acetobutylicum*	Patakova et al. 2011
Sorghum bagasse	*Acetobutylicum*	Zhang et al. 2012
Rice Bran	*Saccharoperbutylacetonicum*	Al-Shorgani 2012a

In addition to naturally occurring, selected strains, much research is ongoing with genetic modifications of bacterial strains. These efforts are focused on increasing the tolerance to butanol (Thormann et al. 2002, Allcock et al. 1981), increasing the selectivity to butanol compared to acetone or ethanol (Green et al. 1996, Tummala et al. 2003), alleviate the pronounced two phase metabolic cycle (Young et al. 1989, Papoutsakis and Bennett 1993, Blaschek and White 1995), and increase resistance to attack by phages (Jones et al. 2000).

Some of the difficulties in achieving high butanol productivity can be engineered away by designs of fermentation, *in situ* removal of inhibitory products, and increasing the effective specific cell mass in the reactor. Lost productivity in batch fermentation due to start up, shut down, and batch preparation can be minimized by operating continuous reactors. In continuous fermentation systems where production is coupled to cell growth, the productivity of a continuous stirred tank fermenter increases with feed rate until it reaches a maximum value. As the feed rate is further increased, the productivity decreases abruptly as cells are washed out of the reactor because cell generation is less than cell loss in the outlet stream from the reactor. There are two generally accepted methods for increasing productivity beyond this maximum, cell immobilization and cell recycle. Cell immobilization is a technique for retaining cells inside the reactor through attachment to a surface (Hu and Dodge 1985), entrapment within porous matrices (Cheetham et al. 1979), and containment behind a barrier or self-aggregation (Karel et al. 1985). Cell recycle is a technique for separating the cells from the product stream by centrifugation, filtration or settling in a conical tank, followed by returning the cells back to the reactor (Shuler and Kargi 2002). Of these two methods, cell immobilization is generally restricted to the laboratory because of significant fouling. In assessing cell recycle technologies, centrifugation to remove cells can be cost prohibitive, and simple settling with or without the addition of flocculating agents requires large tanks because of the similarity in densities between cells and the fermentation broth. Many improvements have been made in axial flow filtration, which have helped to reduce the cost of commercial application of these systems. Cell recycle on a two litre reactor allowed eight weeks of continuous fermentation with *C. Beijerinckii* when coupled with *in situ* removal of product (Potts et al. 2012). Membrane fouling, a common problem with cell recycle reactors was kept in check with an automated, time-based backflush system. Cell immobilization in packed bed reactors has been studied by several, with bonechar, clay brick particles, and polymeric substrates (Qureshi et al. 2004, Qureshi et al. 2005, Napoli et al. 2011). When coupled with one of several butanol removal techniques, described in the butanol separation section of this chapter, operation of continuous reactors for ABE production is a viable technical process.

Typical yields for ABE fermentation are in the range of 0.15–0.25 g of butanol per gram of glucose with productivities of 15.8 g/Lh (Qureshi et al. 2001), 9.5 g/Lh (Ramey 1998), 6.5–15.8 g/Lh (Qureshi and Ezeji 2008), 6.7 g/Lh (Ezeji et al. 2007), 0.2 g/Lh (Golueke et al. 1957), 0.2 g/Lh (Jesse et al. 2002). Unfortunately, both yield and productivity are limited by the presence of butanol in the broth, which is inhibitory to the fermentation. Some researchers report increasing yields substantially by employing *in situ* removal of the butanol during fermentation. *In situ* removal of the butanol can be coupled with cell recycle for additional gains, and in some cases, with carefully defined medium, individuals have reported achieving yields close to the theoretical or stoichiometric yield of 38% fermentation of glucose to solvent (Ramey 1998, Qureshi and Blaschek 2000). The many techniques used for *in situ* removal of the butanol are discussed elsewhere in this paper. It may be noted that those who report very high butanol productivities typically operate small reactors at highly optimized conditions. Operations in pilot plant and industrial scale reactors do not achieve these high levels, but 18–22 g/L is apparently the norm when high sugar substrates are fed to the bacteria (Ni and Sun 2009).

While much progress has been made in the enhancement of the economics of ABE processing, the difficulties of the cost of feedstock materials and the energy cost of separation of the butanol from the large mass of water inherent in the fermentation process. It appears to us that the efforts toward establishing low cost lignocellulose for other biofuels, such as bioethanol, are applicable to the biobutanol process as well. On the basis of feedstock costs, biobutanol looks like a very attractive replacement for bioethanol, due to several advantages butanol has as a vehicular fuel. The cost of butanol isolation and purification, thus, remains the most troublesome impediment for the replacement of ethanol with butanol. If some technical improvement can be made to ABE processing to relieve the purification costs, the future of biobutanol looks rosy, if the purification costs remain high, the future of biobutanol looks bleak. Separations will be discussed later in this book chapter.

Two step process

An interesting approach to improving the ABE fermentation was presented by Ramey (1998), who described a continuous process for producing butanol from sugars by using two different strains of bacteria. In this approach, C. *tyrobutyricum* (or other similar strains) is used to produce butyric acid from sugars and C. *acetobutylicum* (or other similar strains) is used to produce butanol from the butyric acid. These two steps are usually accomplished in two separate fermenters. Butyric acid from the first fermenter can be added to the second fermenter without concentration (Ramey 1998, Bahl and Gottschalk 1984) or by separation of the butyric acid from the broth of the first fermenter followed by addition to the second fermenter (Du et al. 2012). Comparable yields of butyric acid from glucose were obtained in using either *C. tyrobutyricum* or *C. thermobutyricum* (Weigel et al. 1989, Wu and Yang 2003). Liu et al. (2005) developed *C. tyrobutyricum* mutants which gave higher butyrate yields (>0.4 g/g) and concentrations (43 g/L). In producing butanol from sugars by either the direct or indirect fermentation routes, Ramey (1998) noted that 38% of the carbohydrate was converted to butanol by the indirect route using *C. tyrobutyricum* and *C. acetobutylicum*, while only 25% of the carbohydrate was converted to butanol using the direct route with *C. acetobutylicum* alone. Furthermore, Ramey (1998) noted that the butanol productivity increased by 78% when using the indirect fermentation route.

C. *tyrobutyricum* converts both glucose and xylose to butyric, acetic and lactic acids. Elevated pH (>6.3) is favorable for the production of butyric acid, and lower pH (<5.7) is more favorable for the production of acetic and lactic acids (Zhu and Yang 2004). Higher total acid yields are attained at reduced pH, but higher butyrate selectivities and concentrations are attained at increased pH. Table 5 shows a summary of acid production from glucose by *C. tyrobutyricum* as obtained by Wu and Yang (2003) and Du et al. (2012). As expected, immobilized cell systems outperformed free cell systems, and extractive fermentation systems outperformed both the free cell and immobilized cell systems because the solvent removes the inhibitory product butyrate. However, acid extraction can only occur at lower pH levels since most solvents extract products only in the free acid form, and many solvent systems are not particularly selective. In the work by Du et al. (2012), the separation is done at a neutral pH with high selectivity and productivity of butyric acid.

Table 5. Comparison of Fermentation Results from Free Cell, Immobilized Cell and Extractive Fermentations Using *C. tyrobutyricum* ATCC 25755 (Wu and Yang 2003, Du et al. 2012).

	Free Cells (pH 6.0)	Immobilized Cells (pH 6.0)	Immobilized Cells (pH 5.5)	Extractive Fermentation (pH 5.5)	EDI Separation (pH 6.3)
OD$_{max}$	5.8	11.5	8.2	8.1	14.5
Butyrate concentration (gl^{-1})	16.3	43.4	20.4	301	>150
Butyrate yield (gg^{-1})	0.34	0.42	0.38	0.45	.45
Acetate yield (gg^{-1})	0.120	0.095	0.115	0.111	N/A
Butyrate productivity (gl^{-1}h^{-1})	0.193	6.77	5.11	7.37	
Product selectivity	0.74	0.81	0.77	0.80	.92
Product purity				0.91	

By feeding a mixture of glucose and butyric acid (from *C. tyrobutyricum*), *C. acetobutylicum* can remain in the solvent producing stage and produce higher yields and concentrations of butanol than when feeding sugars to *C. acetobutylicum*. Huang et al. (2004) fed a mixture of glucose and butyrate to a fibrous bed bioreactor containing *C. acetobutylicum* ATCC 55025 at 35°C and pH 3.5–5.5. An optimal butanol productivity of 4.6 g/Lh and a butanol yield of 0.42 g/g were obtained at a dilution rate of 0.9 h^{-1} and a pH of 4.3 with 54 g/L glucose and 3.6 g/L butyric acid in the feed stream. The concentration of butanol was 5.1 g/L on average, and the conversions of glucose and butyric acid were 19% and 31%, respectively. The optimum solvent (ABE) yield was 0.53 g/g, under the same process conditions. By contrast, the optimum single step (conventional) ABE fermentation has an optimum butanol yield of 0.25 g/g and a productivity of 4.5 g/Lh.

As described in the ABE process section, the product concentration in the two fermenters can be increased with cell recycle or immobilization. Concentrations of butyric acid in the first fermenter and of butanol in the second fermenter are limited by product inhibition. Product inhibition can be minimized by *in situ* removal of the butyric acid from the fermenter and *in situ* removal of butanol from the second reactor. The techniques used for these separations are discussed in the section on separation and purification. Once butyrate is concentrated, it must then be converted into butanol via reaction or a second fermentation step (Green and Crow 2009).

Separations and Purification

In situ product removal is designed to increase the yield and productivity of a fermentation process by (Freeman et al. 1993):

1. Minimizing the effects of product inhibition on the producing cell, thus allowing for continuous expression at the maximum production level;
2. Minimizing product losses resulting from cross-interaction with the producing cell, environmental conditions or uncontrolled removal from the system (e.g., by evaporation); or
3. Reducing the number of subsequent downstream processing steps.

The product yield is set by overall stoichiometry, the production of cells and cell maintenance. However, in fermentation systems that produce multiple liquid phase products, selective *in situ* removal of one of the products may cause the fermentation system to overproduce that product, and thereby increase the yield of that product relative to the other products in the product matrix. This phenomenon is illustrated in the following examples. Wu and Yang (2003), in fermenting glucose to butyric and acetic acids using *C. tyrobutyricum* with and without *in situ* removal of products by solvent extraction, utilized an amine-based solvent system that preferentially (but not totally) extracted butyric acid over acetic acid. Without product removal, their fed-batch system gave a butyric acid yield of 0.34 g/g and an acetic acid yield of 0.12 g/g, for a product selectivity of 0.74. With product extraction, the overall butyric acid yield was 0.45 g/g and the acetic acid yield was 0.11g/g, for a product selectivity of 0.80. Thus, the fermentation system produced more butyric acid than acetic acid as the butyric acid was preferentially removed by extraction.

Similarly, Grobben et al. (1993), in fermenting potato wastes to acetone, butanol and ethanol using *C. acetobutylicum* with and without *in situ* removal of products by perstraction, utilized a solvent system that preferentially removed butanol (K=3.5) over acetone (K=0.65) and ethanol (K=0.2). Without product removal, their fed-batch system steadied at 12 g/L of butanol, 4 g/L of acetone and just under 1 g/L of ethanol. With product removal, the butanol concentration (both extracted and in the fermenter) reached 39 g/L and the acetone concentration reached 11.5 g/L. In both of these fermentation systems, the preferentially extracted product (butyric acid in the *C. tyrobutyricum* system and butanol in the *C. acetobutylicum* system) was preferentially produced over the lesser extracted product. If a separation system could be developed that removed only butanol (or butyric acid), perhaps only minimal amounts of acetone and ethanol (or acetic acid) would be produced and nearly all of the sugar substrate would be diverted to the preferred product.

In situ butyric acid separation for two step production of butanol

The two most common technologies proposed for separating organic acids from water are liquid-liquid extraction (Mamade and Pastore 2006, Matsumoto et al. 2004, Zigova and Sturdik 2000, Ramey 1998) and electrodialysis (Wang et al. 2006, Hestekin et al. 2002, Nghiem et al. 2001, Huang et al. 2007, Lee 2005). Although liquid-liquid extraction works well, the acid typically has to be protonated for efficient extraction. The pKa of butyric acid is 4.82 and this low pKa requires lowering the pH of the solution from optimal fermentation range (near neutral) in order to remove the organic acids.

Electrodialysis (ED) has a significant advantage with pH flexibility—the removal of products from fermentation broth can occur at a pH that enables the formation of products in their ionized form, a condition that is more desirable for fermentation processes. Huang et al. (2007) published a review on ED techniques, costs and production capacities, showing that ED has been effectively used in concentrating amino acids, lactic acid, citric acid, formic acids, and butyric acid. However, ED is not typically selective for one organic acid over another. In contrast, electrodeionization (EDI) may be selective for particular organic acids because this technology combines selectivity in the membrane with selectivity of the ion exchange resins. Thus, EDI is more effective for *in situ* product recovery from fermentation broths than ED (Datta et al. 2002).

EDI has been used for concentration of organic acids. Specifically, Widiasa et al. (2004) concentrated citric acid from 2,000 to 60,000 ppm. Arora et al. (2007) and Lin et al. (2004) separated organic acids from fermentation broths with initial concentrations below 1 g/L to final concentrations of less than 1 mg/L. Unfortunately, neither of these studies addressed selective organic acid separation. Interestingly, Semmens and Gregory (1974) demonstrated that ion exchange beads become more selective for organic acids as chain length increases. Takahashi et al. (2003) reported that the selectivity of organic acids with anion exchange membranes could be described by the nature of the ion exchange selectivity but, as the length of the chain increases, hydrophobic interactions increase and thus the selectivity of the membrane starts to increase for that acid.

In situ butanol separation

There are many techniques that have been used for the separation of butanol from aqueous streams including liquid-liquid extraction (Ezeji et al. 2007), distillation (Skouras and Skogestad 2004), and gas stripping (Ezeji et al. 2004). However, when separation is used in conjunction with a fermentation process, preservation of substrate, organisms and medium components must be considered in applying any separation technique. Liquid-liquid extraction using organic solvents, although quite useful in performing the separation, often results in the contamination of the recycled fermentation medium by ppt or ppm levels of the solvent which can result in the inhibition of the fermentation or even cell death. Distillation, although highly effective in recovering a number of fermentation products, can thermally degrade sugar substrates or other medium components or produce thermal by-products. In addition, azeotropes can develop during distillation, which limit the separation. For these reasons, the most commonly proposed method for selective extractions of butanol from fermentation broths is pervaporation (Ezeji et al. 2007, Liu et al. 2011, Vane 2005, Yeom et al. 2000, Jitesh et al. 2000, Liu et al. 2005, Thongsukmak and Sirkar 2007, Olsson et al. 2001, Qureshi et al. 1999, Park and Geng 1996).

Distillation

For butanol, the most common form of separation is multi-step distillation. Butanol obtained from fermentation is fed to a distillation unit at a concentration between 1–3 weight percent, which is much lower than found in ethanol systems where typical concentrations range between 9–13 weights percent (Pfromm et al. 2010). The distillation unit generates a distillate where the concentration of butanol is at the azeotropic point of 55.5 weight percent. Then the partially pure butanol is allowed to phase-separate in a decanter into butanol and water rich phases. Next, the butanol rich phase is fed into a second distillation unit to bring it up to purity while the water rich phase is fed back to the first distillation unit. A schematic of this procedure is given in Fig. 5.

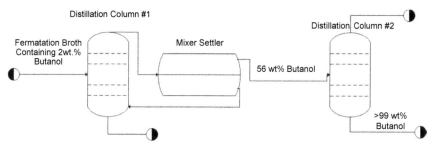

Fig. 5. Process flow diagram of current butanol distillation technology.

Membrane separation

Another method that shows promise in butanol purification is membrane separation (Garcia et al. 2011). Membranes can be used within the fermentation system to continuously draw out product while keeping the bacterial environment stable, effectively maximizing productivity without dealing with bacterial tolerance. Membranes can also be used to effectively separate out butanol without extracting out acetone and ethanol side products. This would encourage bacteria to break down these side products into butanol, further improving yields. Membrane extraction (perstaction) is another effective method for butanol purification. This is similar to general membrane extraction except the use of a chemical potential gradient is applied instead of a pressure driving force. An extractive solvent is placed on one side of the membrane system and fermentation broth is flowed past the membrane. Butanol's affinity for the solvent allows it to move through the membrane into the solvent. This allows for fast and efficient production separation for fermentation systems. Studies using membrane based extraction techniques have proven to reduce product inhibition while boosting glucose consumption and overall productivity (Tanaka et al. 2012, Jeon and Lee 1989).

Pervaporation

Another membrane separation for biobutanol production is pervaporation. Fermentation broth is flowed past a membrane with a vacuum imposed on the other side. Volatile components are pulled through the membrane and vaporized by the low pressure on the vacuum side of the membrane. The products are then condensed and collected. A schematic of this process is shown in Fig. 6. Figure 6 also shows a significant advantage of pervaporation that it can separate outside the vapor-liquid equilibrium of butanol water allowing for much higher selectivity to be obtained. Many studies have been conducted in order to model pervaporation technology for butanol recovery (Li and Srivastava 2011b, El-Zanati et al. 2006, Wijmans and Baker 1995). Other studies focused on membrane materials such as zeolites (Bowen et al. 2002, Bowen et al. 2003, Pera-Titus et al. 2006), polymeric resins (Nielsen and Prather 2008), mix matrix membranes (Wang et al. 2009), and silica based membranes (Hickey and Slater 1991, Fouad and Feng 2009). Some studies focused on studying the effects of fouling and concentration polarization in order to reduce these adverse effects (Qureshi and Blaschek 1999, Fouad and Feng 2008). Consensus of results implies that pervaporation is a very suitable butanol separation technique and is capable of moving butanol fermentation away from batch schemes into continuous processes. Pervaporation systems involving butanol can be operated at low temperatures while still performing selective separation of butanol, and can be employed on both laboratory and commercial scales (Vane 2005). In most of these pervaporation applications, butanol is the most selective component because it is the most hydrophobic component and has the highest solubility in the membrane (Olsson et al. 2001). As an example, Liu et al. (2011) used polydimethylsiloxane based membranes on a quaternary mixture of acetone (1.57%), ethanol (0.9%) and butanol (1.11%) in water and found selectivities to water of 2.4, 3.8 and 9.6 respectively. When the concentration of butanol was increased relative to ethanol and acetone, the selectivity of butanol to the other components also increased. When more selective membranes such as a liquid pervaporation membrane (Thongsukmak and Sirkar 2007) or a silicalite/silicone membrane (Qureshi et al. 1999) were used, butanol selectivities ranging from 40–100 were obtained. Park and Geng (1996) showed that increasing the pervaporation membrane area in a fed-batch butanol production system yielded higher glucose consumption rates and higher production rates.

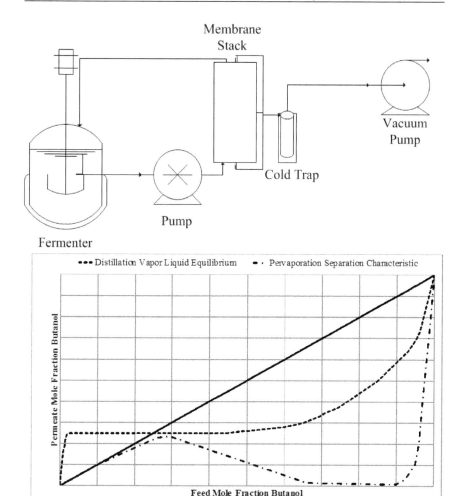

Fig. 6. (Top) Process Schematic for Butanol Pervaporation (Bottom) Comparison of Pervaporation and Distillation Separation Characteristics.

Gas sparging

This process involves bubbling an inert gas (typically nitrogen or CO_2) through the fermentation broth. Volatile components in the fermentation broth (butanol) are stripped from solution and into the vapor phase where it is carried out of the fermentation system. The vapor is then condensed and collected as product. Results from gas sparging implementation show that high flowrates of inert gas can maximize product recovery (Ezeji et al. 2005b). However, inhibitory effects were increased, resulting in lower glucose consumption rates (Park et al. 1991).

Liquid-liquid extraction (LLE)

In LLE a third component is added to a binary mixture in order to create a second liquid phase, one which has a high concentration of product and one with a high concentration of the original solvent. Common practices involve adding an organic solvent to an aqueous solution in order to extract out an organic product. The second phase is then processed through another separation step (which is generally simple and cheap) in order to obtain pure product. Recently ionic liquids have been studied for use in LLE of butanol fermentation systems. Their chemical versatility makes them ideal solvents for butanol extraction (Fadeev and Meagher 2001). Eleven ionic liquids were studied for butanol selectivity and extraction efficiency. Results showed that butanol selectivity can exceed 100 with good extraction efficiencies ranging between 60–80% (Ha et al. 2010). Other studies have focused on using hexane (Gomis et al. 2012), other alcohols (Takriff et al. 2008), surfactants (Dhamole et al. 2012), and biodiesel as an extractant (Adhami et al. 2009).

A summary of butanol separation techniques is provided in Table 6. Each technique has their advantages and limitations, but each can be an effective method for separation with the proper process in place. As research continues, each method will become more cost effective and efficient in preferentially separating butanol from other unwanted fermentation products.

Table 6. Summary of Butanol Separation Methods.

Separation Method	Key Advantages	Major Limitations	Approximate Cost ($)
Distillation	Fast, efficient, low capital cost	High energy and operating cost	$$$$
Membrane Separations/Filtration	Selectivity and Low energy requirements	Time intensive and high capital costs	$$$$
Perstaction	Selectivity and simple process	Fouling, slow, high capital cost	$$$
Pervaporation	Selectivity	Fouling, Slow, high capital & operating costs	$$$
Electrodialysis	Very high selectivity	High capital cost, slow	$$$
LLE	Inexpensive	Not well suited for continuous processing	$$
Gas Sparging	Inexpensive	Low selectivity, high processing cost (compressed gas)	$$

Biofuel Production Case Studies

In order to create a fair industrial perspective of current biofuel technology, case studies were developed for butanol and ethanol production. Each case study focused on the capital and operating costs of producing 10 million gallons of biofuel per year. This was done with a combination of literature, scaling using well known chemical engineering practices, and simulation (when necessary). Figure 7 shows the major steps in producing ethanol and butanol respectively. As shown, the front end of both processes are the same with significant differences upon production of sugars. Commentary on the major limitations and potential for each biofuel is also presented.

Bioethanol case study

Bioethanol is currently the largest biofuel being producing globally. With feedstocks and technology readily available, production of ethanol has steadily increased over the past decade. For this case study, costs reported are higher than estimates for larger scale plants on a per gallon basis. This is due to lower economies of scale values than what is expected for full scale production facilities. The major costs are broken down in six main areas: raw materials, milling, saccharification, fermentation, distillation, and dehydration. Utilities, additional products, and depreciation are all considered. For ethanol biofuels, the main raw material cost is the feedstock, corn. In a 10 million gallon plant, approximately 3.6 million bushels of corn are needed since the productivity of corn ethanol production is 2.8 gallons per bushel (Wu 2007). The price of corn is at approximately $7.38 dollars per bushel, resulting in a feedstock cost of $26.3 million per year (www. quotecorn.com). This high cost makes producing solely ethanol from corn a non-profitable endeavor. However, dried distillers grain with solubles (DDGS) is also produced in corn ethanol plants, improving the economics of the process and making plants feasible. About 18 pounds of DDGS is produced per bushel of corn, resulting in approximately 30,000 metric tons of DDGS produced. Additional raw materials required include bacteria, enzymes, and chemicals for saccharification and fermentation. Total cost for these materials are approximately $27.5 million per year (Whims 2002).

The next costs to consider are costs from milling and saccharification of corn to extract the fermentable sugars. There are two options for this process, dry milling and wet milling. In most ethanol plants, dry milling is used due to the lower capital costs (Whims 2002). Wet milling only becomes profitable for large ethanol production facilities (> 50 million gallons ethanol produce per year). The costs of corn milling are approximately $2.56 million in capital. Saccharification and liquefaction costs mainly include

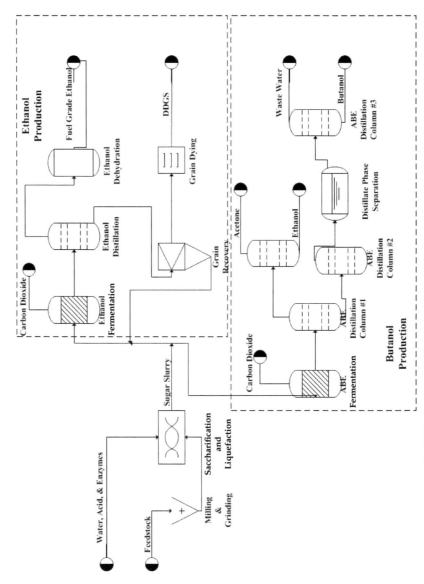

Fig. 7. Comparison of Ethanol and Butanol Current Production Methods.

requirements for enzymes, acids, and water for the breakdown of corn kernels and extraction of sugars. Sulfuric acid is commonly used with an amylase based enzyme (Kwiatkowski et al. 2006). Costs for saccharification are $4.01 million.

The next step is the fermentation of sugars. Yeast is the most commonly used microorganism for this process. Fermentation occurs in batches and generally takes 45 hours per batch (Pfromm et al. 2010). Capital costs for fermentation are $7.92 million. Separation of the products is the final step in production. These costs are divided into distillation and dehydration of ethanol and purification of solids into DDGS. These costs are approximately $6.04 million and $13.44 million respectively. Costs for DDGS processing is high due to the centrifugation and drying steps required in removing water from the DDGS (Kwiatkowski et al. 2006). From this information, total capital and operating costs were obtained. All values were obtained by various case studies of ethanol economics and time correct to meet today's price expectations. Tables 7 and 8 show the estimated capital and operating costs associated with the 10 million gallon facility.

Economic feasibility can be determined by examination of cash flows expected from this facility. The current price of ethanol and DDGS is approximately $2.50/gallon and $350/MT respectively. From these prices the expected revenue from a 10 million gallon ethanol facility is approximately 35.5 million dollars (nasdaq.com, grains.org). The revenue generated from this production facility would be insufficient for the investment required.

Table 7. Estimated capital costs for 10 million gallon ethanol production facility.

Plant Section	Capital Costs (US$ million)
Milling	2.56
Saccharification	4.01
Fermentation	7.92
Ethanol Purification	6.04
Byproduct Processing	13.44
Support Systems	1.28
Total	35.25

Table 8. Estimated Operating costs for 10 million gallon ethanol production facility.

Category	Cost (US$ million)
Raw Materials	27.5
Fuel Costs	1.83
Electricity	1.68
Labor and Maintenance	1.63
Administrative and Other	0.93
Total	33.57

In order for this process to become economically feasible, the process would need to have sufficient scale-up such that economies of scale allow revenue to overtake the increase in capital and operating costs. This has been demonstrated in larger ethanol production facilities. Another possibility is the use of government subsidies in the form of tax credits to mitigate costs and allow the plant to be profitable. Historically small scale facilities were subsidized by tax credits to allow proper scale-up so that plants could become self-sustaining once it reached set productivity levels. With the loss of ethanol subsidies, small scale facilities are no longer economically viable. Current large scale facilities operate at profit. However, as next generation biofuels continue to develop, ethanol production will level off and newer fuels like butanol will see a major boost in production.

Biobutanol case study

The similar costs for the butanol estimates, found in Tables 9 and 10, were gleaned from several sources and recalculated as needed to match the 10 million gallons per year butanol production rate (Durre 2007, Gapes 2000, Qureshi 2013, Zhu and Yang 2004). The raw materials cost were estimated by assuming $100/ton for wheat straw and $7.38/bushel for the corn (www.quotecorn.com 2012). The productivity of wheat straw to butanol was assumed to be 147 kg of butanol from 751 kg of wheat straw (Qureshi

Table 9. Estimated capital costs for 10 million gallon butanol production facility.

Plant Section	Capital Costs (US$ million) Corn Starch	Capital Costs (US$ million) Wheat Straw
Milling	3.91	2.86
Saccharification	6.12	4.49
Fermentation	14.56	19.87
Butanol Purification	4.00	4.00
Byproduct Processing	13.44	18.34
Support Systems	1.28	1.28
Total	43.32	50.84

Table 10. Estimated Operating costs for 10 million gallon butanol production facility.

Category	Cost (US$ million) Corn Starch	Cost (US$ million) Wheat Straw
Raw Materials	29.50	15.80
Utility Costs	2.20	2.20
Labor and Maintenance	1.32	1.32
Administrative and Other	0.98	0.98
Total	34.00	20.30

2012). The productivity of corn starch to butanol was assumed to be 2.5 bushels of corn to one gallon of butanol (Ramey 2004). Capital equipment costs for milling and saccharification were assumed to be similar, regardless of feedstock. Capital costs for purification were calculated using standard estimating techniques, as were the operating costs for fuels, labor and maintenance, and administrative and other. For a plant 5 times larger, Zhu estimates the cost per gallon for butanol from corn starch at $2.56/gallon and from wheat straw at $2.15/gallon. These numbers are consistent with those of Tables 8 and 9 if you do not have to amortize capital.

Comparing the relative costs of ethanol production to butanol production a few conclusions can be made. First, neither process is economical for fuel production at this scale without government subsidies or a major price increase. Second, when comparing the capital costs of the two processes, butanol is much higher due to the much larger fermenters required because of low yield organisms. Thus, the continued research in high yield organisms is justified. However, it should be noted that separations research can lead to lower fermenter sizes as well. In the operating costs, surprisingly, there doesn't seem to be much difference in butanol or ethanol production. This has been reported elsewhere (Zhu and Yang 2009).

Conclusions

- Figure 8 shows a spike in the amount of butanol related scientific literature in the last 5 years. This demonstrates that it is clearly being researched as a biofuel as well as a solvent.

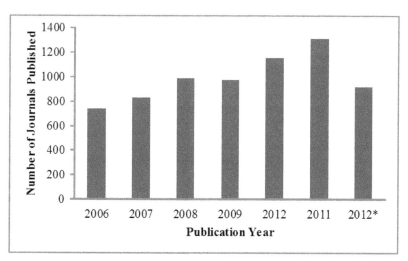

Fig. 8. Butanol Journal Publications by Year. Data Obtained from Web of Science. *As of November 2012.

- Research is taking place on isobutanol production, one step 1-butanol production, and two step 1-butanol production. At present, one step 1-butanol production is the only process that is large scale commercialized.
- Costs of butanol remain high because of low organism productivity and difficulty of operation.

References

Adhami, L., B. Griggs, P. Himebrook and K. Taconi. 2009. Liquid-Liquid Extraction of Butanol from Dilute Aqueous Solutions Using Soybean-Derived Biodiesel. J. Am. Oil Chem. Soc. 86: 1123–1128.

Allcock, E. R., S. J. Reid, D. T. Jones and D. R. Woods. 1981. Autolytic activity and an autolysis-deficient mutant of *Clostridium acetobutylicum*. Appl. Environ. Microbiol. 42: 929–35.

Al-Shorgani, N. K. N., M. S. Kalil and W. M. W. Yusoff. 2011. The Effect of Different Carbon Sources on Biobutanol Production using *Clostridium saccharoperbutylacetonicum* N1-4. Biotechnol. 10: 280–285.

Al-Shorgani, N. K. N., M. S. Kalil and W. M. W. Yusoff. 2012a. Fermentation of sago starch to biobutanol in a batch culture using *Clostridium saccharoperbutylacetonicum* N1-4 (ATCC 13564). Ann. Microbiol. (Heidelberg, Ger). 62: 1059–70.

Al-Shorgani, N. K. N., M. S. Kalil and W. M. W. Yusoff. 2012b. Biobutanol production from rice bran and de-oiled rice bran by *Clostridium saccharoperbutylacetonicum* N1-4. Bioprocess Biosyst. Eng. 35: 817–26.

Altun, S., C. Oner, F. Yasar and H. Adin. 2011. Effect of n-butanol blending with a blend of diesel and biodiesel on performance and exhaust emissions of a diesel engine. Ind. Eng. Chem. Res. 50: 9425–30.

Andersch, W., H. Bahl and G. Gottschalk. 1982. Acetone-butanol production by *Clostridium acetobutylicum* in an ammonium-limited chemostat at low pH values. Biotechnol. Lett. 4: 29–32.

Arora, M. B., J. A. Hestekin, S. W. Snyder, E. J. St. Martin, M. I. Donnelly, C. Sanville-Millard and Y. J. Lin. 2007. The Separative Bioreactor: A Continuous Separation Process for the Simultaneous Production and Direct Capture of Organic Acids. Sep. Sci. Tech. 42: 2519–2538.

Atsumi, S., T. Hanai and J. C. Liao. 2008. Non-fermentative pathways for synthesis of branched-chain higher alcohols as biofuels. Nature. 451: 86–90.

Atsumi, S., T. Y. Wu, E. M. Eckl, S. D. Hawkins, T. Buelter and J. C. Liao. 2009. Engineering the isobutanol biosynthetic pathway in *Escherichia coli* by comparison of three aldehyde reductase/alcohol dehydrogenase genes. Appl. Microbiol. Biotechnol. 85: 651–657.

Atsumi, S., T. Y. Wu, I. M. P. Machado, W. C. Huang, P. Y. Chen, M. Pellegrini and J. C. Liao. 2010. Evolution, genomic analysis, and reconstruction of isobutanol tolerance in *Escherichia coli*. Mol. Syst. Biol. 6: 449.

Awang, G. M., G. A. Jones and W. M. Ingledew. 1988. The acetone-butanol-ethanol fermentation. Crit. Rev. Microbiol. 15 Suppl. 1: S33–67.

Baez, Antonio, K. M. Cho and J. C. Liao. 2011. High-flux isobutanol production using engineered *Escherichia coli*: a bioreactor study with *in situ* product removal. Appl. Microbiol. Biotechnol. 90: 1681–1690.

Bahl, H. and G. Gottschalk. 1984. Parameters affecting solvent production by *Clostridium acetobutylicum* in continuous culture. Biotechnol. Bioeng. Symp. 14: 215–23.

Bastian, S., X. Liu, J. T. Meyerowitz, C. D. Snow, M. M. Y. Chen, and F. H. Arnold. 2011. Engineered ketol-acid reductoisomerase and alcohol dehydrogenase enable anaerobic

2-methylpropan-1-ol production at theoretical yield in *Escherichia coli*. Metab. Eng. 13: 345–352.

Blaschek, H. P. and B. A. White. 1995. Genetic systems development in the clostridia. FEMS Microbiol. Rev. 17: 349–56.

Blombach, B. and B. J. Eikmanns. 2011. Current knowledge on isobutanol production with *Escherichia coli, Bacillus subtilis* and *Corynebacterium glutamicum*. Bioeng. Bugs. 2: 346–350.

Blombach, B., T. Riester, S. Wieschalka, C. Ziert, J. W. Youn, V. F. Wendisch and B. J. Eikmanns. 2011. *Corynebacterium glutamicum* Tailored for Efficient Isobutanol Production. Appl. Environ. Microbiol. 77: 3300–3310.

Bowen, T. C., H. Kalipcilar, J. L. Falconer and R. D. Noble. 2002. Separation of C4 and C6 isomer mixtures and alcohol-water solutions by monolith supported B-ZSM-5 membranes. Desalination. 147: 331–332.

Bowen, T. C., S. Li, R. D. Noble and J. L. Falconer. 2003. Driving force for pervaporation through zeolite membranes. J. Membrane Sci. 225: 165–176.

Cheetham, P. S. J., K. W. Blunt and C. Bucke. 1979. Physical Studies on Cell Immobilization Using Calcium Alginate Beads. Biotechnol. Bioeng. 21: 2155–2168.

Chen, C. and H. P. Blaschek. 1999. Acetate enhances solvent production and prevents degeneration in *Clostridium beijerinckii* BA101. Appl. Microbiol. Biotechnol. 52: 170–3.

Chen, C., Z. Tsai and B. Chen. 2011. Potential bacterial strains for biobutanol production. Shiyou Jikan. 47: 85–98.

Chen, X., K. F. Nielsen, I. Borodina, M. C. Kielland-Brandt and K. Karhumaa. 2011. Increased isobutanol production in *Saccharomyces cerevisiae* by overexpression of genes in valine metabolism. Biotechnol. for Biofuels. 4: 21.

Chiao, J. and Z. Sun. 2007. History of acetone-butanol-ethanol fermentation industry in China: Development of continuous production technology. J. Mol. Microbiol. Biotechnol. 13: 12–4.

Chin, C. S., W. Z. Hwang and H. C. Lee. 1991. Isolation and characterization of mutants of *Clostridium saccharoperbutylacetonicum* fermenting bagasse hydrolyzate. J. Ferment. Bioeng. 72: 249–53.

Datta, R., Y. Lin, D. Burke and S. Tsai. 2002. Electrodeionization Substrate, and Device for Electrodeionization Treatment. U.S. Patent. 6,495,014.

Dhamole, P. B., Z. Wang, Y. Liu, B. Wang and H. Feng. 2012. Extractive fermentation with non-ionic surfactants to enhance butanol production. Biomass Bioenerg. 40: 112–119.

Doelle, H. W. 1975. Bacterial metabolism. 2nd ed. 736.

Dogan, O. 2011. The influence of n-butanol/diesel fuel blends utilization on a small diesel engine performance and emissions. Fuel. 90: 2467–72.

Du, J., N. Lorenz, R. R. Beitle and J. A. Hestekin. 2012. Application of wafer-enhanced electrodeionization in a continuous fermentation process to produce butyric acid with *Clostridium tyrobutyricum*. Sep. Sci. Technol. 47: 43–51.

Durre, P. 2007. Biobutanol: An attractive Biofuel. Biotechnol. 2: 1525–1534.

Edmunds. 2007. http://www.edmunds.com/fuel-economy/e85-vs-gasoline-comparison-test.html.

El-Zanati, E., E. Abdel-Hakim, O. El-Ardi and M. Fahmy. 2006. Modeling and simulation of butanol separation from aqueous solutions using pervaporation. J. Membrane Sci. 280: 278–283.

Ezeji, T. C., N. Qureshi and H. P. Blaschek. 2004. Butanol Fermentation Research: Upstream and Downstream Manipulations. The Chemical Record. 4: 305–314.

Ezeji, T. C., N. Qureshi and H. P. Blaschek. 2005a. Continuous butanol fermentation and feed starch retrogradation: Butanol fermentation sustainability using *Clostridium beijerinckii* BA101. J. Biotechnol. 115: 179–87.

Ezeji, T. C., N. Qureshi and H. P. Blaschek. 2005b. Industrially relevant fermentations. Handb Clostridia. 797–812.

Ezeji, T. C., N. Qureshi and H. P. Blaschek. 2007. Bioproduction of butanol from biomass: from genes to bioreactors. Curr. Opin. in Biotech. 18: 220–227.

Fadeev, A. G. and M. M. Meagher. 2001. Opportunities for ionic liquids in recovery of biofuels. Chem. Commun. 295–296.

Fernbach, A. and E. H. Strange. 1912. Acetone and higher alcohols.

Fouad, E. A. and X. Feng. 2008. Use of pervaporation to separate butanol from dilute aqueous solutions: Effects of operating conditions and concentration polarization. J. Membrane Sci. 323: 428–435.

Fouad, E. A. and X. Feng. 2009. Pervaporative separation of n-butanol from dilute aqueous solutions using silicalite-filled poly(dimethyl siloxane) membranes. J. Membrane Sci. 339: 120–125.

Freeman, A., J. M. Woodley and M. D. Lilly. 1993. *In Situ* Product Removal as a Tool for Bioprocessing. Biotechnol. 11: 1007–1012.

Gapes, J. R. 2000. The economics of acetone-butanol fermentation: Theoretical and market considerations. J. Mol. Microbiol. Biotechnol. 2: 27–32.

Garcia, V., J. Paekkilae, H. Ojamo, E. Muurinen and R. L. Keiski. 2011. Challenges in biobutanol production: How to improve the efficiency?. Renewable Sustainable Energy Rev. 15: 964–80.

Gevo. 2012. Isobutanol—A renewable solution for the transportation fuels value chain. Transportation Fuels.

Glassner, D. A. 2009. Hydrocarbon Fuels from Plant Biomass. Gevo Presentation.

Gomis, V., A. Font, M. D. Saquete and J. Garcia-Cano. 2012. LLE,VLE, and VLLE data for the water-n-butanol-n-hexane system at atmospheric pressure. Fluid Phase Equilibr. 316: 135–140.

Green, E. and M. Crow. 2009. Butanol and other alcohol production process. UK Patent Application 0903501.

Green, E. M. 2011. Fermentative production of butanol—the industrial perspective. Curr. Opin. Biotech. 22: 337–343.

Green, E. M., Z. L. Boynton, L. M. Harris, F. B. Rudolph, E. T. Papoutsakis and G. N. Bennett. 1996. Genetic manipulation of acid formation pathways by gene inactivation in *Clostridium acetobutylicum* ATCC 824. Microbiology (Reading, U K). 142: 2079–86.

Grobben, N. G., G. Eggink, F. P. Cuperus and H. J. Huizing. 1993. Production of acetone, butanol and ethanol (ABE) from potato wastes: Fermentation with integrated membrane extraction. Appl. Microbiol. Biotechnol. 39: 494–8.

Gutierrez, N. A., I. S. Maddox, K. C. Schuster, H. Swoboda and J. R. Gapes. 1998. Strain comparison and medium preparation for the acetone-butanol-ethanol (ABE) fermentation process using a substrate of potato. Bioresour. Technol. 66: 263–5.

Ha, S. H., N. L. Mai and Y. M. Koo. 2010. Butanol recovery from aqueous solution into ionic liquids by liquid-liquid extraction. Process Biochem. 45: 1899–1903.

Herndon, A. 2012. Cobalt, Solvay to Convert Brazil Sugar-Cane Waste Into Chemicals. Bloomberg. http://www.bloomberg.com/news/2012-08-01/cobalt-solvay-to-convert-brazil-sugar-cane-waste-into-chemicals.html.

Hestekin, J. A., S. Snyder and B. Davison. 2002. Direct Product Capture from Biotransformations. ChemPlus Report, Chemical Vision 2020.

Hickey, P. J. and C. S. Slater. 1991. The Separation of Butanol-Water Solutions by Pervaporation: A Comparison of Silicone Based Membranes. Conference Proceedings. 447–459.

Hu, W. and T. C. Dodge. 1985. Cultivation of Mammalian Cells in Bioreactors. Biotechnol. Prog. 1: 4–10.

Huang, W. C., D. E. Ramey and S. T. Yang. 2004. Continuous Production of Butanol by *Clostridium acetobutylicum* Immobilized in a Fibrous Bed Bioreactor. App. Biochem. Biotech. 113–116, 887–898.

Huang, C., T. Xu, Y. Zhang, Y. Xue and G. Chen. 2007. Application of Electrodialysis to the Production of Organic Acids; State of the Art and Recent Developments. J. Membrane Sci. 288: 1–12.

Huntington, H. G. 1986. The US dollar and the world oil market. Energy Policy. 14: 299–306.
Jaffe, A., K. Medlock and R. Soligo. 2011. The status of world oil reserves: conventional and unconventional resources in the future supply mix. James A. Baker III Institute for Public Policy, Rice University, October 2011.
Jeon, Y. J. and Y. Y. Lee. 1989. *In situ* product separation in butanol fermentation by membrane-assisted extraction. Enzyme Microb. Tech. 11: 575–582.
Jesse, T. W., T. C. Ezeji, N. Qureshi and H. P. Blaschek. 2002. Production of butanol from starch-based waste packing peanuts and agricultural waste. J. Ind. Microbiol. Biotechnol. 29: 117–23.
Jia, X., S. Li, S. Xie and J. Wen. 2012. Engineering a Metabolic Pathway for Isobutanol Biosynthesis in *Bacillus subtilis*. Appl. Biochem. Biotechnol. 168: 1–9.
Jones, D. T. 2001. Applied acetone-butanol fermentation. Clostridia. 125–68.
Jones, D. T., M. Shirley, X. Wu and S. Keis. 2000. Bacteriophage infections in the industrial acetone butanol (AB) fermentation process. J. Mol. Microbiol. Biotechnol. 2: 21–8.
Karel, S. F., S. B. Libicki and C. R. Robertson. 1985. The Immobilization of Whole Cells: Engineering Principles. Chem. Eng. Sci. 40: 1321–1354.
Killeffer, D. H. 1927. Butanol and acetone from corn. J. Ind. Eng. Chem. (Washington, D.C.). 19: 46–50.
Koepke, M. and P. Duerre. 2011. Biochemical production of biobutanol. Woodhead Publ. Ser. Energy. 15: 221–57.
Kondo, T., H. Tezuka, J. Ishii, F. Matsuda, C. Ogino and A. Kondo. 2012. Genetic engineering to enhance the Ehrlich pathways and alter carbon flux for increased isobutanol production from glucose by *Saccharomyces cerevisiae*. J. Biotechnol. 159: 32–27.
Kwiatkowski, J. R., A. J. McAloon, F. Taylor and D. B. Johnston. 2006. Modeling the process and costs of fuel ethanol production by the corn dry-grind process. Ind. Crops Prod. 23: 288–96.
Lane, J. 2012. Butamax and Gevo: Bio's Montagues and Capulets get it on, and on, and on. www.altenergystocks.com/archives/biofuels/ethanol.
Lee, H., S. Seo, Y. Bae, H. Nan, Y. Jin and J. Seo. 2012. Isobutanol production in engineered *Saccharomyces cerevisiae* by overexpresion of 2-ketoisovalerate decarboxylase and valine biosynthetic enzyme. Bioprocess and Biosystems Engineering. 35(9): 1467–75.
Lee, K. 2005. A Media Design Program for Lactic Acid Production Coupled with Extraction by Electrodialysis. Bioresource Technol. 96: 1505–1510.
Li, H., P. Opgenorth, D. Wernick, S. Rogers, T. Wu, W. Higashide, P. Malati, Y. Huo, K. Cho and J. Liao. 2012. Integrated electromicrobial conversion of CO_2 to higher alcohols. Nature. 335: 1596.
Li, S., J. Wen and X. Jia. 2011a. Engineering *Bacillus subtilis* for isobutanol production by heterologous Ehrlich pathways construction and the biosynthetic 2-ketoisovalerate precursor pathway overexpression. Appl. Microbiol. Biotechnol. 91: 577–589.
Li, S. and R. Srivastava. 2011b. Study of *in situ* 1-butanol pervaporation from ABE Fermentation using a PDMS composite membrane: Validity of solution-diffusion model for pervaporative ABE fermentation. Biotechnol. Prog. 27: 111–120.
Lin, Y. J., J. A. Hestekin, M. B. Arora and E. J. St. Martin. 2004. Electrodeionization Method. U.S. Patent. 6,797,140.
Liu, F., L. Liu and X. Feng. 2005. Separation of acetone, butanol, ethanol (ABE) from dilute aqueous solutions by pervaporation. Separation and Purification Technology. 42(3): 273–282.
Lu, J. 2012. Studies on the production of branched-chain alcohols in engineered *Ralpha eutropha*. App. Microbiol. Biotechnol. 96: 283–297.
Machado, I. M. P. and S. Atsumi. 2012. Cyanobacterial biofuel production. J. Biotechnol. 162: 50–56.
Maddox, I. S., N. Qureshi and N. A. Gutierrez. 1993. Utilization of whey by clostridia and process technology. Biotechnol. Ser. 25: 343–69.

Mamede, M. E. O. and G. M. Pastore. 2006. Study of Methods for the Extraction of Volatile Compounds from Fermented Grape Must. Food Chem. 96: 586–590.

Marchal, R., D. Blanchet and J. P. Vandecasteele. 1985. Industrial optimization of acetone-butanol fermentation: A study of the utilization of jerusalem artichokes. Appl. Microbiol. Biotechnol. 23: 92–8.

Mariano, A. P., M. O. S. Dias, T. L. Junqueira, M. P. Cunha, A. Bonomi and R. M. Filho. 2012. In Press. Butanol Production in a first-generation Brazilian sugarcane biorefinery: technical aspects and economics of greenfield projects. Bioresource Technol.

Martin, C. 2012. Butamax Taps Fagen for First Ethanol-to-Butanol Conversion Plant. Bloomberg. http://www.bloomberg.com/news/2012-04-19/butamax-taps-fagen-for-first-ethanol-to-butanol-conversion-plant.html.

Matsumoto, M., K. Mochiduki, K. Fukunishi and K. Kondo. 2004. Extraction of Organic Acids using Imidazolium-based Ionic Liquids and their Toxicity to Lactobacillus rhamnosus. Sep. Purif. Technol. 40: 97–101.

Monot, F. and J. M. Engasser. 1983. Production of acetone and butanol by batch and continuous culture of *Clostridium acetobutylicum* under nitrogen limitation. Biotechnol. Lett. 5: 213–8.

Monot, F., J. R. Martin, H. Petitdemange and R. Gay. 1982. Acetone and butanol production by *Clostridium acetobutylicum* in a synthetic medium. Appl. Environ. Microbiol. 44: 1318–24.

Nakas, J. P., M. Schaedle, C. M. Parkinson, C. E. Coonley and S. W. Tanenbaum. 1983. System development for linked-fermentation production of solvents from algal biomass. Appl. Environ. Microbiol. 46: 1017–23.

Napoli, F., G. Olivieri, M. Russo, A. Marzocchella and P. Salatino. 2011. Continuous lactose fermentation by *Clostridium acetobutylicum*—assessment of acidogenesis kinetics. Bioresource Technology. 102: 1608–1614.

Nghiem, N., B. H. Davison, M. I. Donnelly, S. P. Tsai and J. G. Frye. 2001. An Integrated Process for the Production of Chemicals from Biologically Derived Succinic Acid. ACS Symposium Series. 784: 160–173.

Ni, Y. and Z. Sun. 2009. Recent progress on industrial fermentative production of acetone-butanol-ethanol by clostridium acetobutylicum in china. Appl. Microbiol. Biotechnol. 83: 415–23.

Nielsen, D. R. and K. J. Prather. 2008. *In Situ* Product Recovery of n-Butanol using Polymeric Resins. Biotechnol. Bioeng. 102: 811–821.

Nielsen, S. 2012. Green Biologics to Raise $6 Million for Brazil Butanol Facility. Bloomberg. http://www.bloomberg.com/news/2012-02-24/green-biologics-to-raise-6-million-for-brazil-butanol-facility.html.

Olsson, J., G. Tragardh and C. Tragardh. 2001. Pervaporation of Volatile Organics from Water II. Influence of Permeate Pressure on Partial Fluxes. J. Membrane Sci. 186: 239–247.

Ounine, K., H. Petitdemange, G. Raval and R. Gay. 1983. Acetone-butanol production from pentoses by clostridium acetobutylicum. Biotechnol. Lett. 5: 605–10.

Papoutsakis, E. T. and G. N. Bennett. 1993. Cloning, structure, and expression of acid and solvent pathway genes of *Clostridium acetobutylicum*. Biotechnol. Ser. 25: 157–99.

Park, C., M. R. Okos and P. C. Wankat. 1991. Acetone-Butanol-Ethanol (ABE) Fermentation and Simultaneous Separation in a Trickle Bed Reactor. Biotechnol. Prog. 7: 185–194.

Park, C. H. and C. Q. Geng. 1996. Mathematical Modeling of Fed-Batch Butanol Fermentation with Simultaneous Pervaporation. Korean J. Chem. Eng. 13: 612–619.

Patakova, P., M. Pospisil, J. Lipovsky, P. Fribert, M. Linhova, S. S. M. Toure, M. Rychtera, K. Melzoch and G. Sebor. 2010. Prospects for biobutanol production and use in the czech republic. CHEMagazin. 20: 13–5.

Pera-Titus, M., J. Llorens, J. Tejero and F. Cunill. 2006. Description of the pervaporative dehydration performance of A-type zeolite membranes: A modeling approach based on the Maxwell-Stefan theory. Catal. Today. 118: 73–84.

Pfromm, P. H., V. Amanor-Boadu, R. Nelson, P. Vadlani and R. Madl. 2010. Bio-butanol vs. bio-ethanol: A technical and economic assessment for corn and switchgrass fermented by yeast or *Clostridium acetobutylicum*. Biomass Bioenergy. 34: 515–24.

Potts, T., J. Du, M. Paul, P. May, R. Beitle and J. Hestekin. 2012. The production of butanol from Jamaica Bay macro algae. Environ. Prog. Sustainable Energy. 31: 29–36.

Qureshi, N., M. M. Meagher and R. W. Hutkins. 1999. Recovery of Butanol from Model Solutions and Fermentation Broth using a Silicalite/Silicone Membrane. J. Membrane Sci. 158: 115–125.

Qureshi, N. and H. P. Blaschek. 1999. Fouling Studies of a Pervaporation Membrane with Commercial Fermentation Media and Fermentation Broth of Hyper-Butanol-Producing *Clostridium beijerinckii* BA101. Sep. Sci. and Tech. 34: 2803–2815.

Qureshi, N. and T. C. Ezeji. 2008. Butanol, 'a superior biofuel' production from agricultural residues (renewable biomass): Recent progress in technology. Biofuels, Bioprod. Biorefin. 2: 319–30.

Qureshi, N. and H. Blaschek. 2000. Butanol production using *Clostridium beijerinckii* BA101 hyper-butanol producing mutant strain recovery by pervaporation. Applied Biochemistry and Biotechnology. 84–86: 225–235.

Qureshi, N., A. Lolas and H. P. Blaschek. 2001. Soy molasses as fermentation substrate for production of butanol using *Clostridium beijerinckii* BA101. J. Ind. Microbiol. Biotechnol. 26: 290–5.

Qureshi, N., P. Karcher, M. Cotta and H. Blaschek. 2004. High productivity continuous biofilm reactor for butanol production. Applied Biochemistry and Biotechnology. 113–116: 713–721.

Qureshi, N., B. A. Annous, T. C. Ezeji, P. Karcher and I. S. Maddox. 2005. Biofilm reactors for industrial bioconversion processes: Employing potential of enhanced reaction rates. Microb. Cell Fact. 4: 24.

Qureshi, N., B. Saha, M. Cotta and V. Singh. 2013. An economic evaluation of biological conversion of wheat straw to butanol: a biofuel. Energy Conversion and Management. 65: 456–462.

Ramey, D. E. 1998. Continuous two stage, dual path anaerobic fermentation of butanol and other organic solvents using two different strains of bacteria. U.S. 1996-771065; 1996-771065:8.

Ramey, D. E. and S. T. Yang. 2004. Production of butyric acid and butanol from biomass. Final Report to US Department of Energy, Contract No.: DE-F-G02-00ER86106.

Rosgaard, L., A. J. Porcellinis, J. H. Jacobsen, N. U. Frigaard and Y. Sakuragi. 2012. Bioengineering of carbon fixation, biofuels, and biochemical in cyanobacteria and plants. J. Biotechnol. 162: 134–147.

Semmens, M. and J. Gregory. 1974. Selectivity of Strongly Basic Anion Exchange Resins for Organic Anions. Environ. Sci. Technol. 8: 834–839.

Shuler, M. L. and F. Kargi. 2002. Bioprocess Engineering: Basic Concepts, 2nd ed. Prentice-Hall, Inc. Upper Saddle River, NY, p. 248.

Skouras, S. and S. Skogestad. 2004. Separation of Ternary Heteroazeotropic Mixtures in a Closed Multivessel Batch-Distillation Decanter Hybrid. Chem. Eng. Process. 43: 291–304.

Smith, K. M. and J. C. Liao. 2011. An evolutionary strategy for isobutanol production strain development in *Escherichia coli*. Metab. Eng. 13: 674–681.

Smith, K. M., K. M. Cho and J. C. Liao. 2010. Engineering *Corynebacterium glutamicum* for isobutanol production. Appl. Microbiol. Biotechnol. 87: 1045–1055.

Stanier, R., E. A. Adelberg and J. L. Ingraham. 1976. The Microbial World, 4th Edition. Prentice-Hill. N.J.

Stoeberl, M., R. Werkmeister, M. Faulstich and W. Russ. 2011. Biobutanol from food wastes —fermentative production, use as biofuel an the influence on the emissions. Procedia Food Sci. 1: 1867–74.

Szulczyk, K. R. 2010. Which is a better transportation fuel-butanol or ethanol? Int. J. Energy Environ. 1: 501–12.

Takahashi, H., K. Ohba and K. I. Kikuchi. 2003. Sorption of Mono-Carboxylic Acids by an Anion-Exchange Membrane. Biochem. Eng. J. 16: 311–315.

Takriff, M. S., S. J. H. M. Yusof, A. A. H. Kadhum, J. Jahim and A. W. Mohammad. 2008. Recovery of acetone-butanol-ethanol from fermentation broth by liquid-liquid extraction. J. Biotechnol. 5: 46.

Tanaka, S., Y. Tashiro, G. Kobayashi, T. Ikegami, H. Negishi and K. Sakaki. 2012. Membrane-assisted extractive butanol fermentation by *Clostridium saccharoperbutylacetonicum* N1-4 with 1-dodecanol as the extractant. Bioresearch Technol. 116: 448–452.

Tashiro, Y., K. Takeda, G. Kobayashi and K. Sonomoto. 2005. High Production of acetone-butanol-ethanol with high cell density by cell-recycling and bleeding. J. Biotechnol. 120: 197–206.

Tashiro, Y., T. Yoshida and K. Sonomoto. 2010. Bioethanol production by designed biomass. Kagaku to Kogyo (Tokyo, Jpn). 63: 894–6.

Thang, V. H., K. Kanda and G. Kobayashi. 2010. Production of acetone-butanol-ethanol (ABE) in direct fermentation of cassava by *Clostridium saccharoperbutylacetonicum* N1-4. Appl. Biochem. Biotechnol. 161: 157–70.

Thongsukmak, A. and K. K. Sirkar. 2007. Pervaporation Membranes Highly Selective for Solvents Present in Fermentation Broths. J. Membrane Sci. 302: 45–58.

Thormann, K., L. Feustel, K. Lorenz, S. Nakotte and P. Durre. 2002. Control of butanol formation in *Clostridium acetobutylicum* by transcriptional activation. J. Bacteriol. 184: 1966–73.

Trinh, C. T., J. Li, H. W. Blanch and D. S. Clark. 2011. Redesigning *Escherichia coli*. Metabolism for Anaerobic Production of Isobutanol. Appl. Environ. Microbiol. 77: 4894–4904.

Trinh, C. T. 2012. Elucidating and reprogramming *Escherichia coli* metabolisms for obligate anaerobic n-butanol and isobutanol production. Appl. Microbiol. And Biotechnol. 95: 1083–1094.

Tummala, S. B., S. G. Junne and E. T. Papoutsakis. 2003. Antisense RNA downregulation of coenzyme A transferase combined with alcohol-aldehyde dehydrogenase overexpression leads to predominantly alcohologenic clostridium acetobutylicum fermentations. J. Bacteriol. 185: 3644–53.

US Department of Energy. Annual Energy Outlook 2012 with projections to 2035. DOE/EIA 0383(2012). http://www.eia.gov/forecasts/aeo/pdf/038328(2012).pdf.

Vane, L. 2005. A Review of Pervaporation for Product Recovery from Biomass Fermentation Processes. J. Chem. Technol. Biotechnol. 80: 603–629.

Voget, C. E., C. F. Mignone and R. J. Ertola. 1985a. Butanol production from apple pomace. Biotechnol. Lett. 7: 43–6.

Wang, Y., T. S. Chung, H. Wang and S. H. Goh. 2009. Butanol isomer separation using polyamide-imide/CD mixed matrix membranes via pervaporation. Chem. Eng. Sci. 64: 5198–5209.

Wang, Z., Y. Luo and P. Yu. 2006. Recovery of Organic Acids from Waste Salt Solutions Derived from the Manufacture of Cyclohexanone by Electrodialysis. J. Membrane Sci. 280: 134–137.

Weast, R. C. 1978. Handbook of Chemistry and Physics. CRC Press, Inc. Boca Raton, FL.

Weigel, J., K. U. K. Seung-Uk and G. W. Kohring. 1989. *Clostridium thermobutyricum* sp. nov., a Moderate Thermophile Isolated from a Cellulolytic Culture, that Produces Butyrate as the Major Product. Int. J. Syst. Bacteriol. 39: 199–204.

Weizmann, C. 1915 (applied). Acetone and butyl alcohol. US Patent 1,315,484, September 9, 1919.

Whims, J. 2002. Corn based ethanol costs and margins, attachment 1. Department of Agricultural Economics, Kansas State University.

Widiasa, I. N., P. D. Sutrisna and I. G. Wenten. 2004. Performance of a Novel Electrodeionization Technique during Citric Acid Recovery. Sep. Purif. Technol. 39: 89–97.

Wijmans, J. G. and R. W. Baker. 1995. The solution-diffusion model: a review. J. Membrane Sci. 107: 1–21.

Wu, M. Analysis of the Efficiency of the U.S. Ethanol Industry 2007. Argonne National Laboratory, Center for Transportation Research. Sponsored by Renewable Fuels Association. [cited 2012 November 1]. Available from: http://ww1.eere.gov/biomass/pdfs/anl_ethanol_analysis_007.pdf.

Wu, Z. and S. T. Yang. 2003. Extractive Fermentation for Butyric Acid Production from Glucose by *Clostridium tyrobutyricum*. Biotechnol. Bioeng. 82: 93–102.

Yao, M., H. Wang, Z. Zheng and Y. Yue. 2010. Experimental study of n-butanol additive and multi-injection on HD diesel engine performance and emissions. Fuel. 89: 2191–201.

Yeom, C. K., S. H. Lee and J. M. Lee. 2000. Pervaporative Permeations of Homologous Series of Alcohol Aqueous Mixtures through a Hydrophilic Membrane. J. App. Polymer Sci. 79: 703–713.

Yerushalmi, L., B. Volesky and T. Szczesny. 1985. Effect of increased hydrogen partial pressure on the acetone-butanol fermentation by *Clostridium acetobutylicum*. Appl. Microbiol. Biotechnol. 22: 103–7.

Young, J., P. Lee and A. DiGioia. 2007. Fast HPLC analysis for fermentation ethanol process. Waters Application Note. 4.

Yuan, L. and X. Hui-feng. 2012. OXO Market Supply and Demand Forecase & Investment Economic Analysis. Finan. Res. 1: 4–10.

Zhang, H., H. Chong, C. B. Ching, H. Song and R. Jiang. 2012. Engineering global transcription factor cyclic AMP receptor protein of *Escherichia coli* for improved 1-butanol tolerance. Appl. Microbiol. Biotechnol. 94: 1107–17.

Zhu, J. and F. Yang. 2009. Biological process for Butanol Production. *In:* J. Cheng. Biomass to renewable energy processes/Biological process for butanol production. CRC Press/Taylor & Francis, Boca Raton, FL. pp. 271–329.

Zhu, Y. and S. T. Yang. 2004. Effect of pH on Metabolic Pathway Shift in Fermentation of Xylose by *Clostridium tyrobutyricum*. J. Biotechnol. 110: 143–157.

Zigova, J. and E. Sturdik. 2000. Advances in Biotechnological Production of Butyric Acid. J. Ind. Microbiol. Biot. 24: 153–160.

Index

Color Plate Section

Chapter 4

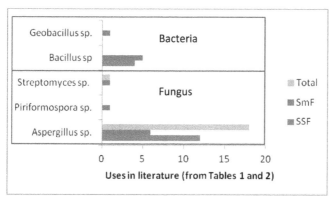

Fig. 1. Microorganisms used for amylases production.

Chapter 8

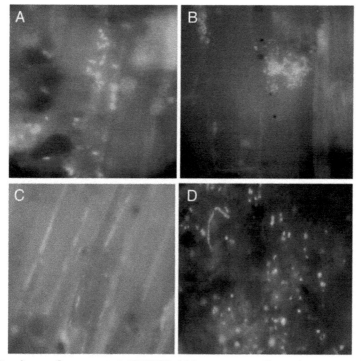

Fig. 1. *Pseudomonas fluorescens* strain F113PCB *gfp* biosensor responding to the presence of PCBs in the rhizosphere of *Pisum sativum* (reproduced from Liu et al. 2010b with permission).

Chapter 11

Fig. 1. Photograph representing the fruit bodies of the white-rot fungus *F. fomentarius* in its natural habitat (dead plants in the forest) (adapted with permission of Etienne Charles, http://champignons.moselle.free.fr).

9 780367 379834